John E. Freund's
Mathematical Statistics with Applications

Eighth Edition

Irwin Miller

Marylees Miller

PEARSON

Authorized adaptation from the United States edition, entitled *John E. Freund's Mathematical Statistics with Applications*, *Eighth Edition,* ISBN 9780321807090, by Irwin Miller and Marylees Miller published by Pearson Education Inc., © 2014, Pearson Education Inc.

Indian Subcontinent Version
© 2014 Dorling Kindersley (India) Pvt. Ltd

ISBN: 978-93-325-1905-3

First Impression, 2014
Second Impression

Published by Dorling Kindersley (India) Pvt. Ltd., licensees of Pearson Education in South Asia.

Head Office: 7th Floor, Knowledge Boulevard, A-8(A), Sector 62, Noida 201 309, UP, India.
Registered Office: 11 Community Centre, Panchsheel Park, New Delhi 110 017, India.

Printed in India by Repro India Ltd.

Table of Contents

INTRODUCTION

1 Introduction

In recent years, the growth of statistics has made itself felt in almost every phase of human activity. Statistics no longer consists merely of the collection of data and their presentation in charts and tables; it is now considered to encompass the science of basing inferences on observed data and the entire problem of making decisions in the face of uncertainty. This covers considerable ground since uncertainties are met when we flip a coin, when a dietician experiments with food additives, when an actuary determines life insurance premiums, when a quality control engineer accepts or rejects manufactured products, when a teacher compares the abilities of students, when an economist forecasts trends, when a newspaper predicts an election, and even when a physicist describes quantum mechanics.

It would be presumptuous to say that statistics, in its present state of development, can handle all situations involving uncertainties, but new techniques are constantly being developed and modern statistics can, at least, provide the framework for looking at these situations in a logical and systematic fashion. In other words, statistics provides the models that are needed to study situations involving uncertainties, in the same way as calculus provides the models that are needed to describe, say, the concepts of Newtonian physics.

The beginnings of the mathematics of statistics may be found in mid-eighteenth-century studies in probability motivated by interest in games of chance. The theory thus developed for "heads or tails" or "red or black" soon found applications in situations where the outcomes were "boy or girl," "life or death," or "pass or fail," and scholars began to apply probability theory to actuarial problems and some aspects of the social sciences. Later, probability and statistics were introduced into physics by L. Boltzmann, J. Gibbs, and J. Maxwell, and by this century they have found applications in all phases of human endeavor that in some way involve an element of uncertainty or risk. The names that are connected most prominently with the growth of mathematical statistics in the first half of the twentieth century are those of R. A. Fisher, J. Neyman, E. S. Pearson, and A. Wald. More recently, the work of R. Schlaifer, L. J. Savage, and others has given impetus to statistical theories based essentially on methods that date back to the eighteenth-century English clergyman Thomas Bayes.

Mathematical statistics is a recognized branch of mathematics, and it can be studied for its own sake by students of mathematics. Today, the theory of statistics is applied to engineering, physics and astronomy, quality assurance and reliability, drug development, public health and medicine, the design of agricultural or industrial experiments, experimental psychology, and so forth. Those wishing to participate in such applications or to develop new applications will do well to understand the mathematical theory of statistics. For only through such an understanding can applications proceed without the serious mistakes that sometimes occur. The applications are illustrated by means of examples and a separate set of applied exercises, many of them involving the use of computers. To this end, we have added at the end of the chapter a discussion of how the theory of the chapter can be applied in practice.

We begin with a brief review of combinatorial methods and binomial coefficients.

2 Combinatorial Methods

In many problems of statistics we must list all the alternatives that are possible in a given situation, or at least determine how many different possibilities there are. In connection with the latter, we often use the following theorem, sometimes called the **basic principle of counting**, the **counting rule for compound events**, or the **rule for the multiplication of choices.**

> **THEOREM 1.** If an operation consists of two steps, of which the first can be done in n_1 ways and for each of these the second can be done in n_2 ways, then the whole operation can be done in $n_1 \cdot n_2$ ways.

Here, "operation" stands for any kind of procedure, process, or method of selection.

To justify this theorem, let us define the ordered pair (x_i, y_j) to be the outcome that arises when the first step results in possibility x_i and the second step results in possibility y_j. Then, the set of all possible outcomes is composed of the following $n_1 \cdot n_2$ pairs:

$$
\begin{array}{c}
(x_1, y_1), (x_1, y_2), \ldots, (x_1, y_{n_2}) \\
(x_2, y_1), (x_2, y_2), \ldots, (x_2, y_{n_2}) \\
\ldots \\
\ldots \\
\ldots \\
(x_{n_1}, y_1), (x_{n_1}, y_2), \ldots, (x_{n_1}, y_{n_2})
\end{array}
$$

EXAMPLE 1

Suppose that someone wants to go by bus, train, or plane on a week's vacation to one of the five East North Central States. Find the number of different ways in which this can be done.

Solution

The particular state can be chosen in $n_1 = 5$ ways and the means of transportation can be chosen in $n_2 = 3$ ways. Therefore, the trip can be carried out in $5 \cdot 3 = 15$ possible ways. If an actual listing of all the possibilities is desirable, a **tree diagram** like that in Figure 1 provides a systematic approach. This diagram shows that there are $n_1 = 5$ branches (possibilities) for the number of states, and for each of these branches there are $n_2 = 3$ branches (possibilities) for the different means of transportation. It is apparent that the 15 possible ways of taking the vacation are represented by the 15 distinct paths along the branches of the tree.

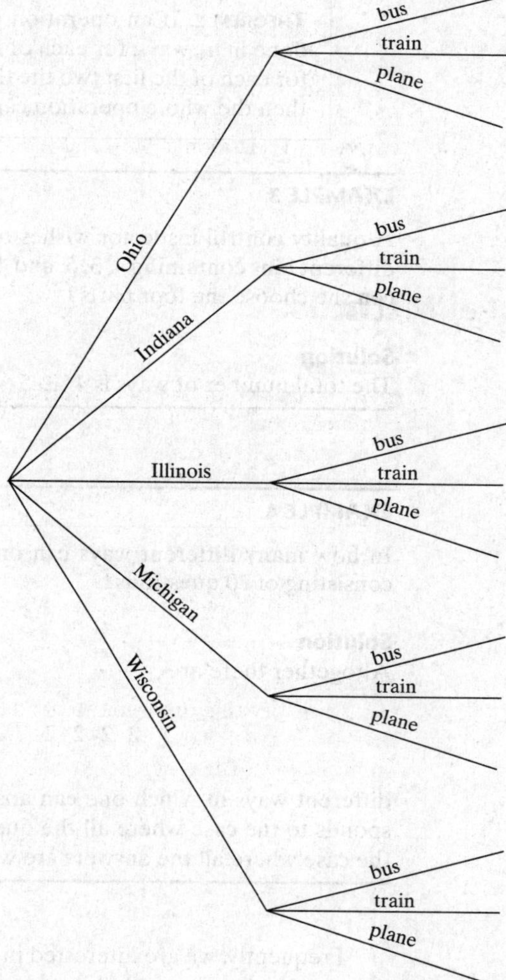

Figure 1. Tree diagram.

EXAMPLE 2

How many possible outcomes are there when we roll a pair of dice, one red and one green?

Solution

The red die can land in any one of six ways, and for each of these six ways the green die can also land in six ways. Therefore, the pair of dice can land in $6 \cdot 6 = 36$ ways.

Theorem 1 may be extended to cover situations where an operation consists of two or more steps. In this case, we have the following theorem.

> **THEOREM 2.** If an operation consists of k steps, of which the first can be done in n_1 ways, for each of these the second step can be done in n_2 ways, for each of the first two the third step can be done in n_3 ways, and so forth, then the whole operation can be done in $n_1 \cdot n_2 \cdot \ldots \cdot n_k$ ways.

EXAMPLE 3

A quality control inspector wishes to select a part for inspection from each of four different bins containing 4, 3, 5, and 4 parts, respectively. In how many different ways can she choose the four parts?

Solution

The total number of ways is $4 \cdot 3 \cdot 5 \cdot 4 = 240$.

EXAMPLE 4

In how many different ways can one answer all the questions of a true–false test consisting of 20 questions?

Solution

Altogether there are

$$2 \cdot 2 \cdot 2 \cdot 2 \cdot \ldots \cdot 2 \cdot 2 = 2^{20} = 1{,}048{,}576$$

different ways in which one can answer all the questions; only one of these corresponds to the case where all the questions are correct and only one corresponds to the case where all the answers are wrong.

Frequently, we are interested in situations where the outcomes are the different ways in which a group of objects can be ordered or arranged. For instance, we might want to know in how many different ways the 24 members of a club can elect a president, a vice president, a treasurer, and a secretary, or we might want to know in how many different ways six persons can be seated around a table. Different arrangements like these are called **permutations**.

> **DEFINITION 1. PERMUTATIONS.** *A* ***permutation*** *is a distinct arrangement of n different elements of a set.*

EXAMPLE 5

How many permutations are there of the letters a, b, and c?

Solution

The possible arrangements are *abc*, *acb*, *bac*, *bca*, *cab*, and *cba*, so the number of distinct permutations is six. Using Theorem 2, we could have arrived at this answer without actually listing the different permutations. Since there are three choices to select a letter for the first position, then two for the second position, leaving only one letter for the third position, the total number of permutations is $3 \cdot 2 \cdot 1 = 6$.

Generalizing the argument used in the preceding example, we find that n distinct objects can be arranged in $n(n-1)(n-2) \cdot \ldots \cdot 3 \cdot 2 \cdot 1$ different ways. To simplify our notation, we represent this product by the symbol $n!$, which is read "n factorial." Thus, $1! = 1, 2! = 2 \cdot 1 = 2, 3! = 3 \cdot 2 \cdot 1 = 6, 4! = 4 \cdot 3 \cdot 2 \cdot 1 = 24, 5! = 5 \cdot 4 \cdot 3 \cdot 2 \cdot 1 = 120$, and so on. Also, by definition we let $0! = 1$.

THEOREM 3. The number of permutations of n distinct objects is $n!$.

EXAMPLE 6

In how many different ways can the five starting players of a basketball team be introduced to the public?

Solution

There are $5! = 5 \cdot 4 \cdot 3 \cdot 2 \cdot 1 = 120$ ways in which they can be introduced.

EXAMPLE 7

The number of permutations of the four letters a, b, c, and d is 24, but what is the number of permutations if we take only two of the four letters or, as it is usually put, if we take the four letters two at a time?

Solution

We have two positions to fill, with four choices for the first and then three choices for the second. Therefore, by Theorem 1, the number of permutations is $4 \cdot 3 = 12$.

Generalizing the argument that we used in the preceding example, we find that n distinct objects taken r at a time, for $r > 0$, can be arranged in $n(n-1) \cdot \ldots \cdot (n-r+1)$ ways. We denote this product by ${}_nP_r$, and we let ${}_nP_0 = 1$ by definition. Therefore, we can state the following theorem.

THEOREM 4. The number of permutations of n distinct objects taken r at a time is

$$ {}_nP_r = \frac{n!}{(n-r)!} $$

for $r = 0, 1, 2, \ldots, n$.

Proof The formula ${}_nP_r = n(n-1) \cdot \ldots \cdot (n-r+1)$ cannot be used for $r = 0$, but we do have

$$ {}_nP_0 = \frac{n!}{(n-0)!} = 1 $$

For $r = 1, 2, \ldots, n$, we have

$$
\begin{aligned}
{}_nP_r &= n(n-1)(n-2)\cdot \ldots \cdot (n-r-1) \\
&= \frac{n(n-1)(n-2)\cdot \ldots \cdot (n-r-1)(n-r)!}{(n-r)!} \\
&= \frac{n!}{(n-r)!}
\end{aligned}
$$

In problems concerning permutations, it is usually easier to proceed by using Theorem 2 as in Example 7, but the factorial formula of Theorem 4 is somewhat easier to remember. Many statistical software packages provide values of ${}_nP_r$ and other combinatorial quantities upon simple commands. Indeed, these quantities are also preprogrammed in many hand-held statistical (or scientific) calculators.

EXAMPLE 8

Four names are drawn from among the 24 members of a club for the offices of president, vice president, treasurer, and secretary. In how many different ways can this be done?

Solution

The number of permutations of 24 distinct objects taken four at a time is

$$
{}_{24}P_4 = \frac{24!}{20!} = 24\cdot 23\cdot 22\cdot 21 = 255{,}024
$$

EXAMPLE 9

In how many ways can a local chapter of the American Chemical Society schedule three speakers for three different meetings if they are all available on any of five possible dates?

Solution

Since we must choose three of the five dates and the order in which they are chosen (assigned to the three speakers) matters, we get

$$
{}_5P_3 = \frac{5!}{2!} = \frac{120}{2} = 60
$$

We might also argue that the first speaker can be scheduled in five ways, the second speaker in four ways, and the third speaker in three ways, so that the answer is $5\cdot 4\cdot 3 = 60$.

Permutations that occur when objects are arranged in a circle are called **circular permutations**. Two circular permutations are not considered different (and are counted only once) if corresponding objects in the two arrangements have the same objects to their left and to their right. For example, if four persons are playing bridge, we do not get a different permutation if everyone moves to the chair at his or her right.

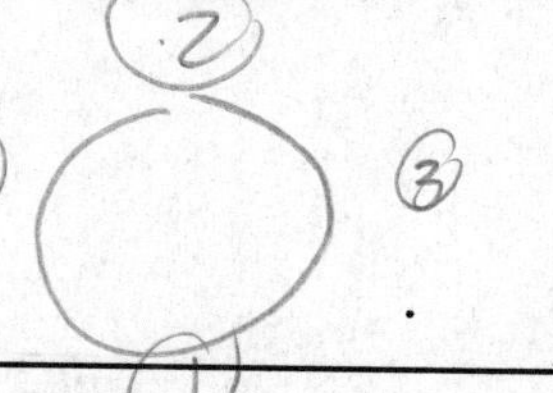

EXAMPLE 10

How many circular permutations are there of four persons playing bridge?

Solution

If we arbitrarily consider the position of one of the four players as fixed, we can seat (arrange) the other three players in $3! = 6$ different ways. In other words, there are six different circular permutations.

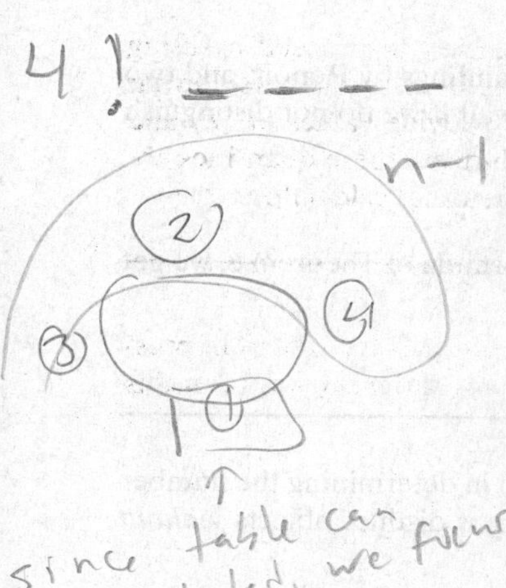

Generalizing the argument used in the preceding example, we obtain the following theorem.

> **THEOREM 5.** The number of permutations of n distinct objects arranged in a circle is $(n-1)!$.

We have been assuming until now that the n objects from which we select r objects and form permutations are all distinct. Thus, the various formulas cannot be used, for example, to determine the number of ways in which we can arrange the letters in the word "book," or the number of ways in which three copies of one novel and one copy each of four other novels can be arranged on a shelf.

EXAMPLE 11

How many different permutations are there of the letters in the word "book"?

Solution

If we distinguish for the moment between the two o's by labeling them o_1 and o_2, there are $4! = 24$ different permutations of the symbols b, o_1, o_2, and k. However, if we drop the subscripts, then bo_1ko_2 and bo_2ko_1, for instance, both yield $boko$, and since each pair of permutations with subscripts yields but one arrangement without subscripts, the total number of arrangements of the letters in the word "book" is $\frac{24}{2} = 12$.

EXAMPLE 12

In how many different ways can three copies of one novel and one copy each of four other novels be arranged on a shelf?

Solution

If we denote the three copies of the first novel by a_1, a_2, and a_3 and the other four novels by b, c, d, and e, we find that *with subscripts* there are $7!$ different permutations of a_1, a_2, a_3, b, c, d, and e. However, since there are $3!$ permutations of a_1, a_2, and a_3 that lead to the same permutation of a, a, a, b, c, d, and e, we find that there are only $\frac{7!}{3!} = 7 \cdot 6 \cdot 5 \cdot 4 = 840$ ways in which the seven books can be arranged on a shelf.

Generalizing the argument that we used in the two preceding examples, we obtain the following theorem.

THEOREM 6. The number of permutations of n objects of which n_1 are of one kind, n_2 are of a second kind, ..., n_k are of a kth kind, and $n_1 + n_2 + \cdots + n_k = n$ is

$$\frac{n!}{n_1! \cdot n_2! \cdot \ldots \cdot n_k!}$$

EXAMPLE 13

In how many ways can two paintings by Monet, three paintings by Renoir, and two paintings by Degas be hung side by side on a museum wall if we do not distinguish between the paintings by the same artists?

Solution

Substituting $n = 7$, $n_1 = 2$, $n_2 = 3$, and $n_3 = 2$ into the formula of Theorem 6, we get

$$\frac{7!}{2! \cdot 3! \cdot 2!} = 210$$

There are many problems in which we are interested in determining the number of ways in which r objects can be selected from among n distinct objects *without regard to the order in which they are selected.*

DEFINITION 2. COMBINATIONS. *A* ***combination*** *is a selection of* r *objects taken from* n *distinct objects without regard to the order of selection.*

EXAMPLE 14

In how many different ways can a person gathering data for a market research organization select three of the 20 households living in a certain apartment complex?

Solution

If we care about the order in which the households are selected, the answer is

$$_{20}P_3 = 20 \cdot 19 \cdot 18 = 6{,}840$$

but each set of three households would then be counted $3! = 6$ times. If we do not care about the order in which the households are selected, there are only $\frac{6{,}840}{6} = 1{,}140$ ways in which the person gathering the data can do his or her job.

Actually, "combination" means the same as "subset," and when we ask for the number of combinations of r objects selected from a set of n distinct objects, we are simply asking for the total number of subsets of r objects that can be selected from a set of n distinct objects. In general, there are $r!$ permutations of the objects in a subset of r objects, so that the $_nP_r$ permutations of r objects selected from a set of n distinct objects contain each subset $r!$ times. Dividing $_nP_r$ by $r!$ and denoting the result by the symbol $\binom{n}{r}$, we thus have the following theorem.

THEOREM 7. The number of combinations of n distinct objects taken r at a time is

$$\binom{n}{r} = \frac{n!}{r!(n-r)!}$$

for $r = 0, 1, 2, \ldots, n$.

EXAMPLE 15

In how many different ways can six tosses of a coin yield two heads and four tails?

Solution

This question is the same as asking for the number of ways in which we can select the two tosses on which heads is to occur. Therefore, applying Theorem 7, we find that the answer is

$$\binom{6}{2} = \frac{6!}{2! \cdot 4!} = 15$$

This result could also have been obtained by the rather tedious process of enumerating the various possibilities, HHTTTT, TTHTHT, HTHTTT, ..., where H stands for head and T for tail.

EXAMPLE 16

How many different committees of two chemists and one physicist can be formed from the four chemists and three physicists on the faculty of a small college?

Solution

Since two of four chemists can be selected in $\binom{4}{2} = \frac{4!}{2! \cdot 2!} = 6$ ways and one of three physicists can be selected in $\binom{3}{1} = \frac{3!}{1! \cdot 2!} = 3$ ways, Theorem 1 shows that the number of committees is $6 \cdot 3 = 18$.

A combination of r objects selected from a set of n distinct objects may be considered a **partition** of the n objects into two subsets containing, respectively, the r objects that are selected and the $n - r$ objects that are left. Often, we are concerned with the more general problem of partitioning a set of n distinct objects into k subsets, which requires that each of the n objects must belong to one and only one of the subsets.[†] The order of the objects within a subset is of no importance.

EXAMPLE 17

In how many ways can a set of four objects be partitioned into three subsets containing, respectively, two, one, and one of the objects?

[†] Symbolically, the subsets $A_1, A_2, \ldots, A_k$ constitute a partition of set A if $A_1 \cup A_2 \cup \cdots \cup A_k = A$ and $A_i \cap A_j = \emptyset$ for all $i \neq j$.

Solution

Denoting the four objects by a, b, c, and d, we find by enumeration that there are the following 12 possibilities:

$ab\|c\|d$	$ab\|d\|c$	$ac\|b\|d$	$ac\|d\|b$
$ad\|b\|c$	$ad\|c\|b$	$bc\|a\|d$	$bc\|d\|a$
$bd\|a\|c$	$bd\|c\|a$	$cd\|a\|b$	$cd\|b\|a$

The number of partitions for this example is denoted by the symbol

$$\binom{4}{2,1,1} = 12$$

where the number at the top represents the total number of objects and the numbers at the bottom represent the number of objects going into each subset.

Had we not wanted to enumerate all the possibilities in the preceding example, we could have argued that the two objects going into the first subset can be chosen in $\binom{4}{2} = 6$ ways, the object going into the second subset can then be chosen in $\binom{2}{1} = 2$ ways, and the object going into the third subset can then be chosen in $\binom{1}{1} = 1$ way. Thus, by Theorem 2 there are $6 \cdot 2 \cdot 1 = 12$ partitions. Generalizing this argument, we have the following theorem.

THEOREM 8. The number of ways in which a set of n distinct objects can be partitioned into k subsets with n_1 objects in the first subset, n_2 objects in the second subset, ..., and n_k objects in the kth subset is

$$\binom{n}{n_1, n_2, \ldots, n_k} = \frac{n!}{n_1! \cdot n_2! \cdot \ldots \cdot n_k!}$$

Proof Since the n_1 objects going into the first subset can be chosen in $\binom{n}{n_1}$ ways, the n_2 objects going into the second subset can then be chosen in $\binom{n-n_1}{n_2}$ ways, the n_3 objects going into the third subset can then be chosen in $\binom{n-n_1-n_2}{n_3}$ ways, and so forth, it follows by Theorem 2 that the total number of partitions is

$$\begin{aligned}
\binom{n}{n_1, n_2, \ldots, n_k} &= \binom{n}{n_1} \cdot \binom{n-n_1}{n_2} \cdot \ldots \cdot \binom{n-n_1-n_2-\cdots-n_{k-1}}{n_k} \\
&= \frac{n!}{n_1! \cdot (n-n_1)!} \cdot \frac{(n-n_1)!}{n_2! \cdot (n-n_1-n_2)!} \\
&\quad \cdot \ldots \cdot \frac{(n-n_1-n_2-\cdots-n_{k-1})!}{n_k! \cdot 0!} \\
&= \frac{n!}{n_1! \cdot n_2! \cdot \ldots \cdot n_k!}
\end{aligned}$$

EXAMPLE 18

In how many ways can seven businessmen attending a convention be assigned to one triple and two double hotel rooms?

Solution

Substituting $n = 7, n_1 = 3, n_2 = 2$, and $n_3 = 2$ into the formula of Theorem 8, we get

$$\binom{7}{3,\ 2,\ 2} = \frac{7!}{3! \cdot 2! \cdot 2!} = 210$$

3 Binomial Coefficients

If n is a positive integer and we multiply out $(x+y)^n$ term by term, each term will be the product of x's and y's, with an x or a y coming from each of the n factors $x+y$. For instance, the expansion

$$\begin{aligned}(x+y)^3 &= (x+y)(x+y)(x+y)\\ &= x \cdot x \cdot x + x \cdot x \cdot y + x \cdot y \cdot x + x \cdot y \cdot y\\ &\quad + y \cdot x \cdot x + y \cdot x \cdot y + y \cdot y \cdot x + y \cdot y \cdot y\\ &= x^3 + 3x^2y + 3xy^2 + y^3\end{aligned}$$

yields terms of the form x^3, x^2y, xy^2, and y^3. Their coefficients are 1, 3, 3, and 1, and the coefficient of xy^2, for example, is $\binom{3}{2} = 3$, the number of ways in which we can choose the two factors providing the y's. Similarly, the coefficient of x^2y is $\binom{3}{1} = 3$, the number of ways in which we can choose the one factor providing the y, and the coefficients of x^3 and y^3 are $\binom{3}{0} = 1$ and $\binom{3}{3} = 1$.

More generally, if n is a positive integer and we multiply out $(x+y)^n$ term by term, the coefficient of $x^{n-r}y^r$ is $\binom{n}{r}$, the number of ways in which we can choose the r factors providing the y's. Accordingly, we refer to $\binom{n}{r}$ as a **binomial coefficient**. Values of the binomial coefficients for $n = 0, 1, \ldots, 20$ and $r = 0, 1, \ldots, 10$ are given in table Factorials and Binomial Coefficients of "Statistical Tables." We can now state the following theorem.

THEOREM 9.

$$(x+y)^n = \sum_{r=0}^{n} \binom{n}{r} x^{n-r}y^r \quad \text{for any positive integer } n$$

DEFINITION 3. BINOMIAL COEFFICIENTS. *The coefficient of* $x^{n-r}y^r$ *in the binomial expansion of* $(x+y)^n$ *is called the* ***binomial coefficient*** $\binom{n}{r}$.

The calculation of binomial coefficients can often be simplified by making use of the three theorems that follow.

THEOREM 10. For any positive integers n and $r = 0, 1, 2, \ldots, n$,

$$\binom{n}{r} = \binom{n}{n-r}$$

Proof We might argue that when we select a subset of r objects from a set of n distinct objects, we leave a subset of $n-r$ objects; hence, there are as many ways of selecting r objects as there are ways of leaving (or selecting) $n-r$ objects. To prove the theorem algebraically, we write

$$\binom{n}{n-r} = \frac{n!}{(n-r)![n-(n-r)]!} = \frac{n!}{(n-r)!r!}$$

$$= \frac{n!}{r!(n-r)!} = \binom{n}{r}$$

Theorem 10 implies that if we calculate the binomial coefficients for $r = 0, 1, \ldots, \frac{n}{2}$ when n is even and for $r = 0, 1, \ldots, \frac{n-1}{2}$ when n is odd, the remaining binomial coefficients can be obtained by making use of the theorem.

EXAMPLE 19

Given $\binom{4}{0} = 1$, $\binom{4}{1} = 4$, and $\binom{4}{2} = 6$, find $\binom{4}{3}$ and $\binom{4}{4}$.

Solution

$$\binom{4}{3} = \binom{4}{4-3} = \binom{4}{1} = 4 \text{ and } \binom{4}{4} = \binom{4}{4-4} = \binom{4}{0} = 1$$

EXAMPLE 20

Given $\binom{5}{0} = 1$, $\binom{5}{1} = 5$, and $\binom{5}{2} = 10$, find $\binom{5}{3}$, $\binom{5}{4}$, and $\binom{5}{5}$.

Solution

$$\binom{5}{3} = \binom{5}{5-3} = \binom{5}{2} = 10,\ \binom{5}{4} = \binom{5}{5-4} = \binom{5}{1} = 5, \text{ and}$$

$$\binom{5}{5} = \binom{5}{5-5} = \binom{5}{0} = 1$$

It is precisely in this fashion that Theorem 10 may have to be used in connection with table Factorials and Binomial Coefficients of "Statistical Tables."

EXAMPLE 21

Find $\binom{20}{12}$ and $\binom{17}{10}$.

Solution

Since $\binom{20}{12}$ is not given in the table, we make use of the fact that $\binom{20}{12} = \binom{20}{8}$, look up $\binom{20}{8}$, and get $\binom{20}{12} = 125{,}970$. Similarly, to find $\binom{17}{10}$, we make use of the fact that $\binom{17}{10} = \binom{17}{7}$, look up $\binom{17}{7}$, and get $\binom{17}{10} = 19{,}448$.

THEOREM 11. For any positive integer n and $r = 1, 2, \ldots, n-1$,

$$\binom{n}{r} = \binom{n-1}{r} + \binom{n-1}{r-1}$$

Proof Substituting $x = 1$ into $(x+y)^n$, let us write $(1+y)^n = (1+y)(1+y)^{n-1} = (1+y)^{n-1} + y(1+y)^{n-1}$ and equate the coefficient of y^r in $(1+y)^n$ with that in $(1+y)^{n-1} + y(1+y)^{n-1}$. Since the coefficient of y^r in $(1+y)^n$ is $\binom{n}{r}$ and the coefficient of y^r in $(1+y)^{n-1} + y(1+y)^{n-1}$ is the sum of the coefficient of y^r in $(1+y)^{n-1}$, that is, $\binom{n-1}{r}$, and the coefficient of y^{r-1} in $(1+y)^{n-1}$, that is, $\binom{n-1}{r-1}$, we obtain

$$\binom{n}{r} = \binom{n-1}{r} + \binom{n-1}{r-1}$$

which completes the proof.

Alternatively, take any one of the n objects. If it is not to be included among the r objects, there are $\binom{n-1}{r}$ ways of selecting the r objects; if it is to be included, there are $\binom{n-1}{r-1}$ ways of selecting the other $r-1$ objects. Therefore, there are $\binom{n-1}{r} + \binom{n-1}{r-1}$ ways of selecting the r objects, that is,

$$\binom{n}{r} = \binom{n-1}{r} + \binom{n-1}{r-1}$$

Theorem 11 can also be proved by expressing the binomial coefficients on both sides of the equation in terms of factorials and then proceeding algebraically, but we shall leave this to the reader in Exercise 12.

An important application of Theorem 11 is a construct known as **Pascal's triangle**. When no table is available, it is sometimes convenient to determine binomial coefficients by means of a simple construction. Applying Theorem 11, we can generate Pascal's triangle as follows:

$$\begin{array}{ccccccccccc} &&&&&1&&&&& \\ &&&&1&&1&&&& \\ &&&1&&2&&1&&& \\ &&1&&3&&3&&1&& \\ &1&&4&&6&&4&&1& \\ 1&&5&&10&&10&&5&&1 \\ &&&&&\cdots&&&&& \end{array}$$

In this triangle, the first and last entries of each row are the numeral "1" each other entry in any given row is obtained by adding the two entries in the preceding row immediately to its left and to its right.

To state the third theorem about binomial coefficients, let us make the following definition: $\binom{n}{r} = 0$ whenever n is a positive integer and r is a positive integer greater than n. (Clearly, there is no way in which we can select a subset that contains more elements than the whole set itself.)

THEOREM 12.

$$\sum_{r=0}^{k} \binom{m}{r}\binom{n}{k-r} = \binom{m+n}{k}$$

Proof Using the same technique as in the proof of Theorem 11, let us prove this theorem by equating the coefficients of y^k in the expressions on both sides of the equation

$$(1+y)^{m+n} = (1+y)^m(1+y)^n$$

The coefficient of y^k in $(1+y)^{m+n}$ is $\binom{m+n}{k}$, and the coefficient of y^k in

$$(1+y)^m(1+y)^n = \left[\binom{m}{0} + \binom{m}{1}y + \cdots + \binom{m}{m}y^m\right] \times \left[\binom{n}{0} + \binom{n}{1}y + \cdots + \binom{n}{n}y^n\right]$$

is the sum of the products that we obtain by multiplying the constant term of the first factor by the coefficient of y^k in the second factor, the coefficient of y in the first factor by the coefficient of y^{k-1} in the second factor, ..., and the coefficient of y^k in the first factor by the constant term of the second factor. Thus, the coefficient of y^k in $(1+y)^m(1+y)^n$ is

$$\binom{m}{0}\binom{n}{k} + \binom{m}{1}\binom{n}{k-1} + \binom{m}{2}\binom{n}{k-2} + \cdots + \binom{m}{k}\binom{n}{0}$$

$$= \sum_{r=0}^{k} \binom{m}{r}\binom{n}{k-r}$$

and this completes the proof.

EXAMPLE 22

Verify Theorem 12 numerically for $m = 2, n = 3$, and $k = 4$.

Solution

Substituting these values, we get

$$\binom{2}{0}\binom{3}{4} + \binom{2}{1}\binom{3}{3} + \binom{2}{2}\binom{3}{2} + \binom{2}{3}\binom{3}{1} + \binom{2}{4}\binom{3}{0} = \binom{5}{4}$$

and since $\binom{3}{4}$, $\binom{2}{3}$, and $\binom{2}{4}$ equal 0 according to the definition on the previous page, the equation reduces to

$$\binom{2}{1}\binom{3}{3} + \binom{2}{2}\binom{3}{2} = \binom{5}{4}$$

which checks, since $2 \cdot 1 + 1 \cdot 3 = 5$.

Using Theorem 8, we can extend our discussion to **multinomial coefficients**, that is, to the coefficients that arise in the expansion of $(x_1 + x_2 + \cdots + x_k)^n$. The multinomial coefficient of the term $x_1^{r_1} \cdot x_2^{r_2} \cdot \ldots \cdot x_k^{r_k}$ in the expansion of $(x_1 + x_2 + \cdots + x_k)^n$ is

$$\binom{n}{r_1, r_2, \ldots, r_k} = \frac{n!}{r_1! \cdot r_2! \cdot \ldots \cdot r_k!}$$

EXAMPLE 23

What is the coefficient of $x_1^3 x_2 x_3^2$ in the expansion of $(x_1 + x_2 + x_3)^6$?

Solution

Substituting $n = 6, r_1 = 3, r_2 = 1$, and $r_3 = 2$ into the preceding formula, we get

$$\frac{6!}{3! \cdot 1! \cdot 2!} = 60$$

Exercises

1. An operation consists of two steps, of which the first can be made in n_1 ways. If the first step is made in the ith way, the second step can be made in n_{2i} ways.†

(a) Use a tree diagram to find a formula for the total number of ways in which the total operation can be made.

(b) A student can study 0, 1, 2, or 3 hours for a history test on any given day. Use the formula obtained in part (a) to verify that there are 13 ways in which the student can study at most 4 hours for the test on two consecutive days.

2. With reference to Exercise 1, verify that if n_{2i} equals the constant n_2, the formula obtained in part (a) reduces to that of Theorem 1.

3. With reference to Exercise 1, suppose that there is a third step, and if the first step is made in the ith way and the second step in the jth way, the third step can be made in n_{3ij} ways.

(a) Use a tree diagram to verify that the whole operation can be made in

† The first subscript denotes the row to which a particular element belongs, and the second subscript denotes the column.

$$\sum_{i=1}^{n_1}\sum_{j=1}^{n_{2i}} n_{3ij}$$

different ways.

(b) With reference to part (b) of Exercise 1, use the formula of part (a) to verify that there are 32 ways in which the student can study at most 4 hours for the test on three consecutive days.

4. Show that if n_{2i} equals the constant n_2 and n_{3ij} equals the constant n_3, the formula of part (a) of Exercise 3 reduces to that of Theorem 2.

5. In a two-team basketball play-off, the winner is the first team to win m games.

(a) Counting separately the number of play-offs requiring $m, m+1, \ldots,$ and $2m-1$ games, show that the total number of different outcomes (sequences of wins and losses by one of the teams) is

$$2\left[\binom{m-1}{m-1}+\binom{m}{m-1}+\cdots+\binom{2m-2}{m-1}\right]$$

(b) How many different outcomes are there in a "2 out of 3" play-off, a "3 out of 5" play-off, and a "4 out of 7" play-off?

6. When n is large, $n!$ can be approximated by means of the expression

$$\sqrt{2\pi n}\left(\frac{n}{e}\right)^n$$

called **Stirling's formula**, where e is the base of natural logarithms. (A derivation of this formula may be found in the book by W. Feller cited among the references at the end of this chapter.)

(a) Use Stirling's formula to obtain approximations for 10! and 12!, and find the percentage errors of these approximations by comparing them with the exact values given in table Factorials and Binomial Coefficients of "Statistical Tables."

(b) Use Stirling's formula to obtain an approximation for the number of 13-card bridge hands that can be dealt with an ordinary deck of 52 playing cards.

7. Using Stirling's formula (see Exercise 6) to approximate $2n!$ and $n!$, show that

$$\frac{\binom{2n}{n}\sqrt{\pi n}}{2^{2n}} \approx 1$$

8. In some problems of **occupancy theory** we are concerned with the number of ways in which certain *distinguishable* objects can be distributed among individuals, urns, boxes, or cells. Find an expression for the number of ways in which r *distinguishable* objects can be distributed among n cells, and use it to find the number of ways in which three different books can be distributed among the 12 students in an English literature class.

9. In some problems of occupancy theory we are concerned with the number of ways in which certain *indistinguishable* objects can be distributed among individuals, urns, boxes, or cells. Find an expression for the number of ways in which r *indistinguishable* objects can be distributed among n cells, and use it to find the number of ways in which a baker can sell five (indistinguishable) loaves of bread to three customers. (*Hint*: We might argue that L|LLL|L represents the case where the three customers buy one loaf, three loaves, and one loaf, respectively, and that LLLL||L represents the case where the three customers buy four loaves, none of the loaves, and one loaf. Thus, we must look for the number of ways in which we can arrange the five L's and the two vertical bars.)

10. In some problems of occupancy theory we are concerned with the number of ways in which certain *indistinguishable* objects can be distributed among individuals, urns, boxes, or cells with at least one in each cell. Find an expression for the number of ways in which r *indistinguishable* objects can be distributed among n cells with at least one in each cell, and rework the numerical part of Exercise 9 with each of the three customers getting at least one loaf of bread.

11. Construct the seventh and eighth rows of Pascal's triangle and write the binomial expansions of $(x+y)^6$ and $(x+y)^7$.

12. Prove Theorem 11 by expressing all the binomial coefficients in terms of factorials and then simplifying algebraically.

13. Expressing the binomial coefficients in terms of factorials and simplifying algebraically, show that

(a) $\binom{n}{r} = \frac{n-r+1}{r}\cdot\binom{n}{r-1}$;

(b) $\binom{n}{r} = \frac{n}{n-r}\cdot\binom{n-1}{r}$;

(c) $n\binom{n-1}{r} = (r+1)\binom{n}{r+1}$.

14. Substituting appropriate values for x and y into the formula of Theorem 9, show that

(a) $\sum_{r=0}^{n}\binom{n}{r} = 2^n$;

(b) $\sum_{r=0}^{n}(-1)^r\binom{n}{r} = 0$;

(c) $\sum_{r=0}^{n}\binom{n}{r}(a-1)^r = a^n$.

15. Repeatedly applying Theorem 11, show that

$$\binom{n}{r} = \sum_{i=1}^{r+1} \binom{n-i}{r-i+1}$$

16. Use Theorem 12 to show that

$$\sum_{r=0}^{n} \binom{n}{r}^2 = \binom{2n}{n}$$

17. Show that $\sum_{r=0}^{n} r\binom{n}{r} = n2^{n-1}$ by setting $x = 1$ in Theorem 9, then differentiating the expressions on both sides with respect to y, and finally substituting $y = 1$.

18. Rework Exercise 17 by making use of part (a) of Exercise 14 and part (c) of Exercise 13.

19. If n is not a positive integer or zero, the binomial expansion of $(1+y)^n$ yields, for $-1 < y < 1$, the infinite series

$$1 + \binom{n}{1}y + \binom{n}{2}y^2 + \binom{n}{3}y^3 + \cdots + \binom{n}{r}y^r + \cdots$$

where $\binom{n}{r} = \dfrac{n(n-1)\cdot\ldots\cdot(n-r+1)}{r!}$ for $r = 1, 2, 3, \ldots$.

Use this **generalized definition of binomial coefficients** to evaluate

(a) $\binom{\frac{1}{2}}{4}$ and $\binom{-3}{3}$;

(b) $\sqrt{5}$ writing $\sqrt{5} = 2(1+\frac{1}{4})^{1/2}$ and using the first four terms of the binomial expansion of $(1+\frac{1}{4})^{1/2}$.

20. With reference to the generalized definition of binomial coefficients in Exercise 19, show that

(a) $\binom{-1}{r} = (-1)^r$;

(b) $\binom{-n}{r} = (-1)^r\binom{n+r-1}{r}$ for $n > 0$.

21. Find the coefficient of $x^2y^3z^3$ in the expansion of $(x+y+z)^8$.

22. Find the coefficient of $x^3y^2z^3w$ in the expansion of $(2x+3y-4z+w)^9$.

23. Show that

$$\binom{n}{n_1, n_2, \ldots, n_k} = \binom{n-1}{n_1-1, n_2, \ldots, n_k} + \binom{n-1}{n_1, n_2-1, \ldots, n_k} + \cdots + \binom{n-1}{n_1, n_2, \ldots, n_k-1}$$

by expressing all these multinomial coefficients in terms of factorials and simplifying algebraically.

4 The Theory in Practice

Applications of the preceding theory of combinatorial methods and binomial coefficients are quite straightforward, and a variety of them have been given in Sections 2 and 3. The following examples illustrate further applications of this theory.

EXAMPLE 24

An assembler of electronic equipment has 20 integrated-circuit chips on her table, and she must solder three of them as part of a larger component. In how many ways can she choose the three chips for assembly?

Solution

Using Theorem 6, we obtain the result

$${}_{20}P_3 = 20!/17! = 20 \cdot 19 \cdot 18 = 6{,}840$$

EXAMPLE 25

A lot of manufactured goods, presented for sampling inspection, contains 16 units. In how many ways can 4 of the 16 units be selected for inspection?

Solution
According to Theorem 7,

$$\binom{16}{4} = 16!/4!12! = 16 \cdot 15 \cdot 14 \cdot 13/4 \cdot 3 \cdot 2 \cdot 1 = 1{,}092 \text{ ways}$$

Applied Exercises SECS. 1–4

24. A thermostat will call for heat 0, 1, or 2 times a night. Construct a tree diagram to show that there are 10 different ways that it can turn on the furnace for a total of 6 times over 4 nights.

25. On August 31 there are five wild-card terms in the American League that can make it to the play-offs, and only two will win spots. Draw a tree diagram which shows the various possible play-off wild-card teams.

26. There are four routes, A, B, C, and D, between a person's home and the place where he works, but route B is one-way, so he cannot take it on the way to work, and route C is one-way, so he cannot take it on the way home.
(a) Draw a tree diagram showing the various ways the person can go to and from work.
(b) Draw a tree diagram showing the various ways he can go to and from work without taking the same route both ways.

27. A person with $2 in her pocket bets $1, even money, on the flip of a coin, and she continues to bet $1 as long as she has any money. Draw a tree diagram to show the various things that can happen during the first four flips of the coin. After the fourth flip of the coin, in how many of the cases will she be
(a) exactly even;
(b) exactly $2 ahead?

28. The pro at a golf course stocks two identical sets of women's clubs, reordering at the end of each day (for delivery early the next morning) if and only if he has sold them both. Construct a tree diagram to show that if he starts on a Monday with two sets of the clubs, there are altogether eight different ways in which he can make sales on the first two days of that week.

29. Suppose that in a baseball World Series (in which the winner is the first team to win four games) the National League champion leads the American League champion three games to two. Construct a tree diagram to show the number of ways in which these teams may win or lose the remaining game or games.

30. If the NCAA has applications from six universities for hosting its intercollegiate tennis championships in two consecutive years, in how many ways can they select the hosts for these championships
(a) if they are not both to be held at the same university;
(b) if they may both be held at the same university?

31. Counting the number of outcomes in games of chance has been a popular pastime for many centuries. This was of interest not only because of the gambling that was involved, but also because the outcomes of games of chance were often interpreted as divine intent. Thus, it was just about a thousand years ago that a bishop in what is now Belgium determined that there are 56 different ways in which three dice can fall *provided one is interested only in the overall result and not in which die does what.* He assigned a virtue to each of these possibilities and each sinner had to concentrate for some time on the virtue that corresponded to his cast of the dice.
(a) Find the number of ways in which three dice can all come up with the same number of points.
(b) Find the number of ways in which two of the three dice can come up with the same number of points, while the third comes up with a different number of points.
(c) Find the number of ways in which all three of the dice can come up with a different number of points.
(d) Use the results of parts (a), (b), and (c) to verify the bishop's calculations that there are altogether 56 possibilities.

32. In a primary election, there are four candidates for mayor, five candidates for city treasurer, and two candidates for county attorney.
(a) In how many ways can a voter mark his ballot for all three of these offices?
(b) In how many ways can a person vote if he exercises his option of not voting for a candidate for any or all of these offices?

33. The five finalists in the Miss Universe contest are Miss Argentina, Miss Belgium, Miss U.S.A., Miss Japan, and Miss Norway. In how many ways can the judges choose
(a) the winner and the first runner-up;
(b) the winner, the first runner-up, and the second runner-up?

34. A multiple-choice test consists of 15 questions, each permitting a choice of three alternatives. In how many different ways can a student check off her answers to these questions?

35. Determine the number of ways in which a distributor can choose 2 of 15 warehouses to ship a large order.

36. The price of a European tour includes four stopovers to be selected from among 10 cities. In how many different ways can one plan such a tour
(a) if the order of the stopovers matters;
(b) if the order of the stopovers does not matter?

37. A carton of 15 light bulbs contains one that is defective. In how many ways can an inspector choose 3 of the bulbs and
(a) get the one that is defective;
(b) not get the one that is defective?

38. In how many ways can a television director schedule a sponsor's six different commercials during the six time slots allocated to commercials during a two-hour program?

39. In how many ways can the television director of Exercise 38 fill the six time slots for commercials if there are three different sponsors and the commercial for each is to be shown twice?

40. In how many ways can five persons line up to get on a bus? In how many ways can they line up if two of the persons refuse to follow each other?

41. In how many ways can eight persons form a circle for a folk dance?

42. How many permutations are there of the letters in the word
(a) "great";
(b) "greet"?

43. How many distinct permutations are there of the letters in the word "statistics"? How many of these begin and end with the letter s?

44. A college team plays 10 football games during a season. In how many ways can it end the season with five wins, four losses, and one tie?

45. If eight persons are having dinner together, in how many different ways can three order chicken, four order steak, and one order lobster?

46. In Example 4 we showed that a true–false test consisting of 20 questions can be marked in 1,048,576 different ways. In how many ways can each question be marked true or false so that
(a) 7 are right and 13 are wrong;
(b) 10 are right and 10 are wrong;
(c) at least 17 are right?

47. Among the seven nominees for two vacancies on a city council are three men and four women. In how many ways can these vacancies be filled
(a) with any two of the seven nominees;
(b) with any two of the four women;
(c) with one of the men and one of the women?

48. A shipment of 10 television sets includes three that are defective. In how many ways can a hotel purchase four of these sets and receive at least two of the defective sets?

49. Ms. Jones has four skirts, seven blouses, and three sweaters. In how many ways can she choose two of the skirts, three of the blouses, and one of the sweaters to take along on a trip?

50. How many different bridge hands are possible containing five spades, three diamonds, three clubs, and two hearts?

51. Find the number of ways in which one A, three B's, two C's, and one F can be distributed among seven students taking a course in statistics.

52. An art collector, who owns 10 paintings by famous artists, is preparing her will. In how many different ways can she leave these paintings to her three heirs?

53. A baseball fan has a pair of tickets for six different home games of the Chicago Cubs. If he has five friends who like baseball, in how many different ways can he take one of them along to each of the six games?

54. At the end of the day, a bakery gives everything that is unsold to food banks for the needy. If it has 12 apple pies left at the end of a given day, in how many different ways can it distribute these pies among six food banks for the needy?

55. With reference to Exercise 54, in how many different ways can the bakery distribute the 12 apple pies if each of the six food banks is to receive at least one pie?

56. On a Friday morning, the pro shop of a tennis club has 14 identical cans of tennis balls. If they are all sold by Sunday night and we are interested only in how many were sold on each day, in how many different ways could the tennis balls have been sold on Friday, Saturday, and Sunday?

57. Rework Exercise 56 given that at least two of the cans of tennis balls were sold on each of the three days.

References

Among the books on the history of statistics there are

WALKER, H. M., *Studies in the History of Statistical Method*. Baltimore: The Williams & Wilkins Company, 1929,

WESTERGAARD, H., *Contributions to the History of Statistics*. London: P. S. King & Son, 1932,

and the more recent publications

KENDALL, M. G., and PLACKETT, R. L., eds., *Studies in the History of Statistics and Probability*, Vol. II. New York: Macmillan Publishing Co., Inc., 1977,

PEARSON, E. S., and KENDALL, M. G., eds., *Studies in the History of Statistics and Probability*. Darien, Conn.: Hafner Publishing Co., Inc., 1970,

PORTER, T. M., *The Rise of Statistical Thinking, 1820–1900*. Princeton, N.J.: Princeton University Press, 1986,

and

STIGLER, S. M., *The History of Statistics*. Cambridge, Mass.: Harvard University Press, 1986.

A wealth of material on combinatorial methods can be found in

COHEN, D. A., *Basic Techniques of Combinatorial Theory*. New York: John Wiley & Sons, Inc., 1978,

EISEN, M., *Elementary Combinatorial Analysis*. New York: Gordon and Breach, Science Publishers, Inc., 1970,

FELLER, W., *An Introduction to Probability Theory and Its Applications*, Vol. I, 3rd ed. New York: John Wiley & Sons, Inc., 1968,

NIVEN, J., *Mathematics of Choice*. New York: Random House, Inc., 1965,

ROBERTS, F. S., *Applied Combinatorics*. Upper Saddle River, N.J.: Prentice Hall, 1984,

and

WHITWORTH, W. A., *Choice and Chance*, 5th ed. New York: Hafner Publishing Co., Inc., 1959, which has become a classic in this field.

More advanced treatments may be found in

BECKENBACH, E. F., ed., *Applied Combinatorial Mathematics*. New York: John Wiley & Sons, Inc., 1964,

DAVID, F. N., and BARTON, D. E., *Combinatorial Chance*. New York: Hafner Publishing Co., Inc., 1962,

and

RIORDAN, J., *An Introduction to Combinatorial Analysis*. New York: John Wiley & Sons, Inc., 1958.

Answers to Odd-Numbered Exercises

1 **(a)** $\sum_{i=1}^{n} n_{2n_i}$.

5 **(b)** 6, 20, and 70.

9 $\binom{r+n-1}{r}$ and 21.

11 Seventh row: 1, 6, 15, 20, 15, 6, 1; Eighth row: 1, 7, 21, 35, 35, 27, 7, 1.

19 **(a)** $\frac{-15}{384}$ and -10; **(b)** 2.230.

21 560.

27 **(a)** 5; **(b)** 4.

31 **(a)** 6; **(b)** 30; **(c)** 20; **(d)** 56.

33 **(a)** 20; **(b)** 60.

35 **(a)** 105.

37 **(a)** 91; **(b)** 364.

39 90.

41 5040.

43 50,400 and 3360.

45 280.

47 **(a)** 21; **(b)** 6; **(c)** 12.

49 630.

51 420.

53 15,625.

55 462.

57 45.

PROBABILITY

1 Introduction

Historically, the oldest way of defining probabilities, the **classical probability concept**, applies when *all possible outcomes are equally likely*, as is presumably the case in most games of chance. We can then say that *if there are* N *equally likely possibilities, of which one must occur and* n *are regarded as favorable, or as a "success," then the probability of a "success" is given by the ratio* $\frac{n}{N}$.

EXAMPLE 1

What is the probability of drawing an ace from an ordinary deck of 52 playing cards?

Solution

Since there are $n = 4$ aces among the $N = 52$ cards, the probability of drawing an ace is $\frac{4}{52} = \frac{1}{13}$. (It is assumed, of course, that each card has the same chance of being drawn.)

Although equally likely possibilities are found mostly in games of chance, the classical probability concept applies also in a great variety of situations where gambling devices are used to make random selections—when office space is assigned to teaching assistants by lot, when some of the families in a township are chosen in such a way that each one has the same chance of being included in a sample study, when machine parts are chosen for inspection so that each part produced has the same chance of being selected, and so forth.

A major shortcoming of the classical probability concept is its limited applicability, for there are many situations in which the possibilities that arise cannot all be regarded as equally likely. This would be the case, for instance, if we are concerned with the question whether it will rain on a given day, if we are concerned with the outcome of an election, or if we are concerned with a person's recovery from a disease.

Among the various probability concepts, most widely held is the **frequency interpretation**, according to which *the probability of an event* (*outcome or happening*) is the proportion of the time that events of the same kind will occur in the long run. If we say that the probability is 0.84 that a jet from Los Angeles to San Francisco will arrive on time, we mean (in accordance with the frequency interpretation) that such flights arrive on time 84 percent of the time. Similarly, if the weather bureau

From Chapter 2 of *John E. Freund's Mathematical Statistics with Applications*, Eighth Edition. Irwin Miller, Marylees Miller.

predicts that there is a 30 percent chance for rain (that is, a probability of 0.30), this means that under the same weather conditions it will rain 30 percent of the time. More generally, we say that an event has a probability of, say, 0.90, in the same sense in which we might say that our car will start in cold weather 90 percent of the time. We cannot guarantee what will happen on any particular occasion—the car may start and then it may not—but if we kept records over a long period of time, we should find that the proportion of "successes" is very close to 0.90.

The approach to probability that we shall use in this chapter is the **axiomatic approach**, in which probabilities are defined as "mathematical objects" that behave according to certain well-defined rules. Then, any one of the preceding probability concepts, or interpretations, can be used in applications as long as it is consistent with these rules.

2 Sample Spaces

Since all probabilities pertain to the occurrence or nonoccurrence of events, let us explain first what we mean here by *event* and by the related terms *experiment, outcome*, and *sample space*.

It is customary in statistics to refer to any process of observation or measurement as an **experiment**. In this sense, an experiment may consist of the simple process of checking whether a switch is turned on or off; it may consist of counting the imperfections in a piece of cloth; or it may consist of the very complicated process of determining the mass of an electron. The results one obtains from an experiment, whether they are instrument readings, counts, "yes" or "no" answers, or values obtained through extensive calculations, are called the **outcomes** of the experiment.

DEFINITION 1. SAMPLE SPACE. *The set of all possible outcomes of an experiment is called the **sample space** and it is usually denoted by the letter* S. *Each outcome in a sample space is called an **element** of the sample space, or simply a **sample point**.*

If a sample space has a finite number of elements, we may list the elements in the usual set notation; for instance, the sample space for the possible outcomes of one flip of a coin may be written

$$S = \{\mathrm{H}, \mathrm{T}\}$$

where H and T stand for head and tail. Sample spaces with a large or infinite number of elements are best described by a statement or rule; for example, if the possible outcomes of an experiment are the set of automobiles equipped with satellite radios, the sample space may be written

$$S = \{x|x \text{ is an automobile with a satellite radio}\}$$

This is read "S is the set of all x such that x is an automobile with a satellite radio." Similarly, if S is the set of odd positive integers, we write

$$S = \{2k+1|k = 0, 1, 2, \ldots\}$$

How we formulate the sample space for a given situation will depend on the problem at hand. If an experiment consists of one roll of a die and we are interested in which face is turned up, we would use the sample space

$$S_1 = \{1, 2, 3, 4, 5, 6\}$$

However, if we are interested only in whether the face turned up is even or odd, we would use the sample space

$$S_2 = \{\text{even, odd}\}$$

This demonstrates that different sample spaces may well be used to describe an experiment. In general, *it is desirable to use sample spaces whose elements cannot be divided (partitioned or separated) into more primitive or more elementary kinds of outcomes.* In other words, *it is preferable that an element of a sample space not represent two or more outcomes that are distinguishable in some way.* Thus, in the preceding illustration S_1 would be preferable to S_2.

EXAMPLE 2

Describe a sample space that might be appropriate for an experiment in which we roll a pair of dice, one red and one green. (The different colors are used to emphasize that the dice are distinct from one another.)

Solution

The sample space that provides the most information consists of the 36 points given by

$$S_1 = \{(x, y) | x = 1, 2, \ldots, 6; y = 1, 2, \ldots, 6\}$$

where x represents the number turned up by the red die and y represents the number turned up by the green die. A second sample space, adequate for most purposes (though less desirable in general as it provides less information), is given by

$$S_2 = \{2, 3, 4, \ldots, 12\}$$

where the elements are the totals of the numbers turned up by the two dice.

Sample spaces are usually classified according to the number of elements that they contain. In the preceding example the sample spaces S_1 and S_2 contained a **finite** number of elements; but if a coin is flipped until a head appears for the first time, this could happen on the first flip, the second flip, the third flip, the fourth flip, ..., and there are infinitely many possibilities. For this experiment we obtain the sample space

$$S = \{\text{H, TH, TTH, TTTH, TTTTH}, \ldots\}$$

with an unending sequence of elements. But even here the number of elements can be matched one-to-one with the whole numbers, and in this sense the sample space is said to be **countable**. If a sample space contains a finite number of elements or an infinite though countable number of elements, it is said to be **discrete**.

The outcomes of some experiments are neither finite nor countably infinite. Such is the case, for example, when one conducts an investigation to determine the distance that a certain make of car will travel over a prescribed test course on 5 liters of gasoline. If we assume that distance is a variable that can be measured to any desired degree of accuracy, there is an infinity of possibilities (distances) that cannot be matched one-to-one with the whole numbers. Also, if we want to measure the amount of time it takes for two chemicals to react, the amounts making up the sample space are infinite in number and not countable. Thus, sample spaces need not be discrete. If a sample space consists of a continuum, such as all the points of a line segment or all the points in a plane, it is said to be **continuous**. Continuous sample spaces arise in practice whenever the outcomes of experiments are measurements of physical properties, such as temperature, speed, pressure, length, ..., that are measured on continuous scales.

3 Events

In many problems we are interested in results that are not given directly by a specific element of a sample space.

EXAMPLE 3

With reference to the first sample space S_1 on the previous page, describe the event A that the number of points rolled with the die is divisible by 3.

Solution

Among 1, 2, 3, 4, 5, and 6, only 3 and 6 are divisible by 3. Therefore, A is represented by the subset $\{3, 6\}$ of the sample space S_1.

EXAMPLE 4

With reference to the sample space S_1 of Example 2, describe the event B that the total number of points rolled with the pair of dice is 7.

Solution

Among the 36 possibilities, only (1, 6), (2, 5), (3, 4), (4, 3), (5, 2), and (6, 1) yield a total of 7. So, we write

$$B = \{(1,6), (2,5), (3,4), (4,3), (5,2), (6,1)\}$$

Note that in Figure 1 the event of rolling a total of 7 with the two dice is represented by the set of points inside the region bounded by the dotted line.

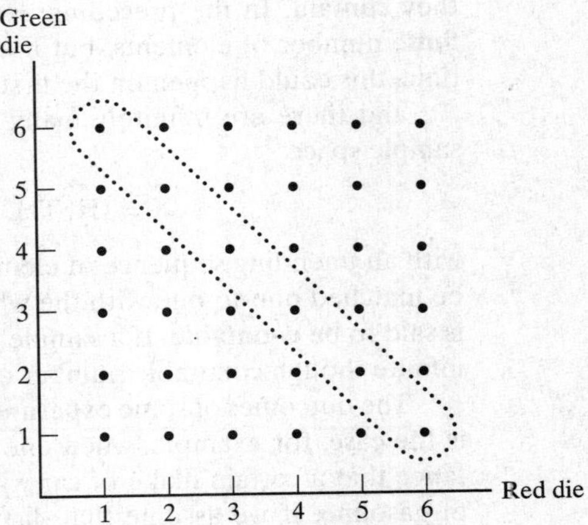

Figure 1. Rolling a total of 7 with a pair of dice.

In the same way, any event (outcome or result) can be identified with a collection of points, which constitute a subset of an appropriate sample space. Such a subset consists of all the elements of the sample space for which the event occurs, and in probability and statistics we identify the subset with the event.

DEFINITION 2. EVENT. *An **event** is a subset of a sample space.*

EXAMPLE 5

If someone takes three shots at a target and we care only whether each shot is a hit or a miss, describe a suitable sample space, the elements of the sample space that constitute event M that the person will miss the target three times in a row, and the elements of event N that the person will hit the target once and miss it twice.

Solution

If we let 0 and 1 represent a miss and a hit, respectively, the eight possibilities (0, 0, 0), (1, 0, 0), (0, 1, 0), (0, 0, 1), (1, 1, 0), (1, 0, 1), (0, 1, 1), and (1, 1, 1) may be displayed as in Figure 2. Thus, it can be seen that

$$M = \{(0, 0, 0)\}$$

and

$$N = \{(1, 0, 0), (0, 1, 0), (0, 0, 1)\}$$

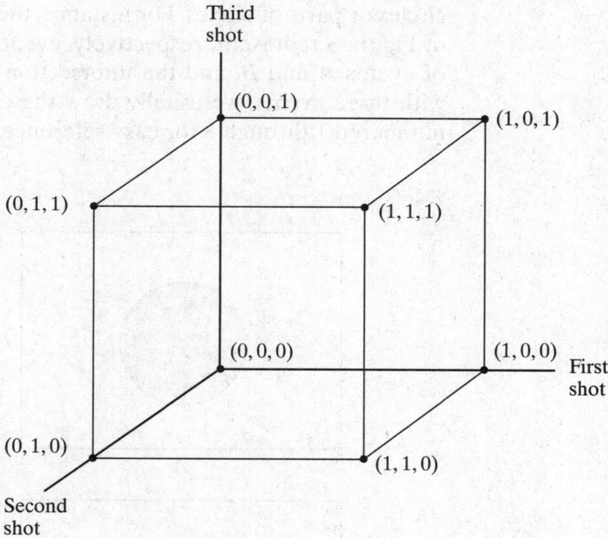

Figure 2. Sample space for Example 5.

EXAMPLE 6

Construct a sample space for the length of the useful life of a certain electronic component and indicate the subset that represents the event F that the component fails before the end of the sixth year.

Solution

If t is the length of the component's useful life in years, the sample space may be written $S = \{t|t \geqq 0\}$, and the subset $F = \{t|0 \leqq t < 6\}$ is the event that the component fails before the end of the sixth year.

According to our definition, any event is a subset of an appropriate sample space, but it should be observed that the converse is not necessarily true. For discrete sample spaces, all subsets are events, but in the continuous case some rather abstruse point sets must be excluded for mathematical reasons. This is discussed further in some of the more advanced texts listed among the references at the end of this chapter.

In many problems of probability we are interested in events that are actually combinations of two or more events, formed by taking **unions**, **intersections**, and **complements**. Although the reader must surely be familiar with these terms, let us review briefly that, if A and B are any two subsets of a sample space S, their union $A \cup B$ is the subset of S that contains all the elements that are either in A, in B, or in both; their intersection $A \cap B$ is the subset of S that contains all the elements that are in both A and B; and the complement A' of A is the subset of S that contains all the elements of S that are not in A. Some of the rules that control the formation of unions, intersections, and complements may be found in Exercises 1 through 4.

Sample spaces and events, particularly relationships among events, are often depicted by means of **Venn diagrams**, in which the sample space is represented by a rectangle, while events are represented by regions within the rectangle, usually by circles or parts of circles. For instance, the shaded regions of the four Venn diagrams of Figure 3 represent, respectively, event A, the complement of event A, the union of events A and B, and the intersection of events A and B. When we are dealing with three events, we usually draw the circles as in Figure 4. Here, the regions are numbered 1 through 8 for easy reference.

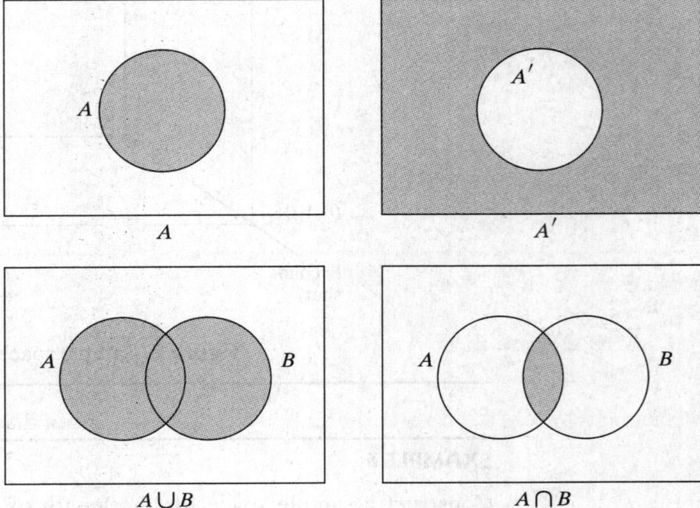

Figure 3. Venn diagrams.

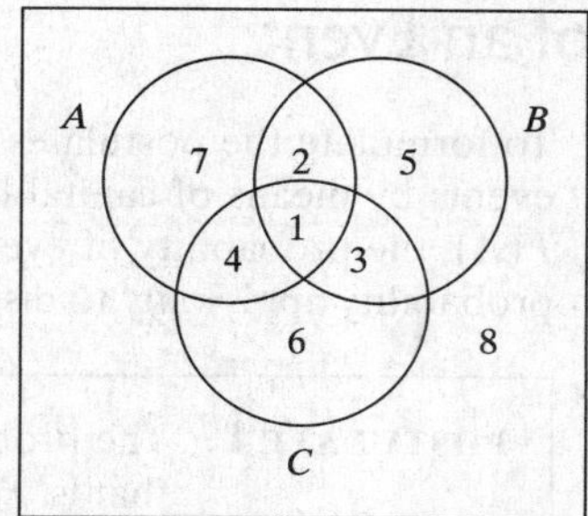

Figure 4. Venn diagram.

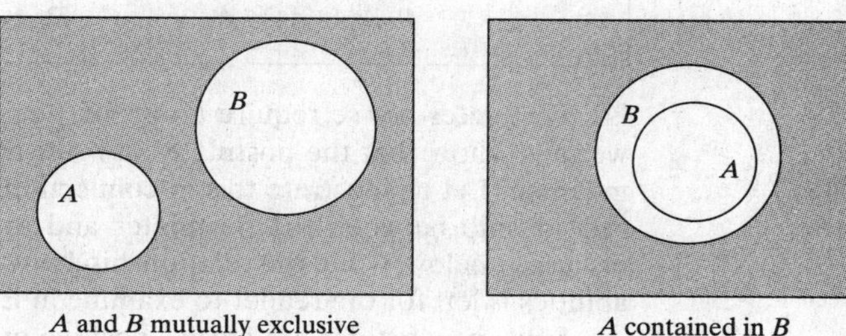

Figure 5. Diagrams showing special relationships among events.

To indicate special relationships among events, we sometimes draw diagrams like those of Figure 5. Here, the one on the left serves to indicate that events A and B are **mutually exclusive**.

> **DEFINITION 3. MUTUALLY EXCLUSIVE EVENTS.** *Two events having no elements in common are said to be* ***mutually exclusive.***

When A and B are mutually exclusive, we write $A \cap B = \emptyset$, where $\emptyset$ denotes the **empty set**, which has no elements at all. The diagram on the right serves to indicate that A is contained in B, and symbolically we express this by writing $A \subset B$.

Exercises

1. Use Venn diagrams to verify that
(a) $(A \cup B) \cup C$ is the same event as $A \cup (B \cup C)$;
(b) $A \cap (B \cup C)$ is the same event as $(A \cap B) \cup (A \cap C)$;
(c) $A \cup (B \cap C)$ is the same event as $(A \cup B) \cap (A \cup C)$.

2. Use Venn diagrams to verify the two **De Morgan laws**:
(a) $(A \cap B)' = A' \cup B'$; **(b)** $(A \cup B)' = A' \cap B'$.

3. Use Venn diagrams to verify that
(a) $(A \cap B) \cup (A \cap B') = A$;
(b) $(A \cap B) \cup (A \cap B') \cup (A' \cap B) = A \cup B$;
(c) $A \cup (A' \cap B) = A \cup B$.

4. Use Venn diagrams to verify that if A is contained in B, then $A \cap B = A$ and $A \cap B' = \emptyset$.

4 The Probability of an Event

To formulate the postulates of probability, we shall follow the practice of denoting events by means of capital letters, and we shall write the probability of event A as $P(A)$, the probability of event B as $P(B)$, and so forth. The following postulates of probability apply only to discrete sample spaces, S.

POSTULATE 1 The probability of an event is a nonnegative real number; that is, $P(A) \geqq 0$ for any subset A of S.

POSTULATE 2 $P(S) = 1$.

POSTULATE 3 If $A_1, A_2, A_3, \ldots$, is a finite or infinite sequence of mutually exclusive events of S, then

$$P(A_1 \cup A_2 \cup A_3 \cup \cdots) = P(A_1) + P(A_2) + P(A_3) + \cdots$$

Postulates per se require no proof, but if the resulting theory is to be applied, we must show that the postulates are satisfied when we give probabilities a "real" meaning. Let us illustrate this in connection with the frequency interpretation; the relationship between the postulates and the classical probability concept will be discussed below, while the relationship between the postulates and subjective probabilities is left for the reader to examine in Exercises 16 and 82.

Since proportions are always positive or zero, the first postulate is in complete agreement with the frequency interpretation. The second postulate states indirectly that certainty is identified with a probability of 1; after all, it is always assumed that one of the possibilities in S must occur, and it is to this certain event that we assign a probability of 1. As far as the frequency interpretation is concerned, a probability of 1 implies that the event in question will occur 100 percent of the time or, in other words, that it is certain to occur.

Taking the third postulate in the simplest case, that is, for two mutually exclusive events A_1 and A_2, it can easily be seen that it is satisfied by the frequency interpretation. If one event occurs, say, 28 percent of the time, another event occurs 39 percent of the time, and the two events cannot both occur at the same time (that is, they are mutually exclusive), then one or the other will occur $28 + 39 = 67$ percent of the time. Thus, the third postulate is satisfied, and the same kind of argument applies when there are more than two mutually exclusive events.

Before we study some of the immediate consequences of the postulates of probability, let us emphasize the point that the three postulates do not tell us how to assign probabilities to events; they merely restrict the ways in which it can be done.

EXAMPLE 7

An experiment has four possible outcomes, A, B, C, and D, that are mutually exclusive. Explain why the following assignments of probabilities are not permissible:

(a) $P(A) = 0.12, P(B) = 0.63, P(C) = 0.45, P(D) = -0.20$;
(b) $P(A) = \frac{9}{120}, P(B) = \frac{45}{120}, P(C) = \frac{27}{120}, P(D) = \frac{46}{120}$.

Solution

(a) $P(D) = -0.20$ violates Postulate 1;
(b) $P(S) = P(A \cup B \cup C \cup D) = \frac{9}{120} + \frac{45}{120} + \frac{27}{120} + \frac{46}{120} = \frac{127}{120} \neq 1$, and this violates Postulate 2.

Of course, in actual practice probabilities are assigned on the basis of past experience, on the basis of a careful analysis of all underlying conditions, on the basis of subjective judgments, or on the basis of assumptions—sometimes the assumption that all possible outcomes are equiprobable.

To assign a probability measure to a sample space, it is not necessary to specify the probability of each possible subset. This is fortunate, for a sample space with as few as 20 possible outcomes has already $2^{20} = 1{,}048{,}576$ subsets, and the number of subsets grows very rapidly when there are 50 possible outcomes, 100 possible outcomes, or more. Instead of listing the probabilities of all possible subsets, we often list the probabilities of the individual outcomes, or sample points of S, and then make use of the following theorem.

THEOREM 1. If A is an event in a discrete sample space S, then $P(A)$ equals the sum of the probabilities of the individual outcomes comprising A.

Proof Let $O_1, O_2, O_3, \ldots$, be the finite or infinite sequence of outcomes that comprise the event A. Thus,

$$A = O_1 \cup O_2 \cup O_3 \cdots$$

and since the individual outcomes, the O's, are mutually exclusive, the third postulate of probability yields

$$P(A) = P(O_1) + P(O_2) + P(O_3) + \cdots$$

This completes the proof.

To use this theorem, we must be able to assign probabilities to the individual outcomes of experiments. How this is done in some special situations is illustrated by the following examples.

EXAMPLE 8

If we twice flip a balanced coin, what is the probability of getting at least one head?

Solution

The sample space is $S = \{HH, HT, TH, TT\}$, where H and T denote head and tail. Since we assume that the coin is balanced, these outcomes are equally likely and we assign to each sample point the probability $\frac{1}{4}$. Letting A denote the event that we will get at least one head, we get $A = \{HH, HT, TH\}$ and

$$\begin{aligned} P(A) &= P(HH) + P(HT) + P(TH) \\ &= \frac{1}{4} + \frac{1}{4} + \frac{1}{4} \\ &= \frac{3}{4} \end{aligned}$$

EXAMPLE 9

A die is loaded in such a way that each odd number is twice as likely to occur as each even number. Find $P(G)$, where G is the event that a number greater than 3 occurs on a single roll of the die.

Solution

The sample space is $S = \{1, 2, 3, 4, 5, 6\}$. Hence, if we assign probability w to each even number and probability $2w$ to each odd number, we find that $2w + w + 2w + w + 2w + w = 9w = 1$ in accordance with Postulate 2. It follows that $w = \frac{1}{9}$ and

$$P(G) = \frac{1}{9} + \frac{2}{9} + \frac{1}{9} = \frac{4}{9}$$

If a sample space is countably infinite, probabilities will have to be assigned to the individual outcomes by means of a mathematical rule, preferably by means of a formula or equation.

EXAMPLE 10

If, for a given experiment, $O_1, O_2, O_3, \ldots$, is an infinite sequence of outcomes, verify that

$$P(O_i) = \left(\frac{1}{2}\right)^i \qquad \text{for } i = 1, 2, 3, \ldots$$

is, indeed, a probability measure.

Solution

Since the probabilities are all positive, it remains to be shown that $P(S) = 1$. Getting

$$P(S) = \frac{1}{2} + \frac{1}{4} + \frac{1}{8} + \frac{1}{16} + \cdots$$

and making use of the formula for the sum of the terms of an infinite geometric progression, we find that

$$P(S) = \frac{\frac{1}{2}}{1 - \frac{1}{2}} = 1$$

In connection with the preceding example, the word "sum" in Theorem 1 will have to be interpreted so that it includes the value of an infinite series.

The probability measure of Example 10 would be appropriate, for example, if O_i is the event that a person flipping a balanced coin will get a tail for the first time on the ith flip of the coin. Thus, the probability that the first tail will come on the third, fourth, or fifth flip of the coin is

$$\left(\frac{1}{2}\right)^3 + \left(\frac{1}{2}\right)^4 + \left(\frac{1}{2}\right)^5 = \frac{7}{32}$$

and the probability that the first tail will come on an odd-numbered flip of the coin is

$$\left(\frac{1}{2}\right)^1 + \left(\frac{1}{2}\right)^3 + \left(\frac{1}{2}\right)^5 + \cdots = \frac{\frac{1}{2}}{1 - \frac{1}{4}} = \frac{2}{3}$$

Here again we made use of the formula for the sum of the terms of an infinite geometric progression.

If an experiment is such that we can assume equal probabilities for all the sample points, as was the case in Example 8, we can take advantage of the following special case of Theorem 1.

THEOREM 2. If an experiment can result in any one of N different equally likely outcomes, and if n of these outcomes together constitute event A, then the probability of event A is

$$P(A) = \frac{n}{N}$$

Proof Let $O_1, O_2, \ldots, O_N$ represent the individual outcomes in S, each with probability $\frac{1}{N}$. If A is the union of n of these mutually exclusive outcomes, and it does not matter which ones, then

$$\begin{aligned} P(A) &= P(O_1 \cup O_2 \cup \cdots \cup O_n) \\ &= P(O_1) + P(O_2) + \cdots + P(O_n) \\ &= \underbrace{\frac{1}{N} + \frac{1}{N} + \cdots + \frac{1}{N}}_{n \text{ terms}} \\ &= \frac{n}{N} \end{aligned}$$

Observe that the formula $P(A) = \frac{n}{N}$ of Theorem 2 is identical with the one for the classical probability concept (see below). Indeed, what we have shown here is that the classical probability concept is consistent with the postulates of probability—it follows from the postulates in the special case where the individual outcomes are all equiprobable.

EXAMPLE 11

A five-card poker hand dealt from a deck of 52 playing cards is said to be a full house if it consists of three of a kind and a pair. If all the five-card hands are equally likely, what is the probability of being dealt a full house?

Solution

The number of ways in which we can be dealt a particular full house, say three kings and two aces, is $\binom{4}{3}\binom{4}{2}$. Since there are 13 ways of selecting the face value for the three of a kind and for each of these there are 12 ways of selecting the face value for the pair, there are altogether

$$n = 13 \cdot 12 \cdot \binom{4}{3}\binom{4}{2}$$

different full houses. Also, the total number of equally likely five-card poker hands is

$$N = \binom{52}{5}$$

and it follows by Theorem 2 that the probability of getting a full house is

$$P(A) = \frac{n}{N} = \frac{13 \cdot 12 \binom{4}{3}\binom{4}{2}}{\binom{52}{5}} = 0.0014$$

5 Some Rules of Probability

Based on the three postulates of probability, we can derive many other rules that have important applications. Among them, the next four theorems are immediate consequences of the postulates.

THEOREM 3. If A and A' are complementary events in a sample space S, then

$$P(A') = 1 - P(A)$$

Proof In the second and third steps of the proof that follows, we make use of the definition of a complement, according to which A and A' are mutually exclusive and $A \cup A' = S$. Thus, we write

$$\begin{aligned} 1 &= P(S) && \text{(by Postulate 2)} \\ &= P(A \cup A') \\ &= P(A) + P(A') && \text{(by Postulate 3)} \end{aligned}$$

and it follows that $P(A') = 1 - P(A)$.

In connection with the frequency interpretation, this result implies that if an event occurs, say, 37 percent of the time, then it does not occur 63 percent of the time.

THEOREM 4. $P(\emptyset) = 0$ for any sample space S.

Proof Since S and $\emptyset$ are mutually exclusive and $S \cup \emptyset = S$ in accordance with the definition of the empty set $\emptyset$, it follows that

$$\begin{aligned} P(S) &= P(S \cup \emptyset) \\ &= P(S) + P(\emptyset) && \text{(by Postulate 3)} \end{aligned}$$

and, hence, that $P(\emptyset) = 0$.

It is important to note that it does not necessarily follow from $P(A) = 0$ that $A = \emptyset$. In practice, we often assign 0 probability to events that, in colloquial terms, would not happen in a million years. For instance, there is the classical example that we assign a probability of 0 to the event that a monkey set loose on a typewriter will type Plato's *Republic* word for word without a mistake. The fact that $P(A) = 0$ does not imply that $A = \emptyset$ is of relevance, especially, in the continuous case.

THEOREM 5. If A and B are events in a sample space S and $A \subset B$, then $P(A) \leqq P(B)$.

Proof Since $A \subset B$, we can write

$$B = A \cup (A' \cap B)$$

as can easily be verified by means of a Venn diagram. Then, since A and $A' \cap B$ are mutually exclusive, we get

$$\begin{aligned} P(B) &= P(A) + P(A' \cap B) \qquad \text{(by Postulate 3)} \\ &\geqq P(A) \qquad \text{(by Postulate 1)} \end{aligned}$$

In words, this theorem states that if A is a subset of B, then $P(A)$ cannot be greater than $P(B)$. For instance, the probability of drawing a heart from an ordinary deck of 52 playing cards cannot be greater than the probability of drawing a red card. Indeed, the probability of drawing a heart is $\frac{1}{4}$, compared with $\frac{1}{2}$, the probability of drawing a red card.

THEOREM 6. $0 \leqq P(A) \leqq 1$ for any event A.

Proof Using Theorem 5 and the fact that $\emptyset \subset A \subset S$ for any event A in S, we have

$$P(\emptyset) \leqq P(A) \leqq P(S)$$

Then, $P(\emptyset) = 0$ and $P(S) = 1$ leads to the result that

$$0 \leqq P(A) \leqq 1$$

The third postulate of probability is sometimes referred to as the **special addition rule**; it is special in the sense that events $A_1, A_2, A_3, \ldots$, must all be mutually exclusive. For *any* two events A and B, there exists the **general addition rule**, or the **inclusion–exclusion principle**:

THEOREM 7. If A and B are any two events in a sample space S, then

$$P(A \cup B) = P(A) + P(B) - P(A \cap B)$$

Proof Assigning the probabilities a, b, and c to the mutually exclusive events $A \cap B, A \cap B'$, and $A' \cap B$ as in the Venn diagram of Figure 6, we find that

$$\begin{aligned} P(A \cup B) &= a + b + c \\ &= (a + b) + (c + a) - a \\ &= P(A) + P(B) - P(A \cap B) \end{aligned}$$

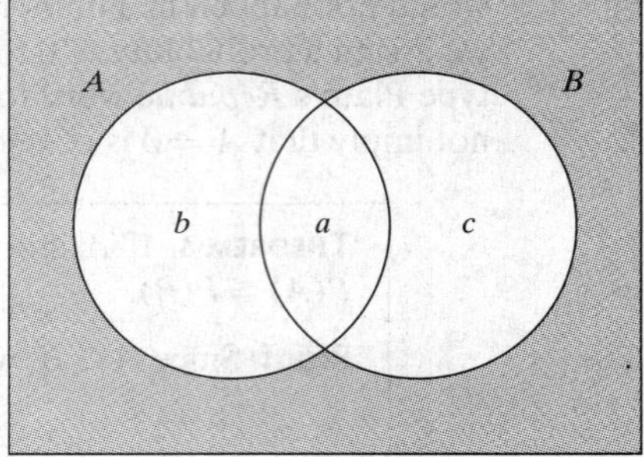

Figure 6. Venn diagram for proof of Theorem 7.

EXAMPLE 12

In a large metropolitan area, the probabilities are 0.86, 0.35, and 0.29, respectively, that a family (randomly chosen for a sample survey) owns a color television set, a HDTV set, or both kinds of sets. What is the probability that a family owns either or both kinds of sets?

Solution

If A is the event that a family in this metropolitan area owns a color television set and B is the event that it owns a HDTV set, we have $P(A) = 0.86, P(B) = 0.35$, and $P(A \cap B) = 0.29$; substitution into the formula of Theorem 7 yields

$$\begin{aligned} P(A \cup B) &= 0.86 + 0.35 - 0.29 \\ &= 0.92 \end{aligned}$$

EXAMPLE 13

Near a certain exit of I-17, the probabilities are 0.23 and 0.24, respectively, that a truck stopped at a roadblock will have faulty brakes or badly worn tires. Also, the probability is 0.38 that a truck stopped at the roadblock will have faulty brakes and/or badly worn tires. What is the probability that a truck stopped at this roadblock will have faulty brakes as well as badly worn tires?

Solution

If B is the event that a truck stopped at the roadblock will have faulty brakes and T is the event that it will have badly worn tires, we have $P(B) = 0.23, P(T) = 0.24$, and $P(B \cup T) = 0.38$; substitution into the formula of Theorem 7 yields

$$0.38 = 0.23 + 0.24 - P(B \cap T)$$

Solving for $P(B \cap T)$, we thus get

$$P(B \cap T) = 0.23 + 0.24 - 0.38 = 0.09$$

Repeatedly using the formula of Theorem 7, we can generalize this addition rule so that it will apply to any number of events. For instance, for three events we obtain the following theorem.

THEOREM 8. If A, B, and C are any three events in a sample space S, then

$$P(A \cup B \cup C) = P(A) + P(B) + P(C) - P(A \cap B) - P(A \cap C) - P(B \cap C) + P(A \cap B \cap C)$$

Proof Writing $A \cup B \cup C$ as $A \cup (B \cup C)$ and using the formula of Theorem 7 twice, once for $P[A \cup (B \cup C)]$ and once for $P(B \cup C)$, we get

$$\begin{aligned} P(A \cup B \cup C) &= P[A \cup (B \cup C)] \\ &= P(A) + P(B \cup C) - P[A \cap (B \cup C)] \\ &= P(A) + P(B) + P(C) - P(B \cap C) \\ &\quad - P[A \cap (B \cup C)] \end{aligned}$$

Then, using the distributive law that the reader was asked to verify in part (b) of Exercise 1, we find that

$$\begin{aligned} P[A \cap (B \cup C)] &= P[(A \cap B) \cup (A \cap C)] \\ &= P(A \cap B) + P(A \cap C) - P[(A \cap B) \cap (A \cap C)] \\ &= P(A \cap B) + P(A \cap C) - P(A \cap B \cap C) \end{aligned}$$

and hence that

$$P(A \cup B \cup C) = P(A) + P(B) + P(C) - P(A \cap B) - P(A \cap C) - P(B \cap C) + P(A \cap B \cap C)$$

(In Exercise 12 the reader will be asked to give an alternative proof of this theorem, based on the method used in the text to prove Theorem 7.)

EXAMPLE 14

If a person visits his dentist, suppose that the probability that he will have his teeth cleaned is 0.44, the probability that he will have a cavity filled is 0.24, the probability that he will have a tooth extracted is 0.21, the probability that he will have his teeth cleaned and a cavity filled is 0.08, the probability that he will have his teeth cleaned and a tooth extracted is 0.11, the probability that he will have a cavity filled and a tooth extracted is 0.07, and the probability that he will have his teeth cleaned, a cavity filled, and a tooth extracted is 0.03. What is the probability that a person visiting his dentist will have at least one of these things done to him?

Solution

If C is the event that the person will have his teeth cleaned, F is the event that he will have a cavity filled, and E is the event that he will have a tooth extracted, we are given $P(C) = 0.44$, $P(F) = 0.24$, $P(E) = 0.21$, $P(C \cap F) = 0.08$, $P(C \cap E) = 0.11$, $P(F \cap E) = 0.07$, and $P(C \cap F \cap E) = 0.03$, and substitution into the formula of Theorem 8 yields

$$\begin{aligned} P(C \cup F \cup E) &= 0.44 + 0.24 + 0.21 - 0.08 - 0.11 - 0.07 + 0.03 \\ &= 0.66 \end{aligned}$$

Exercises

5. Use parts (a) and (b) of Exercise 3 to show that
(a) $P(A) \geqq P(A \cap B)$;
(b) $P(A) \leqq P(A \cup B)$.

6. Referring to Figure 6, verify that

$$P(A \cap B') = P(A) - P(A \cap B)$$

7. Referring to Figure 6 and letting $P(A' \cap B') = d$, verify that

$$P(A' \cap B') = 1 - P(A) - P(B) + P(A \cap B)$$

8. The event that "A or B but not both" will occur can be written as

$$(A \cap B') \cup (A' \cap B)$$

Express the probability of this event in terms of $P(A)$, $P(B)$, and $P(A \cap B)$.

9. Use the formula of Theorem 7 to show that
(a) $P(A \cap B) \leqq P(A) + P(B)$;
(b) $P(A \cap B) \geqq P(A) + P(B) - 1$.

10. Use the Venn diagram of Figure 7 with the probabilities a, b, c, d, e, f, and g assigned to $A \cap B \cap C$, $A \cap B \cap C', \ldots,$ and $A \cap B' \cap C'$ to show that if $P(A) = P(B) = P(C) = 1$, then $P(A \cap B \cap C) = 1$. [*Hint*: Start with the argument that since $P(A) = 1$, it follows that $e = c = f = 0$.]

11. Give an alternative proof of Theorem 7 by making use of the relationships $A \cup B = A \cup (A' \cap B)$ and $B = (A \cap B) \cup (A' \cap B)$.

12. Use the Venn diagram of Figure 7 and the method by which we proved Theorem 7 to prove Theorem 8.

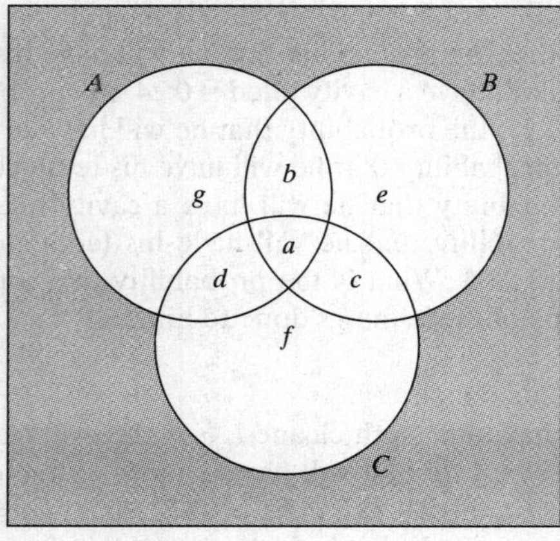

Figure 7. Venn diagram for Exercises 10, 12, and 13.

13. Duplicate the method of proof used in Exercise 12 to show that

$$\begin{aligned} P(A \cup B \cup C \cup D) = {} & P(A) + P(B) + P(C) + P(D) \\ & - P(A \cap B) - P(A \cap C) - P(A \cap D) \\ & - P(B \cap C) - P(B \cap D) - P(C \cap D) \\ & + P(A \cap B \cap C) + P(A \cap B \cap D) \\ & + P(A \cap C \cap D) + P(B \cap C \cap D) \\ & - P(A \cap B \cap C \cap D) \end{aligned}$$

(*Hint:* With reference to the Venn diagram of Figure 7, divide each of the eight regions into two parts, designating one to be inside D and the other outside D and letting $a, b, c, d, e, f, g, h, i, j, k, l, m, n, o$, and p be the probabilities associated with the resulting 16 regions.)

14. Prove by induction that

$$P(E_1 \cup E_2 \cup \cdots \cup E_n) \leqq \sum_{i=1}^{n} P(E_i)$$

for any finite sequence of events $E_1, E_2, \ldots,$ and E_n.

15. The **odds** that an event will occur are given by the ratio of the probability that the event will occur to the probability that it will not occur, provided neither probability is zero. Odds are usually quoted in terms of positive integers having no common factor. Show that if the odds are A to B that an event will occur, its probability is

$$p = \frac{A}{A+B}$$

16. Subjective probabilities may be determined by exposing persons to risk-taking situations and finding the odds at which they would consider it fair to bet on the outcome. The odds are then converted into probabilities by means of the formula of Exercise 15. For instance, if a person feels that 3 to 2 are fair odds that a business venture will succeed (or that it would be fair to bet \$30 against \$20 that it will succeed), the probability is $\frac{3}{3+2} = 0.6$ that the business venture will succeed. Show that if subjective probabilities are determined in this way, they satisfy
(a) Postulate 1;
(b) Postulate 2.

See also Exercise 82.

6 Conditional Probability

Difficulties can easily arise when probabilities are quoted without specification of the sample space. For instance, if we ask for the probability that a lawyer makes more than $75,000 per year, we may well get several different answers, and they may all be correct. One of them might apply to all those who are engaged in the private practice of law, another might apply to lawyers employed by corporations, and so forth. Since the choice of the sample space (that is, the set of all possibilities under consideration) is by no means always self-evident, it often helps to use the symbol $P(A|S)$ to denote the **conditional probability** of event A relative to the sample space S or, as we also call it, "the probability of A given S." The symbol $P(A|S)$ makes it explicit that we are referring to a particular sample space S, and it is preferable to the abbreviated notation $P(A)$ unless the tacit choice of S is clearly understood. It is also preferable when we want to refer to several sample spaces in the same example. If A is the event that a person makes more than $75,000 per year, G is the event that a person is a law school graduate, L is the event that a person is licensed to practice law, and E is the event that a person is actively engaged in the practice of law, then $P(A|G)$ is the probability that a law school graduate makes more than $75,000 per year, $P(A|L)$ is the probability that a person licensed to practice law makes more than $75,000 per year, and $P(A|E)$ is the probability that a person actively engaged in the practice of law makes more than $75,000 per year.

Some ideas connected with conditional probabilities are illustrated in the following example.

EXAMPLE 15

A consumer research organization has studied the services under warranty provided by the 50 new-car dealers in a certain city, and its findings are summarized in the following table.

	Good service under warranty	*Poor service under warranty*
In business 10 years or more	16	4
In business less than 10 years	10	20

If a person randomly selects one of these new-car dealers, what is the probability that he gets one who provides good service under warranty? Also, if a person randomly selects one of the dealers who has been in business for 10 years or more, what is the probability that he gets one who provides good service under warranty?

Solution

By "randomly" we mean that, in each case, all possible selections are equally likely, and we can therefore use the formula of Theorem 2. If we let G denote the selection of a dealer who provides good service under warranty, and if we let $n(G)$ denote the number of elements in G and $n(S)$ the number of elements in the whole sample space, we get

$$P(G) = \frac{n(G)}{n(S)} = \frac{16+10}{50} = 0.52$$

This answers the first question.

For the second question, we limit ourselves to the reduced sample space, which consists of the first line of the table, that is, the $16+4=20$ dealers who have been in business 10 years or more. Of these, 16 provide good service under warranty, and we get

$$P(G|T) = \frac{16}{20} = 0.80$$

where T denotes the selection of a dealer who has been in business 10 years or more. This answers the second question and, as should have been expected, $P(G|T)$ is considerably higher than $P(G)$.

Since the numerator of $P(G|T)$ is $n(T \cap G) = 16$ in the preceding example, the number of dealers who have been in business for 10 years or more and provide good service under warranty, and the denominator is $n(T)$, the number of dealers who have been in business 10 years or more, we can write symbolically

$$P(G|T) = \frac{n(T \cap G)}{n(T)}$$

Then, if we divide the numerator and the denominator by $n(S)$, the total number of new-car dealers in the given city, we get

$$P(G|T) = \frac{\frac{n(T \cap G)}{n(S)}}{\frac{n(T)}{n(S)}} = \frac{P(T \cap G)}{P(T)}$$

and we have, thus, expressed the conditional probability $P(G|T)$ in terms of two probabilities defined for the whole sample space S.

Generalizing from the preceding, let us now make the following definition of conditional probability.

DEFINITION 4. CONDITIONAL PROBABILITY. *If* A *and* B *are any two events in a sample space* S *and* $P(A) \neq 0$, *the* ***conditional probability*** *of* B *given* A *is*

$$P(B|A) = \frac{P(A \cap B)}{P(A)}$$

EXAMPLE 16

With reference to Example 15, what is the probability that one of the dealers who has been in business less than 10 years will provide good service under warranty?

Solution

Since $P(T' \cap G) = \frac{10}{50} = 0.20$ and $P(T') = \frac{10+20}{50} = 0.60$, substitution into the formula yields

$$P(G|T') = \frac{P(T' \cap G)}{P(T')} = \frac{0.20}{0.60} = \frac{1}{3}$$

Although we introduced the formula for $P(B|A)$ by means of an example in which the possibilities were all equally likely, this is not a requirement for its use.

EXAMPLE 17

With reference to the loaded die of Example 9, what is the probability that the number of points rolled is a perfect square? Also, what is the probability that it is a perfect square given that it is greater than 3?

Solution

If A is the event that the number of points rolled is greater than 3 and B is the event that it is a perfect square, we have $A = \{4, 5, 6\}$, $B = \{1, 4\}$, and $A \cap B = \{4\}$. Since the probabilities of rolling a 1, 2, 3, 4, 5, or 6 with the die are $\frac{2}{9}, \frac{1}{9}, \frac{2}{9}, \frac{1}{9}, \frac{2}{9}$, and $\frac{1}{9}$, we find that the answer to the first question is

$$P(B) = \frac{2}{9} + \frac{1}{9} = \frac{1}{3}$$

To determine $P(B|A)$, we first calculate

$$P(A \cap B) = \frac{1}{9} \quad \text{and} \quad P(A) = \frac{1}{9} + \frac{2}{9} + \frac{1}{9} = \frac{4}{9}$$

Then, substituting into the formula of Definition 4, we get

$$P(B|A) = \frac{P(A \cap B)}{P(A)} = \frac{\frac{1}{9}}{\frac{4}{9}} = \frac{1}{4}$$

To verify that the formula of Definition 4 has yielded the "right" answer in the preceding example, we have only to assign probability v to the two even numbers in the reduced sample space A and probability $2v$ to the odd number, such that the sum of the three probabilities is equal to 1. We thus have $v + 2v + v = 1$, $v = \frac{1}{4}$, and, hence, $P(B|A) = \frac{1}{4}$ as before.

EXAMPLE 18

A manufacturer of airplane parts knows from past experience that the probability is 0.80 that an order will be ready for shipment on time, and it is 0.72 that an order will be ready for shipment on time and will also be delivered on time. What is the probability that such an order will be delivered on time given that it was ready for shipment on time?

Solution

If we let R stand for the event that an order is ready for shipment on time and D be the event that it is delivered on time, we have $P(R) = 0.80$ and $P(R \cap D) = 0.72$, and it follows that

$$P(D|R) = \frac{P(R \cap D)}{P(R)} = \frac{0.72}{0.80} = 0.90$$

Thus, 90 percent of the shipments will be delivered on time provided they are shipped on time. Note that $P(R|D)$, the probability that a shipment that is delivered on time was also ready for shipment on time, cannot be determined without further information; for this purpose we would also have to know $P(D)$.

If we multiply the expressions on both sides of the formula of Definition 4 by $P(A)$, we obtain the following **multiplication rule.**

THEOREM 9. If A and B are any two events in a sample space S and $P(A) \neq 0$, then

$$P(A \cap B) = P(A) \cdot P(B|A)$$

In words, the probability that A and B will both occur is the product of the probability of A and the conditional probability of B given A. Alternatively, if $P(B) \neq 0$, the probability that A and B will both occur is the product of the probability of B and the conditional probability of A given B; symbolically,

$$P(A \cap B) = P(B) \cdot P(A|B)$$

To derive this alternative multiplication rule, we interchange A and B in the formula of Theorem 9 and make use of the fact that $A \cap B = B \cap A$.

EXAMPLE 19

If we randomly pick two television sets in succession from a shipment of 240 television sets of which 15 are defective, what is the probability that they will both be defective?

Solution

If we assume equal probabilities for each selection (which is what we mean by "randomly" picking the sets), the probability that the first set will be defective is $\frac{15}{240}$, and the probability that the second set will be defective given that the first set is defective is $\frac{14}{239}$. Thus, the probability that both sets will be defective is $\frac{15}{240} \cdot \frac{14}{239} = \frac{7}{1,912}$. This assumes that we are **sampling without replacement**; that is, the first set is not replaced before the second set is selected.

EXAMPLE 20

Find the probabilities of randomly drawing two aces in succession from an ordinary deck of 52 playing cards if we sample

(a) without replacement;

(b) with replacement.

Solution

(a) If the first card is not replaced before the second card is drawn, the probability of getting two aces in succession is

$$\frac{4}{52} \cdot \frac{3}{51} = \frac{1}{221}$$

(b) If the first card is replaced before the second card is drawn, the corresponding probability is

$$\frac{4}{52} \cdot \frac{4}{52} = \frac{1}{169}$$

In the situations described in the two preceding examples there is a definite temporal order between the two events A and B. In general, this need not be the case when we write $P(A|B)$ or $P(B|A)$. For instance, we could ask for the probability that the first card drawn was an ace given that the second card drawn (without replacement) is an ace—the answer would also be $\frac{3}{51}$.

Theorem 9 can easily be generalized so that it applies to more than two events; for instance, for three events we have the following theorem.

THEOREM 10. If A, B, and C are any three events in a sample space S such that $P(A \cap B) \neq 0$, then

$$P(A \cap B \cap C) = P(A) \cdot P(B|A) \cdot P(C|A \cap B)$$

Proof Writing $A \cap B \cap C$ as $(A \cap B) \cap C$ and using the formula of Theorem 9 twice, we get

$$\begin{aligned} P(A \cap B \cap C) &= P[(A \cap B) \cap C] \\ &= P(A \cap B) \cdot P(C|A \cap B) \\ &= P(A) \cdot P(B|A) \cdot P(C|A \cap B) \end{aligned}$$

EXAMPLE 21

A box of fuses contains 20 fuses, of which 5 are defective. If 3 of the fuses are selected at random and removed from the box in succession without replacement, what is the probability that all 3 fuses are defective?

Solution

If A is the event that the first fuse is defective, B is the event that the second fuse is defective, and C is the event that the third fuse is defective, then $P(A) = \frac{5}{20}$, $P(B|A) = \frac{4}{19}$, $P(C|A \cap B) = \frac{3}{18}$, and substitution into the formula yields

$$\begin{aligned} P(A \cap B \cap C) &= \frac{5}{20} \cdot \frac{4}{19} \cdot \frac{3}{18} \\ &= \frac{1}{114} \end{aligned}$$

Further generalization of Theorems 9 and 10 to k events is straightforward, and the resulting formula can be proved by mathematical induction.

7 Independent Events

Informally speaking, two events A and B are **independent** if the occurrence or nonoccurrence of either one does not affect the probability of the occurrence of the other. For instance, in the preceding example the selections would all have been independent had each fuse been replaced before the next one was selected; the probability of getting a defective fuse would have remained $\frac{5}{20}$.

Symbolically, two events A and B are independent if $P(B|A) = P(B)$ and $P(A|B) = P(A)$, and it can be shown that either of these equalities implies the other when both of the conditional probabilities exist, that is, when neither $P(A)$ nor $P(B)$ equals zero (see Exercise 21).

Now, if we substitute $P(B)$ for $P(B|A)$ into the formula of Theorem 9, we get

$$\begin{aligned} P(A \cap B) &= P(A) \cdot P(B|A) \\ &= P(A) \cdot P(B) \end{aligned}$$

and we shall use this as our formal definition of independence.

DEFINITION 5. INDEPENDENCE. *Two events* A *and* B *are* ***independent*** *if and only if*

$$P(A \cap B) = P(A) \cdot P(B)$$

Reversing the steps, we can also show that Definition 5 implies the definition of independence that we gave earlier.

If two events are not independent, they are said to be **dependent**. In the derivation of the formula of Definition 5, we assume that $P(B|A)$ exists and, hence, that $P(A) \neq 0$. For mathematical convenience, we shall let the definition apply also when $P(A) = 0$ and/or $P(B) = 0$.

EXAMPLE 22

A coin is tossed three times and the eight possible outcomes, HHH, HHT, HTH, THH, HTT, THT, TTH, and TTT, are assumed to be equally likely. If A is the event that a head occurs on each of the first two tosses, B is the event that a tail occurs on the third toss, and C is the event that exactly two tails occur in the three tosses, show that

(a) events A and B are independent;

(b) events B and C are dependent.

Solution

Since

$$\begin{aligned} A &= \{\text{HHH, HHT}\} \\ B &= \{\text{HHT, HTT, THT, TTT}\} \\ C &= \{\text{HTT, THT, TTH}\} \\ A \cap B &= \{\text{HHT}\} \\ B \cap C &= \{\text{HTT, THT}\} \end{aligned}$$

the assumption that the eight possible outcomes are all equiprobable yields $P(A) = \frac{1}{4}, P(B) = \frac{1}{2}, P(C) = \frac{3}{8}, P(A \cap B) = \frac{1}{8}$, and $P(B \cap C) = \frac{1}{4}$.

(a) Since $P(A) \cdot P(B) = \frac{1}{4} \cdot \frac{1}{2} = \frac{1}{8} = P(A \cap B)$, events A and B are independent.

(b) Since $P(B) \cdot P(C) = \frac{1}{2} \cdot \frac{3}{8} = \frac{3}{16} \neq P(B \cap C)$, events B and C are not independent.

In connection with Definition 5, it can be shown that if A and B are independent, then so are A and B', A' and B, and A' and B'. For instance, consider the following theorem.

THEOREM 11. If A and B are independent, then A and B' are also independent.

Proof Since $A = (A \cap B) \cup (A \cap B')$, as the reader was asked to show in part (a) of Exercise 3, $A \cap B$ and $A \cap B'$ are mutually exclusive, and A and B are independent by assumption, we have

$$\begin{aligned} P(A) &= P[(A \cap B) \cup (A \cap B')] \\ &= P(A \cap B) + P(A \cap B') \\ &= P(A) \cdot P(B) + P(A \cap B') \end{aligned}$$

It follows that

$$\begin{aligned} P(A \cap B') &= P(A) - P(A) \cdot P(B) \\ &= P(A) \cdot [1 - P(B)] \\ &= P(A) \cdot P(B') \end{aligned}$$

and hence that A and B' are independent.

In Exercises 22 and 23 the reader will be asked to show that if A and B are independent, then A' and B are independent and so are A' and B', and if A and B are dependent, then A and B' are dependent.

To extend the concept of independence to more than two events, let us make the following definition.

DEFINITION 6. INDEPENDENCE OF MORE THAN TWO EVENTS. *Events* $A_1, A_2, \ldots,$ *and* A_k *are* ***independent*** *if and only if the probability of the intersections of any 2, 3, ..., or k of these events equals the product of their respective probabilities.*

For three events A, B, and C, for example, independence requires that

$$P(A \cap B) = P(A) \cdot P(B)$$

$$P(A \cap C) = P(A) \cdot P(C)$$

$$P(B \cap C) = P(B) \cdot P(C)$$

and

$$P(A \cap B \cap C) = P(A) \cdot P(B) \cdot P(C)$$

It is of interest to note that three or more events can be **pairwise independent** without being independent.

EXAMPLE 23

Figure 8 shows a Venn diagram with probabilities assigned to its various regions. Verify that A and B are independent, A and C are independent, and B and C are independent, but A, B, and C are not independent.

Solution

As can be seen from the diagram, $P(A) = P(B) = P(C) = \frac{1}{2}$, $P(A \cap B) = P(A \cap C) = P(B \cap C) = \frac{1}{4}$, and $P(A \cap B \cap C) = \frac{1}{4}$. Thus,

$$P(A) \cdot P(B) = \frac{1}{4} = P(A \cap B)$$

$$P(A) \cdot P(C) = \frac{1}{4} = P(A \cap C)$$

$$P(B) \cdot P(C) = \frac{1}{4} = P(B \cap C)$$

but

$$P(A) \cdot P(B) \cdot P(C) = \frac{1}{8} \neq P(A \cap B \cap C)$$

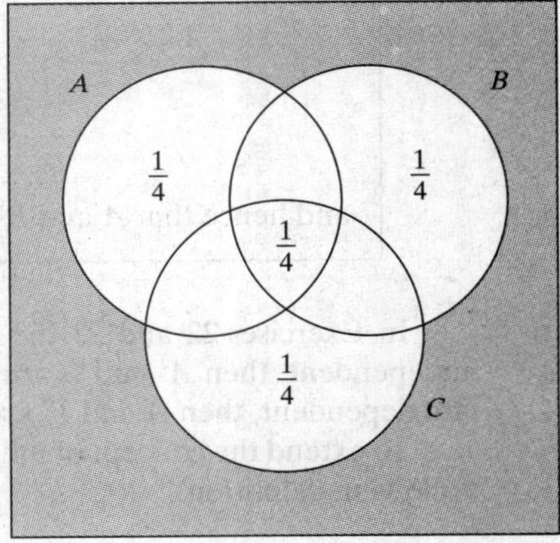

Figure 8. Venn diagram for Example 23.

Incidentally, the preceding example can be given a "real" interpretation by considering a large room that has three separate switches controlling the ceiling lights. These lights will be on when all three switches are "up" and hence also when one of the switches is "up" and the other two are "down." If A is the event that the first switch is "up," B is the event that the second switch is "up," and C is the event that the third switch is "up," the Venn diagram of Figure 8 shows a possible set of probabilities associated with the switches being "up" or "down" when the ceiling lights are on.

It can also happen that $P(A \cap B \cap C) = P(A) \cdot P(B) \cdot P(C)$ without A, B, and C being pairwise independent—this the reader will be asked to verify in Exercise 24.

Of course, if we are given that certain events are independent, the probability that they will all occur is simply the product of their respective probabilities.

EXAMPLE 24

Find the probabilities of getting

(a) three heads in three random tosses of a balanced coin;

(b) four sixes and then another number in five random rolls of a balanced die.

Solution

(a) The probability of a head on each toss is $\frac{1}{2}$ and the three outcomes are independent. Thus we can multiply, obtaining

$$\frac{1}{2} \cdot \frac{1}{2} \cdot \frac{1}{2} = \frac{1}{8}$$

(b) The probability of a six on each toss is $\frac{1}{6}$; thus the probability of tossing a number other than 6 is $\frac{5}{6}$. Inasmuch as the tosses are independent, we can multiply the respective probabilities to obtain

$$\frac{1}{6} \cdot \frac{1}{6} \cdot \frac{1}{6} \cdot \frac{1}{6} \cdot \frac{5}{6} = \frac{5}{7,776}$$

---

8 Bayes' Theorem

In many situations the outcome of an experiment depends on what happens in various intermediate stages. The following is a simple example in which there is one intermediate stage consisting of two alternatives:

EXAMPLE 25

The completion of a construction job may be delayed because of a strike. The probabilities are 0.60 that there will be a strike, 0.85 that the construction job will be completed on time if there is no strike, and 0.35 that the construction job will be completed on time if there is a strike. What is the probability that the construction job will be completed on time?

Solution

If A is the event that the construction job will be completed on time and B is the event that there will be a strike, we are given $P(B) = 0.60$, $P(A|B') = 0.85$, and $P(A|B) = 0.35$. Making use of the formula of part (a) of Exercise 3, the fact that $A \cap B$ and $A \cap B'$ are mutually exclusive, and the alternative form of the multiplication rule, we can write

$$\begin{aligned} P(A) &= P[(A \cap B) \cup (A \cap B')] \\ &= P(A \cap B) + P(A \cap B') \\ &= P(B) \cdot P(A|B) + P(B') \cdot P(A|B') \end{aligned}$$

Then, substituting the given numerical values, we get

$$\begin{aligned} P(A) &= (0.60)(0.35) + (1 - 0.60)(0.85) \\ &= 0.55 \end{aligned}$$

An immediate generalization of this kind of situation is the case where the intermediate stage permits k different alternatives (whose occurrence is denoted by $B_1, B_2, \ldots, B_k$). It requires the following theorem, sometimes called the **rule of total probability** or the **rule of elimination**.

THEOREM 12. If the events $B_1, B_2, \ldots,$ and B_k constitute a partition of the sample space S and $P(B_i) \neq 0$ for $i = 1, 2, \ldots, k$, then for any event A in S

$$P(A) = \sum_{i=1}^{k} P(B_i) \cdot P(A|B_i)$$

The B's constitute a partition of the sample space if they are pairwise mutually exclusive and if their union equals S. A formal proof of Theorem 12 consists, essentially, of the same steps we used in Example 25, and it is left to the reader in Exercise 32.

EXAMPLE 26

The members of a consulting firm rent cars from three rental agencies: 60 percent from agency 1, 30 percent from agency 2, and 10 percent from agency 3. If 9 percent of the cars from agency 1 need an oil change, 20 percent of the cars from agency 2 need an oil change, and 6 percent of the cars from agency 3 need an oil change, what is the probability that a rental car delivered to the firm will need an oil change?

Solution

If A is the event that the car needs an oil change, and B_1, B_2, and B_3 are the events that the car comes from rental agencies 1, 2, or 3, we have $P(B_1) = 0.60$, $P(B_2) = 0.30$, $P(B_3) = 0.10$, $P(A|B_1) = 0.09$, $P(A|B_2) = 0.20$, and $P(A|B_3) = 0.06$. Substituting these values into the formula of Theorem 12, we get

$$\begin{aligned} P(A) &= (0.60)(0.09) + (0.30)(0.20) + (0.10)(0.06) \\ &= 0.12 \end{aligned}$$

Thus, 12 percent of all the rental cars delivered to this firm will need an oil change.

With reference to the preceding example, suppose that we are interested in the following question: If a rental car delivered to the consulting firm needs an oil change, what is the probability that it came from rental agency 2? To answer questions of this kind, we need the following theorem, called **Bayes' theorem**:

THEOREM 13. If $B_1, B_2, \ldots, B_k$ constitute a partition of the sample space S and $P(B_i) \neq 0$ for $i = 1, 2, \ldots, k$, then for any event A in S such that $P(A) \neq 0$

$$P(B_r|A) = \frac{P(B_r) \cdot P(A|B_r)}{\sum_{i=1}^{k} P(B_i) \cdot P(A|B_i)}$$

for $r = 1, 2, \ldots, k$.

In words, the probability that event A was reached via the rth branch of the tree diagram of Figure 9, given that it was reached via one of its k branches, is the *ratio* of the probability associated with the rth branch to the sum of the probabilities associated with all k branches of the tree.

Proof Writing $P(B_r|A) = \frac{P(A \cap B_r)}{P(A)}$ in accordance with the definition of conditional probability, we have only to substitute $P(B_r) \cdot P(A|B_r)$ for $P(A \cap B_r)$ and the formula of Theorem 12 for $P(A)$.

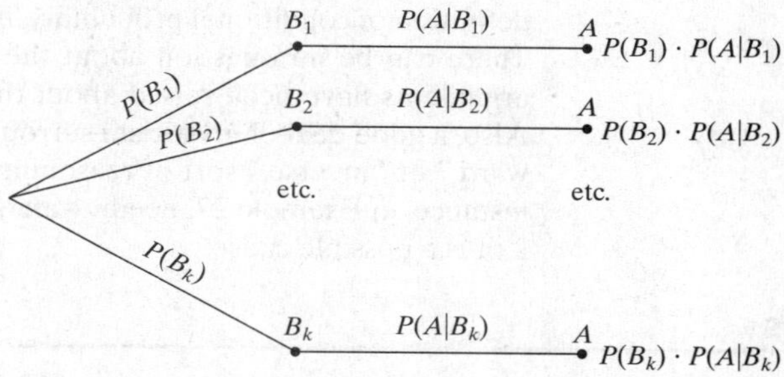

Figure 9. Tree diagram for Bayes' theorem.

EXAMPLE 27

With reference to Example 26, if a rental car delivered to the consulting firm needs an oil change, what is the probability that it came from rental agency 2?

Solution

Substituting the probabilities on the previous page into the formula of Theorem 13, we get

$$\begin{aligned} P(B_2|A) &= \frac{(0.30)(0.20)}{(0.60)(0.09) + (0.30)(0.20) + (0.10)(0.06)} \\ &= \frac{0.060}{0.120} \\ &= 0.5 \end{aligned}$$

Observe that although only 30 percent of the cars delivered to the firm come from agency 2, 50 percent of those requiring an oil change come from that agency.

EXAMPLE 28

A rare but serious disease, D, has been found in 0.01 percent of a certain population. A test has been developed that will be positive, p, for 98 percent of those who have the disease and be positive for only 3 percent of those who do not have the disease. Find the probability that a person tested as positive does not have the disease.

Solution

Let $\overline{D}$ and $\overline{p}$ represent the events that a person randomly selected from the given population, respectively, does not have the disease and is found negative for the disease by the test. Substituting the given probabilities into the formula of Theorem 13, we get

$$P(\overline{D}|p) = \frac{P(\overline{D})P(p|\overline{D})}{P(D)P(p|D)+P(\overline{D})P(p|\overline{D})} = \frac{0.9999 \cdot 0.03}{0.0001 \cdot 0.98 + 0.9999 \cdot 0.03} = 0.997$$

This example demonstrates the near impossibility of finding a test for a rare disease that does not have an unacceptably high probability of false positives.

Although Bayes' theorem follows from the postulates of probability and the definition of conditional probability, it has been the subject of extensive controversy. There can be no question about the validity of Bayes' theorem, but considerable arguments have been raised about the assignment of the **prior probabilities** $P(B_i)$. Also, a good deal of mysticism surrounds Bayes' theorem because it entails a "backward," or "inverse," sort of reasoning, that is, reasoning "from effect to cause." For instance, in Example 27, needing an oil change is the effect and coming from agency 2 of is a possible cause.

Exercises

17. Show that the postulates of probability are satisfied by conditional probabilities. In other words, show that if $P(B) \neq 0$, then
(a) $P(A|B) \geqq 0$;
(b) $P(B|B) = 1$;
(c) $P(A_1 \cup A_2 \cup \ldots |B) = P(A_1|B) + P(A_2|B) + \cdots$ for any sequence of mutually exclusive events $A_1, A_2, \ldots$.

18. Show by means of numerical examples that $P(B|A) + P(B|A')$
(a) may be equal to 1;
(b) need not be equal to 1.

19. Duplicating the method of proof of Theorem 10, show that $P(A \cap B \cap C \cap D) = P(A) \cdot P(B|A) \cdot P(C|A \cap B) \cdot P(D|A \cap B \cap C)$ provided that $P(A \cap B \cap C) \neq 0$.

20. Given three events A, B, and C such that $P(A \cap B \cap C) \neq 0$ and $P(C|A \cap B) = P(C|B)$, show that $P(A|B \cap C) = P(A|B)$.

21. Show that if $P(B|A) = P(B)$ and $P(B) \neq 0$, then $P(A|B) = P(A)$.

22. Show that if events A and B are independent, then
(a) events A' and B are independent;
(b) events A' and B' are independent.

23. Show that if events A and B are dependent, then events A and B' are dependent.

24. Refer to Figure 10 to show that $P(A \cap B \cap C) = P(A) \cdot P(B) \cdot P(C)$ does not necessarily imply that A, B, and C are all pairwise independent.

25. Refer to Figure 10 to show that if A is independent of B and A is independent of C, then B is not necessarily independent of C.

26. Refer to Figure 10 to show that if A is independent of B and A is independent of C, then A is not necessarily independent of $B \cup C$.

27. If events A, B, and C are independent, show that
(a) A and $B \cap C$ are independent;
(b) A and $B \cup C$ are independent.

28. If $P(A|B) < P(A)$, prove that $P(B|A) < P(B)$.

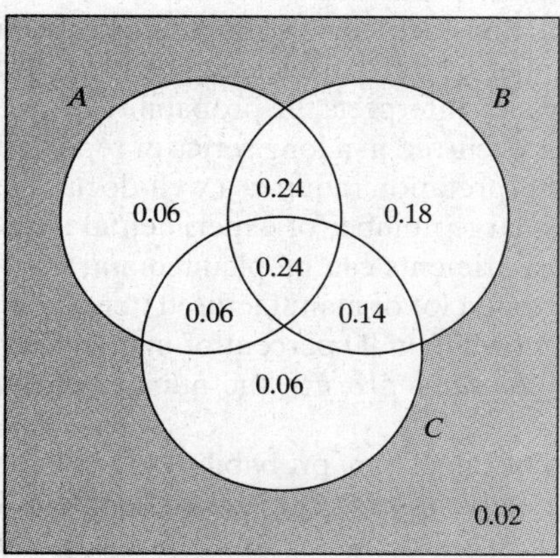

Figure 10. Diagram for Exercises 24, 25, and 26.

29. If $A_1, A_2, \ldots, A_n$ are independent events, prove that

$$P(A_1 \cup A_2 \cup \cdots \cup A_n) = 1 - \{1 - P(A_1)\} \cdot \{1 - P(A_2)\} \ldots \{1 - P(A_n)\}$$

30. Show that $2^k - k - 1$ conditions must be satisfied for k events to be independent.

31. For any event A, show that A and $\emptyset$ are independent.

32. Prove Theorem 12 by making use of the following generalization of the distributive law given in part (b) of Exercise 1:

$$A \cap (B_1 \cup B_2 \cup \cdots \cup B_k) = (A \cap B_1) \cup (A \cap B_2) \cup \cdots \cup (A \cap B_k)$$

33. Suppose that a die has n sides numbered $i = 1, 2, \ldots, n$. Assume that the probability of it coming up on the side numbered i is the same for each value of i. The die is rolled n times (assume independence) and a "match" is defined to be the occurrence of side i on the ith roll. Prove that the probability of at least one match is given by

$$1 - \left(\frac{n-1}{n}\right)^n = 1 - \left(1 - \frac{1}{n}\right)^n$$

34. Show that $P(A \cup B) \geq 1 - P(A') - P(B')$ for any two events A and B defined in the sample space S. (*Hint*: Use Venn diagrams.)

9 The Theory in Practice

The word "probability" is a part of everyday language, but it is difficult to define this word without using the word "probable" or its synonym "likely" in the definition.* To illustrate, Webster's *Third New International Dictionary* defines "probability" as "the quality or state of being probable." If the concept of probability is to be used in mathematics and scientific applications, we require a more exact, less circular, definition.

The postulates of probability given in Section 4 satisfy this criterion. Together with the rules given in Section 5, this definition lends itself to calculations of probabilities that "make sense" and that can be verified experimentally. The entire theory of statistics is based on the notion of probability. It seems remarkable that the entire structure of probability and, therefore of statistics, can be built on the relatively straightforward foundation given in this chapter.

Probabilities were first considered in games of chance, or gambling. Players of various games of chance observed that there seemed to be "rules" that governed the roll of dice or the results of spinning a roulette wheel. Some of them went as far as to postulate some of these rules entirely on the basis of experience. But differences arose among gamblers about probabilities, and they brought their questions to the noted mathematicians of their day. With this motivation, the modern theory of probability began to be developed.

Motivated by problems associated with games of chance, the theory of probability first was developed under the assumption of **equal likelihood**, expressed in Theorem 2. Under this assumption one only had to count the number of "successful" outcomes and divide by the total number of "possible" outcomes to arrive at the probability of an event.

The assumption of equal likelihood fails when we attempt, for example, to find the probability that a trifecta at the race track will pay off. Here, the different horses have different probabilities of winning, and we are forced to rely on a different method of evaluating probabilities. It is common to take into account the various horses' records in previous races, calculating each horse's probability of winning by dividing its number of wins by the number of starts. This idea gives rise to the **frequency interpretation** of probabilities, which interprets the probability of an event to be the proportion of times the event has occurred in a long series of repeated experiments. Application of the frequency interpretation requires a well-documented history of the outcomes of an event over a large number of experimental trials. In the absence of such a history, a series of experiments can be planned and their results observed. For example, the probability that a lot of manufactured items will have at most three defectives is estimated to be 0.90 if, in 90 percent of many previous lots *produced to the same specifications by the same process*, the number of defectives was three or less.

A more recently employed method of calculating probabilities is called the **subjective method**. Here, a personal or subjective assessment is made of the probability of an event which is difficult or impossible to estimate in any other way. For example, the probability that the major stock market indexes will go up in a given future period of time cannot be estimated very well using the frequency interpretation because economic and world conditions rarely replicate themselves very closely. As another example, juries use this method when determining the guilt or innocence of a defendant "beyond a reasonable doubt." Subjective probabilities should be used only when all other methods fail, and then only with a high level of skepticism.

An important application of probability theory relates to the theory of **reliability**. The reliability of a component or system can be defined as follows.

DEFINITION 7. RELIABILITY. *The **reliability** of a product is the probability that it will function within specified limits for a specified period of time under specified environmental conditions.*

Thus, the reliability of a "standard equipment" automobile tire is close to 1 for 10,000 miles of operation on a passenger car traveling within the speed limits on paved roads, but it is close to zero for even short distances at the Indianapolis "500."

The reliability of a system of components can be calculated from the reliabilities of the individual components if the system consists entirely of components connected in series, or in parallel, or both. A **series system** is one in which all components are so interrelated that the entire system will fail if any one (or more) of its components fails. A **parallel system** will fail only if all its components fail. An example of a series system is a string of Christmas lights connected electrically "in series." If one bulb fails, the entire string will fail to light. Parallel systems are sometimes called "redundant" systems. For example, if the hydraulic system on a commercial aircraft that lowers the landing wheels fails, they may be lowered manually with a crank.

We shall assume that the components connected in a series system are independent; that is, the performance of one part does not affect the reliability of the others. Under this assumption, the reliability of a parallel system is given by an extension of Definition 5. Thus, we have the following theorem.

THEOREM 14. The **reliability of a series system** consisting of n independent components is given by

$$R_s = \prod_{i=1}^{n} R_i$$

where R_i is the reliability of the ith component.

Proof The proof follows immediately by iterating in Definition 5.

Theorem 14 vividly demonstrates the effect of increased complexity on reliability. For example, if a series system has 5 components, each with a reliability of 0.970, the reliability of the entire system is only $(0.970)^5 = 0.859$. If the system complexity were increased so it now has 10 such components, the reliability would be reduced to $(0.970)^{10} = 0.738$.

One way to improve the reliability of a series system is to introduce parallel redundancy by replacing some or all of its components by several components connected in parallel. If a system consists of n independent components connected in parallel, it will fail to function only if all components fail. Thus, for the ith component, the probability of failure is $F_i = 1 - R_i$, called the "unreliability" of the component. Again applying Definition 5, we obtain the following theorem.

THEOREM 15. The **reliability of a parallel system** consisting of n independent components is given by

$$R_P = 1 - \prod_{i=1}^{n}(1 - R_i)$$

Proof The proof of this theorem is identical to that of Theorem 14, with $(1 - R_i)$ replacing R_i.

EXAMPLE 29

Consider the system diagramed in Figure 11, which consists of eight components having the reliabilities shown in the figure. Find the reliability of the system.

Solution

The parallel subsystem C, D, E can be replaced by an equivalent component, C' having the reliability $1 - (1 - 0.70)^3 = 0.973$. Likewise, F, G can be replaced by F' having the reliability $1 - (1 - 0.75)^2 = 0.9375$. Thus, the system is reduced to the parallel system A, B, C', F', H, having the reliability $(0.95)(0.99)(0.973)(0.9375)(0.90) = 0.772$.

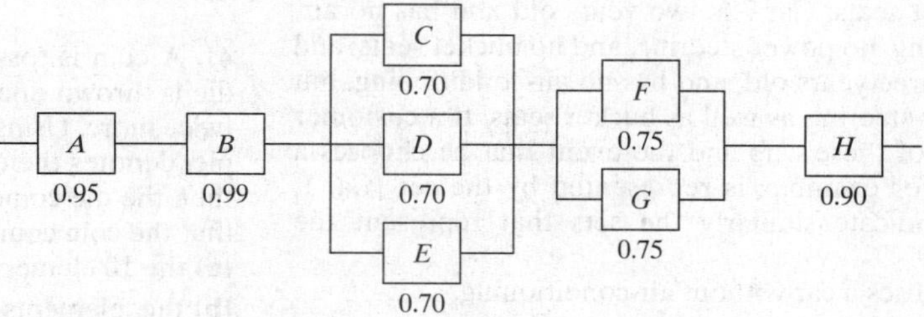

Figure 11. Combination of series and parallel systems.

Applied Exercises

SECS. 1–3

35. If $S = \{1,2,3,4,5,6,7,8,9\}$, $A = \{1,3,5,7\}$, $B = \{6,7,8,9\}$, $C = \{2,4,8\}$, and $D = \{1,5,9\}$, list the elements of the subsets of S corresponding to the following events:

(a) $A' \cap B$; **(b)** $(A' \cap B) \cap C$; **(c)** $B' \cup C$;
(d) $(B' \cup C) \cap D$; **(e)** $A' \cap C$; **(f)** $(A' \cap C) \cap D$.

36. An electronics firm plans to build a research laboratory in Southern California, and its management has to decide between sites in Los Angeles, San Diego, Long Beach, Pasadena, Santa Barbara, Anaheim, Santa Monica, and Westwood. If A represents the event that they will choose a site in San Diego or Santa Barbara, B represents the event that they will choose a site in San Diego or Long Beach, C represents the event that they will choose a site in Santa Barbara or Anaheim, and D represents the event that they will choose a site in Los Angeles or Santa Barbara, list the elements of each of the following subsets of the sample space, which consists of the eight site selections:

(a) A'; **(b)** D'; **(c)** $C \cap D$;
(d) $B \cap C$; **(e)** $B \cup C$; **(f)** $A \cup B$;
(g) $C \cup D$; **(h)** $(B \cup C)'$; **(i)** $B' \cap C'$.

37. Among the eight cars that a dealer has in his showroom, Car 1 is new and has air-conditioning, power steering, and bucket seats; Car 2 is one year old and has air-conditioning, but neither power steering nor bucket seats; Car 3 is two years old and has air-conditioning and power steering, but no bucket seats; Car 4 is three years old and has air-conditioning, but neither power steering nor bucket seats; Car 5 is new and has no air-conditioning, no power steering, and no bucket seats; Car 6 is one year old and has power steering, but neither air-conditioning nor bucket seats; Car 7 is two years old and has no air-conditioning, no power steering, and no bucket seats; and Car 8 is three years old, and has no air-conditioning, but has power steering as well as bucket seats. If a customer buys one of these cars and the event that he chooses a new car, for example, is represented by the set {Car 1, Car 5}, indicate similarly the sets that represent the events that

(a) he chooses a car without air-conditioning;
(b) he chooses a car without power steering;
(c) he chooses a car with bucket seats;
(d) he chooses a car that is either two or three years old.

38. With reference to Exercise 37, state in words what kind of car the customer will choose, if his choice is given by

(a) the complement of the set of part (a);
(b) the union of the sets of parts (b) and (c);
(c) the intersection of the sets of parts (c) and (d);
(d) the intersection of parts (b) and (c) of this exercise.

39. If Ms. Brown buys one of the houses advertised for sale in a Seattle newspaper (on a given Sunday), T is the event that the house has three or more baths, U is the event that it has a fireplace, V is the event that it costs more than \$200,000, and W is the event that it is new, describe (in words) each of the following events:

(a) T'; **(b)** U'; **(c)** V';
(d) W'; **(e)** $T \cap U$; **(f)** $T \cap V$;
(g) $U' \cap V$; **(h)** $V \cup W$; **(i)** $V' \cup W$;
(j) $T \cup U$; **(k)** $T \cup V$; **(l)** $V \cap W$.

40. A resort hotel has two station wagons, which it uses to shuttle its guests to and from the airport. If the larger of the two station wagons can carry five passengers and the smaller can carry four passengers, the point (0, 3) represents the event that at a given moment the larger station wagon is empty while the smaller one has three passengers, the point (4, 2) represents the event that at the given moment the larger station wagon has four passengers while the smaller one has two passengers, ..., draw a figure showing the 30 points of the corresponding sample space. Also, if E stands for the event that at least one of the station wagons is empty, F stands for the event that together they carry two, four, or six passengers, and G stands for the event that each carries the same number of passengers, list the points of the sample space that correspond to each of the following events:

(a) E; **(b)** F; **(c)** G;
(d) $E \cup F$; **(e)** $E \cap F$; **(f)** $F \cup G$;
(g) $E \cup F'$; **(h)** $E \cap G'$; **(i)** $F' \cap E'$.

41. A coin is tossed once. Then, if it comes up heads, a die is thrown once; if the coin comes up tails, it is tossed twice more. Using the notation in which (H, 2), for example, denotes the event that the coin comes up heads and then the die comes up 2, and (T, T, T) denotes the event that the coin comes up tails three times in a row, list

(a) the 10 elements of the sample space S;
(b) the elements of S corresponding to event A that exactly one head occurs;
(c) the elements of S corresponding to event B that at least two tails occur or a number greater than 4 occurs.

42. An electronic game contains three components arranged in the series–parallel circuit shown in Figure 12. At any given time, each component may or may not be operative, and the game will operate only if there is a continuous circuit from P to Q. Let A be the event that the game will operate; let B be the event that

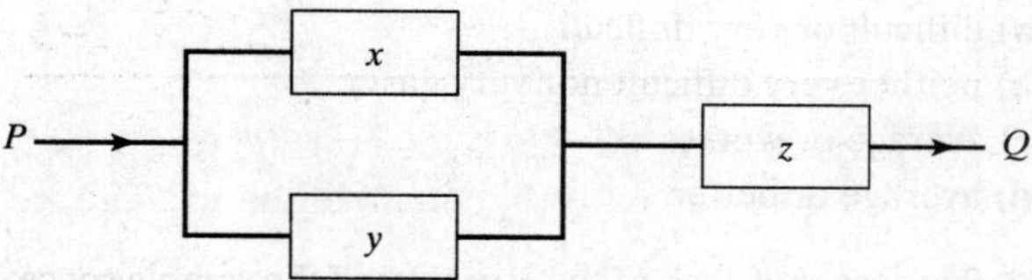

Figure 12. Diagram for Exercise 42.

the game will operate though component x is not operative; and let C be the event that the game will operate though component y is not operative. Using the notation in which (0, 0, 1), for example, denotes that component z is operative but components x and y are not,
(a) list the elements of the sample space S and also the elements of S corresponding to events A, B, and C;
(b) determine which pairs of events, A and B, A and C, or B and C, are mutually exclusive.

43. An experiment consists of rolling a die until a 3 appears. Describe the sample space and determine
(a) how many elements of the sample space correspond to the event that the 3 appears on the kth roll of the die;
(b) how many elements of the sample space correspond to the event that the 3 appears not later than the kth roll of the die.

44. Express symbolically the sample space S that consists of all the points (x, y) on or in the circle of radius 3 centered at the point $(2, -3)$.

45. If $S = \{x|0 < x < 10\}$, $M = \{x|3 < x \leqq 8\}$, and $N = \{x|5 < x < 10\}$, find
(a) $M \cup N$; **(b)** $M \cap N$;
(c) $M \cap N'$; **(d)** $M' \cup N$.

46. In Figure 13, L is the event that a driver has liability insurance and C is the event that she has collision insurance. Express in words what events are represented by regions 1, 2, 3, and 4.

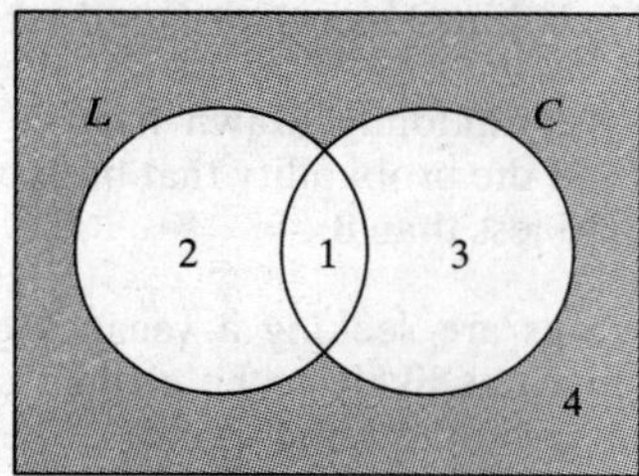

Figure 13. Venn diagram for Exercise 46.

47. With reference to Exercise 46 and Figure 13, what events are represented by
(a) regions 1 and 2 together;
(b) regions 2 and 4 together;
(c) regions 1, 2, and 3 together;
(d) regions 2, 3, and 4 together?

48. In Figure 14, E, T, and N are the events that a car brought to a garage needs an engine overhaul, transmission repairs, or new tires. Express in words the events represented by
(a) region 1;
(b) region 3;
(c) region 7;
(d) regions 1 and 4 together;
(e) regions 2 and 5 together;
(f) regions 3, 5, 6, and 8 together.

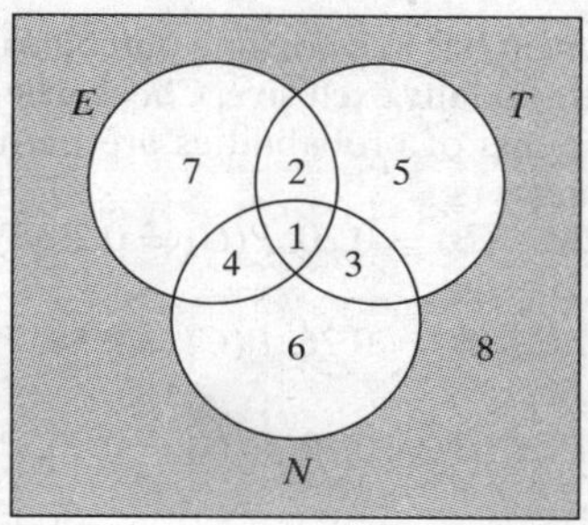

Figure 14. Venn diagram for Exercise 48.

49. With reference to Exercise 48 and Figure 14, list the region or combinations of regions representing the events that a car brought to the garage needs
(a) transmission repairs, but neither an engine overhaul nor new tires;
(b) an engine overhaul and transmission repairs;
(c) transmission repairs or new tires, but not an engine overhaul;
(d) new tires.

50. A market research organization claims that, among 500 shoppers interviewed, 308 regularly buy Product X, 266 regularly buy Product Y, 103 regularly buy both, and 59 buy neither on a regular basis. Using a Venn diagram and filling in the number of shoppers associated with the various regions, check whether the results of this study should be questioned.

51. In a group of 200 college students, 138 are enrolled in a course in psychology, 115 are enrolled in a course in sociology, and 91 are enrolled in both. How many of these students are not enrolled in either course? (*Hint:* Draw a suitable Venn diagram and fill in the numbers associated with the various regions.)

52. Among 120 visitors to Disneyland, 74 stayed for at least 3 hours, 86 spent at least \$20, 64 went on the Matterhorn ride, 60 stayed for at least 3 hours and spent at

least \$20, 52 stayed for at least 3 hours and went on the Matterhorn ride, 54 spent at least \$20 and went on the Matterhorn ride, and 48 stayed for at least 3 hours, spent at least \$20, and went on the Matterhorn ride. Drawing a Venn diagram with three circles (like that of Figure 4) and filling in the numbers associated with the various regions, find how many of the 120 visitors to Disneyland
(a) stayed for at least 3 hours, spent at least \$20, but did not go on the Matterhorn ride;
(b) went on the Matterhorn ride, but stayed less than 3 hours and spent less than \$20;
(c) stayed less than 3 hours, spent at least \$20, but did not go on the Matterhorn ride.

SECS. 4–5

53. An experiment has five possible outcomes, A, B, C, D, and E, that are mutually exclusive. Check whether the following assignments of probabilities are permissible and explain your answers:
(a) $P(A) = 0.20$, $P(B) = 0.20$, $P(C) = 0.20$, $P(D) = 0.20$, and $P(E) = 0.20$;
(b) $P(A) = 0.21$, $P(B) = 0.26$, $P(C) = 0.58$, $P(D) = 0.01$, and $P(E) = 0.06$;
(c) $P(A) = 0.18$, $P(B) = 0.19$, $P(C) = 0.20$, $P(D) = 0.21$, and $P(E) = 0.22$;
(d) $P(A) = 0.10$, $P(B) = 0.30$, $P(C) = 0.10$, $P(D) = 0.60$, and $P(E) = -0.10$;
(e) $P(A) = 0.23$, $P(B) = 0.12$, $P(C) = 0.05$, $P(D) = 0.50$, and $P(E) = 0.08$.

54. If A and B are mutually exclusive, $P(A) = 0.37$, and $P(B) = 0.44$, find
(a) $P(A')$; **(b)** $P(B')$; **(c)** $P(A \cup B)$;
(d) $P(A \cap B)$; **(e)** $P(A \cap B')$; **(f)** $P(A' \cap B')$.

55. Explain why there must be a mistake in each of the following statements:
(a) The probability that Jean will pass the bar examination is 0.66 and the probability that she will not pass is −0.34.
(b) The probability that the home team will win an upcoming football game is 0.77, the probability that it will tie the game is 0.08, and the probability that it will win or tie the game is 0.95.
(c) The probabilities that a secretary will make 0, 1, 2, 3, 4, or 5 *or more* mistakes in typing a report are, respectively, 0.12, 0.25, 0.36, 0.14, 0.09, and 0.07.
(d) The probabilities that a bank will get 0, 1, 2, or 3 *or more* bad checks on any given day are, respectively, 0.08, 0.21, 0.29, and 0.40.

56. The probabilities that the serviceability of a new X-ray machine will be rated very difficult, difficult, average, easy, or very easy are, respectively, 0.12, 0.17, 0.34, 0.29, and 0.08. Find the probabilities that the serviceability of the machine will be rated
(a) difficult or very difficult;
(b) neither very difficult nor very easy;
(c) average or worse;
(d) average or better.

57. Suppose that each of the 30 points of the sample space of Exercise 40 is assigned the probability $\frac{1}{30}$. Find the probabilities that at a given moment
(a) at least one of the station wagons is empty;
(b) each of the two station wagons carries the same number of passengers;
(c) the larger station wagon carries more passengers than the smaller station wagon;
(d) together they carry at least six passengers.

58. A hat contains 20 white slips of paper numbered from 1 through 20, 10 red slips of paper numbered from 1 through 10, 40 yellow slips of paper numbered from 1 through 40, and 10 blue slips of paper numbered from 1 through 10. If these 80 slips of paper are thoroughly shuffled so that each slip has the same probability of being drawn, find the probabilities of drawing a slip of paper that is
(a) blue or white;
(b) numbered 1, 2, 3, 4, or 5;
(c) red or yellow and also numbered 1, 2, 3, or 4;
(d) numbered 5, 15, 25, or 35;
(e) white and numbered higher than 12 or yellow and numbered higher than 26.

59. A police department needs new tires for its patrol cars and the probabilities are 0.15, 0.24, 0.03, 0.28, 0.22, and 0.08, respectively, that it will buy Uniroyal tires, Goodyear tires, Michelin tires, General tires, Goodrich tires, or Armstrong tires. Find the probabilities that it will buy
(a) Goodyear or Goodrich tires;
(b) Uniroyal, Michelin, or Goodrich tires;
(c) Michelin or Armstrong tires;
(d) Uniroyal, Michelin, General, or Goodrich tires.

60. Two cards are randomly drawn from a deck of 52 playing cards. Find the probability that both cards will be greater than 3 and less than 8.

61. Four candidates are seeking a vacancy on a school board. If A is twice as likely to be elected as B, and B and C are given about the same chance of being elected, while C is twice as likely to be elected as D, what are the probabilities that
(a) C will win;
(b) A will not win?

62. In a poker game, five cards are dealt at random from an ordinary deck of 52 playing cards. Find the probabilities of getting

(a) two pairs (any two distinct face values occurring exactly twice);
(b) four of a kind (four cards of equal face value).

63. In a game of Yahtzee, five balanced dice are rolled simultaneously. Find the probabilities of getting
(a) two pairs;
(b) three of a kind;
(c) a full house (three of a kind and a pair);
(d) four of a kind.

64. Explain on the basis of the various rules of Exercises 5 through 9 why there is a mistake in each of the following statements:
(a) The probability that it will rain is 0.67, and the probability that it will rain or snow is 0.55.
(b) The probability that a student will get a passing grade in English is 0.82, and the probability that she will get a passing grade in English and French is 0.86.
(c) The probability that a person visiting the San Diego Zoo will see the giraffes is 0.72, the probability that he will see the bears is 0.84, and the probability that he will see both is 0.52.

65. Among the 78 doctors on the staff of a hospital, 64 carry malpractice insurance, 36 are surgeons, and 34 of the surgeons carry malpractice insurance. If one of these doctors is chosen by lot to represent the hospital staff at an A.M.A. convention (that is, each doctor has a probability of $\frac{1}{78}$ of being selected), what is the probability that the one chosen is not a surgeon and does not carry malpractice insurance?

66. A line segment of length l is divided by a point selected at random within the segment. What is the probability that it will divide the line segment in a ratio greater than 1:2? What is the probability that it will divide the segment exactly in half?

67. A right triangle has the legs 3 and 4 units, respectively. Find the probability that a line segment, drawn at random parallel to the hypotenuse and contained entirely in the triangle, will divide the triangle so that the area between the line and the vertex opposite the hypotenuse will equal at least half the area of the triangle.

68. For married couples living in a certain suburb, the probability that the husband will vote in a school board election is 0.21, the probability that the wife will vote in the election is 0.28, and the probability that they will both vote is 0.15. What is the probability that at least one of them will vote?

69. Given $P(A) = 0.59$, $P(B) = 0.30$, and $P(A \cap B) = 0.21$, find
(a) $P(A \cup B)$; **(b)** $P(A \cap B')$;
(c) $P(A' \cup B')$; **(d)** $P(A' \cap B')$.

70. At Roanoke College it is known that $\frac{1}{3}$ of the students live off campus. It is also known that $\frac{5}{9}$ of the students are from within the state of Virginia and that $\frac{3}{4}$ of the students are from out of state or live on campus. What is the probability that a student selected at random from Roanoke College is from out of state and lives on campus?

71. A biology professor has two graduate assistants helping her with her research. The probability that the older of the two assistants will be absent on any given day is 0.08, the probability that the younger of the two will be absent on any given day is 0.05, and the probability that they will both be absent on any given day is 0.02. Find the probabilities that
(a) either or both of the graduate assistants will be absent on any given day;
(b) at least one of the two graduate assistants will not be absent on any given day;
(c) only one of the two graduate assistants will be absent on any given day.

72. Suppose that if a person visits Disneyland, the probability that he will go on the Jungle Cruise is 0.74, the probability that he will ride the Monorail is 0.70, the probability that he will go on the Matterhorn ride is 0.62, the probability that he will go on the Jungle Cruise and ride the Monorail is 0.52, the probability that he will go on the Jungle Cruise as well as the Matterhorn ride is 0.46, the probability that he will ride the Monorail and go on the Matterhorn ride is 0.44, and the probability that he will go on all three of these rides is 0.34. What is the probability that a person visiting Disneyland will go on at least one of these three rides?

73. Suppose that if a person travels to Europe for the first time, the probability that he will see London is 0.70, the probability that he will see Paris is 0.64, the probability that he will see Rome is 0.58, the probability that he will see Amsterdam is 0.58, the probability that he will see London and Paris is 0.45, the probability that he will see London and Rome is 0.42, the probability that he will see London and Amsterdam is 0.41, the probability that he will see Paris and Rome is 0.35, the probability that he will see Paris and Amsterdam is 0.39, the probability that he will see Rome and Amsterdam is 0.32, the probability that he will see London, Paris, and Rome is 0.23, the probability that he will see London, Paris, and Amsterdam is 0.26, the probability that he will see London, Rome, and Amsterdam is 0.21, the probability that he will see Paris, Rome, and Amsterdam is 0.20, and the probability that he will see all four of these cities is 0.12. What is the probability that a person traveling to Europe for the first time will see at least one of these four cities? (*Hint*: Use the formula of Exercise 13.)

74. Use the formula of Exercise 15 to convert each of the following odds to probabilities:
(a) If three eggs are randomly chosen from a carton of 12 eggs of which 3 are cracked, the odds are 34 to 21 that at least one of them will be cracked.
(b) If a person has eight \$1 bills, five \$5 bills, and one \$20 bill, and randomly selects three of them, the odds are 11 to 2 that they will not all be \$1 bills.
(c) If we arbitrarily arrange the letters in the word "nest," the odds are 5 to 1 that we will not get a meaningful word in the English language.

75. Use the definition of "odds" given in Exercise 15 to convert each of the following probabilities to odds:
(a) The probability that the last digit of a car's license plate is a 2, 3, 4, 5, 6, or 7 is $\frac{6}{10}$.
(b) The probability of getting at least two heads in four flips of a balanced coin is $\frac{11}{16}$.
(c) The probability of rolling "7 or 11" with a pair of balanced dice is $\frac{2}{9}$.

SECS. 6–8

76. There are 90 applicants for a job with the news department of a television station. Some of them are college graduates and some are not; some of them have at least three years' experience and some have not, with the exact breakdown being

	College graduates	*Not college graduates*
At least three years' experience	18	9
Less than three years' experience	36	27

If the order in which the applicants are interviewed by the station manager is random, G is the event that the first applicant interviewed is a college graduate, and T is the event that the first applicant interviewed has at least three years' experience, determine each of the following probabilities directly from the entries and the row and column totals of the table:
(a) $P(G)$; **(b)** $P(T')$; **(c)** $P(G \cap T)$;
(d) $P(G' \cap T')$; **(e)** $P(T|G)$; **(f)** $P(G'|T')$.

77. Use the results of Exercise 76 to verify that
(a) $P(T|G) = \frac{P(G \cap T)}{P(G)}$;
(b) $P(G'|T') = \frac{P(G' \cap T')}{P(T')}$.

78. With reference to Exercise 65, what is the probability that the doctor chosen to represent the hospital staff at the convention carries malpractice insurance given that he or she is a surgeon?

79. With reference to Exercise 68, what is the probability that a husband will vote in the election given that his wife is going to vote?

80. With reference to Exercise 70, what is the probability that one of the students will be living on campus given that he or she is from out of state?

81. A bin contains 100 balls, of which 25 are red, 40 are white, and 35 are black. If two balls are selected from the bin without replacement, what is the probability that one will be red and one will be white?

82. If subjective probabilities are determined by the method suggested in Exercise 16, the third postulate of probability may not be satisfied. However, proponents of the subjective probability concept usually impose this postulate as a **consistency criterion**; in other words, they regard subjective probabilities that do not satisfy the postulate as inconsistent.
(a) A high school principal feels that the odds are 7 to 5 against her getting a \$1,000 raise and 11 to 1 against her getting a \$2,000 raise. Furthermore, she feels that it is an even-money bet that she will get one of these raises or the other. Discuss the consistency of the corresponding subjective probabilities.
(b) Asked about his political future, a party official replies that the odds are 2 to 1 that he will not run for the House of Representatives and 4 to 1 that he will not run for the Senate. Furthermore, he feels that the odds are 7 to 5 that he will run for one or the other. Are the corresponding probabilities consistent?

83. If we let $x =$ the number of spots facing up when a pair of dice is cast, then we can use the sample space S_2 of Example 2 to describe the outcomes of the experiment.
(a) Find the probability of each outcome in S_2.
(b) Verify that the sum of these probabilities is 1.

84. There are two Porsches in a road race in Italy, and a reporter feels that the odds against their winning are 3 to 1 and 5 to 3. To be consistent (see Exercise 82), what odds should the reporter assign to the event that either car will win?

85. Using a computer program that can generate random integers on the interval (0, 9) with equal probabilities, generate 1,000 such integers and use the frequency interpretation to estimate the probability that such a randomly chosen integer will have a value less than 1.

86. Using the method of Exercise 85, generate a second set of 1,000 random integers on (0, 9). Estimate the probability that A: an integer selected at random from the first set will be less than 1 or B: an integer selected at random from the second set will be less than 1
(a) using the frequency interpretation of probabilities;
(b) using Theorem 7 and $P(A \cap B) = \frac{1}{81}$.

87. It is felt that the probabilities are 0.20, 0.40, 0.30, and 0.10 that the basketball teams of four universities, T, U, V, and W, will win their conference championship. If university U is placed on probation and declared ineligible for the championship, what is the probability that university T will win the conference championship?

88. With reference to Exercise 72, find the probabilities that a person who visits Disneyland will
(a) ride the Monorail given that he will go on the Jungle Cruise;
(b) go on the Matterhorn ride given that he will go on the Jungle Cruise and ride the Monorail;
(c) not go on the Jungle Cruise given that he will ride the Monorail and/or go on the Matterhorn ride;
(d) go on the Matterhorn ride and the Jungle Cruise given that he will not ride the Monorail.

(*Hint*: Draw a Venn diagram and fill in the probabilities associated with the various regions.)

89. Crates of eggs are inspected for blood clots by randomly removing three eggs in succession and examining their contents. If all three eggs are good, the crate is shipped; otherwise it is rejected. What is the probability that a crate will be shipped if it contains 120 eggs, of which 10 have blood clots?

90. The probability of surviving a certain transplant operation is 0.55. If a patient survives the operation, the probability that his or her body will reject the transplant within a month is 0.20. What is the probability of surviving both of these critical stages?

91. Suppose that in Vancouver, B.C., the probability that a rainy fall day is followed by a rainy day is 0.80 and the probability that a sunny fall day is followed by a rainy day is 0.60. Find the probabilities that a rainy fall day is followed by
(a) a rainy day, a sunny day, and another rainy day;
(b) two sunny days and then a rainy day;
(c) two rainy days and then two sunny days;
(d) rain two days later.

[*Hint*: In part (c) use the formula of Exercise 19.]

92. Use the formula of Exercise 19 to find the probability of randomly choosing (without replacement) four healthy guinea pigs from a cage containing 20 guinea pigs, of which 15 are healthy and 5 are diseased.

93. A sharpshooter hits a target with probability 0.75. Assuming independence, find the probabilities of getting
(a) a hit followed by two misses;
(b) two hits and a miss in any order.

94. A balanced die is tossed twice. If A is the event that an even number comes up on the first toss, B is the event that an even number comes up on the second toss, and C is the event that both tosses result in the same number, are the events A, B, and C
(a) pairwise independent;
(b) independent?

95. A shipment of 1,000 parts contains 1 percent defective parts. Find the probability that
(a) the first four parts chosen arbitrarily for inspection are nondefective;
(b) the first defective part found will be on the fourth inspection.

96. A coin is loaded so that the probabilities of heads and tails are 0.52 and 0.48, respectively. If the coin is tossed three times, what are the probabilities of getting
(a) all heads;
(b) two tails and a head in that order?

97. If 5 of a company's 10 delivery trucks do not meet emission standards and 3 of them are chosen for inspection, what is the probability that none of the trucks chosen will meet emission standards?

98. Medical records show that one out of 10 persons in a certain town has a thyroid deficiency. If 12 persons in this town are randomly chosen and tested, what is the probability that at least one of them will have a thyroid deficiency?

99. If a person randomly picks 4 of the 15 gold coins a dealer has in stock, and 6 of the coins are counterfeits, what is the probability that the coins picked will all be counterfeits?

100. A department store that bills its charge-account customers once a month has found that if a customer pays promptly one month, the probability is 0.90 that he or she will also pay promptly the next month; however, if a customer does not pay promptly one month, the probability that he or she will pay promptly the next month is only 0.40. (Assume that the probability of paying or not paying on any given month depends only on the outcome of the previous month.)
(a) What is the probability that a customer who pays promptly one month will also pay promptly the next three months?
(b) What is the probability that a customer who does not pay promptly one month will also not pay promptly the next two months and then make a prompt payment the month after that?

101. With reference to Figure 15, verify that events A, B, C, and D are independent. Note that the region representing A consists of two circles, and so do the regions representing B and C.

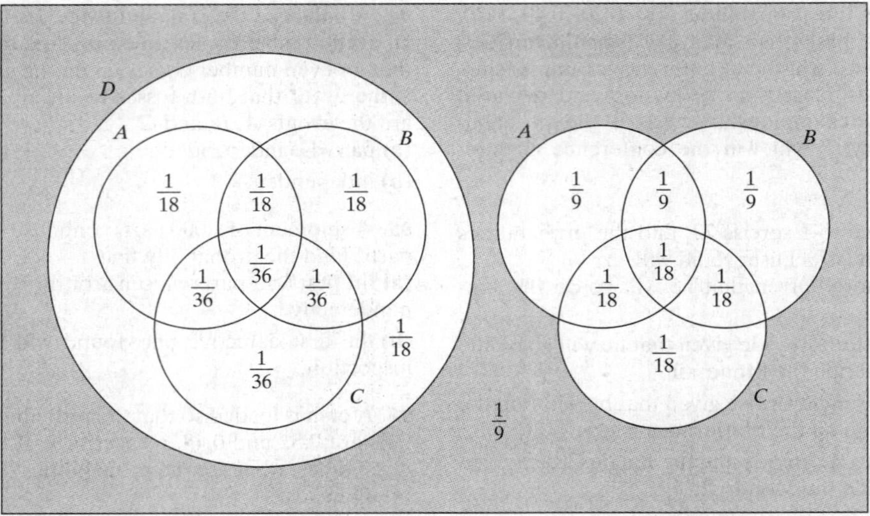

Figure 15. Diagram for Exercise 101.

102. At an electronics plant, it is known from past experience that the probability is 0.84 that a new worker who has attended the company's training program will meet the production quota, and that the corresponding probability is 0.49 for a new worker who has not attended the company's training program. If 70 percent of all new workers attend the training program, what is the probability that a new worker will meet the production quota?

103. It is known from experience that in a certain industry 60 percent of all labor–management disputes are over wages, 15 percent are over working conditions, and 25 percent are over fringe issues. Also, 45 percent of the disputes over wages are resolved without strikes, 70 percent of the disputes over working conditions are resolved without strikes, and 40 percent of the disputes over fringe issues are resolved without strikes. What is the probability that a labor–management dispute in this industry will be resolved without a strike?

104. In a T-maze, a rat is given food if it turns left and an electric shock if it turns right. On the first trial there is a 50–50 chance that a rat will turn either way; then, if it receives food on the first trial, the probability is 0.68 that it will turn left on the next trial, and if it receives a shock on the first trial, the probability is 0.84 that it will turn left on the next trial. What is the probability that a rat will turn left on the second trial?

105. With reference to Exercise 103, what is the probability that if a labor–management dispute in this industry is resolved without a strike, it was over wages?

106. The probability that a one-car accident is due to faulty brakes is 0.04, the probability that a one-car accident is correctly attributed to faulty brakes is 0.82, and the probability that a one-car accident is incorrectly attributed to faulty brakes is 0.03. What is the probability that

(a) a one-car accident will be attributed to faulty brakes;

(b) a one-car accident attributed to faulty brakes was actually due to faulty brakes?

107. With reference to Example 25, suppose that we discover later that the job was completed on time. What is the probability that there had been a strike?

108. In a certain community, 8 percent of all adults over 50 have diabetes. If a health service in this community correctly diagnoses 95 percent of all persons with diabetes as having the disease and incorrectly diagnoses 2 percent of all persons without diabetes as having the disease, find the probabilities that

(a) the community health service will diagnose an adult over 50 as having diabetes;

(b) a person over 50 diagnosed by the health service as having diabetes actually has the disease.

109. An explosion at a construction site could have occurred as the result of static electricity, malfunctioning of equipment, carelessness, or sabotage. Interviews with construction engineers analyzing the risks involved led to the estimates that such an explosion would occur with probability 0.25 as a result of static electricity, 0.20 as a result of malfunctioning of equipment, 0.40 as a result of carelessness, and 0.75 as a result of sabotage. It is also felt that the prior probabilities of the four causes of the explosion are 0.20, 0.40, 0.25, and 0.15. Based on all this information, what is

(a) the most likely cause of the explosion;

(b) the least likely cause of the explosion?

110. A mail-order house employs three stock clerks, U, V, and W, who pull items from shelves and assemble them for subsequent verification and packaging. U makes a mistake in an order (gets a wrong item or the wrong quantity) one time in a hundred, V makes a mistake in an order five times in a hundred, and W makes a mistake in an order three times in a hundred. If U, V, and W fill, respectively, 30, 40, and 30 percent of all orders, what are the probabilities that

(a) a mistake will be made in an order;

(b) if a mistake is made in an order, the order was filled by U;

(c) if a mistake is made in an order, the order was filled by V?

111. An art dealer receives a shipment of five old paintings from abroad, and, on the basis of past experience, she feels that the probabilities are, respectively, 0.76, 0.09, 0.02, 0.01, 0.02, and 0.10 that 0, 1, 2, 3, 4, or all 5 of them are forgeries. Since the cost of authentication is fairly high, she decides to select one of the five paintings at random and send it away for authentication. If it turns out that this painting is a forgery, what probability should she now assign to the possibility that all the other paintings are also forgeries?

112. To get answers to sensitive questions, we sometimes use a method called the **randomized response technique**. Suppose, for instance, that we want to determine what percentage of the students at a large university smoke marijuana. We construct 20 flash cards, write "I smoke marijuana at least once a week" on 12 of the cards, where 12 is an arbitrary choice, and "I do not smoke marijuana at least once a week" on the others. Then, we let each student (in the sample interviewed) select one of the 20 cards at random and respond "yes" or "no" without divulging the question.

(a) Establish a relationship between $P(Y)$, the probability that a student will give a "yes" response, and $P(M)$, the probability that a student randomly selected at that university smokes marijuana at least once a week.

(b) If 106 of 250 students answered "yes" under these conditions, use the result of part (a) and $\frac{106}{250}$ as an estimate of $P(Y)$ to estimate $P(M)$.

SEC. 9

113. A series system consists of three components, each having the reliability 0.95, and three components, each having the reliability 0.99. Find the reliability of the system.

114. Find the reliability of a series systems having five components with reliabilities 0.995, 0.990, 0.992, 0.995, and 0.998, respectively.

115. What must be the reliability of each component in a series system consisting of six components that must have a system reliability of 0.95?

116. Referring to Exercise 115, suppose now that there are 10 components, and the system reliability must be 0.90.

117. Suppose a system consists of four components, connected in parallel, having the reliabilities 0.8, 0.7, 0.7, and 0.65, respectively. Find the system reliability.

118. Referring to Exercise 117, suppose now that the system has five components with reliabilities 0.85, 0.80, 0.65, 0.60, and 0.70, respectively. Find the system reliability.

119. A system consists of two components having the reliabilities 0.95 and 0.90, connected in series to two parallel subsystems, the first containing four components, each having the reliability 0.60, and the second containing two components, each having the reliability 0.75. Find the system reliability.

120. A series system consists of two components having the reliabilities 0.98 and 0.99, respectively, connected to a parallel subsystem containing five components having the reliabilities 0.75, 0.60, 0.65, 0.70, and 0.60, respectively. Find the system reliability.

References

Among the numerous textbooks on probability theory published in recent years, one of the most popular is

FELLER, W., *An Introduction to Probability Theory and Its Applications*, Vol. I, 3rd ed. New York: John Wiley & Sons, Inc., 1968.

More elementary treatments may be found in

BARR, D. R., and ZEHNA, P. W., *Probability: Modeling Uncertainty*. Reading, Mass.: Addison-Wesley Publishing Company, Inc., 1983,

DRAPER, N. R., and LAWRENCE, W. E., *Probability: An Introductory Course*. Chicago: Markam Publishing Company, 1970,

FREUND, J. E., *Introduction to Probability*. New York: Dover Publications, Inc., 1993 Reprint,

GOLDBERG, S., *Probability—An Introduction*. Mineola, N.Y.: Dover Publications, Inc. (republication of 1960 edition),

HODGES, J. L., and LEHMANN, E. L., *Elements of Finite Probability*. San Francisco: Holden Day, Inc., 1965,

NOSAL, M., *Basic Probability and Applications*. Philadelphia: W. B. Saunders Company, 1977.

More advanced treatments are given in many texts, for instance,

HOEL, P., PORT, S. C., and STONE, C. J., *Introduction to Probability Theory*. Boston: Houghton Mifflin Company, 1971,

KHAZANIE, R., *Basic Probability Theory and Applications*. Pacific Palisades, Calif.: Goodyear Publishing Company, Inc., 1976,

PARZEN, E., *Modern Probability Theory and Its Applications*. New York: John Wiley & Sons, Inc., 1960,

ROSS, S., *A First Course in Probability*, 3rd ed. New York: Macmillan Publishing Company, 1988.

SOLOMON, F., *Probability and Stochastic Processes*. Upper Saddle River, N.J.: Prentice Hall, 1987.

For more on reliability and related topics, see

JOHNSON, R. A., *Miller and Freund's Probability and Statistics for Engineers*. Upper Saddle River, N.J.: Prentice Hall, 2000,

MILLER, I. and MILLER, M., *Statistical Methods for Quality with Applications to Engineering and Management*. Upper Saddle River, N.J.: Prentice Hall, 1995.

Answers to Odd-Numbered Exercises

35 (a) $\{6,8,9\}$; **(b)** $\{8\}$; **(c)** $\{1,2,3,4,5,8\}$; **(d)** $\{1,5\}$; **(e)** $\{2,4,8\}$; **(f)** $\emptyset$.

37 (a) {Car 5, Car 6, Car 7, Car 8}; **(b)** {Car 2, Car 4, Car 5, Car 7}; **(c)** {Car 1, Car 8}; **(d)** {Car 3, Car 4, Car 7, Car 8}.

39 (a) The house has fewer than three baths. **(b)** The house does not have a fireplace. **(c)** The house does not cost more than \$200,000. **(d)** The house is not new. **(e)** The house has three or more baths and a fireplace. **(f)** The house has three or more baths and costs more than \$200,000. **(g)** The house costs more than \$200,000 but has no fireplace. **(h)** The house is new or costs more than \$200,000. **(i)** The house is new or costs at most \$200,000. **(j)** The house has three or more baths and/or a fireplace. **(k)** The house has three or more baths and/or costs more than \$200,000. **(l)** The house is new and costs more than \$200,000.

41 (a) (H,1), (H,2), (H,3), (H,4), (H,5), (H,6), (T,H,H), (T,H,T), (T,T,H), and (T,T,T); **(b)** (H,1), (H,2), (H,3), (H,4), (H,5), (H,6), (T,H,T), and (T,T,H); **(c)** (H,5), (H,6), (T,H,T), (T,T,H), and (T,T,T).

43 (a) 5^{k-1}; **(b)** $\dfrac{5^k-1}{4}$.

45 (a) $(x|3<x<10)$; **(b)** $(x|15<x\le 8)$; **(c)** $(x|3<x\le 5)$; **(d)** $(x|0<x\le 3)$ or $(5<x<10)$.

47 (a) The event that a driver has liability insurance. **(b)** The event that a driver does not have collision insurance. **(c)** The event that a driver has liability insurance or collision insurance, but not both. **(d)** The event that a driver does not have both kinds of insurance.

49 (a) Region 5; **(b)** regions 1 and 2 together; **(c)** regions 3, 5, and 6 together; **(d)** regions 1, 3, 4, and 6 together.

51 38.

53 (a) Permissible; **(b)** not permissible because the sum of the probabilities exceeds 1; **(c)** permissible; **(d)** not permissible because $P(E)$ is negative; **(e)** not permissible because the sum of the probabilities is less than 1.

55 (a) The probability that she cannot pass cannot be negative. **(b)** $0.77+0.08=0.85\neq 0.95$; **(c)** $0.12+0.25+0.36+0.14+0.09+0.07=1.03>1$; **(d)** $0.08+0.21+0.29+0.40=0.98<1$.

57 (a) $\frac{1}{3}$; **(b)** $\frac{1}{6}$; **(c)** $\frac{1}{2}$; **(d)** $\frac{1}{3}$.

59 (a) 0.46; **(b)** 0.40; **(c)** 0.11 **(d)** 0.68.

61 (a) $\frac{2}{9}$; **(b)** $\frac{5}{9}$.

63 (a) $\frac{25}{108}$; **(b)** $\frac{25}{162}$; **(c)** $\frac{25}{648}$; **(d)** $\frac{25}{1296}$.

65 $\frac{2}{13}$.

67 $1-\frac{\sqrt{2}}{2}$.

69 (a) 0.68; **(b)** 0.38; **(c)** 0.79; **(d)** 0.32.

71 (a) 0.11; **(b)** 0.98; **(c)** 0.09.

73 0.94.

75 (a) 3 to 2; **(b)** 11 to 5; **(c)** 7 to 2 against it.

77 (a) $\frac{1}{3}$; **(b)** $\frac{3}{7}$.

79 $\frac{15}{28}$.

81 (a) 0.2; **(b)** $\frac{20}{99}$.

83

Outcome	2	3	4	5	6	7	8	9	10	11	12
Probability	$\frac{1}{36}$	$\frac{1}{18}$	$\frac{1}{12}$	$\frac{1}{9}$	$\frac{5}{36}$	$\frac{1}{6}$	$\frac{5}{36}$	$\frac{1}{9}$	$\frac{1}{12}$	$\frac{1}{18}$	$\frac{1}{36}$

87 $\frac{1}{3}$.

89 0.7685.

91 (a) 0.096; **(b)** 0.048; **(c)** 0.0512; **(d)** 0.76.

93 (a) $\frac{3}{64}$; **(b)** $\frac{27}{64}$.

95 (a) Required probability $=0.9606$; exact probability $=0.9605$; **(b)** required probability $=0.0097$ (assuming independence); exact probability $=0.0097$.

97 $\frac{1}{12}$.

99 $\frac{1}{91}$.

103 0.475.

105 0.5684.

107 0.3818.

109 (a) Most likely cause is sabotage ($P=0.3285$); **(b)** least likely cause is static electricity ($P=0.1460$).

111 0.6757.

113 0.832.

115 0.991.

117 0.9937.

119 0.781.

Probability Distributions and Probability Densities

1 Random Variables

In most applied problems involving probabilities we are interested only in a particular aspect (or in two or a few particular aspects) of the outcomes of experiments. For instance, when we roll a pair of dice we are usually interested only in the total, and not in the outcome for each die; when we interview a randomly chosen married couple we may be interested in the size of their family and in their joint income, but not in the number of years they have been married or their total assets; and when we sample mass-produced light bulbs we may be interested in their durability or their brightness, but not in their price.

In each of these examples we are interested in numbers that are associated with the outcomes of chance experiments, that is, in the values taken on by **random variables**. In the language of probability and statistics, the total we roll with a pair of dice is a random variable, the size of the family of a randomly chosen married couple and their joint income are random variables, and so are the durability and the brightness of a light bulb randomly picked for inspection.

To be more explicit, consider Figure 1, which pictures the sample space for an experiment in which we roll a pair of dice, and let us assume that each of the 36 possible outcomes has the probability $\frac{1}{36}$. Note, however, that in Figure 1 we have attached a number to each point: For instance, we attached the number 2 to the point (1, 1), the number 6 to the point (1, 5), the number 8 to the point (6, 2), the number 11 to the point (5, 6), and so forth. Evidently, we associated with each point the value of a random variable, that is, the corresponding total rolled with the pair of dice.

Since "associating a number with each point (element) of a sample space" is merely another way of saying that we are "defining a function over the points of a sample space," let us now make the following definition.

From Chapter 3 of *John E. Freund's Mathematical Statistics with Applications*, Eighth Edition. Irwin Miller, Marylees Miller.

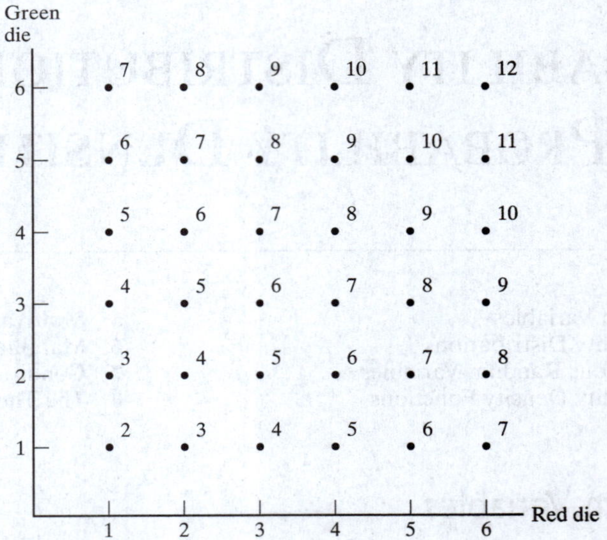

Figure 1. The total number of points rolled with a pair of dice.

> **DEFINITION 1. RANDOM VARIABLE.** *If S is a sample space with a probability measure and X is a real-valued function defined over the elements of S, then X is called a random variable.*†

In this chapter we shall denote random variables by capital letters and their values by the corresponding lowercase letters; for instance, we shall write x to denote a *value* of the random variable X.

With reference to the preceding example and Figure 1, observe that the random variable X takes on the value 9, and we write $X = 9$ for the subset

$$\{(6,3),(5,4),(4,5),(3,6)\}$$

of the sample space S. Thus, $X = 9$ is to be interpreted as the set of elements of S for which the total is 9 and, more generally, $X = x$ is to be interpreted as the set of elements of the sample space for which the random variable X takes on the value x. This may seem confusing, but it reminds one of mathematicians who say "$f(x)$ is a function of x" instead of "$f(x)$ is the value of a function at x."

EXAMPLE 1

Two socks are selected at random and removed in succession from a drawer containing five brown socks and three green socks. List the elements of the sample space, the corresponding probabilities, and the corresponding values w of the random variable W, where W is the number of brown socks selected.

† Instead of "random variable," the terms "chance variable," "stochastic variable," and "variate" are also used in some books.

Solution

If B and G stand for brown and green, the probabilities for BB, BG, GB, and GG are, respectively, $\frac{5}{8} \cdot \frac{4}{7} = \frac{5}{14}$, $\frac{5}{8} \cdot \frac{3}{7} = \frac{15}{56}$, $\frac{3}{8} \cdot \frac{5}{7} = \frac{15}{56}$, and $\frac{3}{8} \cdot \frac{2}{7} = \frac{3}{28}$, and the results are shown in the following table:

Element of sample space	*Probability*	w
BB	$\frac{5}{14}$	2
BG	$\frac{15}{56}$	1
GB	$\frac{15}{56}$	1
GG	$\frac{3}{28}$	0

Also, we can write $P(W = 2) = \frac{5}{14}$, for example, for the probability of the event that the random variable W will take on the value 2.

EXAMPLE 2

A balanced coin is tossed four times. List the elements of the sample space that are presumed to be equally likely, as this is what we mean by a coin being balanced, and the corresponding values x of the random variable X, the total number of heads.

Solution

If H and T stand for heads and tails, the results are as shown in the following table:

Element of sample space	*Probability*	x
HHHH	$\frac{1}{16}$	4
HHHT	$\frac{1}{16}$	3
HHTH	$\frac{1}{16}$	3
HTHH	$\frac{1}{16}$	3
THHH	$\frac{1}{16}$	3
HHTT	$\frac{1}{16}$	2
HTHT	$\frac{1}{16}$	2

Element of sample space	*Probability*	x
HTTH	$\frac{1}{16}$	2
THHT	$\frac{1}{16}$	2
THTH	$\frac{1}{16}$	2
TTHH	$\frac{1}{16}$	2
HTTT	$\frac{1}{16}$	1
THTT	$\frac{1}{16}$	1
TTHT	$\frac{1}{16}$	1
TTTH	$\frac{1}{16}$	1
TTTT	$\frac{1}{16}$	0

Thus, we can write $P(X = 3) = \frac{4}{16}$, for example, for the probability of the event that the random variable X will take on the value 3.

The fact that Definition 1 is limited to real-valued functions does not impose any restrictions. If the numbers we want to assign to the outcomes of an experiment are complex numbers, we can always look upon the real and the imaginary parts separately as values taken on by two random variables. Also, if we want to describe the outcomes of an experiment quantitatively, say, by giving the color of a person's hair, we can arbitrarily make the descriptions real-valued by coding the various colors, perhaps by representing them with the numbers 1, 2, 3, and so on.

In all of the examples of this section we have limited our discussion to discrete sample spaces, and hence to **discrete random variables**, namely, random variables whose range is finite or countably infinite. Continuous random variables defined over continuous sample spaces will be taken up in Section 3.

2 Probability Distributions

As we already saw in Examples 1 and 2, the probability measure defined over a discrete sample space automatically provides the probabilities that a random variable will take on any given value within its range.

For instance, having assigned the probability $\frac{1}{36}$ to each element of the sample space of Figure 1, we immediately find that the random variable X, the total rolled with the pair of dice, takes on the value 9 with probability $\frac{4}{36}$; as described in Section 1, $X = 9$ contains four of the equally likely elements of the sample space. The probabilities associated with all possible values of X are shown in the following table:

x	$P(X = x)$
2	$\frac{1}{36}$
3	$\frac{2}{36}$
4	$\frac{3}{36}$
5	$\frac{4}{36}$
6	$\frac{5}{36}$
7	$\frac{6}{36}$
8	$\frac{5}{36}$
9	$\frac{4}{36}$
10	$\frac{3}{36}$
11	$\frac{2}{36}$
12	$\frac{1}{36}$

Instead of displaying the probabilities associated with the values of a random variable in a table, as we did in the preceding illustration, it is usually preferable to give a formula, that is, to express the probabilities by means of a function such that its values, $f(x)$, equal $P(X = x)$ for each x within the range of the random variable X. For instance, for the total rolled with a pair of dice we could write

$$f(x) = \frac{6 - |x - 7|}{36} \quad \text{for } x = 2, 3, \ldots, 12$$

as can easily be verified by substitution. Clearly,

$$f(2) = \frac{6 - |2 - 7|}{36} = \frac{6 - 5}{36} = \frac{1}{36}$$

$$f(3) = \frac{6 - |3 - 7|}{36} = \frac{6 - 4}{36} = \frac{2}{36}$$

$$\cdots\cdots\cdots\cdots\cdots\cdots\cdots$$

$$f(12) = \frac{6 - |12 - 7|}{36} = \frac{6 - 5}{36} = \frac{1}{36}$$

and all these values agree with the ones shown in the preceding table.

DEFINITION 2. PROBABILITY DISTRIBUTION. *If* X *is a discrete random variable, the function given by* f(x) = P(X = x) *for each* x *within the range of* X *is called the* ***probability distribution*** *of* X.

Based on the postulates of probability, we obtain the following theorem.

THEOREM 1. A function can serve as the probability distribution of a discrete random variable X if and only if its values, $f(x)$, satisfy the conditions

1. $f(x) \geqq 0$ for each value within its domain;
2. $\sum_{x} f(x) = 1$, where the summation extends over all the values within its domain.

EXAMPLE 3

Find a formula for the probability distribution of the total number of heads obtained in four tosses of a balanced coin.

Solution

Based on the probabilities in the table, we find that $P(X = 0) = \frac{1}{16}, P(X = 1) = \frac{4}{16}, P(X = 2) = \frac{6}{16}, P(X = 3) = \frac{4}{16}$, and $P(X = 4) = \frac{1}{16}$. Observing that the numerators of these five fractions, 1, 4, 6, 4, and 1, are the binomial coefficients $\binom{4}{0}, \binom{4}{1}, \binom{4}{2}, \binom{4}{3}$, and $\binom{4}{4}$, we find that the formula for the probability distribution can be written as

$$f(x) = \frac{\binom{4}{x}}{16} \quad \text{for } x = 0, 1, 2, 3, 4$$

EXAMPLE 4

Check whether the function given by

$$f(x) = \frac{x+2}{25} \quad \text{for } x = 1, 2, 3, 4, 5$$

can serve as the probability distribution of a discrete random variable.

Solution

Substituting the different values of x, we get $f(1) = \frac{3}{25}, f(2) = \frac{4}{25}, f(3) = \frac{5}{25}$, $f(4) = \frac{6}{25}$, and $f(5) = \frac{7}{25}$. Since these values are all nonnegative, the first condition of Theorem 1 is satisfied, and since

$$\begin{aligned} f(1)+f(2)+f(3)+f(4)+f(5) &= \frac{3}{25}+\frac{4}{25}+\frac{5}{25}+\frac{6}{25}+\frac{7}{25} \\ &= 1 \end{aligned}$$

the second condition of Theorem 1 is satisfied. Thus, the given function can serve as the probability distribution of a random variable having the range $\{1, 2, 3, 4, 5\}$. Of course, whether any given random variable actually has this probability distribution is an entirely different matter.

In some problems it is desirable to present probability distributions graphically, and two kinds of graphical presentations used for this purpose are shown in Figures 2 and 3. The one shown in Figure 2, called a **probability histogram**, represents the probability distribution of Example 3. The height of each rectangle equals

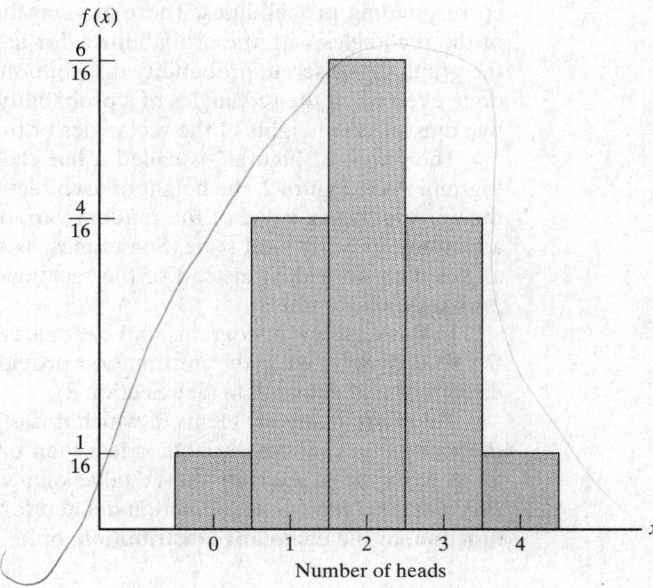

Figure 2. Probability histogram.

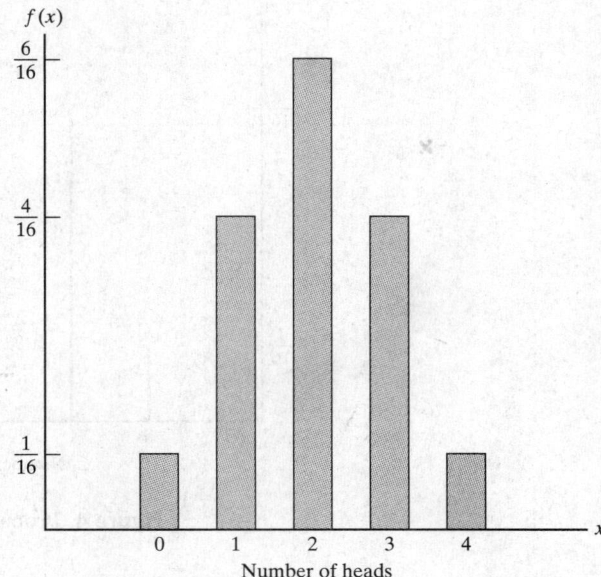

Figure 3. Bar chart.

the probability that X takes on the value that corresponds to the midpoint of its base. By representing 0 with the interval from -0.5 to 0.5, 1 with the interval from 0.5 to 1.5, . . . , and 4 with the interval from 3.5 to 4.5, we are, so to speak, "spreading" the values of the given discrete random variable over a continuous scale.

Since each rectangle of the probability histogram of Figure 2 has unit width, we could have said that the *areas* of the rectangles, rather than their heights, equal the corresponding probabilities. There are certain advantages to identifying the areas of the rectangles with the probabilities, for instance, when we wish to approximate the graph of a discrete probability distribution with a continuous curve. This can be done even when the rectangles of a probability histogram do not all have unit width by adjusting the heights of the rectangles or by modifying the vertical scale.

The graph of Figure 3 is called a **bar chart**, but it is also referred to as a histogram. As in Figure 2, the height of each rectangle, or bar, equals the probability of the corresponding value of the random variable, but there is no pretense of having a continuous horizontal scale. Sometimes, as shown in Figure 4, we use lines (rectangles with no width) instead of the rectangles, but we still refer to the graphs as probability histograms.

In this chapter, histograms and bar charts are used mainly in descriptive statistics to convey visually the information provided by a probability distribution or a distribution of actual data (see Section 8).

There are many problems in which it is of interest to know the probability that the value of a random variable is less than or equal to some real number x. Thus, let us write the probability that X takes on a value less than or equal to x as $F(x) = P(X \leqq x)$ and refer to this function defined for all real numbers x as the **distribution function**, or the **cumulative distribution**, of X.

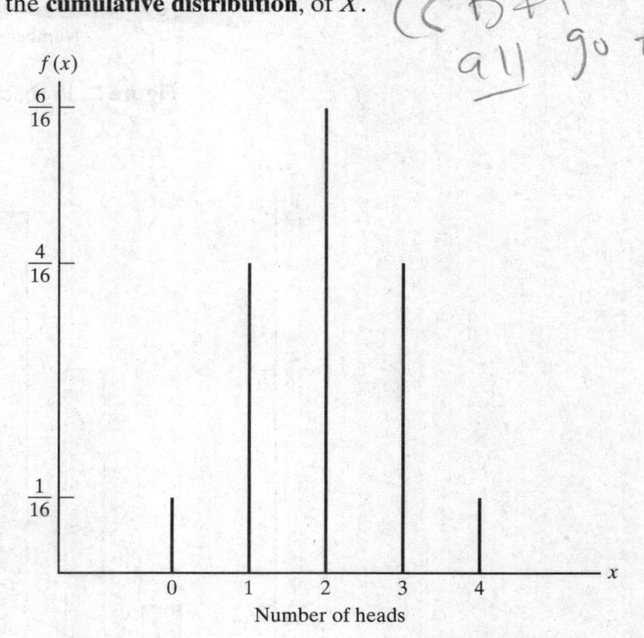

Figure 4. Probability histogram.

DEFINITION 3. DISTRIBUTION FUNCTION. *If X is a discrete random variable, the function given by*

$$F(x) = P(X \leq x) = \sum_{t \leq x} f(t) \qquad \text{for } -\infty < x < \infty$$

where f(t) *is the value of the probability distribution of* X *at* t, *is called the* ***distribution function****, or the* ***cumulative distribution*** *of* X.

Based on the postulates of probability and some of their immediate consequences, we obtain the following theorem.

THEOREM 2. The values $F(x)$ of the distribution function of a discrete random variable X satisfy the conditions

1. $F(-\infty) = 0$ and $F(\infty) = 1$;
2. if $a < b$, then $F(a) \leqq F(b)$ for any real numbers a and b.

If we are given the probability distribution of a discrete random variable, the corresponding distribution function is generally easy to find.

EXAMPLE 5

Find the distribution function of the total number of heads obtained in four tosses of a balanced coin.

Solution

Given $f(0) = \frac{1}{16}, f(1) = \frac{4}{16}, f(2) = \frac{6}{16}, f(3) = \frac{4}{16}$, and $f(4) = \frac{1}{16}$ from Example 3, it follows that

$$F(0) = f(0) = \frac{1}{16}$$

$$F(1) = f(0) + f(1) = \frac{5}{16}$$

$$F(2) = f(0) + f(1) + f(2) = \frac{11}{16}$$

$$F(3) = f(0) + f(1) + f(2) + f(3) = \frac{15}{16}$$

$$F(4) = f(0) + f(1) + f(2) + f(3) + f(4) = 1$$

Hence, the distribution function is given by

$$F(x) = \begin{cases} 0 & \text{for } x < 0 \\ \frac{1}{16} & \text{for } 0 \leqq x < 1 \\ \frac{5}{16} & \text{for } 1 \leqq x < 2 \\ \frac{11}{16} & \text{for } 2 \leqq x < 3 \\ \frac{15}{16} & \text{for } 3 \leqq x < 4 \\ 1 & \text{for } x \geqq 4 \end{cases}$$

Observe that this distribution function is defined not only for the values taken on by the given random variable, but for all real numbers. For instance, we can write $F(1.7) = \frac{5}{16}$ and $F(100) = 1$, although the probabilities of getting "at most 1.7 heads" or "at most 100 heads" in four tosses of a balanced coin may not be of any real significance.

EXAMPLE 6

Find the distribution function of the random variable W of Example 1 and plot its graph.

Solution

Based on the probabilities given in the table in Section 1, we can write $f(0) = \frac{3}{28}, f(1) = \frac{15}{56} + \frac{15}{56} = \frac{15}{28}$, and $f(2) = \frac{5}{14}$, so that

$$F(0) = f(0) = \frac{3}{28}$$

$$F(1) = f(0) + f(1) = \frac{9}{14}$$

$$F(2) = f(0) + f(1) + f(2) = 1$$

Hence, the distribution function of W is given by

$$F(w) = \begin{cases} 0 & \text{for } w < 0 \\ \frac{3}{28} & \text{for } 0 \leqq w < 1 \\ \frac{9}{14} & \text{for } 1 \leqq w < 2 \\ 1 & \text{for } w \geqq 2 \end{cases}$$

The graph of this distribution function, shown in Figure 5, was obtained by first plotting the points $(w, F(w))$ for $w = 0, 1$, and 2 and then completing the step function as indicated. Note that at all points of discontinuity the distribution function takes on the greater of the two values, as indicated by the heavy dots in Figure 5.

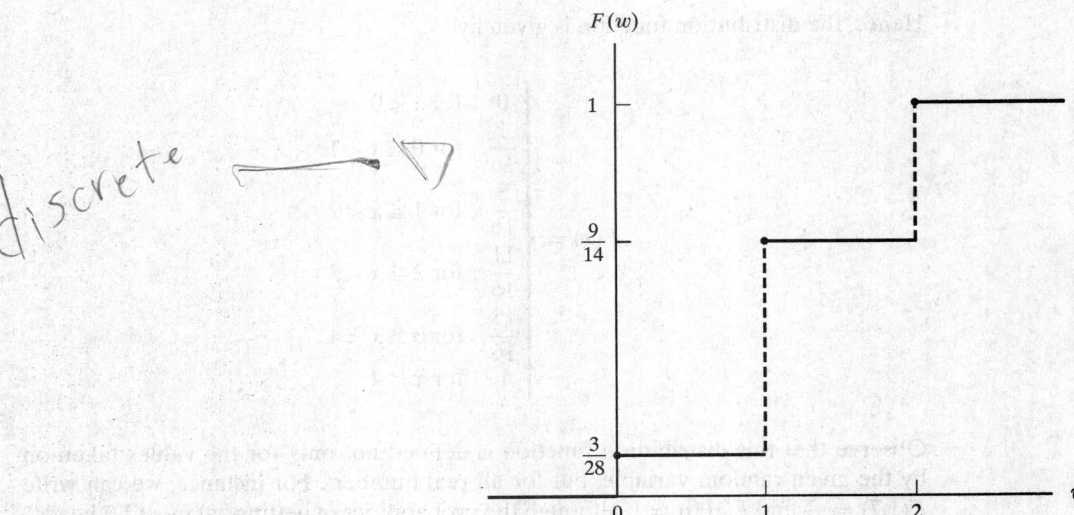

Figure 5. Graph of the distribution function of Example 6.

We can also reverse the process illustrated in the two preceding examples, that is, obtain values of the probability distribution of a random variable from its distribution function. To this end, we use the following result.

THEOREM 3. If the range of a random variable X consists of the values $x_1 < x_2 < x_3 < \cdots < x_n$, then $f(x_1) = F(x_1)$ and

$$f(x_i) = F(x_i) - F(x_{i-1}) \quad \text{for } i = 2, 3, \ldots, n$$

EXAMPLE 7

If the distribution function of X is given by

$$F(x) = \begin{cases} 0 & \text{for } x < 2 \\ \frac{1}{36} & \text{for } 2 \leqq x < 3 \\ \frac{3}{36} & \text{for } 3 \leqq x < 4 \\ \frac{6}{36} & \text{for } 4 \leqq x < 5 \\ \frac{10}{36} & \text{for } 5 \leqq x < 6 \\ \frac{15}{36} & \text{for } 6 \leqq x < 7 \\ \frac{21}{36} & \text{for } 7 \leqq x < 8 \\ \frac{26}{36} & \text{for } 8 \leqq x < 9 \\ \frac{30}{36} & \text{for } 9 \leqq x < 10 \\ \frac{33}{36} & \text{for } 10 \leqq x < 11 \\ \frac{35}{36} & \text{for } 11 \leqq x < 12 \\ 1 & \text{for } x \geqq 12 \end{cases}$$

find the probability distribution of this random variable.

Solution

Making use of Theorem 3, we get $f(2) = \frac{1}{36}, f(3) = \frac{3}{36} - \frac{1}{36} = \frac{2}{36}, f(4) = \frac{6}{36} - \frac{3}{36} = \frac{3}{36}, f(5) = \frac{10}{36} - \frac{6}{36} = \frac{4}{36}, \ldots, f(12) = 1 - \frac{35}{36} = \frac{1}{36}$, and comparison with the probabilities in the table in Section 2 reveals that the random variable with which we are concerned here is the total number of points rolled with a pair of dice.

In the remainder of this chapter we will be concerned with continuous random variables and their distributions and with problems relating to the simultaneous occurrence of the values of two or more random variables.

Exercises

1. For each of the following, determine whether the given values can serve as the values of a probability distribution of a random variable with the range $x = 1, 2, 3$, and 4:

(a) $f(1) = 0.25, f(2) = 0.75, f(3) = 0.25$, and $f(4) = -0.25$;

(b) $f(1) = 0.15, f(2) = 0.27, f(3) = 0.29$, and $f(4) = 0.29$;

(c) $f(1) = \frac{1}{19}, f(2) = \frac{10}{19}, f(3) = \frac{2}{19}$, and $f(4) = \frac{5}{19}$.

2. For each of the following, determine whether the given function can serve as the probability distribution of a random variable with the given range:

(a) $f(x) = \frac{x-2}{5}$ for $x = 1, 2, 3, 4, 5$;

(b) $f(x) = \frac{x^2}{30}$ for $x = 0, 1, 2, 3, 4$;

(c) $f(x) = \frac{1}{5}$ for $x = 0, 1, 2, 3, 4, 5$.

3. Verify that $f(x) = \frac{2x}{k(k+1)}$ for $x = 1, 2, 3, \ldots, k$ can serve as the probability distribution of a random variable with the given range.

4. For each of the following, determine c so that the function can serve as the probability distribution of a random variable with the given range:

(a) $f(x) = cx$ for $x = 1, 2, 3, 4, 5$;

(b) $f(x) = c\binom{5}{x}$ for $x = 0, 1, 2, 3, 4, 5$;

(c) $f(x) = cx^2$ for $x = 1, 2, 3, \ldots, k$;

(d) $f(x) = c\left(\frac{1}{4}\right)^x$ for $x = 1, 2, 3, \ldots$.

5. For what values of k can

$$f(x) = (1-k)k^x$$

serve as the values of the probability distribution of a random variable with the countably infinite range $x = 0, 1, 2, \ldots$?

6. Show that there are no values of c such that

$$f(x) = \frac{c}{x}$$

can serve as the values of the probability distribution of a random variable with the countably infinite range $x = 1, 2, 3, \ldots$.

7. Construct a probability histogram for each of the following probability distributions:

(a) $f(x) = \frac{\binom{2}{x}\binom{4}{3-x}}{\binom{6}{3}}$ for $x = 0, 1, 2$;

(b) $f(x) = \binom{5}{x}\left(\frac{1}{5}\right)^x\left(\frac{4}{5}\right)^{5-x}$ for $x = 0, 1, 2, 3, 4, 5$.

8. Prove Theorem 2.

9. For each of the following, determine whether the given values can serve as the values of a distribution function of a random variable with the range $x = 1, 2, 3$, and 4:

(a) $F(1) = 0.3, F(2) = 0.5, F(3) = 0.8$, and $F(4) = 1.2$;

(b) $F(1) = 0.5, F(2) = 0.4, F(3) = 0.7$, and $F(4) = 1.0$;

(c) $F(1) = 0.25, F(2) = 0.61, F(3) = 0.83$, and $F(4) = 1.0$.

10. Find the distribution function of the random variable of part (a) of Exercise 7 and plot its graph.

11. If X has the distribution function

$$F(x) = \begin{cases} 0 & \text{for } x < 1 \\ \frac{1}{3} & \text{for } 1 \leqq x < 4 \\ \frac{1}{2} & \text{for } 4 \leqq x < 6 \\ \frac{5}{6} & \text{for } 6 \leqq x < 10 \\ 1 & \text{for } x \geqq 10 \end{cases}$$

find

(a) $P(2 < X \leqq 6)$;

(b) $P(X = 4)$;

(c) the probability distribution of X.

12. Find the distribution function of the random variable that has the probability distribution

$$f(x) = \frac{x}{15} \quad \text{for } x = 1, 2, 3, 4, 5$$

13. If X has the distribution function

$$F(x) = \begin{cases} 0 & \text{for } x < -1 \\ \frac{1}{4} & \text{for } -1 \leqq x < 1 \\ \frac{1}{2} & \text{for } 1 \leqq x < 3 \\ \frac{3}{4} & \text{for } 3 \leqq x < 5 \\ 1 & \text{for } x \geqq 5 \end{cases}$$

find

(a) $P(X \leqq 3)$; **(b)** $P(X = 3)$; **(c)** $P(X < 3)$;

(d) $P(X \geqq 1)$; **(e)** $P(-0.4 < X < 4)$; **(f)** $P(X = 5)$.

14. With reference to Example 4, verify that the values of the distribution function are given by

$$F(x) = \frac{x^2 + 5x}{50}$$

for $x = 1, 2, 3, 4$, and 5.

15. With reference to Theorem 3, verify that

(a) $P(X > x_i) = 1 - F(x_i)$ for $i = 1, 2, 3, \ldots, n$;

(b) $P(X \geqq x_i) = 1 - F(x_{i-1})$ for $i = 2, 3, \ldots, n$, and $P(X \geqq x_1) = 1$.

3 Continuous Random Variables

In Section 1 we introduced the concept of a random variable as a real-valued function defined over the points of a sample space with a probability measure, and in Figure 1 we illustrated this by assigning the total rolled with a pair of dice to each of the 36 equally likely points of the sample space. In the continuous case, where random variables can take on values on a continuous scale, the procedure is very much the same. The outcomes of experiments are represented by the points on line segments or lines, and the values of random variables are numbers appropriately assigned to the points by means of rules or equations. When the value of a random variable is given directly by a measurement or observation, we generally do not bother to distinguish between the value of the random variable (the measurement that we obtain) and the outcome of the experiment (the corresponding point on the real axis). Thus, if an experiment consists of determining the actual content of a 230-gram jar of instant coffee, the result itself, say, 225.3 grams, is the value of the random variable with which we are concerned, and there is no real need to add that the sample space consists of a certain continuous interval of points on the positive real axis.

The problem of defining probabilities in connection with continuous sample spaces and continuous random variables involves some complications. To illustrate, let us consider the following situation.

EXAMPLE 8

Suppose that we are concerned with the possibility that an accident will occur on a freeway that is 200 kilometers long and that we are interested in the probability that it will occur at a given location, or perhaps on a given stretch of the road. The sample space of this "experiment" consists of a continuum of points, those on the interval from 0 to 200, and we shall assume, for the sake of argument, that the probability that an accident will occur on any interval of length d is $\frac{d}{200}$, with d measured in kilometers. Note that this assignment of probabilities is consistent with Postulates 1 and 2. (Postulate 1 states that probability of an event is a nonnegative real number; that is, $P(A)$ G 0 for any subset A of S but in Postulate 2 $P(S) = 1$.) The probabilities $\frac{d}{200}$ are all nonnegative and $P(S) = \frac{200}{200} = 1$. So far this assignment of probabilities applies only to intervals on the line segment from 0 to 200, but if we use Postulate 3 (Postulate 3: If $A_1, A_2, A_3, \ldots$, is a finite or infinite sequence of mutually exclusive events of S, then $P(A_1 \cup A_2 \cup A_3 \cup \cdots) = P(A_1) + P(A_2) + P(A_3) + \cdots$), we can also obtain probabilities for the union of any finite or countably infinite sequence of nonoverlapping intervals. For instance, the probability that an accident will occur on either of two nonoverlapping intervals of length d_1 and d_2 is

$$\frac{d_1 + d_2}{200}$$

and the probability that it will occur on any one of a countably infinite sequence of nonoverlapping intervals of length $d_1, d_2, d_3, \ldots$ is

$$\frac{d_1 + d_2 + d_3 + \cdots}{200}$$

With reference to Example 8, observe also that the probability of the accident occurring on a very short interval, say, an interval of 1 centimeter, is only 0.00000005, which is very small. As the length of the interval approaches zero, the probability that an accident will occur on it also approaches zero; indeed, in the continuous case we always assign zero probability to individual points. This does not mean that the corresponding events cannot occur; after all, when an accident occurs on the 200-kilometer stretch of road, it has to occur at some point even though each point has zero probability.

4 Probability Density Functions

The way in which we assigned probabilities in Example 8 is very special, and it is similar in nature to the way in which we assign equal probabilities to the six faces of a die, heads and tails, the 52 playing cards in a standard deck, and so forth. To treat the problem of associating probabilities with values of continuous random variables more generally, suppose that a bottler of soft drinks is concerned about the actual amount of a soft drink that his bottling machine puts into 16-ounce bottles. Evidently, the amount will vary somewhat from bottle to bottle; it is, in fact, a continuous random variable. However, if he rounds the amounts to the nearest tenth of an ounce, he will be dealing with a discrete random variable that has a probability distribution, and this probability distribution may be pictured as a histogram in which the probabilities are given by the areas of rectangles, say, as in the diagram at the top of Figure 6. If he rounds the amounts to the nearest hundredth of an ounce, he will again be dealing with a discrete random variable (a different one) that has a probability distribution, and this probability distribution may be pictured as a probability histogram in which the probabilities

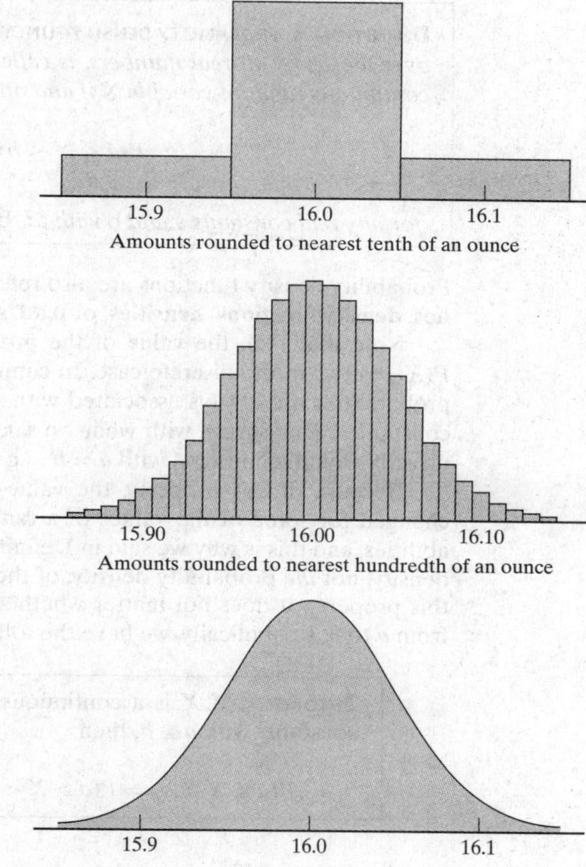

Figure 6. Definition of probability in the continuous case.

are given by the areas of rectangles, say, as in the diagram in the middle of Figure 6.

It should be apparent that if he rounds the amounts to the nearest thousandth of an ounce or to the nearest ten-thousandth of an ounce, the probability histograms of the probability distributions of the corresponding discrete random variables will approach the continuous curve shown in the diagram at the bottom of Figure 6, and the sum of the areas of the rectangles that represent the probability that the amount falls within any specified interval approaches the corresponding area under the curve.

Indeed, the definition of probability in the continuous case presumes for each random variable the existence of a function, called a **probability density function**, such that areas under the curve give the probabilities associated with the corresponding intervals along the horizontal axis. In other words, a probability density function, integrated from a to b (with $a \leqq b$), gives the probability that the corresponding random variable will take on a value on the interval from a to b.

DEFINITION 4. PROBABILITY DENSITY FUNCTION. *A function with values* f(x), *defined over the set of all real numbers, is called a* **probability density function** *of the continuous random variable* X *if and only if*

$$P(a \le X \le b) = \int_a^b f(x)dx$$

for any real constants a *and* b *with* a≤b.

Probability density functions are also referred to, more briefly, as **probability densities, density functions, densities**, or **p.d.f.'s**.

Note that $f(c)$, the value of the probability density of X at c, does not give $P(X = c)$ as in the discrete case. In connection with continuous random variables, probabilities are always associated with intervals and $P(X = c) = 0$ for any real constant c. This agrees with what we said on the previous page and it also follows directly from Definition 4 with $a = b = c$.

Because of this property, the value of a probability density function can be changed for some of the values of a random variable without changing the probabilities, and this is why we said in Definition 4 that $f(x)$ is the value of *a* probability density, not *the* probability density, of the random variable X at x. Also, in view of this property, it does not matter whether we include the endpoints of the interval from a to b; symbolically, we have the following theorem.

THEOREM 4. If X is a continuous random variable and a and b are real constants with $a \leqq b$, then

$$P(a \leqq X \leqq b) = P(a \leqq X < b) = P(a < X \leqq b) = P(a < X < b)$$

Analogous to Theorem 1, let us now state the following properties of probability densities, which again follow directly from the postulates of probability.

THEOREM 5. A function can serve as a probability density of a continuous random variable X if its values, $f(x)$, satisfy the conditions†

1. $f(x) \geqq 0$ for $-\infty < x < \infty$;
2. $\int_{-\infty}^{\infty} f(x)\, dx = 1.$

EXAMPLE 9

If X has the probability density

$$f(x) = \begin{cases} k \cdot e^{-3x} & \text{for } x > 0 \\ 0 & \text{elsewhere} \end{cases}$$

find k and $P(0.5 \leqq X \leqq 1)$.

† The conditions are not "if and only if" as in Theorem 1 because $f(x)$ could be negative for some values of the random variable without affecting any of the probabilities. However, both conditions of Theorem 5 will be satisfied by nearly all the probability densities used in practice and studied in this text.

Solution

To satisfy the second condition of Theorem 5, we must have

$$\int_{-\infty}^{\infty} f(x)\,dx = \int_{0}^{\infty} k \cdot e^{-3x}\,dx = k \cdot \lim_{t\to\infty} \frac{e^{-3x}}{-3}\Big|_0^t = \frac{k}{3} = 1$$

and it follows that $k = 3$. For the probability we get

$$P(0.5 \leqq X \leqq 1) = \int_{0.5}^{1} 3e^{-3x}\,dx = -e^{-3x}\Big|_{0.5}^{1} = -e^{-3} + e^{-1.5} = 0.173$$

Although the random variable of the preceding example cannot take on negative values, we artificially extended the domain of its probability density to include all the real numbers. This is a practice we shall follow throughout this text.

As in the discrete case, there are many problems in which it is of interest to know the probability that the value of a continuous random variable X is less than or equal to some real number x. Thus, let us make the following definition analogous to Definition 3.

DEFINITION 5. DISTRIBUTION FUNCTION. *If* X *is a continuous random variable and the value of its probability density at* t *is* f(t), *then the function given by*

$$F(x) = P(X \leq x) = \int_{-\infty}^{x} f(t)dt \qquad \text{for } -\infty < x < \infty$$

is called the ***distribution function*** *or the* ***cumulative distribution function*** *of* X.

The properties of distribution functions given in Theorem 2 hold also for the continuous case; that is, $F(-\infty) = 0$, $F(\infty) = 1$, and $F(a) \leqq F(b)$ when $a < b$. Furthermore, based on Definition 5, we can state the following theorem.

THEOREM 6. If $f(x)$ and $F(x)$ are the values of the probability density and the distribution function of X at x, then

$$P(a \leqq X \leqq b) = F(b) - F(a)$$

for any real constants a and b with $a \leqq b$, and

$$f(x) = \frac{dF(x)}{dx}$$

where the derivative exists.

EXAMPLE 10

Find the distribution function of the random variable X of Example 9, and use it to reevaluate $P(0.5 \leqq X \leqq 1)$.

Solution

For $x > 0$,

$$F(x) = \int_{-\infty}^{x} f(t)dt = \int_{0}^{x} 3e^{-3t}dt = -e^{-3t}\Big|_{0}^{x} = 1 - e^{-3x}$$

and since $F(x) = 0$ for $x \leqq 0$, we can write

$$F(x) = \begin{cases} 0 & \text{for } x \leqq 0 \\ 1 - e^{-3x} & \text{for } x > 0 \end{cases}$$

To determine the probability $P(0.5 \leqq X \leqq 1)$, we use the first part of Theorem 6, getting

$$\begin{aligned} P(0.5 \leqq X \leqq 1) &= F(1) - F(0.5) \\ &= (1 - e^{-3}) - (1 - e^{-1.5}) \\ &= 0.173 \end{aligned}$$

This agrees with the result obtained by using the probability density directly in Example 9.

EXAMPLE 11

Find a probability density function for the random variable whose distribution function is given by

$$F(x) = \begin{cases} 0 & \text{for } x \leqq 0 \\ x & \text{for } 0 < x < 1 \\ 1 & \text{for } x \geqq 1 \end{cases}$$

and plot its graph.

Solution

Since the given density function is differentiable everywhere except at $x = 0$ and $x = 1$, we differentiate for $x < 0, 0 < x < 1$, and $x > 1$, getting 0, 1, and 0. Thus, according to the second part of Theorem 6, we can write

$$f(x) = \begin{cases} 0 & \text{for } x < 0 \\ 1 & \text{for } 0 < x < 1 \\ 0 & \text{for } x > 1 \end{cases}$$

To fill the gaps at $x = 0$ and $x = 1$, we let $f(0)$ and $f(1)$ both equal zero. Actually, it does not matter how the probability density is defined at these two points, but there are certain advantages for choosing the values in such a way that the probability density is nonzero over an open interval. Thus, we can write the probability density of the original random variable as

$$f(x) = \begin{cases} 1 & \text{for } 0 < x < 1 \\ 0 & \text{elsewhere} \end{cases}$$

Its graph is shown in Figure 7.

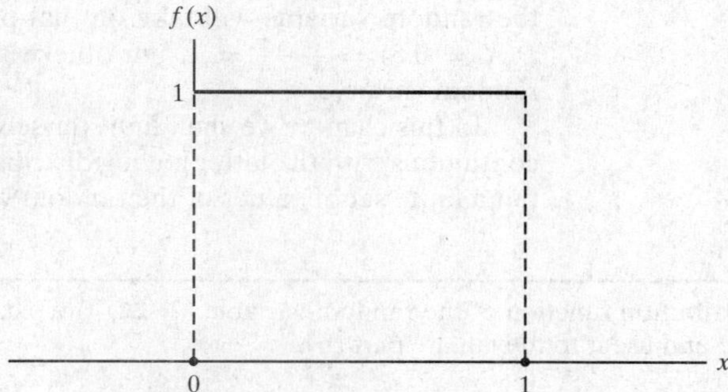

Figure 7. Probability density of Example 11.

In most applications we encounter random variables that are either discrete or continuous, so the corresponding distribution functions have a steplike appearance as in Figure 5, or they are continuous curves or combinations of lines as in Figure 8, which shows the graph of the distribution function of Example 11.

Discontinuous distribution functions like the one shown in Figure 9 arise when random variables are **mixed**. Such a distribution function will be discontinuous at each point having a nonzero probability and continuous elsewhere. As in the discrete case, the height of the step at a point of discontinuity gives the probability that

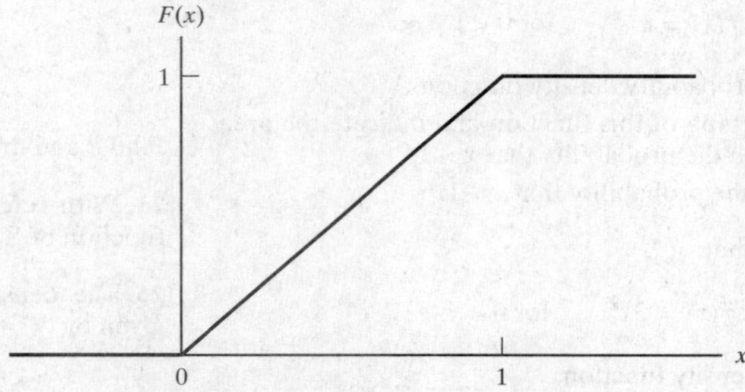

Figure 8. Distribution function of Example 11.

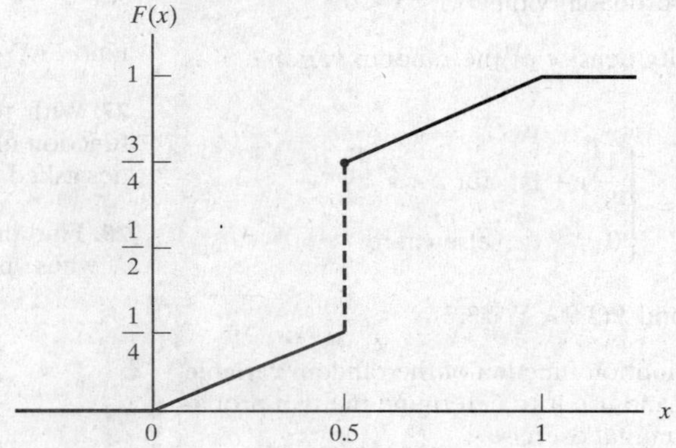

Figure 9. Distribution function of a mixed random variable.

the random variable will take on that particular value. With reference to Figure 9, $P(X = 0.5) = \frac{3}{4} - \frac{1}{4} = \frac{1}{2}$, but otherwise the random variable is like a continuous random variable.

In this chapter we shall limit ourselves to random variables that are discrete or continuous with the latter having distribution functions that are differentiable for all but a finite set of values of the random variables.

Exercises

16. Find the distribution function of the random variable X of Exercise 17 and use it to reevaluate part (b).

17. The probability density of the continuous random variable X is given by

$$f(x) = \begin{cases} \frac{1}{5} & \text{for } 2 < x < 7 \\ 0 & \text{elsewhere} \end{cases}$$

(a) Draw its graph and verify that the total area under the curve (above the x-axis) is equal to 1.
(b) Find $P(3 < X < 5)$.

18. (a) Show that

$$f(x) = e^{-x} \qquad \text{for } 0 < x < \infty$$

represents a probability density function.
(b) Sketch a graph of this function and indicate the area associated with the probability that $x > 1$.
(c) Calculate the probability that $x > 1$.

19. (a) Show that

$$f(x) = 3x^2 \qquad \text{for } 0 < x < 1$$

represents a density function.
(b) Sketch a graph of this function, and indicate the area associated with the probability that $0.1 < x < 0.5$.
(c) Calculate the probability that $0.1 < x < 0.5$.

20. The probability density of the random variable Y is given by

$$f(y) = \begin{cases} \frac{1}{8}(y+1) & \text{for } 2 < y < 4 \\ 0 & \text{elsewhere} \end{cases}$$

Find $P(Y < 3.2)$ and $P(2.9 < Y < 3.2)$.

21. Find the distribution function of the random variable Y of Exercise 20 and use it to determine the two probabilities asked for in that exercise.

22. The p.d.f. of the random variable X is given by

$$f(x) = \begin{cases} \frac{c}{\sqrt{x}} & \text{for } 0 < x < 4 \\ 0 & \text{elsewhere} \end{cases}$$

Find
(a) the value of c;
(b) $P(X < \frac{1}{4})$ and $P(X > 1)$.

23. Find the distribution function of the random variable X of Exercise 22 and use it to determine the two probabilities asked for in part (b) of that exercise.

24. The probability density of the random variable Z is given by

$$f(z) = \begin{cases} kze^{-z^2} & \text{for } z > 0 \\ 0 & \text{for } z \leqq 0 \end{cases}$$

Find k and draw the graph of this probability density.

25. With reference to Exercise 24, find the distribution function of Z and draw its graph.

26. The density function of the random variable X is given by

$$g(x) = \begin{cases} 6x(1-x) & \text{for } 0 < x < 1 \\ 0 & \text{elsewhere} \end{cases}$$

Find $P(X < \frac{1}{4})$ and $P(X > \frac{1}{2})$.

27. With reference to Exercise 26, find the distribution function of X and use it to reevaluate the two probabilities asked for in that exercise.

28. Find the distribution function of the random variable X whose probability density is given by

$$f(x) = \begin{cases} x & \text{for } 0 < x < 1 \\ 2 - x & \text{for } 1 \leqq x < 2 \\ 0 & \text{elsewhere} \end{cases}$$

Also sketch the graphs of the probability density and distribution functions.

29. Find the distribution function of the random variable X whose probability density is given by

$$f(x) = \begin{cases} \frac{1}{3} & \text{for } 0 < x < 1 \\ \frac{1}{3} & \text{for } 2 < x < 4 \\ 0 & \text{elsewhere} \end{cases}$$

Also sketch the graphs of the probability density and distribution functions.

30. With reference to Exercise 28, find $P(0.8 < X < 1.2)$ using

(a) the probability density;

(b) the distribution function.

31. Find the distribution function of the random variable X whose probability density is given by

$$f(x) = \begin{cases} \frac{x}{2} & \text{for } 0 < x \leqq 1 \\ \frac{1}{2} & \text{for } 1 < x \leqq 2 \\ \frac{3-x}{2} & \text{for } 2 < x < 3 \\ 0 & \text{elsewhere} \end{cases}$$

Also sketch the graphs of these probability density and distribution functions.

32. The distribution function of the random variable X is given by

$$F(x) = \begin{cases} 0 & \text{for } x < -1 \\ \frac{x+1}{2} & \text{for } -1 \leqq x < 1 \\ 1 & \text{for } x \geqq 1 \end{cases}$$

Find $P(-\frac{1}{2} < X < \frac{1}{2})$ and $P(2 < X < 3)$.

33. With reference to Exercise 32, find the probability density of X and use it to recalculate the two probabilities.

34. The distribution function of the random variable Y is given by

$$F(y) = \begin{cases} 1 - \frac{9}{y^2} & \text{for } y > 3 \\ 0 & \text{elsewhere} \end{cases}$$

Find $P(Y \leqq 5)$ and $P(Y > 8)$.

35. With reference to Exercise 34, find the probability density of Y and use it to recalculate the two probabilities.

36. With reference to Exercise 34 and the result of Exercise 35, sketch the graphs of the distribution function and the probability density of Y, letting $f(3) = 0$.

37. The distribution function of the random variable X is given by

$$F(x) = \begin{cases} 1 - (1+x)e^{-x} & \text{for } x > 0 \\ 0 & \text{for } x \leqq 0 \end{cases}$$

Find $P(X \leqq 2)$, $P(1 < X < 3)$, and $P(X > 4)$.

38. With reference to Exercise 37, find the probability density of X.

39. With reference to Figure 9, find expressions for the values of the distribution function of the mixed random variable X for

(a) $x \leqq 0$; **(b)** $0 < x < 0.5$;

(c) $0.5 \leqq x < 1$; **(d)** $x \geqq 1$.

40. Use the results of Exercise 39 to find expressions for the values of the probability density of the mixed random variable X for

(a) $x < 0$; **(b)** $0 < x < 0.5$;

(c) $0.5 < x < 1$; **(d)** $x > 1$.

$P(X = 0.5) = \frac{1}{2}$, and $f(0)$ and $f(1)$ are undefined.

41. The distribution function of the mixed random variable Z is given by

$$F(z) = \begin{cases} 0 & \text{for } z < -2 \\ \frac{z+4}{8} & \text{for } -2 \leqq z < 2 \\ 1 & \text{for } z \geqq 2 \end{cases}$$

Find $P(Z = -2)$, $P(Z = 2)$, $P(-2 < Z < 1)$, and $P(0 \leqq Z \leqq 2)$.

5 Multivariate Distributions

In the beginning of this chapter we defined a random variable as a real-valued function defined over a sample space with a probability measure, and it stands to reason that many different random variables can be defined over one and the same sample space. With reference to the sample space of Figure 1, for example, we considered only the random variable whose values were the totals rolled with a pair of dice, but we could also have considered the random variable whose values are the products of the numbers rolled with the two dice, the random variable whose values are the differences between the numbers rolled with the red die and the green die, the random variable whose values are 0, 1, or 2 depending on the number of dice that come up 2, and so forth. Closer to life, an experiment may consist of randomly choosing some of the 345 students attending an elementary school, and the principal may be interested in their I.Q.'s, the school nurse in their weights, their teachers in the number of days they have been absent, and so forth.

In this section we shall be concerned first with the **bivariate case**, that is, with situations where we are interested at the same time in a pair of random variables defined over a joint sample space. Later, we shall extend this discussion to the **multivariate case**, covering any finite number of random variables.

If X and Y are discrete random variables, we write the probability that X will take on the value x and Y will take on the value y as $P(X = x, Y = y)$. Thus, $P(X = x, Y = y)$ is the probability of the intersection of the events $X = x$ and $Y = y$. As in the **univariate case**, where we dealt with one random variable and could display the probabilities associated with all values of X by means of a table, we can now, in the bivariate case, display the probabilities associated with all pairs of values of X and Y by means of a table.

EXAMPLE 12

Two caplets are selected at random from a bottle containing 3 aspirin, 2 sedative, and 4 laxative caplets. If X and Y are, respectively, the numbers of aspirin and sedative caplets included among the 2 caplets drawn from the bottle, find the probabilities associated with all possible pairs of values of X and Y.

Solution

The possible pairs are (0, 0), (0, 1), (1, 0), (1, 1), (0, 2), and (2, 0). To find the probability associated with (1, 0), for example, observe that we are concerned with the event of getting one of the 3 aspirin caplets, none of the 2 sedative caplets, and, hence, one of the 4 laxative caplets. The number of ways in which this can be done is $\binom{3}{1}\binom{2}{0}\binom{4}{1} = 12$, and the total number of ways in which 2 of the 9 caplets can be selected is $\binom{9}{2} = 36$. Since those possibilities are all equally likely by virtue of the assumption that the selection is random, it follows from a theorem (If an experiment can result in any one of N different equally likely outcomes, and if n of these outcomes together constitute event A, then the probability of event A is $P(A) = n/N$) that the probability associated with $(1, 0)$ is $\frac{12}{36} = \frac{1}{3}$. Similarly, the probability associated with $(1, 1)$ is

$$\frac{\binom{3}{1}\binom{2}{1}\binom{4}{0}}{36} = \frac{6}{36} = \frac{1}{6}$$

and, continuing this way, we obtain the values shown in the following table:

		x		
		0	1	2
y	0	$\frac{1}{6}$	$\frac{1}{3}$	$\frac{1}{12}$
	1	$\frac{2}{9}$	$\frac{1}{6}$	
	2	$\frac{1}{36}$		

Actually, as in the univariate case, it is generally preferable to represent probabilities such as these by means of a formula. In other words, it is preferable to express the probabilities by means of a function with the values $f(x, y) = P(X = x, Y = y)$ for any pair of values (x, y) within the range of the random variables X and Y. For instance, for the two random variables of Example 12 we can write

$$f(x, y) = \frac{\binom{3}{x}\binom{2}{y}\binom{4}{2-x-y}}{\binom{9}{2}} \qquad \text{for } x = 0, 1, 2; \quad y = 0, 1, 2; \quad 0 \leqq x + y \leqq 2$$

DEFINITION 6. JOINT PROBABILITY DISTRIBUTION. *If* X *and* Y *are discrete random variables, the function given by* f(x, y) = P(X = x, Y = y) *for each pair of values* (x, y) *within the range of* X *and* Y *is called the* ***joint probability distribution*** *of* X *and* Y.

Analogous to Theorem 1, let us state the following theorem, which follows from the postulates of probability.

THEOREM 7. A bivariate function can serve as the joint probability distribution of a pair of discrete random variables X and Y if and only if its values, $f(x, y)$, satisfy the conditions

1. $f(x, y) \geqq 0$ for each pair of values (x, y) within its domain;
2. $\sum_x \sum_y f(x, y) = 1$, where the double summation extends over all possible pairs (x, y) within its domain.

EXAMPLE 13

Determine the value of k for which the function given by

$$f(x, y) = kxy \qquad \text{for } x = 1, 2, 3; \quad y = 1, 2, 3$$

can serve as a joint probability distribution.

Solution

Substituting the various values of x and y, we get $f(1,1) = k$, $f(1,2) = 2k$, $f(1,3) = 3k$, $f(2,1) = 2k$, $f(2,2) = 4k$, $f(2,3) = 6k$, $f(3,1) = 3k$, $f(3,2) = 6k$, and $f(3,3) = 9k$. To satisfy the first condition of Theorem 7, the constant k must be nonnegative, and to satisfy the second condition,

$$k + 2k + 3k + 2k + 4k + 6k + 3k + 6k + 9k = 1$$

so that $36k = 1$ and $k = \frac{1}{36}$.

As in the univariate case, there are many problems in which it is of interest to know the probability that the values of two random variables are less than or equal to some real numbers x and y.

> **DEFINITION 7. JOINT DISTRIBUTION FUNCTION.** *If* X *and* Y *are discrete random variables, the function given by*
>
> $$F(x,y) = P(X \le x, Y \le y) = \sum_{s \le x} \sum_{t \le y} f(s,t) \qquad \text{for } -\infty < x < \infty \quad -\infty < y < \infty$$
>
> *where* f(s, t) *is the value of the joint probability distribution of* X *and* Y *at* (s, t), *is called the **joint distribution function**, or the **joint cumulative distribution** of* X *and* Y.

In Exercise 48 the reader will be asked to prove properties of joint distribution functions that are analogous to those of Theorem 2.

EXAMPLE 14

With reference to Example 12, find $F(1, 1)$.

Solution

$$\begin{aligned} F(1,1) &= P(X \leqq 1, Y \leqq 1) \\ &= f(0,0) + f(0,1) + f(1,0) + f(1,1) \\ &= \frac{1}{6} + \frac{2}{9} + \frac{1}{3} + \frac{1}{6} \\ &= \frac{8}{9} \end{aligned}$$

As in the univariate case, the joint distribution function of two random variables is defined for all real numbers. For instance, for Example 12 we also get $F(-2,1) = P(X \leqq -2, Y \leqq 1) = 0$ and $F(3.7, 4.5) = P(X \leqq 3.7, Y \leqq 4.5) = 1$.

Let us now extend the various concepts introduced in this section to the continuous case.

DEFINITION 8. JOINT PROBABILITY DENSITY FUNCTION. *A bivariate function with values* f(x, y) *defined over the* xy*-plane is called a* ***joint probability density function*** *of the continuous random variables* X *and* Y *if and only if*

$$P(X,Y) \in A = \iint_A f(x,y)\,dx\,dy$$

for any region A *in the* xy*-plane.*

Analogous to Theorem 5, it follows from the postulates of probability that

THEOREM 8. A bivariate function can serve as a joint probability density function of a pair of continuous random variables X and Y if its values, $f(x,y)$, satisfy the conditions

1. $f(x,y) \geqq 0 \quad \text{for } -\infty < x < \infty, \quad -\infty < y < \infty;$
2. $\displaystyle\int_{-\infty}^{\infty}\int_{-\infty}^{\infty} f(x,y)\,dx\,dy = 1.$

EXAMPLE 15

Given the joint probability density function

$$f(x,y) = \begin{cases} \dfrac{3}{5}x(y+x) & \text{for } 0<x<1, 0<y<2 \\ 0 & \text{elsewhere} \end{cases}$$

of two random variables X and Y, find $P[(X,Y) \in A]$, where A is the region $\{(x,y)|0 < x < \frac{1}{2}, 1 < y < 2\}$.

Solution

$$P[(X,Y) \in A] = P\left(0 < X < \frac{1}{2}, 1 < Y < 2\right)$$

$$= \int_1^2 \int_0^{\frac{1}{2}} \frac{3}{5}x(y+x)\,dx\,dy$$

$$= \int_1^2 \frac{3x^2y}{10} + \frac{3x^3}{15}\Bigg|_{x=0}^{x=\frac{1}{2}} dy$$

$$= \int_1^2 \left(\frac{3y}{40} + \frac{1}{40}\right) dy = \frac{3y^2}{80} + \frac{y}{40}\Bigg|_1^2$$

$$= \frac{11}{80}$$

Analogous to Definition 7, we have the following definition of the joint distribution function of two continuous random variables.

DEFINITION 9. JOINT DISTRIBUTION FUNCTION. *If X and Y are continuous random variables, the function given by*

$$F(x,y) = P(X \leq x, Y \leq y) = \int_{-\infty}^{y}\int_{-\infty}^{x} f(s,t)\,ds\,dt \qquad \text{for } -\infty < x < \infty,\ -\infty < y < \infty$$

where f(s, t) *is the joint probability density of* X *and* Y *at* (s, t), *is called the* ***joint distribution function of*** X ***and*** Y.

Note that the properties of joint distribution functions, which the reader will be asked to prove in Exercise 48 for the discrete case, hold also for the continuous case.

As in Section 4, we shall limit our discussion here to random variables whose joint distribution function is continuous everywhere and partially differentiable with respect to each variable for all but a finite set of values of the two random variables.

Analogous to the relationship $f(x) = \dfrac{dF(x)}{dx}$ of Theorem 6, partial differentiation in Definition 9 leads to

$$f(x,y) = \frac{\partial^2}{\partial x \partial y} F(x,y)$$

wherever these partial derivatives exist. As in Section 4, the joint distribution function of two continuous random variables determines their **joint density** (short for joint probability density function) at all points (x, y) where the joint density is continuous. Also as in Section 4, we generally let the values of joint probability densities equal zero wherever they are not defined by the above relationship.

EXAMPLE 16

If the joint probability density of X and Y is given by

$$f(x,y) = \begin{cases} x+y & \text{for } 0 < x < 1, 0 < y < 1 \\ 0 & \text{elsewhere} \end{cases}$$

find the joint distribution function of these two random variables.

Solution

If either $x < 0$ or $y < 0$, it follows immediately that $F(x,y) = 0$. For $0 < x < 1$ and $0 < y < 1$ (Region I of Figure 10), we get

$$F(x,y) = \int_0^y \int_0^x (s+t)\,ds\,dt = \frac{1}{2}xy(x+y)$$

for $x > 1$ and $0 < y < 1$ (Region II of Figure 10), we get

$$F(x,y) = \int_0^y \int_0^1 (s+t)\,ds\,dt = \frac{1}{2}y(y+1)$$

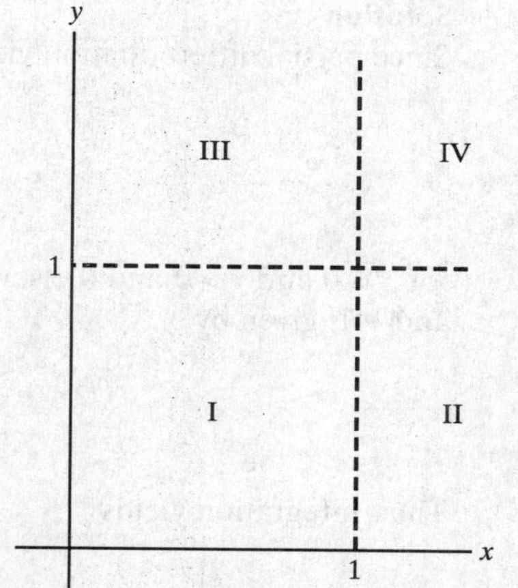

Figure 10. Diagram for Example 16.

for $0 < x < 1$ and $y > 1$ (Region III of Figure 10), we get

$$F(x, y) = \int_0^1 \int_0^x (s+t)\, ds\, dt = \frac{1}{2}x(x+1)$$

and for $x > 1$ and $y > 1$ (Region IV of Figure 10), we get

$$F(x, y) = \int_0^1 \int_0^1 (s+t)\, ds\, dt = 1$$

Since the joint distribution function is everywhere continuous, the boundaries between any two of these regions can be included in either one, and we can write

$$F(x, y) = \begin{cases} 0 & \text{for } x \leqq 0 \text{ or } y \leqq 0 \\ \frac{1}{2}xy(x+y) & \text{for } 0 < x < 1, 0 < y < 1 \\ \frac{1}{2}y(y+1) & \text{for } x \geqq 1, 0 < y < 1 \\ \frac{1}{2}x(x+1) & \text{for } 0 < x < 1, y \geqq 1 \\ 1 & \text{for } x \geqq 1, y \geqq 1 \end{cases}$$

EXAMPLE 17

Find the joint probability density of the two random variables X and Y whose joint distribution function is given by

$$F(x, y) = \begin{cases} (1-e^{-x})(1-e^{-y}) & \text{for } x > 0 \text{ and } y > 0 \\ 0 & \text{elsewhere} \end{cases}$$

Also use the joint probability density to determine $P(1 < X < 3, 1 < Y < 2)$.

Solution

Since partial differentiation yields

$$\frac{\partial^2}{\partial x \partial y} F(x, y) = e^{-(x+y)}$$

for $x > 0$ and $y > 0$ and 0 elsewhere, we find that the joint probability density of X and Y is given by

$$f(x, y) = \begin{cases} e^{-(x+y)} & \text{for } x > 0 \text{ and } y > 0 \\ 0 & \text{elsewhere} \end{cases}$$

Thus, integration yields

$$\begin{aligned} \int_1^2 \int_1^3 e^{-(x+y)}\, dx\, dy &= (e^{-1} - e^{-3})(e^{-1} - e^{-2}) \\ &= e^{-2} - e^{-3} - e^{-4} + e^{-5} \\ &= 0.074 \end{aligned}$$

for $P(1 < X < 3, 1 < Y < 2)$.

For two random variables, the joint probability is, geometrically speaking, a surface, and the probability that we calculated in the preceding example is given by the volume under this surface, as shown in Figure 11.

All the definitions of this section can be generalized to the **multivariate** case, where there are n random variables. Corresponding to Definition 6, the values of the joint probability distribution of n discrete random variables $X_1, X_2, \ldots$, and X_n are given by

$$f(x_1, x_2, \ldots, x_n) = P(X_1 = x_1, X_2 = x_2, \ldots, X_n = x_n)$$

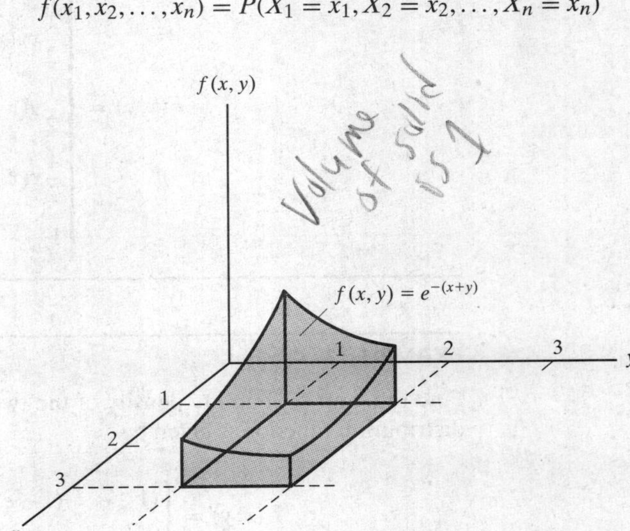

Figure 11. Diagram for Example 17.

for each n-tuple $(x_1, x_2, \ldots, x_n)$ within the range of the random variables; and corresponding to Definition 7, the values of their joint distribution function are given by

$$F(x_1, x_2, \ldots, x_n) = P(X_1 \leqq x_1, X_2 \leqq x_2, \ldots, X_n \leqq x_n)$$

for $-\infty < x_1 < \infty, -\infty < x_2 < \infty, \ldots, -\infty < x_n < \infty$.

EXAMPLE 18

If the joint probability distribution of three discrete random variables X, Y, and Z is given by

$$f(x, y, z) = \frac{(x+y)z}{63} \qquad \text{for } x = 1, 2; \quad y = 1, 2, 3; \quad z = 1, 2$$

find $P(X = 2, Y + Z \leqq 3)$.

Solution

$$\begin{aligned} P(X = 2, Y + Z \leqq 3) &= f(2, 1, 1) + f(2, 1, 2) + f(2, 2, 1) \\ &= \frac{3}{63} + \frac{6}{63} + \frac{4}{63} \\ &= \frac{13}{63} \end{aligned}$$

In the continuous case, probabilities are again obtained by integrating the joint probability density, and the joint distribution function is given by

$$F(x_1, x_2, \ldots, x_n) = \int_{-\infty}^{x_n} \cdots \int_{-\infty}^{x_2} \int_{-\infty}^{x_1} f(t_1, t_2, \ldots, t_n)\, dt_1\, dt_2 \ldots dt_n$$

for $-\infty < x_1 < \infty, -\infty < x_2 < \infty, \ldots, -\infty < x_n < \infty$, analogous to Definition 9. Also, partial differentiation yields

$$f(x_1, x_2, \ldots, x_n) = \frac{\partial^n}{\partial x_1 \partial x_2 \cdots \partial x_n} F(x_1, x_2, \ldots, x_n)$$

wherever these partial derivatives exist.

EXAMPLE 19

If the **trivariate** probability density of X_1, X_2, and X_3 is given by

$$f(x_1, x_2, x_3) = \begin{cases} (x_1 + x_2)e^{-x_3} & \text{for } 0 < x_1 < 1, 0 < x_2 < 1, x_3 > 0 \\ 0 & \text{elsewhere} \end{cases}$$

find $P[(X_1, X_2, X_3) \in A]$, where A is the region

$$\left\{ (x_1, x_2, x_3) | 0 < x_1 < \frac{1}{2}, \frac{1}{2} < x_2 < 1, x_3 < 1 \right\}$$

Solution

$$P[(X_1, X_2, X_3) \in A] = P\left(0 < X_1 < \frac{1}{2}, \frac{1}{2} < X_2 < 1, X_3 < 1\right)$$

$$= \int_0^1 \int_{\frac{1}{2}}^1 \int_0^{\frac{1}{2}} (x_1 + x_2)e^{-x_3}\, dx_1\, dx_2\, dx_3$$

$$= \int_0^1 \int_{\frac{1}{2}}^1 \left(\frac{1}{8} + \frac{x_2}{2}\right) e^{-x_3}\, dx_2\, dx_3$$

$$= \int_0^1 \frac{1}{4} e^{-x_3}\, dx_3$$

$$= \frac{1}{4}(1 - e^{-1}) = 0.158$$

Exercises

42. If the values of the joint probability distribution of X and Y are as shown in the table

y \ x	0	1	2
0	$\frac{1}{12}$	$\frac{1}{6}$	$\frac{1}{24}$
1	$\frac{1}{4}$	$\frac{1}{4}$	$\frac{1}{40}$
2	$\frac{1}{8}$	$\frac{1}{20}$	
3	$\frac{1}{120}$		

find
(a) $P(X = 1, Y = 2)$; **(b)** $P(X = 0, 1 \leqq Y < 3)$;
(c) $P(X + Y \leqq 1)$; **(d)** $P(X > Y)$.

43. With reference to Exercise 42, find the following values of the joint distribution function of the two random variables:
(a) $F(1.2,\ 0.9)$; **(b)** $F(-3, 1.5)$;
(c) $F(2,\ 0)$; **(d)** $F(4,\ 2.7)$.

44. If the joint probability distribution of X and Y is given by

$$f(x, y) = c(x^2 + y^2) \qquad \text{for } x = -1, 0, 1, 3; \quad y = -1, 2, 3$$

find the value of c.

45. With reference to Exercise 44 and the value obtained for c, find
(a) $P(X \leqq 1, Y > 2)$;
(b) $P(X = 0, Y \leqq 2)$;
(c) $P(X + Y > 2)$.

46. Show that there is no value of k for which

$$f(x, y) = ky(2y - x) \qquad \text{for } x = 0, 3; \quad y = 0, 1, 2$$

can serve as the joint probability distribution of two random variables.

47. If the joint probability distribution of X and Y is given by

$$f(x, y) = \frac{1}{30}(x + y) \qquad \text{for } x = 0, 1, 2, 3; \quad y = 0, 1, 2$$

construct a table showing the values of the joint distribution function of the two random variables at the 12 points $(0, 0), (0, 1), \ldots, (3, 2)$.

48. If $F(x, y)$ is the value of the joint distribution function of two discrete random variables X and Y at (x, y), show that
(a) $F(-\infty, -\infty) = 0$;
(b) $F(\infty, \infty) = 1$;
(c) if $a < b$ and $c < d$, then $F(a, c) \leqq F(b, d)$.

49. Determine k so that

$$f(x, y) = \begin{cases} kx(x - y) & \text{for } 0 < x < 1, -x < y < x \\ 0 & \text{elsewhere} \end{cases}$$

can serve as a joint probability density.

50. If the joint probability density of X and Y is given by

$$f(x, y) = \begin{cases} 24xy & \text{for } 0 < x < 1, 0 < y < 1, x + y < 1 \\ 0 & \text{elsewhere} \end{cases}$$

find $P(X + Y < \frac{1}{2})$.

51. If the joint probability density of X and Y is given by

$$f(x,y) = \begin{cases} 2 & \text{for } x > 0, y > 0, x+y < 1 \\ 0 & \text{elsewhere} \end{cases}$$

find
(a) $P(X \leqq \frac{1}{2}, Y \leqq \frac{1}{2})$;
(b) $P(X+Y > \frac{2}{3})$;
(c) $P(X > 2Y)$.

52. With reference to Exercise 51, find an expression for the values of the joint distribution function of X and Y when $x > 0$, $y > 0$, and $x+y < 1$, and use it to verify the result of part (a).

53. If the joint probability density of X and Y is given by

$$f(x,y) = \begin{cases} \dfrac{1}{y} & \text{for } 0 < x < y, 0 < y < 1 \\ 0 & \text{elsewhere} \end{cases}$$

find the probability that the sum of the values of X and Y will exceed $\frac{1}{2}$.

54. Find the joint probability density of the two random variables X and Y whose joint distribution function is given by

$$F(x,y) = \begin{cases} (1-e^{-x^2})(1-e^{-y^2}) & \text{for } x > 0, y > 0 \\ 0 & \text{elsewhere} \end{cases}$$

55. Use the joint probability density obtained in Exercise 54 to find $P(1 < X \leqq 2, 1 < Y \leqq 2)$.

56. Find the joint probability density of the two random variables X and Y whose joint distribution function is given by

$$F(x,y) = \begin{cases} 1-e^{-x}-e^{-y}+e^{-x-y} & \text{for } x > 0, y > 0 \\ 0 & \text{elsewhere} \end{cases}$$

57. Use the joint probability density obtained in Exercise 56 to find $P(X+Y > 3)$.

58. If $F(x,y)$ is the value of the joint distribution function of the two continuous random variables X and Y at (x,y), express $P(a < X \leqq b, c < Y \leqq d)$ in terms of $F(a,c)$, $F(a,d)$, $F(b,c)$, and $F(b,d)$. Observe that the result holds also for discrete random variables.

59. Use the formula obtained in Exercise 58 to verify the result, 0.074, of Example 17.

60. Use the formula obtained in Exercise 58 to verify the result of Exercise 55.

61. Use the formula obtained in Exercise 58 to verify the result of Exercise 57.

62. Find k if the joint probability distribution of X, Y, and Z is given by

$$f(x,y,z) = kxyz$$

for $x = 1, 2$; $y = 1, 2, 3$; $z = 1, 2$.

63. With reference to Exercise 62, find
(a) $P(X = 1, Y \leqq 2, Z = 1)$;
(b) $P(X = 2, Y+Z = 4)$.

64. With reference to Exercise 62, find the following values of the joint distribution function of the three random variables:
(a) $F(2, 1, 2)$;
(b) $F(1, 0, 1)$;
(c) $F(4, 4, 4)$.

65. Find k if the joint probability density of X, Y, and Z is given by

$$f(x,y,z) = \begin{cases} kxy(1-z) & \text{for } 0 < x < 1, 0 < y < 1, \\ & 0 < z < 1, x+y+z < 1 \\ 0 & \text{elsewhere} \end{cases}$$

66. With reference to Exercise 65, find $P(X+Y < \frac{1}{2})$.

67. Use the result of Example 16 to verify that the joint distribution function of the random variables X_1, X_2, and X_3 of Example 19 is given by

$$F(x_1,x_2,x_3) = \begin{cases} 0 & \text{for } x_1 \leqq 0, x_2 \leqq 0, \text{ or } x_3 \leqq 0 \\ \frac{1}{2}x_1x_2(x_1+x_2)(1-e^{-x_3}) & \text{for } 0 < x_1 < 1, 0 < x_2 < 1, x_3 > 0 \\ \frac{1}{2}x_2(x_2+1)(1-e^{-x_3}) & \text{for } x_1 \geqq 1, 0 < x_2 < 1, x_3 > 0 \\ \frac{1}{2}x_1(x_1+1)(1-e^{-x_3}) & \text{for } 0 < x_1 < 1, x_2 \geqq 1, x_3 > 0 \\ 1-e^{-x_3} & \text{for } x_1 \geqq 1, x_2 \geqq 1, x_3 > 0 \end{cases}$$

68. If the joint probability density of X, Y, and Z is given by

$$f(x,y,z) = \begin{cases} \frac{1}{3}(2x+3y+z) & \text{for } 0 < x < 1, 0 < y < 1, \\ & 0 < z < 1 \\ 0 & \text{elsewhere} \end{cases}$$

find
(a) $P(X = \frac{1}{2}, Y = \frac{1}{2}, Z = \frac{1}{2})$;
(b) $P(X < \frac{1}{2}, Y < \frac{1}{2}, Z < \frac{1}{2})$.

6 Marginal Distributions

To introduce the concept of a **marginal distribution**, let us consider the following example.

EXAMPLE 20

In Example 12 we derived the joint probability distribution of two random variables X and Y, the number of aspirin caplets and the number of sedative caplets included among two caplets drawn at random from a bottle containing three aspirin, two sedative, and four laxative caplets. Find the probability distribution of X alone and that of Y alone.

Solution

The results of Example 12 are shown in the following table, together with the marginal totals, that is, the totals of the respective rows and columns:

		x = 0	x = 1	x = 2	
y	0	$\frac{1}{6}$	$\frac{1}{3}$	$\frac{1}{12}$	$\frac{7}{12}$
	1	$\frac{2}{9}$	$\frac{1}{6}$		$\frac{7}{18}$
	2	$\frac{1}{36}$			$\frac{1}{36}$
		$\frac{5}{12}$	$\frac{1}{2}$	$\frac{1}{12}$	

The column totals are the probabilities that X will take on the values 0, 1, and 2. In other words, they are the values

$$g(x) = \sum_{y=0}^{2} f(x, y) \qquad \text{for } x = 0, 1, 2$$

of the probability distribution of X. By the same token, the row totals are the values

$$h(y) = \sum_{x=0}^{2} f(x, y) \qquad \text{for } y = 0, 1, 2$$

of the probability distribution of Y.

We are thus led to the following definition.

DEFINITION 10. MARGINAL DISTRIBUTION. *If* X *and* Y *are discrete random variables and* f(x, y) *is the value of their joint probability distribution at* (x, y), *the function given by*

$$g(x) = \sum_{y} f(x, y)$$

for each x *within the range of* X *is called the* ***marginal distribution of*** X. *Correspondingly, the function given by*

$$h(y) = \sum_{x} f(x, y)$$

for each y *within the range of* Y *is called the* ***marginal distribution of*** Y.

When X and Y are continuous random variables, the probability distributions are replaced by probability densities, the summations are replaced by integrals, and we obtain the following definition.

DEFINITION 11. MARGINAL DENSITY. *If* X *and* Y *are continuous random variables and* f(x, y) *is the value of their joint probability density at* (x, y)*, the function given by*

$$g(x) = \int_{-\infty}^{\infty} f(x, y)\,dy \qquad \text{for } -\infty < x < \infty$$

is called the ***marginal density of*** **X**. *Correspondingly, the function given by*

$$h(y) = \int_{-\infty}^{\infty} f(x, y)\,dx \qquad \text{for } -\infty < y < \infty$$

is called the ***marginal density of*** **Y**.

EXAMPLE 21

Given the joint probability density

$$f(x, y) = \begin{cases} \frac{2}{3}(x + 2y) & \text{for } 0 < x < 1, 0 < y < 1 \\ 0 & \text{elsewhere} \end{cases}$$

find the marginal densities of X and Y.

Solution

Performing the necessary integrations, we get

$$g(x) = \int_{-\infty}^{\infty} f(x, y)\,dy = \int_{0}^{1} \frac{2}{3}(x + 2y)\,dy = \frac{2}{3}(x + 1)$$

for $0 < x < 1$ and $g(x) = 0$ elsewhere. Likewise,

$$h(y) = \int_{-\infty}^{\infty} f(x, y)\,dx = \int_{0}^{1} \frac{2}{3}(x + 2y)\,dx = \frac{1}{3}(1 + 4y)$$

for $0 < y < 1$ and $h(y) = 0$ elsewhere.

When we are dealing with more than two random variables, we can speak not only of the marginal distributions of the individual random variables, but also of the

joint marginal distributions of several of the random variables. If the joint probability distribution of the discrete random variables $X_1, X_2, \ldots,$ and X_n has the values $f(x_1, x_2, \ldots, x_n)$, the marginal distribution of X_1 alone is given by

$$g(x_1) = \sum_{x_2} \cdots \sum_{x_n} f(x_1, x_2, \ldots, x_n)$$

for all values within the range of X_1, the joint marginal distribution of X_1, X_2, and X_3 is given by

$$m(x_1, x_2, x_3) = \sum_{x_4} \cdots \sum_{x_n} f(x_1, x_2, \ldots, x_n)$$

for all values within the range of X_1, X_2, and X_3, and other marginal distributions can be defined in the same way. For the continuous case, probability distributions are replaced by probability densities, summations are replaced by integrals, and if the joint probability density of the continuous random variables $X_1, X_2, \ldots,$ and X_n has the values $f(x_1, x_2, \ldots, x_n)$, the marginal density of X_2 alone is given by

$$h(x_2) = \int_{-\infty}^{\infty} \cdots \int_{-\infty}^{\infty} f(x_1, x_2, \ldots, x_n)\, dx_1\, dx_3 \cdots dx_n$$

for $-\infty < x_2 < \infty$, the joint marginal density of X_1 and X_n is given by

$$\varphi(x_1, x_n) = \int_{-\infty}^{\infty} \cdots \int_{-\infty}^{\infty} f(x_1, x_2, \ldots, x_n)\, dx_2\, dx_3 \cdots dx_{n-1}$$

for $-\infty < x_1 < \infty$ and $-\infty < x_n < \infty$, and so forth.

EXAMPLE 22

Considering again the trivariate probability density of Example 19,

$$f(x_1, x_2, x_3) = \begin{cases} (x_1 + x_2)e^{-x_3} & \text{for } 0 < x_1 < 1, 0 < x_2 < 1, x_3 > 0 \\ 0 & \text{elsewhere} \end{cases}$$

find the joint marginal density of X_1 and X_3 and the marginal density of X_1 alone.

Solution

Performing the necessary integration, we find that the joint marginal density of X_1 and X_3 is given by

$$m(x_1, x_3) = \int_0^1 (x_1 + x_2)e^{-x_3}\, dx_2 = \left(x_1 + \frac{1}{2}\right) e^{-x_3}$$

for $0 < x_1 < 1$ and $x_3 > 0$ and $m(x_1, x_3) = 0$ elsewhere. Using this result, we find that the marginal density of X_1 alone is given by

$$\begin{aligned} g(x_1) &= \int_0^{\infty} \int_0^1 f(x_1, x_2, x_3)\, dx_2\, dx_3 = \int_0^{\infty} m(x_1, x_3)\, dx_3 \\ &= \int_0^{\infty} \left(x_1 + \frac{1}{2}\right) e^{-x_3}\, dx_3 = x_1 + \frac{1}{2} \end{aligned}$$

for $0 < x_1 < 1$ and $g(x_1) = 0$ elsewhere.

Corresponding to the various marginal and joint marginal distributions and densities we have introduced in this section, we can also define **marginal** and **joint marginal distribution functions**. Some problems relating to such distribution functions will be left to the reader in Exercises 72, 79, and 80.

7 Conditional Distributions

In the conditional probability of event A, given event B, as

$$P(A|B) = \frac{P(A \cap B)}{P(B)}$$

provided $P(B) \neq 0$. Suppose now that A and B are the events $X = x$ and $Y = y$ so that we can write

$$P(X = x|Y = y) = \frac{P(X = x, Y = y)}{P(Y = y)}$$

$$= \frac{f(x,y)}{h(y)}$$

provided $P(Y = y) = h(y) \neq 0$, where $f(x,y)$ is the value of the joint probability distribution of X and Y at (x,y), and $h(y)$ is the value of the marginal distribution of Y at y. Denoting the conditional probability by $f(x|y)$ to indicate that x is a variable and y is fixed, let us now make the following definition.

DEFINITION 12. CONDITIONAL DISTRIBUTION. *If* $f(x, y)$ *is the value of the joint probability distribution of the discrete random variables* X *and* Y *at* (x, y) *and* $h(y)$ *is the value of the marginal distribution of* Y *at* y*, the function given by*

$$f(x|y) = \frac{f(x,y)}{h(y)} \qquad h(y) \neq 0$$

for each x *within the range of* X *is called the* **conditional distribution of X given Y = y**. *Correspondingly, if* $g(x)$ *is the value of the marginal distribution of* X *at* x*, the function given by*

$$w(y|x) = \frac{f(x,y)}{g(x)} \qquad g(x) \neq 0$$

for each y *within the range of* Y *is called the* **conditional distribution of Y given X = x**.

EXAMPLE 23

With reference to Examples 12 and 20, find the conditional distribution of X given $Y = 1$.

Solution

Substituting the appropriate values from the table in Example 20, we get

$$f(0|1) = \frac{\frac{2}{9}}{\frac{7}{18}} = \frac{4}{7}$$

$$f(1|1) = \frac{\frac{1}{6}}{\frac{7}{18}} = \frac{3}{7}$$

$$f(2|1) = \frac{0}{\frac{7}{18}} = 0$$

When X and Y are continuous random variables, the probability distributions are replaced by probability densities, and we obtain the following definition.

DEFINITION 13. CONDITIONAL DENSITY. *If* f(x, y) *is the value of the joint density of the continuous random variables* X *and* Y *at* (x, y) *and* h(y) *is the value of the marginal distribution of* Y *at* y, *the function given by*

$$f(x|y) = \frac{f(x, y)}{h(y)} \qquad h(y) \neq 0$$

for $-\infty < x < \infty$, *is called the* ***conditional density of* X *given* Y = y**. *Correspondingly, if* g(x) *is the value of the marginal density of* X *at* x, *the function given by*

$$w(y|x) = \frac{f(x, y)}{g(x)} \qquad g(x) \neq 0$$

for $-\infty < y < \infty$, *is called the* ***conditional density of* Y *given* X = x.**

EXAMPLE 24

With reference to Example 21, find the conditional density of X given $Y = y$, and use it to evaluate $P(X \leq \frac{1}{2} | Y = \frac{1}{2})$.

Solution

Using the results obtained on the previous page, we have

$$f(x|y) = \frac{f(x, y)}{h(y)} = \frac{\frac{2}{3}(x + 2y)}{\frac{1}{3}(1 + 4y)}$$

$$= \frac{2x + 4y}{1 + 4y}$$

for $0 < x < 1$ and $f(x|y) = 0$ elsewhere. Now,

$$f\left(x \middle| \frac{1}{2}\right) = \frac{2x + 4 \cdot \frac{1}{2}}{1 + 4 \cdot \frac{1}{2}}$$

$$= \frac{2x + 2}{3}$$

and we can write

$$P\left(X \leqq \frac{1}{2} \middle| Y = \frac{1}{2}\right) = \int_0^{\frac{1}{2}} \frac{2x+2}{3}\, dx = \frac{5}{12}$$

It is of interest to note that in Figure 12 this probability is given by the ratio of the area of trapezoid $ABCD$ to the area of trapezoid $AEFD$.

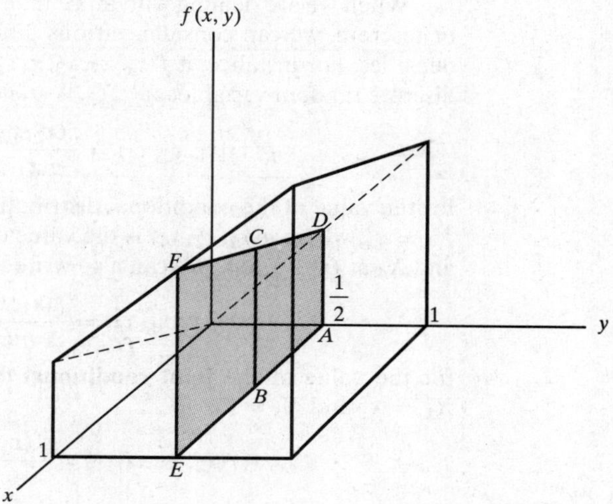

Figure 12. Diagram for Example 24.

EXAMPLE 25

Given the joint probability density

$$f(x,y) = \begin{cases} 4xy & \text{for } 0 < x < 1, 0 < y < 1 \\ 0 & \text{elsewhere} \end{cases}$$

find the marginal densities of X and Y and the conditional density of X given $Y = y$.

Solution

Performing the necessary integrations, we get

$$g(x) = \int_{-\infty}^{\infty} f(x,y)\, dy = \int_0^1 4xy\, dy$$

$$= 2xy^2 \Big|_{y=0}^{y=1} = 2x$$

for $0 < x < 1$, and $g(x) = 0$ elsewhere; also

$$h(y) = \int_{-\infty}^{\infty} f(x,y)\, dx = \int_0^1 4xy\, dx$$

$$= 2x^2y \Big|_{x=0}^{x=1} = 2y$$

for $0 < y < 1$, and $h(y) = 0$ elsewhere. Then, substituting into the formula for a conditional density, we get

$$f(x|y) = \frac{f(x,y)}{h(y)} = \frac{4xy}{2y} = 2x$$

for $0 < x < 1$, and $f(x|y) = 0$ elsewhere.

When we are dealing with more than two random variables, whether continuous or discrete, we can consider various different kinds of conditional distributions or densities. For instance, if $f(x_1, x_2, x_3, x_4)$ is the value of the joint distribution of the discrete random variables X_1, X_2, X_3, and X_4 at (x_1, x_2, x_3, x_4), we can write

$$p(x_3|x_1, x_2, x_4) = \frac{f(x_1, x_2, x_3, x_4)}{g(x_1, x_2, x_4)} \qquad g(x_1, x_2, x_4) \neq 0$$

for the value of the conditional distribution of X_3 at x_3 given $X_1 = x_1$, $X_2 = x_2$, and $X_4 = x_4$, where $g(x_1, x_2, x_4)$ is the value of the joint marginal distribution of X_1, X_2, and X_4 at (x_1, x_2, x_4). We can also write

$$q(x_2, x_4|x_1, x_3) = \frac{f(x_1, x_2, x_3, x_4)}{m(x_1, x_3)} \qquad m(x_1, x_3) \neq 0$$

for the value of the **joint conditional distribution** of X_2 and X_4 at (x_2, x_4) given $X_1 = x_1$ and $X_3 = x_3$, or

$$r(x_2, x_3, x_4|x_1) = \frac{f(x_1, x_2, x_3, x_4)}{b(x_1)} \qquad b(x_1) \neq 0$$

for the value of the joint conditional distribution of X_2, X_3, and X_4 at (x_2, x_3, x_4) given $X_1 = x_1$.

When we are dealing with two or more random variables, questions of **independence** are usually of great importance. In Example 25 we see that $f(x|y) = 2x$ does not depend on the given value $Y = y$, but this is clearly not the case in Example 24, where $f(x|y) = \frac{2x + 4y}{1 + 4y}$. Whenever the values of the conditional distribution of X given $Y = y$ do not depend on y, it follows that $f(x|y) = g(x)$, and hence the formulas of Definitions 12 and 13 yield

$$f(x,y) = f(x|y) \cdot h(y) = g(x) \cdot h(y)$$

That is, the values of the joint distribution are given by the products of the corresponding values of the two marginal distributions. Generalizing from this observation, let us now make the following definition.

DEFINITION 14. INDEPENDENCE OF DISCRETE RANDOM VARIABLES. *If* $f(x_1, x_2, \ldots, x_n)$ *is the value of the joint probability distribution of the discrete random variables* $X_1, X_2, \ldots, X_n$ *at* $(x_1, x_2, \ldots, x_n)$ *and* $f_i(x_i)$ *is the value of the marginal distribution of* X_i *at* x_i *for* $i = 1, 2, \ldots, n$, *then the* n *random variables are* ***independent*** *if and only if*

$$f(x_1, x_2, \ldots, x_n) = f_1(x_1) \cdot f_2(x_2) \cdot \ldots \cdot f_n(x_n)$$

for all $(x_1, x_2, \ldots, x_n)$ *within their range.*

To give a corresponding definition for continuous random variables, we simply substitute the word "density" for the word "distribution."

With this definition of independence, it can easily be verified that the three random variables of Example 22 are not independent, but that the two random variables

X_1 and X_3 and also the two random variables X_2 and X_3 are **pairwise independent** (see Exercise 81).

The following examples serve to illustrate the use of Definition 14 in finding probabilities relating to several independent random variables.

EXAMPLE 26

Considering n independent flips of a balanced coin, let X_i be the number of heads (0 or 1) obtained in the ith flip for $i = 1, 2, \ldots, n$. Find the joint probability distribution of these n random variables.

Solution

Since each of the random variables X_i, for $i = 1, 2, \ldots, n$, has the probability distribution

$$f_i(x_i) = \frac{1}{2} \qquad \text{for } x_i = 0, 1$$

and the n random variables are independent, their joint probability distribution is given by

$$\begin{aligned} f(x_1, x_2, \ldots, x_n) &= f_1(x_1) \cdot f_2(x_2) \cdot \ldots \cdot f_n(x_n) \\ &= \frac{1}{2} \cdot \frac{1}{2} \cdot \ldots \cdot \frac{1}{2} = \left(\frac{1}{2}\right)^n \end{aligned}$$

where $x_i = 0$ or 1 for $i = 1, 2, \ldots, n$.

EXAMPLE 27

Given the independent random variables X_1, X_2, and X_3 with the probability densities

$$f_1(x_1) = \begin{cases} e^{-x_1} & \text{for } x_1 > 0 \\ 0 & \text{elsewhere} \end{cases}$$

$$f_2(x_2) = \begin{cases} 2e^{-2x_2} & \text{for } x_2 > 0 \\ 0 & \text{elsewhere} \end{cases}$$

$$f_3(x_3) = \begin{cases} 3e^{-3x_3} & \text{for } x_3 > 0 \\ 0 & \text{elsewhere} \end{cases}$$

find their joint probability density, and use it to evaluate the probability $P(X_1 + X_2 \leqq 1, X_3 > 1)$.

Solution

According to Definition 14, the values of the joint probability density are given by

$$\begin{aligned} f(x_1, x_2, x_3) &= f_1(x_1) \cdot f_2(x_2) \cdot f_3(x_3) \\ &= e^{-x_1} \cdot 2e^{-2x_2} \cdot 3e^{-3x_3} \\ &= 6e^{-x_1 - 2x_2 - 3x_3} \end{aligned}$$

for $x_1 > 0, x_2 > 0, x_3 > 0$, and $f(x_1, x_2, x_3) = 0$ elsewhere. Thus,

$$\begin{aligned} P(X_1 + X_2 \leqq 1, X_3 > 1) &= \int_1^{\infty} \int_0^1 \int_0^{1-x_2} 6e^{-x_1-2x_2-3x_3}\, dx_1\, dx_2\, dx_3 \\ &= (1 - 2e^{-1} + e^{-2})e^{-3} \\ &= 0.020 \end{aligned}$$

Exercises

69. Given the values of the joint probability distribution of X and Y shown in the table

		x	
		-1	1
	-1	$\frac{1}{8}$	$\frac{1}{2}$
y	0	0	$\frac{1}{4}$
	1	$\frac{1}{8}$	0

find
(a) the marginal distribution of X;
(b) the marginal distribution of Y;
(c) the conditional distribution of X given $Y = -1$.

70. With reference to Exercise 42, find
(a) the marginal distribution of X;
(b) the marginal distribution of Y;
(c) the conditional distribution of X given $Y = 1$;
(d) the conditional distribution of Y given $X = 0$.

71. Given the joint probability distribution

$$f(x, y, z) = \frac{xyz}{108} \quad \text{for } x = 1, 2, 3; \quad y = 1, 2, 3; \quad z = 1, 2$$

find
(a) the joint marginal distribution of X and Y;
(b) the joint marginal distribution of X and Z;
(c) the marginal distribution of X;
(d) the conditional distribution of Z given $X = 1$ and $Y = 2$;
(e) the joint conditional distribution of Y and Z given $X = 3$.

72. With reference to Example 20, find
(a) the **marginal distribution function** of X, that is, the function given by $G(x) = P(X \leqq x)$ for $-\infty < x < \infty$;
(b) the **conditional distribution function** of X given $Y = 1$, that is, the function given by $F(x|1) = P(X \leqq x|Y = 1)$ for $-\infty < x < \infty$.

73. Check whether X and Y are independent if their joint probability distribution is given by
(a) $f(x, y) = \frac{1}{4}$ for $x = -1$ and $y = -1$, $x = -1$ and $y = 1$, $x = 1$ and $y = -1$, and $x = 1$ and $y = 1$;
(b) $f(x, y) = \frac{1}{3}$ for $x = 0$ and $y = 0$, $x = 0$ and $y = 1$, and $x = 1$ and $y = 1$.

74. If the joint probability density of X and Y is given by

$$f(x, y) = \begin{cases} \frac{1}{4}(2x + y) & \text{for } 0 < x < 1, 0 < y < 2 \\ 0 & \text{elsewhere} \end{cases}$$

find
(a) the marginal density of X;
(b) the conditional density of Y given $X = \frac{1}{4}$.

75. With reference to Exercise 74, find
(a) the marginal density of Y;
(b) the conditional density of X given $Y = 1$.

76. If the joint probability density of X and Y is given by

$$f(x, y) = \begin{cases} 24y(1 - x - y) & \text{for } x > 0, y > 0, x + y < 1 \\ 0 & \text{elsewhere} \end{cases}$$

find
(a) the marginal density of X;
(b) the marginal density of Y.

Also determine whether the two random variables are independent.

77. With reference to Exercise 53, find
(a) the marginal density of X;
(b) the marginal density of Y.

Also determine whether the two random variables are independent.

78. With reference to Example 22, find
(a) the conditional density of X_2 given $X_1 = \frac{1}{3}$ and $X_3 = 2$;
(b) the joint conditional density of X_2 and X_3 given $X_1 = \frac{1}{2}$.

79. If $F(x, y)$ is the value of the joint distribution function of X and Y at (x, y), show that the **marginal distribution function** of X is given by

$$G(x) = F(x, \infty) \qquad \text{for } -\infty < x < \infty$$

Use this result to find the marginal distribution function of X for the random variables of Exercise 54.

80. If $F(x_1, x_2, x_3)$ is the value of the joint distribution function of X_1, X_2, and X_3 at (x_1, x_2, x_3), show that the **joint marginal distribution function** of X_1 and X_3 is given by

$$M(x_1, x_3) = F(x_1, \infty, x_3) \quad \text{for } -\infty < x_1 < \infty, -\infty < x_3 < \infty$$

and that the **marginal distribution function** of X_1 is given by

$$G(x_1) = F(x_1, \infty, \infty) \qquad \text{for } -\infty < x_1 < \infty$$

With reference to Example 19, use these results to find

(a) the joint marginal distribution function of X_1 and X_3;
(b) the marginal distribution function of X_1.

81. With reference to Example 22, verify that the three random variables X_1, X_2, and X_3 are not independent, but that the two random variables X_1 and X_3 and also the two random variables X_2 and X_3 are **pairwise independent**.

82. If the independent random variables X and Y have the marginal densities

$$f(x) = \begin{cases} \frac{1}{2} & \text{for } 0 < x < 2 \\ 0 & \text{elsewhere} \end{cases}$$

$$\pi(y) = \begin{cases} \frac{1}{3} & \text{for } 0 < y < 3 \\ 0 & \text{elsewhere} \end{cases}$$

find
(a) the joint probability density of X and Y;
(b) the value of $P(X^2 + Y^2 > 1)$.

8 The Theory in Practice

This chapter has been about how probabilities can group themselves into probability distributions, and how, in the case of continuous random variables, these distributions become probability density functions. In practice, however, all data appear to be discrete. (Even if data arise from continuous random variables, the limitations of measuring instruments and roundoff produce discrete values.) In this section, we shall introduce some applications of the ideas of probability distributions and densities to the exploration of raw data, an important element of what is called **data analysis**.

When confronted with raw data, often consisting of a long list of measurements, it is difficult to understand what the data are informing us about the process, product, or service which gave rise to them. The following data, giving the response times of 30 integrated circuits (in picoseconds), illustrate this point:

Integrated Circuit Response Times (ps)									
4.6	4.0	3.7	4.1	4.1	5.6	4.5	6.0	6.0	3.4
3.4	4.6	3.7	4.2	4.6	4.7	4.1	3.7	3.4	3.3
3.7	4.1	4.5	4.6	4.4	4.8	4.3	4.4	5.1	3.9

Examination of this long list of numbers seems to tell us little other than, perhaps, the response times are greater than 3 ps or less than 7 ps. (If the list contained several hundred numbers, even this information would be difficult to elicit.)

A start at exploring data can be made by constructing a **stem-and-leaf display**. To construct such a display, the first digit of each response time is listed in a column at the left, and the associated second digits are listed to the right of each first digit. For the response-time data, we obtain the following stem-and-leaf display:

3	7 4 4 7 7 4 3 7 9
4	6 0 1 1 5 6 2 6 7 1 1 5 6 4 8 3 4
5	6 1
6	0 0

In this display, each row is a **stem** and the numbers in the column to the left of the vertical line are called **stem labels**. Each number on a stem to the right of the vertical line is called a **leaf**.

The stem-and-leaf display allows examination of the data in a way that would be difficult, if not impossible, from the original listing. For example, it can quickly be seen that there are more response times in the range 4.0 to 4.9 ps than any other, and that the great majority of circuits had response times of less than 5. This method of exploratory data analysis yields another advantage; namely there is no loss of information in a stem-and-leaf display.

The first two stems of this stem-and-leaf display contain the great majority of the observations, and more detail might be desirable. To obtain a finer subdivision of the data in each stem, a **double-stem display** can be constructed by dividing each stem in half so that the leaves in the first half of each stem are 0, 1, 2, 3, and 4, and those in the second half are 5, 6, 7, 8, and 9. The resulting double-stem display looks like this:

3*f*	4 4 4 3
3*s*	7 7 7 7 9
4*f*	0 1 1 2 1 1 4 3 4
4*s*	6 5 6 6 7 5 6 8
5*	6 1
6*	0 0

The stem labels include the letter *f* (for *first*) to denote that the leaves of this stem are 0–4, and *s* (for *second*) to denote that the leaves are 5–9. The asterisk is used with stem labels 5 and 6 to show that all 10 digits are included in these stems.

Numerical data can be grouped according to their values in several other ways in addition to stem-and-leaf displays.

DEFINITION 15. FREQUENCY DISTRIBUTION. *A grouping of numerical data into classes having definite upper and lower limits is called a* ***frequency distribution****.*

The construction of a frequency distribution is easily facilitated with a computer program such as MINITAB. The following discussion may be omitted if a computer program is used to construct frequency distributions.

To construct a frequency distribution, first a decision is made about the number of classes to use in grouping the data. The number of classes can be chosen to make the specification of upper and lower class limits convenient. Generally, the number of classes should increase as the number of observations becomes larger, but it is rarely helpful to use fewer than 5 or more than 15 classes.

The smallest and largest observations that can be put into each class are called the **class limits**. In choosing class limits, it is important that the classes do not overlap, so there is no ambiguity about which class contains any given observation. Also, enough classes should be included to accommodate all observations. Finally, the observations are tallied to determine the **class frequencies**, the number of observations falling into each class.

EXAMPLE 28

Construct a frequency distribution of the following compressive strengths (in psi) of concrete samples, given to the nearest 10 psi:

4890	4830	5490	4820	5230	4860	5040	5060	4500	5260
4610	5100	4730	5250	5540	4910	4430	4850	5040	5000
4600	4630	5330	5160	4950	4480	5310	4730	4700	4390
4710	5160	4970	4710	4430	4260	4890	5110	5030	4850
4820	4550	4970	4740	4840	4910	5200	4880	5150	4890
4900	4990	4570	4790	4480	5060	4340	4830	4670	4750

Solution

Since the smallest observation is 4260 and the largest is 5540, it will be convenient to choose seven classes, having the class limits 4200–4390, 4400–4590, . . ., 5400–5990. (Note that class limits of 4200–4400, 4400–4600, etc., are not used because they would overlap and assignment of 4400, for example, would be ambiguous; it could fit into either of the first two classes.) The following table exhibits the results of tallying the observations, that is, counting the number that fall into each class:

Class Limits	*Tally*	*Frequency*
4200–4390	///	3
4400–4590	~~////~~ //	7
4600–4790	~~////~~ ~~////~~ //	12
4800–4990	~~////~~ ~~////~~ ~~////~~ ////	19
5000–5190	~~////~~ ~~////~~ /	11
5200–5390	~~////~~ /	6
5400–5590	//	2
	Total	60

Note the similarity between frequency distributions and probability distributions. A frequency distribution represents data, but a probability distribution represents a theoretical distribution of probabilities.

The midpoint between the upper class limit of a class and the lower class limit of the next class in a frequency distribution is called a **class boundary**. Class boundaries, rather than class marks, are used in constructing cumulative distributions (Exercise 88). The interval between successive class boundaries is called the **class interval**; it can also be defined as the difference between successive lower class limits or successive upper class limits. (Note that the class interval is *not* obtained by subtracting the lower class limit of a class from its upper class limit.) A class can be represented by a single number, called the **class mark**. This number is calculated for any class by averaging its upper and lower class limits.

Once data have been grouped into a frequency distribution, each observation in a given class is treated as if its value is the class mark of that class. In so doing, its actual value is lost; it is known only that its value lies somewhere between the class limits of its class. Such an approximation is the price paid for the convenience of working with a frequency distribution.

EXAMPLE 29

For the frequency distribution of compressive strengths of concrete given in Example 28, find (a) the class boundaries, (b) the class interval, and (c) the class mark of each class.

Solution

(a) The class boundaries of the first class are 4195–4395. The class boundaries of the second through the sixth classes are 4395–4595, 4595–4795, 4795–4995, 4995–5195, and 5195–5395, respectively. The class boundaries of the last class are 5395–5595. Note that the lower class boundary of the first class is calculated as if there were a class below the first class, and the upper class boundary of the last class is calculated as if there were a class above it. Also note that, unlike class limits, the class boundaries overlap.

(b) The class interval is 200, the difference between the upper and lower class boundaries of any class. It also can be found by subtracting successive lower class limits, for example, $4400 - 4200 = 200$ psi, or by subtracting successive upper class limits, for example, $4590 - 4390 = 200$.

(c) The class mark of the first class is $(4200 + 4390)/2 = 4295$; it is $(4400 + 4590)/2 = 4495$ for the second class; and the class marks are 4695, 4895, 5095, 5295, and 5495 for the remaining five classes. Note that the class interval, 200, also is given by the difference between any two successive class marks.

Histograms are easily constructed using most statistical software packages. Using MINITAB software to construct the histogram of compressive strengths, we obtain the result shown in Figure 13.

EXAMPLE 30

Suppose a wire is soldered to a board and pulled with continuously increasing force until the bond breaks. The forces required to break the solder bonds are as follows:

Force Required to Break Solder Bonds (grams)

19.8	13.9	30.4	16.4	11.6	36.9	14.8	21.1	13.5	5.8
10.0	17.1	14.1	16.6	23.3	12.1	18.8	10.4	9.4	23.8
14.2	26.7	7.8	22.9	12.6	6.8	13.5	10.7	12.2	27.7
9.0	14.9	24.0	12.0	7.1	12.8	18.6	26.0	37.4	13.3

Use MINITAB or other statistical software to obtain a histogram of these data.

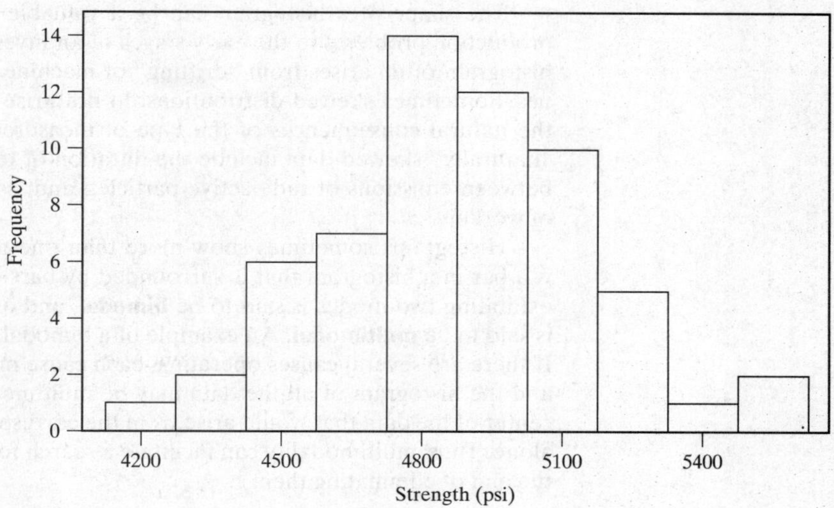

Figure 13. Histogram of compressive strengths.

Solution

The resulting histogram is shown in Figure 14. This histogram exhibits a right-hand "tail," suggesting that while most of the solder bonds have low or moderate breaking strengths, a few had strengths that were much greater than the rest.

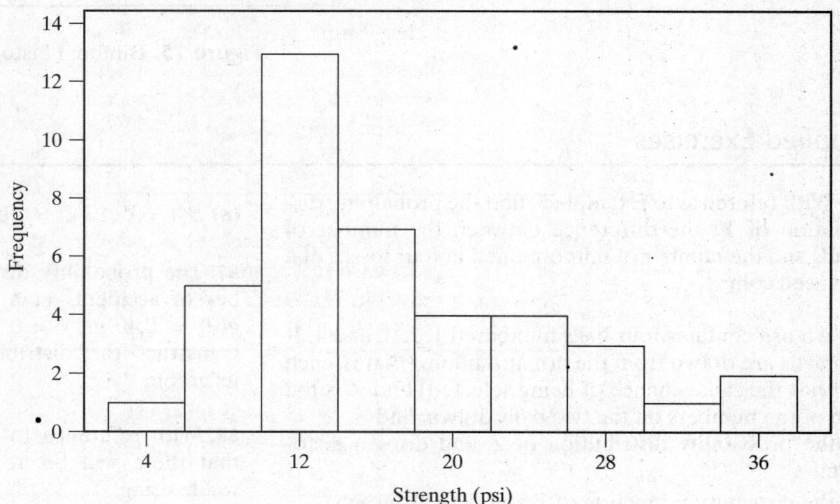

Figure 14. Histogram of solder-bond strengths.

Data having histograms with a long tail on the right or on the left are said to be **skewed.** A histogram exhibiting a long right-hand tail arises when the data have **positive skewness**. Likewise, if the tail is on the left, the data are said the have **negative skewness**. Examples of data that often are skewed include product lifetimes, many kinds of stress tests, workers' incomes, and many weather-related phenomena, such as the proportion of cloudiness on a given day.

The shape of a histogram can be a valuable guide to a search for causes of production problems in the early stages of an investigation. For example, a skewed histogram often arises from "drifting" of machine settings from their nominal values. Sometimes skewed distributions do not arise from underlying causes but are the natural consequences of the type of measurements made. Some examples of "naturally" skewed data include the duration of telephone calls, the time intervals between emissions of radioactive particles, and, as previously mentioned, incomes of workers.

Histograms sometimes show more than one **mode**, or "high points." A mode is a bar in a histogram that is surrounded by bars of lower frequency. A histogram exhibiting two modes is said to be **bimodal**, and one having more than two modes is said to be **multimodal**. An example of a bimodal histogram is shown in Figure 15. If there are several causes operating, each cause may generate its own distribution, and the histogram of all the data may be multimodal, each mode representing the center of the data that would arise from the corresponding cause if it were operating alone. Thus, multimodality can facilitate a search for underlying causes of error with the aim of eliminating them.

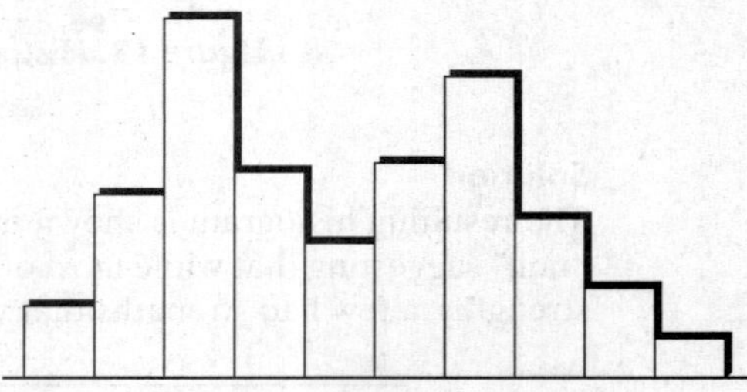

Figure 15. Bimodal histogram.

Applied Exercises SECS. 1–2

83. With reference to Example 3, find the probability distribution of Y, the difference between the number of heads and the number of tails obtained in four tosses of a balanced coin.

84. An urn contains four balls numbered 1, 2, 3, and 4. If two balls are drawn from the urn at random (that is, each pair has the same chance of being selected) and Z is the sum of the numbers on the two balls drawn, find
(a) the probability distribution of Z and draw a histogram;
(b) the distribution function of Z and draw its graph.

85. A coin is biased so that heads is twice as likely as tails. For three independent tosses of the coin, find
(a) the probability distribution of X, the total number of heads;
(b) the probability of getting at most two heads.

86. With reference to Exercise 85, find the distribution function of the random variable X and plot its graph. Use the distribution function of X to find
(a) $P(1 < X \leqq 3)$; **(b)** $P(X > 2)$.

87. The probability distribution of V, the weekly number of accidents at a certain intersection, is given by $g(0) = 0.40, g(1) = 0.30, g(2) = 0.20$, and $g(3) = 0.10$. Construct the distribution function of V and draw its graph.

88. With reference to Exercise 87, find the probability that there will be at least two accidents in any one week, using
(a) the original probabilities;
(b) the values of the distribution function.

89. This question has been intentionally omitted for this edition.

90. With reference to Exercise 80, find the distribution function of the sum of the spots on the dice, that is, the probability that this sum of the spots on the dice will be at most S, where $S = 2, 3, \ldots, 12$.

SECS. 3–4

91. The actual amount of coffee (in grams) in a 230-gram jar filled by a certain machine is a random variable whose probability density is given by

$$f(x) = \begin{cases} 0 & \text{for } x \leqq 227.5 \\ \frac{1}{5} & \text{for } 227.5 < x < 232.5 \\ 0 & \text{for } x \geqq 232.5 \end{cases}$$

Find the probabilities that a 230-gram jar filled by this machine will contain
(a) at most 228.65 grams of coffee;
(b) anywhere from 229.34 to 231.66 grams of coffee;
(c) at least 229.85 grams of coffee.

92. The number of minutes that a flight from Phoenix to Tucson is early or late is a random variable whose probability density is given by

$$f(x) = \begin{cases} \frac{1}{288}(36 - x^2) & \text{for } -6 < x < 6 \\ 0 & \text{elsewhere} \end{cases}$$

where negative values are indicative of the flight's being early and positive values are indicative of its being late. Find the probabilities that one of these flights will be
(a) at least 2 minutes early;
(b) at least 1 minute late;
(c) anywhere from 1 to 3 minutes early;
(d) exactly 5 minutes late.

93. The tread wear (in thousands of kilometers) that car owners get with a certain kind of tire is a random variable whose probability density is given by

$$f(x) = \begin{cases} \frac{1}{30}e^{-\frac{x}{30}} & \text{for } x > 0 \\ 0 & \text{for } x \leqq 0 \end{cases}$$

Find the probabilities that one of these tires will last
(a) at most 18,000 kilometers;
(b) anywhere from 27,000 to 36,000 kilometers;
(c) at least 48,000 kilometers.

94. The shelf life (in hours) of a certain perishable packaged food is a random variable whose probability density function is given by

$$f(x) = \begin{cases} \frac{20,000}{(x+100)^3} & \text{for } x > 0 \\ 0 & \text{elsewhere} \end{cases}$$

Find the probabilities that one of these packages will have a shelf life of
(a) at least 200 hours;
(b) at most 100 hours;
(c) anywhere from 80 to 120 hours.

95. The total lifetime (in years) of five-year-old dogs of a certain breed is a random variable whose distribution function is given by

$$F(x) = \begin{cases} 0 & \text{for } x \leqq 5 \\ 1 - \frac{25}{x^2} & \text{for } x > 5 \end{cases}$$

Find the probabilities that such a five-year-old dog will live
(a) beyond 10 years;
(b) less than eight years;
(c) anywhere from 12 to 15 years.

96. In a certain city the daily consumption of water (in millions of liters) is a random variable whose probability density is given by

$$f(x) = \begin{cases} \frac{1}{9}xe^{-\frac{x}{3}} & \text{for } x > 0 \\ 0 & \text{elsewhere} \end{cases}$$

What are the probabilities that on a given day
(a) the water consumption in this city is no more than 6 million liters;
(b) the water supply is inadequate if the daily capacity of this city is 9 million liters?

SEC. 5

97. Two textbooks are selected at random from a shelf that contains three statistics texts, two mathematics texts, and three physics texts. If X is the number of statistics texts and Y the number of mathematics texts actually chosen, construct a table showing the values of the joint probability distribution of X and Y.

98. Suppose that we roll a pair of balanced dice and X is the number of dice that come up 1, and Y is the number of dice that come up 4, 5, or 6.
(a) Draw a diagram like that of Figure 1 showing the values of X and Y associated with each of the 36 equally likely points of the sample space.
(b) Construct a table showing the values of the joint probability distribution of X and Y.

99. If X is the number of heads and Y the number of heads minus the number of tails obtained in three flips of a balanced coin, construct a table showing the values of the joint probability distribution of X and Y.

100. A sharpshooter is aiming at a circular target with radius 1. If we draw a rectangular system of coordinates

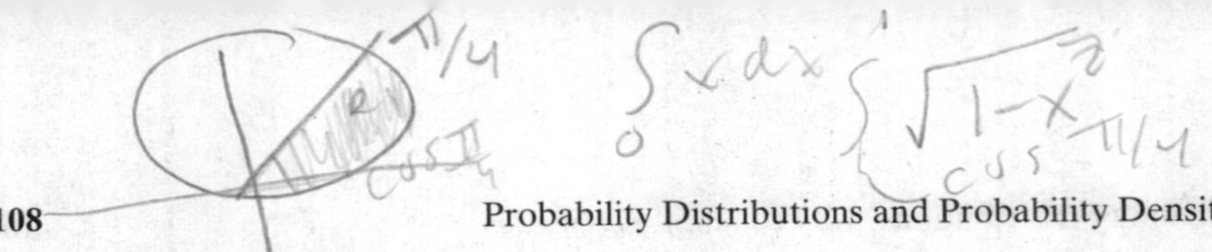

with its origin at the center of the target, the coordinates of the point of impact, (X, Y), are random variables having the joint probability density

$$f(x,y) = \begin{cases} \frac{1}{\pi} & \text{for } 0 < x^2 + y^2 < 1 \\ 0 & \text{elsewhere} \end{cases}$$

Find
(a) $P[(X, Y) \in A]$, where A is the sector of the circle in the first quadrant bounded by the lines $y = 0$ and $y = x$;
(b) $P[(X, Y) \in B]$, where $B = \{(x, y)|0 < x^2 + y^2 < \frac{1}{2}\}$.

101. Suppose that P, the price of a certain commodity (in dollars), and S, its total sales (in 10,000 units), are random variables whose joint probability distribution can be approximated closely with the joint probability density

$$f(p,s) = \begin{cases} 5pe^{-ps} & \text{for } 0.20 < p < 0.40, s > 0 \\ 0 & \text{elsewhere} \end{cases}$$

Find the probabilities that
(a) the price will be less than 30 cents and sales will exceed 20,000 units;
(b) the price will be between 25 cents and 30 cents and sales will be less than 10,000 units.

102. A certain college gives aptitude tests in the sciences and the humanities to all entering freshmen. If X and Y are, respectively, the proportions of correct answers that a student gets on the tests in the two subjects, the joint probability distribution of these random variables can be approximated with the joint probability density

$$f(x,y) = \begin{cases} \frac{2}{5}(2x+3y) & \text{for } 0 < x < 1, 0 < y < 1 \\ 0 & \text{elsewhere} \end{cases}$$

What are the probabilities that a student will get
(a) less than 0.40 on both tests;
(b) more than 0.80 on the science test and less than 0.50 on the humanities test?

SECS. 6–7

103. With reference to Exercise 97, find
(a) the marginal distribution of X;
(b) the conditional distribution of Y given $X = 0$.

104. If X is the proportion of persons who will respond to one kind of mail-order solicitation, Y is the proportion of persons who will respond to another kind of mail-order solicitation, and the joint probability density of X and Y is given by

$$f(x,y) = \begin{cases} \frac{2}{5}(x+4y) & \text{for } 0 < x < 1, 0 < y < 1 \\ 0 & \text{elsewhere} \end{cases}$$

find the probabilities that
(a) at least 30 percent will respond to the first kind of mail-order solicitation;
(b) at most 50 percent will respond to the second kind of mail-order solicitation given that there has been a 20 percent response to the first kind of mail-order solicitation.

105. If two cards are randomly drawn (without replacement) from an ordinary deck of 52 playing cards, Z is the number of aces obtained in the first draw, and W is the total number of aces obtained in both draws, find
(a) the joint probability distribution of Z and W;
(b) the marginal distribution of Z;
(c) the conditional distribution of W given $Z = 1$.

106. With reference to Exercise 101, find
(a) the marginal density of P;
(b) the conditional density of S given $P = p$;
(c) the probability that sales will be less than 30,000 units when $p = 25$ cents.

107. If X is the amount of money (in dollars) that a salesperson spends on gasoline during a day and Y is the corresponding amount of money (in dollars) for which he or she is reimbursed, the joint density of these two random variables is given by

$$f(x,y) = \begin{cases} \frac{1}{25}\left(\frac{20-x}{x}\right) & \text{for } 10 < x < 20, \frac{x}{2} < y < x \\ 0 & \text{elsewhere} \end{cases}$$

find
(a) the marginal density of X;
(b) the conditional density of Y given $X = 12$;
(c) the probability that the salesperson will be reimbursed at least \$8 when spending \$12.

108. Show that the two random variables of Exercise 102 are not independent.

109. The useful life (in hours) of a certain kind of integrated circuit is a random variable having the probability density

$$f(x) = \begin{cases} \frac{20{,}000}{(x+100)^3} & \text{for } x > 0 \\ 0 & \text{elsewhere} \end{cases}$$

If three of these circuits operate independently, find
(a) the joint probability density of X_1, X_2, and X_3, representing the lengths of their useful lives;
(b) the value of $P(X_1 < 100, X_2 < 100, X_3 \geqq 200)$.

SEC. 8

110. The following are the percentages of tin in measurements made on 24 solder joints:

61	63	59	54	65	60	62	61	67	60	55	68
57	64	65	62	59	59	60	62	61	63	58	61

(a) Construct a stem-and-leaf diagram using 5 and 6 as the stem labels.
(b) Construct a double-stem display.
(c) Which is more informative?

111. Suppose the first row of 12 observations in Exercise 110 came from solder connections made at station 105 and the second row came from station 107. Use a pair of stem-and-leaf diagrams to determine whether you should suspect a difference in the soldering process at the two stations.

112. Two different lathes turn shafts to be used in electric motors. Measurements made of their diameters (in cm) are

Lathe A:	1.42	1.38	1.40	1.41	1.39	1.44	1.36	1.42	1.40
Lathe B:	1.47	1.31	1.56	1.33	1.29	1.46	1.28	1.51	

Construct two stem-and-leaf diagrams to see if you should suspect that the two lathes are turning out shafts of different diameters.

113. Use MINITAB or some other computer software to construct a stem-and-leaf display for the following data representing the time to make coke (in hours) in successive runs of a coking oven.

7.8	9.2	6.4	8.2	7.6	5.9	7.4	7.1	6.7	8.5
10.1	8.6	7.7	5.9	9.3	6.4	6.8	7.9	7.2	10.2
6.9	7.4	7.8	6.6	8.1	9.5	6.4	7.6	8.4	9.2

114. Use MINITAB or some other computer software to construct a stem-and-leaf display for the combined data of Exercise 112.

115. The following are the drying times (minutes) of 100 sheets coated with polyurethane under various ambient conditions:

45.6	50.3	55.1	63.0	58.2	65.5	51.1	57.4	60.4	54.9
56.1	62.1	43.5	63.8	64.9	59.9	63.0	67.7	53.8	57.9
61.8	52.2	61.2	51.6	58.6	73.8	53.9	64.1	57.2	75.4
55.9	70.1	46.2	63.6	56.0	48.1	62.2	58.8	50.8	68.1
51.4	73.9	66.7	42.9	71.0	56.1	60.8	58.6	70.6	62.2
59.9	47.5	72.5	62.0	56.8	54.3	61.0	66.3	52.6	63.5
64.3	63.6	53.5	55.1	62.8	63.3	64.7	54.9	54.4	69.6
64.2	59.3	60.6	57.1	68.3	46.7	73.7	56.8	62.9	58.4
68.5	68.9	62.1	62.8	74.4	43.8	40.0	64.4	50.8	49.9
55.8	66.8	67.0	64.8	57.6	68.3	42.5	64.4	48.3	56.5

Construct a frequency distribution of these data, using eight classes.

116. Eighty pilots were tested in a flight simulator and the time for each to take corrective action for a given emergency was measured in seconds, with the following results:

11.1	5.2	3.6	7.6	12.4	6.8	3.8	5.7	9.0	6.0	4.9	12.6
7.4	5.3	14.2	8.0	12.6	13.7	3.8	10.6	6.8	5.4	9.7	6.7
14.1	5.3	11.1	13.4	7.0	8.9	6.2	8.3	7.7	4.5	7.6	5.0
9.4	3.5	7.9	11.0	8.6	10.5	5.7	7.0	5.6	9.1	5.1	4.5
6.2	6.8	4.3	8.5	3.6	6.1	5.8	10.0	6.4	4.0	5.4	7.0
4.1	8.1	5.8	11.8	6.1	9.1	3.3	12.5	8.5	10.8	6.5	7.9
6.8	10.1	4.9	5.4	9.6	8.2	4.2	3.4				

Construct a frequency distribution of these data.

117. Find the class boundaries, the class interval, and the class marks of the frequency distribution constructed in Exercise 115.

118. Find the class boundaries, the class interval, and the class marks of the frequency distribution constructed in Exercise 116.

119. The following are the number of highway accidents reported on 30 successive days in a certain county:

6	4	0	3	5	6	2	0	0	12	3	7	2	1	1
0	4	0	0	0	1	8	0	2	4	7	3	6	2	0

Construct a frequency distribution of these data. Identify the class boundaries, the class marks, and the class interval.

120. A **percentage distribution** is obtained from a frequency distribution by replacing each frequency by 100 times the ratio of that frequency to the total frequency. Construct a percentage distribution using the reaction-time data of Exercise 116.

121. Construct a percentage distribution using the drying-time data of Exercise 115.

122. Percentage distributions are useful in comparing two frequency distributions having different total frequencies. Construct percentage distributions from the following two frequency distributions and determine whether the distributions of daily absences in the two departments follow similar patterns.

	FREQUENCIES	
Class Limits	*Shipping Department*	*Security Department*
0–1	26	18
2–3	18	11
4–5	10	7
6–7	4	3
8–9	2	1
Totals	60	40

123. A **cumulative frequency distribution** is constructed from a frequency distribution by replacing each frequency with the sum of the frequency of the given class and the frequencies of all classes above it, and representing each class by its upper class boundary. Construct a cumulative frequency distribution using the data of Exercise 115.

124. Construct a cumulative frequency distribution using the data of Exercise 116.

125. Construct cumulative percentage distributions from the frequency distributions of absences given in Exercise 122.

126. *Unequal class intervals.* The small number of observations greater than 7 in Exercise 119 may cause some difficulty in constructing a frequency distribution. To keep class intervals equal, one is faced with the dilemma of either creating too many classes for only 30 observations or using a small number of classes with excessive loss of information in the first few classes. In such cases, one is tempted to drop the rule of equal-size classes, using a larger interval for the last class.
(a) If that were done, what would the resulting frequency distribution become?
(b) Is there a unique class interval?

127. The following are the times to failure of 38 light bulbs, given in hours of operation.

150	389	345	310	20	310	175	376	334	340
332	331	327	344	328	341	325	2	311	320
256	315	55	345	111	349	245	367	81	327
355	309	375	316	336	278	396	287		

(a) Dropping the rule that class intervals must be equal, construct a frequency distribution from these data.
(b) Can you find the class mark of every class?

128. (a) Construct a histogram of the reaction times of pilots from the data in Exercise 116.
(b) What can be said about the shape of this histogram?

129. (a) Construct a histogram of the drying times of polyurethane from the data in Exercise 115.
(b) What can be said about the shape of this histogram?

130. Use the data of Exercise 128 to illustrate that class marks are given by the midpoint between successive class boundaries as well as the midpoint between successive class limits.

131. Using the data of Exercise 129, show that the class marks also are given by the midpoint between successive class boundaries.

132. Construct a histogram using the solder-joint data in Exercise 110.

133. (a) Using only the first two rows of the data for the response times given in Section 8, construct a histogram.
(b) How would you describe the shape of the histogram?

134. (a) Combining the data for both lathes in Exercise 112, construct a histogram.
(b) How would you describe the shape of the histogram?

135. Use MINITAB or some other computer software to construct a histogram of the coking-time data given in Exercise 113.

136. Use MINITAB or some other computer software to construct a histogram of the drying-time data in Exercise 115.

137. A plot of the points (x, f), where x represents the class mark of a given class in a frequency distribution and f represents its frequency, is called a **frequency polygon**. Construct a frequency polygon using the data in Exercise 116.

138. Construct a frequency polygon from the data in Exercise 115.

139. A plot of the cumulative frequency (see Exercise 123) on the y-axis and the corresponding upper class boundary on the x-axis is called an **ogive**.

(a) Construct an ogive for the data of Exercise 115.

(b) Using the same set of axes, relabel the y-axis so that the same graph also shows the ogive of the percentage distribution of drying times.

140. (a) Construct an ogive for the reaction times given in Exercise 116.

(b) Construct an ogive representing the cumulative percentage distribution.

References

More advanced or more detailed treatments of the material in this chapter may be found in

BRUNK, H. D., *An Introduction to Mathematical Statistics*, 3rd ed. Lexington, Mass.: Xerox College Publishing, 1975,

DEGROOT, M. H., *Probability and Statistics*, 2nd ed. Reading, Mass.: Addison-Wesley Publishing Company, Inc., 1986,

FRASER, D. A. S., *Probability and Statistics: Theory and Applications*. North Scituate, Mass.: Duxbury Press, 1976,

HOGG, R. V., and CRAIG, A. T., *Introduction to Mathematical Statistics*, 4th ed. New York: Macmillan Publishing Co., Inc., 1978,

KENDALL, M. G., and STUART, A., *The Advanced Theory of Statistics*, Vol. 1, 4th ed. New York: Macmillan Publishing Co., Inc., 1977,

KHAZANIE, R., *Basic Probability Theory and Applications*. Pacific Palisades, Calif.: Goodyear Publishing Company, Inc., 1976.

Answers to Odd-Numbered Exercises

1 (a) no, because $f(4)$ is negative; **(b)** yes; **(c)** no, because the sum of the probabilities is less than 1.

5 $0 < k < 1$.

9 (a) no, because $F(4)$ exceeds 1; **(b)** no, because $F(2)$ is less than $F(1)$; **(c)** yes.

11 (a) $\frac{1}{2}$; **(b)** $\frac{1}{6}$; **(c)** $f(1) = \frac{1}{3}$; $f(4) = \frac{1}{6}$; $f(6) = \frac{2}{3}$; $f(10) = \frac{1}{6}$; $f(x) = 0$ elsewhere.

13 (a) $\frac{3}{4}$; **(b)** $\frac{1}{4}$; **(c)** $\frac{1}{2}$; **(d)** $\frac{3}{4}$; **(e)** $\frac{1}{2}$; **(f)** $\frac{1}{4}$.

17 (b) $\frac{2}{5}$.

19 (c) 0.124.

21 $F(y) = \begin{cases} 0 & \text{for } y \le 2 \\ \frac{1}{16}(y^2 + 2y - 8) & \text{for } 2 < y < 4 \\ 1 & \text{for } y \ge 4 \end{cases}$

The probabilities are 0.454 and 0.1519.

23 $F(x) = \begin{cases} 0 & \text{for } x \le 0 \\ \frac{1}{2}\sqrt{x} & \text{for } 0 < x < 4; \\ 1 & \text{for } x \ge 4 \end{cases}$ **(b)** $\frac{1}{4}$ and $\frac{1}{2}$.

25 $F(z) = \begin{cases} 0 & \text{for } z \le 0 \\ 1 - e^{-z^2} & \text{for } z > 0 \end{cases}$

27 $G(x) = \begin{cases} 0 & \text{for } x \le 0 \\ 3x^2 - 2x^3 & \text{for } 0 < x < 1 \\ 1 & \text{for } x \ge 1 \end{cases}$

The probabilities are $\frac{5}{32}$ and $\frac{1}{2}$.

29 $F(x) = \begin{cases} 0 & \text{for } x \le 0 \\ \frac{x^2}{2} & \text{for } 0 < x < 1 \\ 2x - \frac{x^2}{2} - 1 & \text{for } \le x < 2 \\ 1 & \text{for } x \ge 2 \end{cases}$

31 $F(x) = \begin{cases} 0 & \text{for } x \le 0 \\ \frac{x^2}{4} & \text{for } 0 < x \le 1 \\ \frac{1}{4}(2x - 1) & \text{for } 1 < x \le 2 \\ \frac{1}{4}(6x - x^2 - 5) & \text{for } 2 < x < 3 \\ 1 & \text{for } x \ge 3 \end{cases}$

33 $f(x) = \frac{1}{2}$ for $-1 < x < 1$ and $f(x) = 0$ elsewhere.

35 $f(y) = \dfrac{18}{y^3}$ for $y > 0$ and $f(y) = 0$ elsewhere; the two probabilities are $\frac{16}{25}$ and $\frac{9}{64}$.

37 The three probabilities are $1 - 3e^{-2}, 2e^{-1} - 4e^{-3}$, and $5e^{-5}$.

39 (a) $F(x) = 0$; **(b)** $F(x) = \frac{1}{2}x$; **(c)** $F(x) = \frac{1}{2}(x + 1)$; **(d)** $F(x) = 0$.

41 The probabilities are $\frac{1}{4}$, $\frac{1}{4}$, $\frac{3}{8}$, and $\frac{1}{2}$.

43 (a) $\frac{1}{4}$; **(b)** 0; **(c)** $\frac{7}{24}$; **(d)** $\frac{119}{120}$.

45 (a) $\frac{29}{89}$; **(b)** $\frac{5}{89}$; **(c)** $\frac{55}{89}$.

47

y \ x	0	1	2	3
0	0	$\frac{1}{30}$	$\frac{1}{10}$	$\frac{1}{5}$
1	$\frac{1}{30}$	$\frac{2}{15}$	$\frac{3}{10}$	$\frac{8}{15}$
2	$\frac{1}{10}$	$\frac{3}{10}$	$\frac{3}{5}$	1

49 $k = 2$.

51 **(a)** $\frac{1}{2}$; **(b)** $\frac{5}{9}$; **(c)** $\frac{1}{3}$.

53 $1 - \frac{1}{2}\ln 2 = 0.6354$.

55 $(e^{-1} - e^{-4})^2$.

57 $(e^{-2} - e^{-3})^2$.

63 **(a)** $\frac{1}{18}$; **(b)** $\frac{7}{27}$.

65 $k = 144$.

71 **(a)** $m(x, y) = \frac{xy}{36}$ for $x = 1, 2, 3$ and $y = 1, 2, 3$; **(b)** $n(x, z) = \frac{xz}{18}$ for $x = 1, 2, 3$ and $z = 1, 2$; **(c)** $g(x) = \frac{x}{6}$ for $x = 1, 2, 3$; **(d)** $\phi(z|1, 2) = \frac{z}{3}$ for $z = 1, 2$; **(e)** $\psi(y, z|3) = \frac{yz}{18}$ for $y = 1, 2, 3$ and $z = 1, 2$.

73 **(a)** Independent; **(b)** not independent.

75 **(a)** $h(y) = \frac{1}{4}(1 + y)$ for $0 < y < 2$ and $h(y) = 0$ elsewhere; **(b)** $f(x|1) = \frac{1}{2}(2x + 1)$ for $0 < x < 1$ and $f(x|1) = 0$ elsewhere.

77 **(a)** $g(x) = -\ln x$ for $0 < x < 1$ and $g(x) = 0$ elsewhere; **(b)** $h(y) = 1$ for $0 < y < 1$ and $h(y) = 0$ elsewhere. The two random variables are not independent.

79 $G(x) = 1 - e^{-x^2}$ for $x > 0$ and $G(x) = 0$ elsewhere.

83

Y	−4	−2	0	2	4
$P(Y)$	$\frac{1}{16}$	$\frac{4}{16}$	$\frac{6}{16}$	$\frac{4}{16}$	$\frac{1}{16}$

85 **(a)**

X	0	1	2	3
$P(X)$	$\frac{1}{27}$	$\frac{6}{27}$	$\frac{12}{27}$	$\frac{8}{27}$

(b) $\frac{19}{27}$.

87 $F(V) = \begin{cases} 0 & \text{for } V < 0 \\ 0.40 & \text{for } 0 \le V < 1 \\ 0.70 & \text{for } 1 \le V < 2 \\ 0.90 & \text{for } 2 \le V < 3 \\ 1 & \text{for } V \ge 3 \end{cases}$.

89 Yes; $\sum_{x=2}^{12} f(x) = 1$.

91 **(a)** 0.23; **(b)** 0.464; **(c)** 0.53.

93 **(a)** 0.4512; **(b)** 0.1054; **(c)** 0.2019.

95 **(a)** $\frac{1}{4}$; **(b)** $\frac{39}{64}$; **(c)** $\frac{1}{16}$.

101 **(a)** 0.3038; **(b)** $\frac{1}{221}$.

103 **(a)** $g(0) = \frac{5}{14}$, $g(1) = \frac{15}{28}$, $g(2) = \frac{3}{28}$; **(b)** $\phi(0|0) = \frac{3}{10}$, $\phi(1|0) = \frac{6}{10}$, $\phi(2|0) = \frac{1}{10}$.

105 **(a)** $f(0, 0) = \frac{188}{221}$, $f(0, 1) = \frac{16}{221}$, $f(1, 0) = \frac{16}{221}$, $f(1, 1) = \frac{1}{221}$. **(b)** $g(0) = \frac{204}{221}$, $g(1) = \frac{17}{221}$; **(c)** $\phi(0|0) = \frac{16}{17}$, $\phi(1, 1) = \frac{1}{17}$.

107 **(a)** $g(x) = \frac{20 - x}{50}$ for $10 < x < 20$ and $g(x) = 0$ elsewhere; **(b)** $\phi(y|12) = \frac{1}{6}$ for $6 < y < 12$ and $\phi(y|12) = 0$ elsewhere; **(c)** $\frac{2}{3}$.

109 **(a)** $f(x_1, x_2, x_3) = \frac{(20{,}000)^3}{(x_1 + 100)^3(x_2 + 100)^3(x_3 + 100)^3}$ for $x_1 > 0$, $x_2 > 0$, $x_3 > 0$ and $f(x_1, x_2, x_3) = 0$ elsewhere; **(b)** $\frac{1}{16}$.

111 Station 107 data show less variability than station 105 data.

Mathematical Expectation

1 Introduction

Originally, the concept of a **mathematical expectation** arose in connection with games of chance, and in its simplest form it is the product of the amount a player stands to win and the probability that he or she will win. For instance, if we hold one of 10,000 tickets in a raffle for which the grand prize is a trip worth \$4,800, our mathematical expectation is $4{,}800 \cdot \frac{1}{10{,}000} = \$0.48$. This amount will have to be interpreted in the sense of an average—altogether the 10,000 tickets pay \$4,800, or on the average $\frac{\$4{,}800}{10{,}000} = \$0.48$ per ticket.

If there is also a second prize worth \$1,200 and a third prize worth \$400, we can argue that altogether the 10,000 tickets pay $\$4{,}800 + \$1{,}200 + \$400 = \$6{,}400$, or on the average $\frac{\$6{,}400}{10{,}000} = \$0.64$ per ticket. Looking at this in a different way, we could argue that if the raffle is repeated many times, we would lose 99.97 percent of the time (or with probability 0.9997) and win each of the prizes 0.01 percent of the time (or with probability 0.0001). On the average we would thus win

$$0(0.9997) + 4{,}800(0.0001) + 1{,}200(0.0001) + 400(0.0001) = \$0.64$$

which is the sum of the products obtained by multiplying each amount by the corresponding probability.

2 The Expected Value of a Random Variable

In the illustration of the preceding section, the amount we won was a random variable, and the mathematical expectation of this random variable was the sum of the products obtained by multiplying each value of the random variable by the corresponding probability. Referring to the mathematical expectation of a random variable simply as its **expected value**, and extending the definition to the continuous case by replacing the operation of summation by integration, we thus have the following definition.

DEFINITION 1. EXPECTED VALUE. *If X is a discrete random variable and f(x) is the value of its probability distribution at x, the* ***expected value of X*** *is*

$$E(X) = \sum_{x} x \cdot f(x)$$

Correspondingly, if X is a continuous random variable and f(x) is the value of its probability density at x, the ***expected value of X*** *is*

$$E(X) = \int_{-\infty}^{\infty} x \cdot f(x)dx$$

In this definition it is assumed, of course, that the sum or the integral exists; otherwise, the mathematical expectation is undefined.

EXAMPLE 1

A lot of 12 television sets includes 2 with white cords. If 3 of the sets are chosen at random for shipment to a hotel, how many sets with white cords can the shipper expect to send to the hotel?

Solution

Since x of the 2 sets with white cords and $3-x$ of the 10 other sets can be chosen in $\binom{2}{x}\binom{10}{3-x}$ ways, 3 of the 12 sets can be chosen in $\binom{12}{3}$ ways, and these $\binom{12}{3}$ possibilities are presumably equiprobable, we find that the probability distribution of X, the number of sets with white cords shipped to the hotel, is given by

$$f(x) = \frac{\binom{2}{x}\binom{10}{3-x}}{\binom{12}{3}} \quad \text{for } x = 0, 1, 2$$

or, in tabular form,

x	0	1	2
$f(x)$	$\frac{6}{11}$	$\frac{9}{22}$	$\frac{1}{22}$

Now,

$$E(X) = 0 \cdot \frac{6}{11} + 1 \cdot \frac{9}{22} + 2 \cdot \frac{1}{22} = \frac{1}{2}$$

and since half a set cannot possibly be shipped, it should be clear that the term "expect" is not used in its colloquial sense. Indeed, it should be interpreted as an average pertaining to repeated shipments made under the given conditions.

EXAMPLE 2

Certain coded measurements of the pitch diameter of threads of a fitting have the probability density

$$f(x) = \begin{cases} \dfrac{4}{\pi(1+x^2)} & \text{for } 0 < x < 1 \\ 0 & \text{elsewhere} \end{cases}$$

Find the expected value of this random variable.

Solution

Using Definition 1, we have

$$\begin{aligned} E(X) &= \int_0^1 x \cdot \frac{4}{\pi(1+x^2)}\,dx \\ &= \frac{4}{\pi}\int_0^1 \frac{x}{1+x^2}\,dx \\ &= \frac{\ln 4}{\pi} = 0.4413 \end{aligned}$$

There are many problems in which we are interested not only in the expected value of a random variable X, but also in the expected values of random variables related to X. Thus, we might be interested in the random variable Y, whose values are related to those of X by means of the equation $y = g(x)$; to simplify our notation, we denote this random variable by $g(X)$. For instance, $g(X)$ might be X^3 so that when X takes on the value 2, $g(X)$ takes on the value $2^3 = 8$. If we want to find the expected value of such a random variable $g(X)$, we could first determine its probability distribution or density and then use Definition 1, but generally it is easier and more straightforward to use the following theorem.

THEOREM 1. If X is a discrete random variable and $f(x)$ is the value of its probability distribution at x, the expected value of $g(X)$ is given by

$$E[g(X)] = \sum_x g(x) \cdot f(x)$$

Correspondingly, if X is a continuous random variable and $f(x)$ is the value of its probability density at x, the expected value of $g(X)$ is given by

$$E[g(X)] = \int_{-\infty}^{\infty} g(x) \cdot f(x)\,dx$$

Proof Since a more general proof is beyond the scope of this chapter, we shall prove this theorem here only for the case where X is discrete and has a finite range. Since $y = g(x)$ does not necessarily define a one-to-one correspondence, suppose that $g(x)$ takes on the value g_i when x takes on the values $x_{i1}, x_{i2}, \ldots, x_{in_i}$. Then, the probability that $g(X)$ will take on the value g_i is

$$P[g(X) = g_i] = \sum_{j=1}^{n_i} f(x_{ij})$$

and if $g(x)$ takes on the values $g_1, g_2, \ldots, g_m$, it follows that

$$\begin{aligned} E[g(X)] &= \sum_{i=1}^{m} g_i \cdot P[g(X) = g_i] \\ &= \sum_{i=1}^{m} g_i \cdot \sum_{j=1}^{n_i} f(x_{ij}) \\ &= \sum_{i=1}^{m} \sum_{j=1}^{n_i} g_i \cdot f(x_{ij}) \\ &= \sum_{x} g(x) \cdot f(x) \end{aligned}$$

where the summation extends over all values of X.

EXAMPLE 3

If X is the number of points rolled with a balanced die, find the expected value of $g(X) = 2X^2 + 1$.

Solution

Since each possible outcome has the probability $\frac{1}{6}$, we get

$$\begin{aligned} E[g(X)] &= \sum_{x=1}^{6} (2x^2 + 1) \cdot \frac{1}{6} \\ &= (2 \cdot 1^2 + 1) \cdot \frac{1}{6} + \cdots + (2 \cdot 6^2 + 1) \cdot \frac{1}{6} \\ &= \frac{94}{3} \end{aligned}$$

EXAMPLE 4

If X has the probability density

$$f(x) = \begin{cases} e^x & \text{for } x > 0 \\ 0 & \text{elsewhere} \end{cases}$$

find the expected value of $g(X) = e^{3X/4}$.

Solution
According to Theorem 1, we have

$$\begin{aligned}
E[e^{3X/4}] &= \int_0^\infty e^{3x/4} \cdot e^{-x}\, dx \\
&= \int_0^\infty e^{-x/4}\, dx \\
&= 4
\end{aligned}$$

The determination of mathematical expectations can often be simplified by using the following theorems, which enable us to calculate expected values from other known or easily computed expectations. Since the steps are essentially the same, some proofs will be given for either the discrete case or the continuous case; others are left for the reader as exercises.

THEOREM 2. If a and b are constants, then

$$E(aX + b) = aE(X) + b$$

Proof Using Theorem 1 with $g(X) = aX + b$, we get

$$\begin{aligned}
E(aX + b) &= \int_{-\infty}^{\infty} (ax + b) \cdot f(x)\, dx \\
&= a \int_{-\infty}^{\infty} x \cdot f(x)\, dx + b \int_{-\infty}^{\infty} f(x)\, dx \\
&= aE(X) + b
\end{aligned}$$

If we set $b = 0$ or $a = 0$, we can state the following corollaries to Theorem 2.

COROLLARY 1. If a is a constant, then

$$E(aX) = aE(X)$$

COROLLARY 2. If b is a constant, then

$$E(b) = b$$

Observe that if we write $E(b)$, the constant b may be looked upon as a random variable that always takes on the value b.

THEOREM 3. If $c_1, c_2, \ldots,$ and c_n are constants, then

$$E\left[\sum_{i=1}^{n} c_i g_i(X)\right] = \sum_{i=1}^{n} c_i E[g_i(X)]$$

Proof According to Theorem 1 with $g(X) = \sum_{i=1}^{n} c_i g_i(X)$, we get

$$
\begin{aligned}
E\left[\sum_{i=1}^{n} c_i g_i(X)\right] &= \sum_{x}\left[\sum_{i=1}^{n} c_i g_i(x)\right] f(x) \\
&= \sum_{i=1}^{n}\sum_{x} c_i g_i(x) f(x) \\
&= \sum_{i=1}^{n} c_i \sum_{x} g_i(x) f(x) \\
&= \sum_{i=1}^{n} c_i E[g_i(X)]
\end{aligned}
$$

EXAMPLE 5

Making use of the fact that

$$E(X^2) = (1^2 + 2^2 + 3^2 + 4^2 + 5^2 + 6^2) \cdot \frac{1}{6} = \frac{91}{6}$$

for the random variable of Example 3, rework that example.

Solution

$$E(2X^2 + 1) = 2E(X^2) + 1 = 2 \cdot \frac{91}{6} + 1 = \frac{94}{3}$$

EXAMPLE 6

If the probability density of X is given by

$$f(x) = \begin{cases} 2(1-x) & \text{for } 0 < x < 1 \\ 0 & \text{elsewhere} \end{cases}$$

(a) show that

$$E(X^r) = \frac{2}{(r+1)(r+2)}$$

(b) and use this result to evaluate

$$E[(2X+1)^2]$$

Solution

(a)

$$
\begin{aligned}
E(X^r) &= \int_0^1 x^r \cdot 2(1-x)\, dx = 2\int_0^1 (x^r - x^{r+1})\, dx \\
&= 2\left(\frac{1}{r+1} - \frac{1}{r+2}\right) = \frac{2}{(r+1)(r+2)}
\end{aligned}
$$

(b) Since $E[(2X+1)^2] = 4E(X^2)+4E(X)+1$ and substitution of $r = 1$ and $r = 2$ into the preceding formula yields $E(X) = \frac{2}{2 \cdot 3} = \frac{1}{3}$ and $E(X^2) = \frac{2}{3 \cdot 4} = \frac{1}{6}$, we get

$$E[(2X+1)^2] = 4 \cdot \frac{1}{6} + 4 \cdot \frac{1}{3} + 1 = 3$$

EXAMPLE 7

Show that

$$E[(aX+b)^n] = \sum_{i=0}^{n} \binom{n}{i} a^{n-i} b^i E(X^{n-i})$$

Solution

Since $(ax+b)^n = \sum_{i=0}^{n} \binom{n}{i} (ax)^{n-i} b^i$, it follows that

$$E[(aX+b)^n] = E\left[\sum_{i=0}^{n} \binom{n}{i} a^{n-i} b^i X^{n-i}\right]$$

$$= \sum_{i=0}^{n} \binom{n}{i} a^{n-i} b^i E(X^{n-i})$$

The concept of a mathematical expectation can easily be extended to situations involving more than one random variable. For instance, if Z is the random variable whose values are related to those of the two random variables X and Y by means of the equation $z = g(x, y)$, we can state the following theorem.

THEOREM 4. If X and Y are discrete random variables and $f(x, y)$ is the value of their joint probability distribution at (x, y), the expected value of $g(X, Y)$ is

$$E[g(X, Y)] = \sum_{x} \sum_{y} g(x, y) \cdot f(x, y)$$

Correspondingly, if X and Y are continuous random variables and $f(x, y)$ is the value of their joint probability density at (x, y), the expected value of $g(X, Y)$ is

$$E[g(X, Y)] = \int_{-\infty}^{\infty} \int_{-\infty}^{\infty} g(x, y) f(x, y) \, dx \, dy$$

Generalization of this theorem to functions of any finite number of random variables is straightforward.

EXAMPLE 8

Find the expected value of $g(X, Y) = X + Y$.

Solution

$$E(X+Y) = \sum_{x=0}^{2}\sum_{y=0}^{2}(x+y)\cdot f(x,y)$$

$$= (0+0)\cdot\frac{1}{6} + (0+1)\cdot\frac{2}{9} + (0+2)\cdot\frac{1}{36} + (1+0)\cdot\frac{1}{3}$$

$$+ (1+1)\cdot\frac{1}{6} + (2+0)\cdot\frac{1}{12}$$

$$= \frac{10}{9}$$

EXAMPLE 9

If the joint probability density of X and Y is given by

$$f(x,y) = \begin{cases} \frac{2}{7}(x+2y) & \text{for } 0 < x < 1, 1 < y < 2 \\ 0 & \text{elsewhere} \end{cases}$$

find the expected value of $g(X, Y) = X/Y^3$.

Solution

$$E(X/Y^3) = \int_1^2\int_0^1 \frac{2x(x+2y)}{7y^3}\,dx\,dy$$

$$= \frac{2}{7}\int_1^2\left(\frac{1}{3y^3} + \frac{1}{y^2}\right)dy$$

$$= \frac{15}{84}$$

The following is another theorem that finds useful applications in subsequent work. It is a generalization of Theorem 3, and its proof parallels the proof of that theorem.

THEOREM 5. If $c_1, c_2, \ldots,$ and c_n are constants, then

$$E\left[\sum_{i=1}^{n} c_i g_i(X_1, X_2, \ldots, X_k)\right] = \sum_{i=1}^{n} c_i E[g_i(X_1, X_2, \ldots, X_k)]$$

Exercises

1. To illustrate the proof of Theorem 1, consider the random variable X, which takes on the values -2, -1, 0, 1, 2, and 3 with probabilities $f(-2)$, $f(-1)$, $f(0)$, $f(1)$, $f(2)$, and $f(3)$. If $g(X) = X^2$, find
(a) g_1, g_2, g_3, and g_4, the four possible values of $g(x)$;
(b) the probabilities $P[g(X) = g_i]$ for $i = 1, 2, 3, 4$;
(c) $E[g(X)] = \sum_{i=1}^{4} g_i \cdot P[g(X) = g_i]$, and show that it equals

$$\sum_x g(x) \cdot f(x)$$

2. Prove Theorem 2 for discrete random variables.

3. Prove Theorem 3 for continuous random variables.

4. Prove Theorem 5 for discrete random variables.

5. Given two continuous random variables X and Y, use Theorem 4 to express $E(X)$ in terms of
(a) the joint density of X and Y;
(b) the marginal density of X.

6. Find the expected value of the discrete random variable X having the probability distribution

$$f(x) = \frac{|x-2|}{7} \quad \text{for } x = -1, 0, 1, 3$$

7. Find the expected value of the random variable Y whose probability density is given by

$$f(y) = \begin{cases} \frac{1}{8}(y+1) & \text{for } 2 < y < 4 \\ 0 & \text{elsewhere} \end{cases}$$

8. Find the expected value of the random variable X whose probability density is given by

$$f(x) = \begin{cases} x & \text{for } 0 < x < 1 \\ 2-x & \text{for } 1 \leqq x < 2 \\ 0 & \text{elsewhere} \end{cases}$$

9. (a) If X takes on the values 0, 1, 2, and 3 with probabilities $\frac{1}{125}$, $\frac{12}{125}$, $\frac{48}{125}$, and $\frac{64}{125}$, find $E(X)$ and $E(X^2)$.
(b) Use the results of part (a) to determine the value of $E[(3X+2)^2]$.

10. (a) If the probability density of X is given by

$$f(x) = \begin{cases} \frac{1}{x(\ln 3)} & \text{for } 1 < x < 3 \\ 0 & \text{elsewhere} \end{cases}$$

find $E(X)$, $E(X^2)$, and $E(X^3)$.
(b) Use the results of part (a) to determine $E(X^3 + 2X^2 - 3X + 1)$.

11. If the probability density of X is given by

$$f(x) = \begin{cases} \frac{x}{2} & \text{for } 0 < x \leqq 1 \\ \frac{1}{2} & \text{for } 1 < x \leqq 2 \\ \frac{3-x}{2} & \text{for } 2 < x < 3 \\ 0 & \text{elsewhere} \end{cases}$$

find the expected value of $g(X) = X^2 - 5X + 3$.

12. This has been intentionally omitted for this edition.

13. This has been intentionally omitted for this edition.

14. This has been intentionally omitted for this edition.

15. This has been intentionally omitted for this edition.

16. If the probability distribution of X is given by

$$f(x) = \left(\frac{1}{2}\right)^x \quad \text{for } x = 1, 2, 3, \ldots$$

show that $E(2^X)$ does not exist. This is the famous **Petersburg paradox**, according to which a player's expectation is infinite (does not exist) if he or she is to receive 2^x dollars when, in a series of flips of a balanced coin, the first head appears on the xth flip.

3 Moments

In statistics, the mathematical expectations defined here and in Definition 4, called the **moments** of the distribution of a random variable or simply the **moments** of a random variable, are of special importance.

DEFINITION 2. MOMENTS ABOUT THE ORIGIN. *The **r*th *moment about the origin** of a random variable X, denoted by μ'_r, is the expected value of X^r; symbolically*

$$\mu'_r = E(X^r) = \sum_x x^r \cdot f(x)$$

for $r = 0, 1, 2, \ldots$ *when X is discrete, and*

$$\mu'_r = E(X^r) = \int_{-\infty}^{\infty} x^r \cdot f(x)dx$$

when X is continuous.

It is of interest to note that the term "moment" comes from the field of physics: If the quantities $f(x)$ in the discrete case were point masses acting perpendicularly to the x-axis at distances x from the origin, μ'_1 would be the x-coordinate of the center of gravity, that is, the first moment divided by $\sum f(x) = 1$, and μ'_2 would be the moment of inertia. This also explains why the moments μ'_r are called moments about the origin: In the analogy to physics, the length of the lever arm is in each case the distance from the origin. The analogy applies also in the continuous case, where μ'_1 and μ'_2 might be the x-coordinate of the center of gravity and the moment of inertia of a rod of variable density.

When $r = 0$, we have $\mu'_0 = E(X^0) = E(1) = 1$ by Corollary 2 of Theorem 2. When $r = 1$, we have $\mu'_1 = E(X)$, which is just the expected value of the random variable X, and in view of its importance in statistics we give it a special symbol and a special name.

DEFINITION 3. MEAN OF A DISTRIBUTION. μ'_1 *is called the **mean** of the distribution of X, or simply the **mean of X**, and it is denoted simply by* μ.

The special moments we shall define next are of importance in statistics because they serve to describe the shape of the distribution of a random variable, that is, the shape of the graph of its probability distribution or probability density.

DEFINITION 4. MOMENTS ABOUT THE MEAN. *The **r*th *moment about the mean** of a random variable X, denoted by μ_r, is the expected value of $(X - \mu)^r$, symbolically*

$$\mu_r = E\left[(X-\mu)^r\right] = \sum_x (x-\mu)^r \cdot f(x)$$

for $r = 0, 1, 2, \ldots,$ *when X is discrete, and*

$$\mu_r = E\left[(X-\mu)^r\right] = \int_{-\infty}^{\infty} (x-u)^r \cdot f(x)dx$$

when X is continuous.

Note that $\mu_0 = 1$ and $\mu_1 = 0$ for any random variable for which μ exists (see Exercise 17).

The second moment about the mean is of special importance in statistics because it is indicative of the spread or dispersion of the distribution of a random variable; thus, it is given a special symbol and a special name.

> **DEFINITION 5. VARIANCE.** *μ_2 is called the **variance** of the distribution of X, or simply the **variance of X**, and it is denoted by σ^2, $\sigma_X{}^2$, var(X), or V(X). The positive square root of the variance, σ, is called the **standard deviation of X**.*

Figure 1 shows how the variance reflects the spread or dispersion of the distribution of a random variable. Here we show the histograms of the probability distributions of four random variables with the same mean $\mu = 5$ but variances equaling 5.26, 3.18, 1.66, and 0.88. As can be seen, a small value of σ^2 suggests that we are likely to get a value close to the mean, and a large value of σ^2 suggests that there is a greater probability of getting a value that is not close to the mean. This will be discussed further in Section 4. A brief discussion of how μ_3, the third moment about the mean, describes the **symmetry** or **skewness** (lack of symmetry) of a distribution is given in Exercise 26.

In many instances, moments about the mean are obtained by first calculating moments about the origin and then expressing the μ_r in terms of the μ'_r. To serve this purpose, the reader will be asked to verify a general formula in Exercise 25. Here, let us merely derive the following computing formula for σ^2.

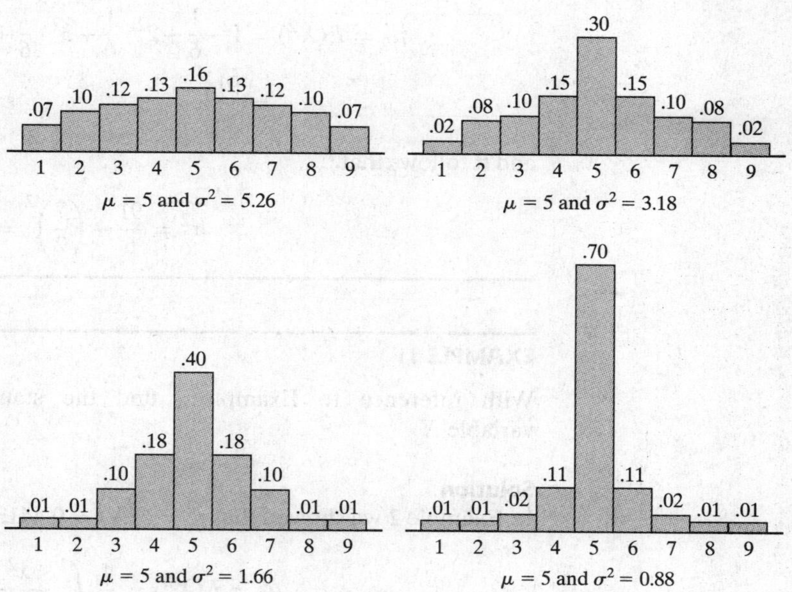

Figure 1. Distributions with different dispersions.

THEOREM 6.

$$\sigma^2 = \mu_2' - \mu^2$$

Proof

$$\begin{aligned}\sigma^2 &= E[(X-\mu)^2] \\ &= E(X^2 - 2\mu X + \mu^2) \\ &= E(X^2) - 2\mu E(X) + E(\mu^2) \\ &= E(X^2) - 2\mu \cdot \mu + \mu^2 \\ &= \mu_2' - \mu^2\end{aligned}$$

EXAMPLE 10

Use Theorem 6 to calculate the variance of X, representing the number of points rolled with a balanced die.

Solution

First we compute

$$\begin{aligned}\mu = E(X) &= 1 \cdot \frac{1}{6} + 2 \cdot \frac{1}{6} + 3 \cdot \frac{1}{6} + 4 \cdot \frac{1}{6} + 5 \cdot \frac{1}{6} + 6 \cdot \frac{1}{6} \\ &= \frac{7}{2}\end{aligned}$$

Now,

$$\begin{aligned}\mu_2' = E(X^2) &= 1^2 \cdot \frac{1}{6} + 2^2 \cdot \frac{1}{6} + 3^2 \cdot \frac{1}{6} + 4^2 \cdot \frac{1}{6} + 5^2 \cdot \frac{1}{6} + 6^2 \cdot \frac{1}{6} \\ &= \frac{91}{6}\end{aligned}$$

and it follows that

$$\sigma^2 = \frac{91}{6} - \left(\frac{7}{2}\right)^2 = \frac{35}{12}$$

EXAMPLE 11

With reference to Example 2, find the standard deviation of the random variable X.

Solution

In Example 2 we showed that $\mu = E(X) = 0.4413$. Now

$$\begin{aligned}\mu_2' = E(X^2) &= \frac{4}{\pi} \int_0^1 \frac{x^2}{1+x^2}\, dx \\ &= \frac{4}{\pi} \int_0^1 \left(1 - \frac{1}{1+x^2}\right) dx\end{aligned}$$

$$= \frac{4}{\pi} - 1$$
$$= 0.2732$$

and it follows that

$$\sigma^2 = 0.2732 - (0.4413)^2 = 0.0785$$

and $\sigma = \sqrt{0.0785} = 0.2802$.

The following is another theorem that is of importance in work connected with standard deviations or variances.

THEOREM 7. If X has the variance σ^2, then

$$\text{var}(aX + b) = a^2\sigma^2$$

The proof of this theorem will be left to the reader, but let us point out the following corollaries: For $a = 1$, we find that the addition of a constant to the values of a random variable, resulting in a shift of all the values of X to the left or to the right, in no way affects the spread of its distribution; for $b = 0$, we find that if the values of a random variable are multiplied by a constant, the variance is multiplied by the square of that constant, resulting in a corresponding change in the spread of the distribution.

4 Chebyshev's Theorem

To demonstrate how σ or σ^2 is indicative of the spread or dispersion of the distribution of a random variable, let us now prove the following theorem, called **Chebyshev's theorem** after the nineteenth-century Russian mathematician P. L. Chebyshev. We shall prove it here only for the continuous case, leaving the discrete case as an exercise.

THEOREM 8. (*Chebyshev's Theorem*) If μ and σ are the mean and the standard deviation of a random variable X, then for any positive constant k the probability is *at least* $1 - \frac{1}{k^2}$ that X will take on a value within k standard deviations of the mean; symbolically,

$$P(|x - \mu| < k\sigma) \geq 1 - \frac{1}{k^2}, \quad \sigma \neq 0$$

Proof According to Definitions 4 and 5, we write

$$\sigma^2 = E[(X - \mu)^2] = \int_{-\infty}^{\infty} (x - \mu)^2 \cdot f(x)\, dx$$

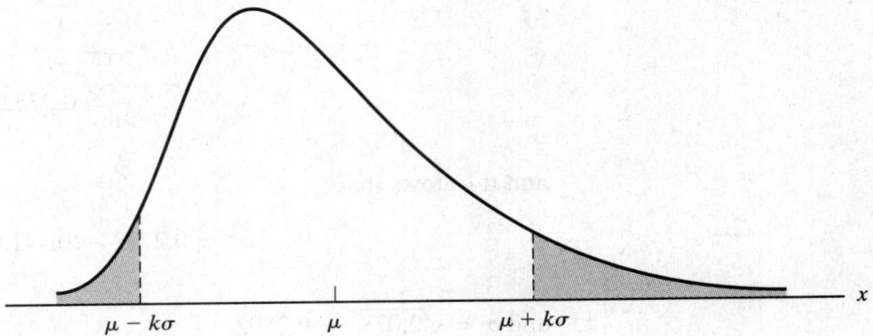

Figure 2. Diagram for proof of Chebyshev's theorem.

Then, dividing the integral into three parts as shown in Figure 2, we get

$$\sigma^2 = \int_{-\infty}^{\mu-k\sigma} (x-\mu)^2 \cdot f(x)\,dx + \int_{\mu-k\sigma}^{\mu+k\sigma} (x-\mu)^2 \cdot f(x)\,dx + \int_{\mu+k\sigma}^{\infty} (x-\mu)^2 \cdot f(x)\,dx$$

Since the integrand $(x-\mu)^2 \cdot f(x)$ is nonnegative, we can form the inequality

$$\sigma^2 \geqq \int_{-\infty}^{\mu-k\sigma} (x-\mu)^2 \cdot f(x)\,dx + \int_{\mu+k\sigma}^{\infty} (x-\mu)^2 \cdot f(x)\,dx$$

by deleting the second integral. Therefore, since $(x-\mu)^2 \geqq k^2\sigma^2$ for $x \leqq \mu - k\sigma$ or $x \geqq \mu + k\sigma$ it follows that

$$\sigma^2 \geqq \int_{-\infty}^{\mu-k\sigma} k^2\sigma^2 \cdot f(x)\,dx + \int_{\mu+k\sigma}^{\infty} k^2\sigma^2 \cdot f(x)\,dx$$

and hence that

$$\frac{1}{k^2} \geqq \int_{-\infty}^{\mu-k\sigma} f(x)\,dx + \int_{\mu+k\sigma}^{\infty} f(x)\,dx$$

provided $\sigma^2 \neq 0$. Since the sum of the two integrals on the right-hand side is the probability that X will take on a value less than or equal to $\mu - k\sigma$ or greater than or equal to $\mu + k\sigma$, we have thus shown that

$$P(|X-\mu| \geqq k\sigma) \leqq \frac{1}{k^2}$$

and it follows that

$$P(|X-\mu| < k\sigma) \geqq 1 - \frac{1}{k^2}$$

For instance, the probability is at least $1 - \frac{1}{2^2} = \frac{3}{4}$ that a random variable X will take on a value within two standard deviations of the mean, the probability is at least $1 - \frac{1}{3^2} = \frac{8}{9}$ that it will take on a value within three standard deviations of the mean, and the probability is at least $1 - \frac{1}{5^2} = \frac{24}{25}$ that it will take on a value within five standard deviations of the mean. It is in this sense that σ controls the spread or dispersion of the distribution of a random variable. Clearly, the probability given by Chebyshev's theorem is only a lower bound; whether the probability that a given random variable will take on a value within k standard deviations of the mean is actually greater than $1 - \frac{1}{k^2}$ and, if so, by how much we cannot say, but Chebyshev's theorem assures us that this probability cannot be less than $1 - \frac{1}{k^2}$. Only when the distribution of a random variable is known can we calculate the exact probability.

EXAMPLE 12

If the probability density of X is given by

$$f(x) = \begin{cases} 630x^4(1-x)^4 & \text{for } 0 < x < 1 \\ 0 & \text{elsewhere} \end{cases}$$

find the probability that it will take on a value within two standard deviations of the mean and compare this probability with the lower bound provided by Chebyshev's theorem.

Solution

Straightforward integration shows that $\mu = \frac{1}{2}$ and $\sigma^2 = \frac{1}{44}$, so that $\sigma = \sqrt{1/44}$ or approximately 0.15. Thus, the probability that X will take on a value within two standard deviations of the mean is the probability that it will take on a value between 0.20 and 0.80, that is,

$$P(0.20 < X < 0.80) = \int_{0.20}^{0.80} 630x^4(1-x)^4\, dx$$
$$= 0.96$$

Observe that the statement "the probability is 0.96" is a much stronger statement than "the probability is at least 0.75," which is provided by Chebyshev's theorem.

---

5 Moment-Generating Functions

Although the moments of most distributions can be determined directly by evaluating the necessary integrals or sums, an alternative procedure sometimes provides considerable simplifications. This technique utilizes **moment-generating functions.**

DEFINITION 6. MOMENT GENERATING FUNCTION. *The **moment generating function** of a random variable X, where it exists, is given by*

$$M_X(t) = E(e^{tX}) = \sum_x e^{tX} \cdot f(x)$$

when X is discrete, and

$$M_X(t) = E(e^{tX}) = \int_{-\infty}^{\infty} e^{tx} \cdot f(x)dx$$

when X is continuous.

The independent variable is t, and we are usually interested in values of t in the neighborhood of 0.

To explain why we refer to this function as a "moment-generating" function, let us substitute for e^{tx} its Maclaurin's series expansion, that is,

$$e^{tx} = 1 + tx + \frac{t^2x^2}{2!} + \frac{t^3x^3}{3!} + \cdots + \frac{t^rx^r}{r!} + \cdots$$

For the discrete case, we thus get

$$\begin{aligned} M_X(t) &= \sum_x \left[1 + tx + \frac{t^2x^2}{2!} + \cdots + \frac{t^rx^r}{r!} + \cdots \right] f(x) \\ &= \sum_x f(x) + t \cdot \sum_x xf(x) + \frac{t^2}{2!} \cdot \sum_x x^2f(x) + \cdots + \frac{t^r}{r!} \cdot \sum_x x^rf(x) + \cdots \\ &= 1 + \mu \cdot t + \mu_2' \cdot \frac{t^2}{2!} + \cdots + \mu_r' \cdot \frac{t^r}{r!} + \cdots \end{aligned}$$

and it can be seen that in the Maclaurin's series of the moment-generating function of X the coefficient of $\frac{t^r}{r!}$ is μ_r', the rth moment about the origin. In the continuous case, the argument is the same.

EXAMPLE 13

Find the moment-generating function of the random variable whose probability density is given by

$$f(x) = \begin{cases} e^{-x} & \text{for } x > 0 \\ 0 & \text{elsewhere} \end{cases}$$

and use it to find an expression for μ_r'.

Solution

By definition

$$\begin{aligned} M_X(t) = E(e^{tX}) &= \int_0^\infty e^{tx} \cdot e^{-x}\, dx \\ &= \int_0^\infty e^{-x(1-t)}\, dx \\ &= \frac{1}{1-t} \quad \text{for } t < 1 \end{aligned}$$

As is well known, when $|t| < 1$ the Maclaurin's series for this moment-generating function is

$$\begin{aligned} M_X(t) &= 1 + t + t^2 + t^3 + \cdots + t^r + \cdots \\ &= 1 + 1! \cdot \frac{t}{1!} + 2! \cdot \frac{t^2}{2!} + 3! \cdot \frac{t^3}{3!} + \cdots + r! \cdot \frac{t^r}{r!} + \cdots \end{aligned}$$

and hence $\mu_r' = r!$ for $r = 0, 1, 2, \ldots$.

The main difficulty in using the Maclaurin's series of a moment-generating function to determine the moments of a random variable is usually *not* that of finding the moment-generating function, but that of expanding it into a Maclaurin's series. If we are interested only in the first few moments of a random variable, say, μ_1' and μ_2', their determination can usually be simplified by using the following theorem.

THEOREM 9.

$$\left.\frac{d^r M_X(t)}{dt^r}\right|_{t=0} = \mu_r'$$

This follows from the fact that if a function is expanded as a power series in t, the coefficient of $\frac{t^r}{r!}$ is the rth derivative of the function with respect to t at $t = 0$.

EXAMPLE 14

Given that X has the probability distribution $f(x) = \frac{1}{8}\binom{3}{x}$ for $x = 0, 1, 2,$ and 3, find the moment-generating function of this random variable and use it to determine μ_1' and μ_2'.

Solution

In accordance with Definition 6,

$$\begin{aligned} M_X(t) = E(e^{tX}) &= \frac{1}{8} \cdot \sum_{x=0}^{3} e^{tx}\binom{3}{x} \\ &= \frac{1}{8}(1 + 3e^t + 3e^{2t} + e^{3t}) \\ &= \frac{1}{8}(1 + e^t)^3 \end{aligned}$$

Then, by Theorem 9,

$$\mu_1' = M_X'(0) = \left.\frac{3}{8}(1 + e^t)^2 e^t\right|_{t=0} = \frac{3}{2}$$

and

$$\mu_2' = M_X''(0) = \left.\frac{3}{4}(1 + e^t)e^{2t} + \frac{3}{8}(1 + e^t)^2 e^t\right|_{t=0} = 3$$

Often the work involved in using moment-generating functions can be simplified by making use of the following theorem.

THEOREM 10. If a and b are constants, then

1. $M_{X+a}(t) = E[e^{(X+a)t}] = e^{at} \cdot M_X(t)$;
2. $M_{bX}(t) = E(e^{bXt}) = M_X(bt)$;
3. $M_{\frac{X+a}{b}}(t) = E[e^{\left(\frac{X+a}{b}\right)t}] = e^{\frac{a}{b}t} \cdot M_X\left(\frac{t}{b}\right)$.

The proof of this theorem is left to the reader in Exercise 39. The first part of the theorem is of special importance when $a = -\mu$, and the third part is of special importance when $a = -\mu$ and $b = \sigma$, in which case

$$M_{\frac{X-\mu}{\sigma}}(t) = e^{-\frac{\mu t}{\sigma}} \cdot M_X\left(\frac{t}{\sigma}\right)$$

Exercises

17. With reference to Definition 4, show that $\mu_0 = 1$ and that $\mu_1 = 0$ for any random variable for which $E(X)$ exists.

18. Find μ, μ_2', and σ^2 for the random variable X that has the probability distribution $f(x) = \frac{1}{2}$ for $x = -2$ and $x = 2$.

19. Find μ, μ_2', and σ^2 for the random variable X that has the probability density

$$f(x) = \begin{cases} \dfrac{x}{2} & \text{for } 0 < x < 2 \\ 0 & \text{elsewhere} \end{cases}$$

20. Find μ_r' and σ^2 for the random variable X that has the probability density

$$f(x) = \begin{cases} \dfrac{1}{\ln 3} \cdot \dfrac{1}{x} & \text{for } 1 < x < 3 \\ 0 & \text{elsewhere} \end{cases}$$

21. Prove Theorem 7.

22. With reference to Exercise 8, find the variance of $g(X) = 2X + 3$.

23. If the random variable X has the mean μ and the standard deviation σ, show that the random variable Z whose values are related to those of X by means of the equation $z = \frac{x-\mu}{\sigma}$ has

$$E(Z) = 0 \quad \text{and} \quad \text{var}(Z) = 1$$

A distribution that has the mean 0 and the variance 1 is said to be in **standard form**, and when we perform the above change of variable, we are said to be **standardizing** the distribution of X.

24. If the probability density of X is given by

$$f(x) = \begin{cases} 2x^{-3} & \text{for } x > 1 \\ 0 & \text{elsewhere} \end{cases}$$

check whether its mean and its variance exist.

25. Show that

$$\mu_r = \mu_r' - \binom{r}{1}\mu_{r-1}' \cdot \mu + \cdots + (-1)^i \binom{r}{i} \mu_{r-i}' \cdot \mu^i + \cdots + (-1)^{r-1}(r-1) \cdot \mu^r$$

for $r = 1, 2, 3, \ldots$, and use this formula to express μ_3 and μ_4 in terms of moments about the origin.

26. The **symmetry** or **skewness** (lack of symmetry) of a distribution is often measured by means of the quantity

$$\alpha_3 = \frac{\mu_3}{\sigma^3}$$

Use the formula for μ_3 obtained in Exercise 25 to determine α_3 for each of the following distributions (which have equal means and standard deviations):

(a) $f(1) = 0.05$, $f(2) = 0.15$, $f(3) = 0.30$, $f(4) = 0.30$, $f(5) = 0.15$, and $f(6) = 0.05$;

(b) $f(1) = 0.05$, $f(2) = 0.20$, $f(3) = 0.15$, $f(4) = 0.45$, $f(5) = 0.10$, and $f(6) = 0.05$.

Also draw histograms of the two distributions and note that whereas the first is symmetrical, the second has a "tail" on the left-hand side and is said to be **negatively skewed**.

27. The extent to which a distribution is peaked or flat, also called the **kurtosis** of the distribution, is often measured by means of the quantity

$$\alpha_4 = \frac{\mu_4}{\sigma^4}$$

Use the formula for μ_4 obtained in Exercise 25 to find α_4 for each of the following symmetrical distributions, of which the first is more peaked (narrow humped) than the second:

(a) $f(-3) = 0.06$, $f(-2) = 0.09$, $f(-1) = 0.10$, $f(0) = 0.50$, $f(1) = 0.10$, $f(2) = 0.09$, and $f(3) = 0.06$;

(b) $f(-3) = 0.04$, $f(-2) = 0.11$, $f(-1) = 0.20$, $f(0) = 0.30$, $f(1) = 0.20$, $f(2) = 0.11$, and $f(3) = 0.04$.

28. Duplicate the steps used in the proof of Theorem 8 to prove Chebyshev's theorem for a discrete random variable X.

29. Show that if X is a random variable with the mean μ for which $f(x) = 0$ for $x < 0$, then for any positive constant a,

$$P(X \geqq a) \leqq \frac{\mu}{a}$$

This inequality is called **Markov's inequality**, and we have given it here mainly because it leads to a relatively simple alternative proof of Chebyshev's theorem.

30. Use the inequality of Exercise 29 to prove Chebyshev's theorem. [*Hint*: Substitute $(X-\mu)^2$ for X.]

31. What is the smallest value of k in Chebyshev's theorem for which the probability that a random variable will take on a value between $\mu - k\sigma$ and $\mu + k\sigma$ is
(a) at least 0.95;
(b) at least 0.99?

32. If we let $k\sigma = c$ in Chebyshev's theorem, what does this theorem assert about the probability that a random variable will take on a value between $\mu - c$ and $\mu + c$?

33. Find the moment-generating function of the continuous random variable X whose probability density is given by

$$f(x) = \begin{cases} 1 & \text{for } 0 < x < 1 \\ 0 & \text{elsewhere} \end{cases}$$

and use it to find μ'_1, μ'_2, and σ^2.

34. Find the moment-generating function of the discrete random variable X that has the probability distribution

$$f(x) = 2\left(\frac{1}{3}\right)^x \quad \text{for } x = 1, 2, 3, \ldots$$

and use it to determine the values of μ'_1 and μ'_2.

35. If we let $R_X(t) = \ln M_X(t)$, show that $R'_X(0) = \mu$ and $R''_X(0) = \sigma^2$. Also, use these results to find the mean and the variance of a random variable X having the moment-generating function

$$M_X(t) = e^{4(e^t - 1)}$$

36. Explain why there can be no random variable for which $M_X(t) = \frac{t}{1-t}$.

37. Show that if a random variable has the probability density

$$f(x) = \frac{1}{2}e^{-|x|} \quad \text{for } -\infty < x < \infty$$

its moment-generating function is given by

$$M_X(t) = \frac{1}{1-t^2}$$

38. With reference to Exercise 37, find the variance of the random variable by
(a) expanding the moment-generating function as an infinite series and reading off the necessary coefficients;
(b) using Theorem 9.

39. Prove the three parts of Theorem 10.

40. Given the moment-generating function $M_X(t) = e^{3t+8t^2}$, find the moment-generating function of the random variable $Z = \frac{1}{4}(X-3)$, and use it to determine the mean and the variance of Z.

6 Product Moments

To continue the discussion of Section 3, let us now present the **product moments** of two random variables.

DEFINITION 7. PRODUCT MOMENTS ABOUT THE ORIGIN. *The* ***rth and sth product moment about the origin*** *of the random variables X and Y, denoted by $\mu'_{r,s}$, is the expected value of X^rY^s; symbolically,*

$$\mu'_{r,s} = E(X^rY^s) = \sum_x \sum_y x^r y^s \cdot f(x, y)$$

for $r = 0, 1, 2, \ldots$ and $s = 0, 1, 2, \ldots$ when X and Y are discrete, and

$$\mu'_{r,s} = E(X^rY^s) = \int_{-\infty}^{\infty}\int_{-\infty}^{\infty} x^r y^s \cdot f(x, y)\,dx\,dy$$

when X and Y are continuous.

In the discrete case, the double summation extends over the entire joint range of the two random variables. Note that $\mu'_{1,0} = E(X)$, which we denote here by μ_X, and that $\mu'_{0,1} = E(Y)$, which we denote here by μ_Y.

Analogous to Definition 4, let us now state the following definition of product moments about the respective means.

DEFINITION 8. PRODUCT MOMENTS ABOUT THE MEAN. *The* ***rth and sth product moment about the means*** *of the random variables X and Y, denoted by $\mu_{r,s}$, is the expected value of $(X - \mu X)^r (Y - \mu_Y)^s$; symbolically,*

$$\mu_{r,s} = E[(X - \mu_X)^r (Y - \mu_Y)^s]$$
$$= \sum_x \sum_y (x - \mu_X)^r (y - \mu_Y)^s \cdot f(x, y)$$

for $r = 0, 1, 2, \ldots$ *and* $s = 0, 1, 2, \ldots$ *when X and Y are discrete, and*

$$\mu_{r,s} = E[(X - \mu_X)^r (Y - \mu_Y)^s]$$
$$= \int_{-\infty}^{\infty} \int_{-\infty}^{\infty} (x - \mu_X)^r (y - \mu_Y)^s \cdot f(x, y) dx dy$$

when X and Y are continuous.

In statistics, $\mu_{1,1}$ is of special importance because it is indicative of the relationship, if any, between the values of X and Y; thus, it is given a special symbol and a special name.

DEFINITION 9. COVARIANCE. $\mu_{1,1}$ *is called the* ***covariance*** *of X and Y, and it is denoted by σ_{XY}, cov(X, Y), or C(X, Y).*

Observe that if there is a high probability that large values of X will go with large values of Y and small values of X with small values of Y, the covariance will be positive; if there is a high probability that large values of X will go with small values of Y, and vice versa, the covariance will be negative. It is in this sense that the covariance measures the relationship, or association, between the values of X and Y.

Let us now prove the following result, analogous to Theorem 6, which is useful in actually determining covariances.

THEOREM 11.

$$\sigma_{XY} = \mu'_{1,1} - \mu_X \mu_Y$$

Proof Using the various theorems about expected values, we can write

$$\begin{aligned}
\sigma_{XY} &= E[(X - \mu_X)(Y - \mu_Y)] \\
&= E(XY - X\mu_Y - Y\mu_X + \mu_X\mu_Y) \\
&= E(XY) - \mu_Y E(X) - \mu_X E(Y) + \mu_X\mu_Y \\
&= E(XY) - \mu_Y\mu_X - \mu_X\mu_Y + \mu_X\mu_Y \\
&= \mu'_{1,1} - \mu_X\mu_Y
\end{aligned}$$

EXAMPLE 15

The joint and marginal probabilities of X and Y, the numbers of aspirin and sedative caplets among two caplets drawn at random from a bottle containing three aspirin, two sedative, and four laxative caplets, are recorded as follows:

		x = 0	x = 1	x = 2	
	0	$\frac{1}{6}$	$\frac{1}{3}$	$\frac{1}{12}$	$\frac{7}{12}$
y	1	$\frac{2}{9}$	$\frac{1}{6}$		$\frac{7}{18}$
	2	$\frac{1}{36}$			$\frac{1}{36}$
		$\frac{5}{12}$	$\frac{1}{2}$	$\frac{1}{12}$	

Find the covariance of X and Y.

Solution

Referring to the joint probabilities given here, we get

$$\begin{aligned}\mu'_{1,1} &= E(XY)\\ &= 0\cdot 0\cdot\frac{1}{6}+0\cdot 1\cdot\frac{2}{9}+0\cdot 2\cdot\frac{1}{36}+1\cdot 0\cdot\frac{1}{3}+1\cdot 1\cdot\frac{1}{6}+2\cdot 0\cdot\frac{1}{12}\\ &= \frac{1}{6}\end{aligned}$$

and using the marginal probabilities, we get

$$\mu_X = E(X) = 0\cdot\frac{5}{12}+1\cdot\frac{1}{2}+2\cdot\frac{1}{12} = \frac{2}{3}$$

and

$$\mu_Y = E(Y) = 0\cdot\frac{7}{12}+1\cdot\frac{7}{18}+2\cdot\frac{1}{36} = \frac{4}{9}$$

It follows that

$$\sigma_{XY} = \frac{1}{6}-\frac{2}{3}\cdot\frac{4}{9} = -\frac{7}{54}$$

The negative result suggests that the more aspirin tablets we get the fewer sedative tablets we will get, and vice versa, and this, of course, makes sense.

EXAMPLE 16

Find the covariance of the random variables whose joint probability density is given by

$$f(x,y) = \begin{cases} 2 & \text{for } x>0, y>0, x+y<1\\ 0 & \text{elsewhere}\end{cases}$$

Solution
Evaluating the necessary integrals, we get

$$\mu_X = \int_0^1 \int_0^{1-x} 2x \, dy \, dx = \frac{1}{3}$$

$$\mu_Y = \int_0^1 \int_0^{1-x} 2y \, dy \, dx = \frac{1}{3}$$

and

$$\sigma'_{1,1} = \int_0^1 \int_0^{1-x} 2xy \, dy \, dx = \frac{1}{12}$$

It follows that

$$\sigma_{XY} = \frac{1}{12} - \frac{1}{3} \cdot \frac{1}{3} = -\frac{1}{36}$$

As far as the relationship between X and Y is concerned, observe that if X and Y are independent, their covariance is zero; symbolically, we have the following theorem.

THEOREM 12. If X and Y are independent, then $E(XY) = E(X) \cdot E(Y)$ and $\sigma_{XY} = 0$.

Proof For the discrete case we have, by definition,

$$E(XY) = \sum_x \sum_y xy \cdot f(x, y)$$

Since X and Y are independent, we can write $f(x, y) = g(x) \cdot h(y)$, where $g(x)$ and $h(y)$ are the values of the marginal distributions of X and Y, and we get

$$\begin{aligned} E(XY) &= \sum_x \sum_y xy \cdot g(x)h(y) \\ &= \left[\sum_x x \cdot g(x)\right]\left[\sum_y y \cdot h(y)\right] \\ &= E(X) \cdot E(Y) \end{aligned}$$

Hence,

$$\begin{aligned} \sigma_{XY} &= \mu'_{1,1} - \mu_X \mu_Y \\ &= E(X) \cdot E(Y) - E(X) \cdot E(Y) \\ &= 0 \end{aligned}$$

It is of interest to note that the independence of two random variables implies a zero covariance, but a zero covariance does not necessarily imply their independence. This is illustrated by the following example (see also Exercises 46 and 47).

EXAMPLE 17

If the joint probability distribution of X and Y is given by

		x			
		-1	0	1	
	-1	$\frac{1}{6}$	$\frac{1}{3}$	$\frac{1}{6}$	$\frac{2}{3}$
y	0	0	0	0	0
	1	$\frac{1}{6}$	0	$\frac{1}{6}$	$\frac{1}{3}$
		$\frac{1}{3}$	$\frac{1}{3}$	$\frac{1}{3}$	

show that their covariance is zero even though the two random variables are not independent.

Solution

Using the probabilities shown in the margins, we get

$$\mu_X = (-1)\cdot\frac{1}{3} + 0\cdot\frac{1}{3} + 1\cdot\frac{1}{3} = 0$$

$$\mu_Y = (-1)\cdot\frac{2}{3} + 0\cdot 0 + 1\cdot\frac{1}{3} = -\frac{1}{3}$$

and

$$\mu'_{1,1} = (-1)(-1)\cdot\frac{1}{6} + 0(-1)\cdot\frac{1}{3} + 1(-1)\cdot\frac{1}{6} + (-1)1\cdot\frac{1}{6} + 1\cdot 1\cdot\frac{1}{6}$$
$$= 0$$

Thus, $\sigma_{XY} = 0 - 0(-\frac{1}{3}) = 0$, the covariance is zero, but the two random variables are not independent. For instance, $f(x, y) \neq g(x)\cdot h(y)$ for $x = -1$ and $y = -1$.

Product moments can also be defined for the case where there are more than two random variables. Here let us merely state the important result, in the following theorem.

THEOREM 13. If $X_1, X_2, \ldots, X_n$ are independent, then

$$E(X_1X_2\cdot\ldots\cdot X_n) = E(X_1)\cdot E(X_2)\cdot\ldots\cdot E(X_n)$$

This is a generalization of the first part of Theorem 12.

7 Moments of Linear Combinations of Random Variables

In this section we shall derive expressions for the mean and the variance of a linear combination of n random variables and the covariance of two linear combinations of n random variables. Applications of these results will be important in our later discussion of sampling theory and problems of statistical inference.

THEOREM 14. If $X_1, X_2, \ldots, X_n$ are random variables and

$$Y = \sum_{i=1}^{n} a_i X_i$$

where $a_1, a_2, \ldots, a_n$ are constants, then

$$E(Y) = \sum_{i=1}^{n} a_i E(X_i)$$

and

$$\text{var}(Y) = \sum_{i=1}^{n} a_i^2 \cdot \text{var}(X_i) + 2 \sum\sum_{i<j} a_i a_j \cdot \text{cov}(X_i X_j)$$

where the double summation extends over all values of i and j, from 1 to n, for which $i < j$.

Proof From Theorem 5 with $g_i(X_1, X_2, \ldots, X_k) = X_i$ for $i = 0, 1, 2, \ldots, n$, it follows immediately that

$$E(Y) = E\left(\sum_{i=1}^{n} a_i X_i\right) = \sum_{i=1}^{n} a_i E(X_i)$$

and this proves the first part of the theorem. To obtain the expression for the variance of Y, let us write μ_i for $E(X_i)$ so that we get

$$\text{var}(Y) = E\left([Y - E(Y)]^2\right) = E\left\{\left[\sum_{i=1}^{n} a_i X_i - \sum_{i=1}^{n} a_i E(X_i)\right]^2\right\}$$

$$= E\left\{\left[\sum_{i=1}^{n} a_i (X_i - \mu_i)\right]^2\right\}$$

Then, expanding by means of the multinomial theorem, according to which $(a+b+c+d)^2$, for example, equals $a^2+b^2+c^2+d^2+2ab+2ac+2ad+2bc+2bd+2cd$, and again referring to Theorem 5, we get

$$\text{var}(Y) = \sum_{i=1}^{n} a_i^2 E[(X_i - \mu_i)^2] + 2 \sum\sum_{i<j} a_i a_j E[(X_i - \mu_i)(X_j - \mu_j)]$$

$$= \sum_{i=1}^{n} a_i^2 \cdot \text{var}(X_i) + 2 \sum\sum_{i<j} a_i a_j \cdot \text{cov}(X_i, X_j)$$

Note that we have tacitly made use of the fact that $\text{cov}(X_i, X_j) = \text{cov}(X_j, X_i)$.

Since $\text{cov}(X_i, X_j) = 0$ when X_i and X_j are independent, we obtain the following corollary.

COROLLARY 3. If the random variables $X_1, X_2, \ldots, X_n$ are independent and $Y = \sum_{i=1}^{n} a_i X_i$, then

$$\text{var}(Y) = \sum_{i=1}^{n} a_i^2 \cdot \text{var}(X_i)$$

EXAMPLE 18

If the random variables X, Y, and Z have the means $\mu_X = 2, \mu_Y = -3$, and $\mu_Z = 4$, the variances $\sigma_X^2 = 1, \sigma_Y^2 = 5$, and $\sigma_Z^2 = 2$, and the covariances $\text{cov}(X, Y) = -2$, $\text{cov}(X, Z) = -1$, and $\text{cov}(Y, Z) = 1$, find the mean and the variance of $W = 3X - Y + 2Z$.

Solution

By Theorem 14, we get

$$\begin{aligned} E(W) &= E(3X - Y + 2Z) \\ &= 3E(X) - E(Y) + 2E(Z) \\ &= 3 \cdot 2 - (-3) + 2 \cdot 4 \\ &= 17 \end{aligned}$$

and

$$\begin{aligned} \text{var}(W) &= 9\,\text{var}(X) + \text{var}(Y) + 4\,\text{var}(Z) - 6\,\text{cov}(X, Y) \\ &\quad + 12\,\text{cov}(X, Z) - 4\,\text{cov}(Y, Z) \\ &= 9 \cdot 1 + 5 + 4 \cdot 2 - 6(-2) + 12(-1) - 4 \cdot 1 \\ &= 18 \end{aligned}$$

The following is another important theorem about linear combinations of random variables; it concerns the covariance of two linear combinations of n random variables.

THEOREM 15. If $X_1, X_2, \ldots, X_n$ are random variables and

$$Y_1 = \sum_{i=1}^{n} a_i X_i \quad \text{and} \quad Y_2 = \sum_{i=1}^{n} b_i X_i$$

where $a_1, a_2, \ldots, a_n, b_1, b_2, \ldots, b_n$ are constants, then

$$\text{cov}(Y_1, Y_2) = \sum_{i=1}^{n} a_i b_i \cdot \text{var}(X_i) + \sum_{i<j}\sum (a_i b_j + a_j b_i) \cdot \text{cov}(X_i, X_j)$$

The proof of this theorem, which is very similar to that of Theorem 14, will be left to the reader in Exercise 52.

Since $\text{cov}(X_i, X_j) = 0$ when X_i and X_j are independent, we obtain the following corollary.

COROLLARY 4. If the random variables $X_1, X_2, \ldots, X_n$ are independent, $Y_1 = \sum_{i=1}^{n} a_i X_i$ and $Y_2 = \sum_{i=1}^{n} b_i X_i$, then

$$\text{cov}(Y_1, Y_2) = \sum_{i=1}^{n} a_i b_i \cdot \text{var}(X_i)$$

EXAMPLE 19

If the random variables X, Y, and Z have the means $\mu_X = 3, \mu_Y = 5$, and $\mu_Z = 2$, the variances $\sigma_X^2 = 8, \sigma_Y^2 = 12$, and $\sigma_Z^2 = 18$, and $\text{cov}(X,Y) = 1, \text{cov}(X,Z) = -3$, and $\text{cov}(Y,Z) = 2$, find the covariance of

$$U = X + 4Y + 2Z \quad \text{and} \quad V = 3X - Y - Z$$

Solution

By Theorem 15, we get

$$\begin{aligned}
\text{cov}(U,V) &= \text{cov}(X + 4Y + 2Z, 3X - Y - Z) \\
&= 3\,\text{var}(X) - 4\,\text{var}(Y) - 2\,\text{var}(Z) + 11\,\text{cov}(X,Y) \\
&\quad + 5\,\text{cov}(X,Z) - 6\,\text{cov}(Y,Z) \\
&= 3 \cdot 8 - 4 \cdot 12 - 2 \cdot 18 + 11 \cdot 1 + 5(-3) - 6 \cdot 2 \\
&= -76
\end{aligned}$$

---

8 Conditional Expectations

Conditional probabilities are obtained by adding the values of conditional probability distributions, or integrating the values of conditional probability densities. **Conditional expectations** of random variables are likewise defined in terms of their conditional distributions.

DEFINITION 10. CONDITIONAL EXPECTATION. *If* X *is a discrete random variable, and* f(x|y) *is the value of the conditional probability distribution of* X *given* Y = y *at* x, *the* ***conditional expectation of* u(X) *given* Y = y** *is*

$$E[u(X)|y)] = \sum_x u(x) \cdot f(x|y)$$

Correspondingly, if X *is a continuous variable and* f(x|y) *is the value of the conditional probability distribution of* X *given* Y = y *at* x, *the* ***conditional expectation of* u(X) *given* Y = y** *is*

$$E[(u(X)|y)] = \int_{-\infty}^{\infty} u(x) \cdot f(x|y)dx$$

Similar expressions based on the conditional probability distribution or density of Y given $X = x$ define the conditional expectation of $v(Y)$ given $X = x$.

If we let $u(X) = X$ in Definition 10, we obtain the **conditional mean** of the random variable X given $Y = y$, which we denote by

$$\mu_{X|y} = E(X|y)$$

Correspondingly, the **conditional variance** of X given $Y = y$ is

$$\sigma^2_{X|y} = E[(X - \mu_{X|y})^2|y]$$
$$= E(X^2|y) - \mu^2_{X|y}$$

where $E(X^2|y)$ is given by Definition 10 with $u(X) = X^2$. The reader should not find it difficult to generalize Definition 10 for conditional expectations involving more than two random variables.

EXAMPLE 20

If the joint probability density of X and Y is given by

$$f(x, y) = \begin{cases} \frac{2}{3}(x+2y) & \text{for } 0 < x < 1, 0 < y < 1 \\ 0 & \text{elsewhere} \end{cases}$$

find the conditional mean and the conditional variance of X given $Y = \frac{1}{2}$.

Solution

For these random variables the conditional density of X given $Y = y$ is

$$f(x|y) = \begin{cases} \dfrac{2x+4y}{1+4y} & \text{for } 0 < x < 1 \\ 0 & \text{elsewhere} \end{cases}$$

so that

$$f\left(x \middle| \frac{1}{2}\right) = \begin{cases} \frac{2}{3}(x+1) & \text{for } 0 < x < 1 \\ 0 & \text{elsewhere} \end{cases}$$

Thus, $\mu_{X|\frac{1}{2}}$ is given by

$$E\left(X \middle| \frac{1}{2}\right) = \int_0^1 \frac{2}{3}x(x+1)\,dx$$
$$= \frac{5}{9}$$

Next we find

$$E\left(X^2 \middle| \frac{1}{2}\right) = \int_0^1 \frac{2}{3}x^2(x+1)\,dx$$
$$= \frac{7}{18}$$

and it follows that

$$\sigma^2_{X|\frac{1}{2}} = \frac{7}{18} - \left(\frac{5}{9}\right)^2 = \frac{13}{162}$$

Exercises

41. Prove that $\text{cov}(X, Y) = \text{cov}(Y, X)$ for both discrete and continuous random variables X and Y.

42. If X and Y have the joint probability distribution $f(x, y) = \frac{1}{4}$ for $x = -3$ and $y = -5$, $x = -1$ and $y = -1$, $x = 1$ and $y = 1$, and $x = 3$ and $y = 5$, find $\text{cov}(X, Y)$.

43. This has been intentionally omitted for this edition.

44. This has been intentionally omitted for this edition.

45. This has been intentionally omitted for this edition.

46. If X and Y have the joint probability distribution $f(-1, 0) = 0$, $f(-1, 1) = \frac{1}{4}$, $f(0, 0) = \frac{1}{6}$, $f(0, 1) = 0$, $f(1, 0) = \frac{1}{12}$, and $f(1, 1) = \frac{1}{2}$, show that

(a) $\text{cov}(X, Y) = 0$;

(b) the two random variables are not independent.

47. If the probability density of X is given by

$$f(x) = \begin{cases} 1+x & \text{for } -1 < x \leqq 0 \\ 1-x & \text{for } 0 < x < 1 \\ 0 & \text{elsewhere} \end{cases}$$

and $U = X$ and $V = X^2$, show that

(a) $\text{cov}(U, V) = 0$;

(b) U and V are dependent.

48. For k random variables $X_1, X_2, \ldots, X_k$, the values of their **joint moment-generating function** are given by

$$E\left(e^{t_1X_1 + t_2X_2 + \cdots + t_kX_k}\right)$$

(a) Show for either the discrete case or the continuous case that the partial derivative of the joint moment-generating function with respect to t_i at $t_1 = t_2 = \cdots = t_k = 0$ is $E(X_i)$.

(b) Show for either the discrete case or the continuous case that the second partial derivative of the joint moment-generating function with respect to t_i and t_j, $i \neq j$, at $t_1 = t_2 = \cdots = t_k = 0$ is $E(X_iX_j)$.

(c) If two random variables have the joint density given by

$$f(x, y) = \begin{cases} e^{-x-y} & \text{for } x > 0,\ y > 0 \\ 0 & \text{elsewhere} \end{cases}$$

find their joint moment-generating function and use it to determine the values of $E(XY)$, $E(X)$, $E(Y)$, and $\text{cov}(X, Y)$.

49. If X_1, X_2, and X_3 are independent and have the means 4, 9, and 3 and the variances 3, 7, and 5, find the mean and the variance of

(a) $Y = 2X_1 - 3X_2 + 4X_3$;

(b) $Z = X_1 + 2X_2 - X_3$.

50. Repeat both parts of Exercise 49, dropping the assumption of independence and using instead the information that $\text{cov}(X_1, X_2) = 1$, $\text{cov}(X_2, X_3) = -2$, and $\text{cov}(X_1, X_3) = -3$.

51. If the joint probability density of X and Y is given by

$$f(x, y) = \begin{cases} \frac{1}{3}(x+y) & \text{for } 0 < x < 1,\ 0 < y < 2 \\ 0 & \text{elsewhere} \end{cases}$$

find the variance of $W = 3X + 4Y - 5$.

52. Prove Theorem 15.

53. Express $\text{var}(X+Y)$, $\text{var}(X-Y)$, and $\text{cov}(X+Y, X-Y)$ in terms of the variances and covariance of X and Y.

54. If $\text{var}(X_1) = 5$, $\text{var}(X_2) = 4$, $\text{var}(X_3) = 7$, $\text{cov}(X_1, X_2) = 3$, $\text{cov}(X_1, X_3) = -2$, and X_2 and X_3 are independent, find the covariance of $Y_1 = X_1 - 2X_2 + 3X_3$ and $Y_2 = -2X_1 + 3X_2 + 4X_3$.

55. With reference to Exercise 49, find $\text{cov}(Y, Z)$.

56. This question has been intentionally omitted for this edition.

57. This question has been intentionally omitted for this edition.

58. This question has been intentionally omitted for this edition.

59. This question has been intentionally omitted for this edition.

60. (a) Show that the conditional distribution function of the continuous random variable X, given $a < X \leqq b$, is given by

$$F(x|a < X \leqq b) = \begin{cases} 0 & \text{for } x \leqq a \\ \dfrac{F(x) - F(a)}{F(b) - F(a)} & \text{for } a < x \leqq b \\ 1 & \text{for } x > b \end{cases}$$

(b) Differentiate the result of part (a) with respect to x to find the conditional probability density of X given $a < X \leqq b$, and show that

$$E[u(X)|a < X \leqq b] = \frac{\int_a^b u(x)f(x)\,dx}{\int_a^b f(x)\,dx}$$

9 The Theory in Practice

Empirical distributions, those arising from data, can be described by their shape. We will discuss **descriptive measures**, calculated from data, that extend the methodology of describing data. These descriptive measures are based on the ideas of moments, given in Section 3.

The analog of the first moment, $\mu_1' = \mu$, is the **sample mean**, $\bar{x}$, defined as

$$\bar{x} = \sum_{i=1}^{n} x_i/n$$

where $i = 1, 2, \ldots, n$ and n is the number of observations.

The usefulness of the sample mean as a description of data can be envisioned by imagining that the histogram of a data distribution has been cut out of a piece of cardboard and balanced by inserting a fulcrum along the horizontal axis. This balance point corresponds to the mean of the data. Thus, the mean can be thought of as the centroid of the data and, as such, it describes its **location**.

The mean is an excellent measure of location for symmetric or nearly symmetric distributions. But it can be misleading when used to measure the location of highly skewed data. To give an example, suppose, in a small company, the annual salaries of its 10 employees (rounded to the nearest $1,000) are 25, 18, 36, 28, 16, 20, 29, 32, 41, and 150. The mean of these observations is $39,500. One of the salaries, namely $150,000, is much higher than the others (it's what the owner pays himself) and only one other employee earns as much as $39,500. Suppose the owner, in a recruiting ad, claimed that "Our company pays an average salary of $39,500." He would be technically correct, but very misleading.

Other descriptive measures for the location of data should be used in cases like the one just described. The **median** describes the center of the data as the middle point of the observations. If the data are ranked from, say, smallest to largest, the median becomes observation number $n/2$ if n is an even integer, and it is defined as the mean value of observations $\frac{(n-1)}{2}$ and $\frac{(n+1)}{2}$ if n is an odd integer. The median of the 10 observations given in the preceding example is $28,000, and it is a much better description of what an employee of this company can expect to earn. You may very well have heard the term "median income" for, say, the incomes of American families. The median is used instead of the mean here because it is well known that the distribution of family incomes in the United States is highly skewed—the great majority of families earn low to moderate incomes, but a relatively few have very high incomes.

The **dispersion** of data also is important in its description. Give the location of data, one reasonably wants to know how closely the observations are grouped around this value. A reasonable measure of dispersion can be based on the square root of the second moment about the mean, σ. The **sample standard deviation,** s, is calculated analogously to the second moment, as follows:

$$s = \sqrt{\frac{\sum_{i=1}^{n}(x - \bar{x})^2}{n-1}}$$

Since this formula requires first the calculation of the mean, then subtraction of the mean from each observation before squaring and adding, it is much easier to use the following *calculating formula* for s:

$$s = \sqrt{\frac{n\sum_{i=1}^{n} x_i^2 - \left(\sum_{i=1}^{n} x_i\right)^2}{n(n-1)}}$$

Note that in both formulas we divide by $n - 1$ instead of n. Using either formula for the calculation of s requires tedious calculation, but every statistical computer program in common use will calculate both the sample mean and the sample standard deviation once the data have been inputted.

EXAMPLE 21

The following are the lengths (in feet) of 10 steel beams rolled in a steel mill and cut to a nominal length of 12 feet:

11.8 12.1 12.5 11.7 11.9 12.0 12.2 11.5 11.9 12.2

Calculate the mean length and its standard deviation. Is the mean a reasonable measure of the location of the data? Why or why not?

Solution

The mean is given by the sum of the observations, $11.8 + 12.1 + \ldots 12.2 = 119.8$, divided by 10, or $\bar{x} = 11.98$ feet. To calculate the standard deviation, we first calculate the sum of the squares of the observations, $(11.8)^2 + (12.1)^2 + \ldots + (12.2)^2 = 1,435.94$. Then substituting into the formula for s, we obtain $s^2 = (10)(1435.94) - (119.8)^2/(10)(9) = 0.082$ foot. Taking the square root, we obtain $s = 0.29$. The mean, 11.98 feet, seems to be a reasonable measure of location inasmuch as the data seem to be approximately symmetrically distributed.

The standard deviation is not the only measure of the dispersion, or variability of data. The **sample range** sometimes is used for this purpose. To calculate the range, we find the largest and the smallest observations, x_l and x_s, defining the range to be

$$r = x_l - x_s$$

This measure of dispersion is used only for small samples; for larger and larger sample sizes, the range becomes a poorer and poorer measure of dispersion.

Applied Exercises SECS. 1–2

61. This question has been intentionally omitted for this edition.

62. The probability that Ms. Brown will sell a piece of property at a profit of \$3,000 is $\frac{3}{20}$, the probability that she will sell it at a profit of \$1,500 is $\frac{7}{20}$, the probability that she will break even is $\frac{7}{20}$, and the probability that she will lose \$1,500 is $\frac{3}{20}$. What is her expected profit?

63. A game of chance is considered **fair**, or **equitable**, if each player's expectation is equal to zero. If someone pays us \$10 each time that we roll a 3 or a 4 with a balanced die, how much should we pay that person when we roll a 1, 2, 5, or 6 to make the game equitable?

64. The manager of a bakery knows that the number of chocolate cakes he can sell on any given day is a random variable having the probability distribution $f(x) = \frac{1}{6}$ for $x = 0, 1, 2, 3, 4,$ and 5. He also knows that there is a profit of \$1.00 for each cake that he sells and a loss (due to spoilage) of \$0.40 for each cake that he does not sell. Assuming that each cake can be sold only on the day it is

made, find the baker's expected profit for a day on which he bakes
(a) one of the cakes;
(b) two of the cakes;
(c) three of the cakes;
(d) four of the cakes;
(e) five of the cakes.

How many should he bake in order to maximize his expected profit?

65. If a contractor's profit on a construction job can be looked upon as a continuous random variable having the probability density

$$f(x) = \begin{cases} \frac{1}{18}(x+1) & \text{for } -1 < x < 5 \\ 0 & \text{elsewhere} \end{cases}$$

where the units are in \$1,000, what is her expected profit?

66. This question has been intentionally omitted for this edition.

67. This question has been intentionally omitted for this edition.

68. This question has been intentionally omitted for this edition.

69. Mr. Adams and Ms. Smith are betting on repeated flips of a coin. At the start of the game Mr. Adams has a dollars and Ms. Smith has b dollars, at each flip the loser pays the winner one dollar, and the game continues until either player is "ruined." Making use of the fact that in an equitable game each player's mathematical expectation is zero, find the probability that Mr. Adams will win Ms. Smith's b dollars before he loses his a dollars.

SECS. 3–5

70. With reference to Example 1, find the variance of the number of television sets with white cords.

71. The amount of time it takes a person to be served at a given restaurant is a random variable with the probability density

$$f(x) = \begin{cases} \frac{1}{4}e^{-\frac{x}{4}} & \text{for } x > 0 \\ 0 & \text{elsewhere} \end{cases}$$

Find the mean and the variance of this random variable.

72. This question has been intentionally omitted for this edition.

73. This question has been intentionally omitted for this edition.

74. The following are some applications of the Markov inequality of Exercise 29:
(a) The scores that high school juniors get on the verbal part of the PSAT/NMSQT test may be looked upon as values of a random variable with the mean $\mu = 41$. Find an upper bound to the probability that one of the students will get a score of 65 or more.
(b) The weight of certain animals may be looked upon as a random variable with a mean of 212 grams. If none of the animals weighs less than 165 grams, find an upper bound to the probability that such an animal will weigh at least 250 grams.

75. The number of marriage licenses issued in a certain city during the month of June may be looked upon as a random variable with $\mu = 124$ and $\sigma = 7.5$. According to Chebyshev's theorem, with what probability can we assert that between 64 and 184 marriage licenses will be issued there during the month of June?

76. A study of the nutritional value of a certain kind of bread shows that the amount of thiamine (vitamin B_1) in a slice may be looked upon as a random variable with $\mu = 0.260$ milligram and $\sigma = 0.005$ milligram. According to Chebyshev's theorem, between what values must be the thiamine content of
(a) at least $\frac{35}{36}$ of all slices of this bread;
(b) at least $\frac{143}{144}$ of all slices of this bread?

77. With reference to Exercise 71, what can we assert about the amount of time it takes a person to be served at the given restaurant if we use Chebyshev's theorem with $k = 1.5$? What is the corresponding probability rounded to four decimals?

SECS. 6–9

78. A quarter is bent so that the probabilities of heads and tails are 0.40 and 0.60. If it is tossed twice, what is the covariance of Z, the number of heads obtained on the first toss, and W, the total number of heads obtained in the two tosses of the coin?

79. The inside diameter of a cylindrical tube is a random variable with a mean of 3 inches and a standard deviation of 0.02 inch, the thickness of the tube is a random variable with a mean of 0.3 inch and a standard deviation of 0.005 inch, and the two random variables are independent. Find the mean and the standard deviation of the outside diameter of the tube.

80. The length of certain bricks is a random variable with a mean of 8 inches and a standard deviation of 0.1 inch, and the thickness of the mortar between two bricks is a random variable with a mean of 0.5 inch and a standard deviation of 0.03 inch. What is the mean and the standard deviation of the length of a wall made of 50 of these bricks laid side by side, if we can assume that all the random variables involved are independent?

81. If heads is a success when we flip a coin, getting a six is a success when we roll a die, and getting an ace is a success when we draw a card from an ordinary deck of 52 playing cards, find the mean and the standard deviation of the total number of successes when we
(a) flip a balanced coin, roll a balanced die, and then draw a card from a well-shuffled deck;
(b) flip a balanced coin three times, roll a balanced die twice, and then draw a card from a well-shuffled deck.

82. If we alternately flip a balanced coin and a coin that is loaded so that the probability of getting heads is 0.45, what are the mean and the standard deviation of the number of heads that we obtain in 10 flips of these coins?

83. This question has been intentionally omitted for this edition.

84. This question has been intentionally omitted for this edition.

85. The amount of time (in minutes) that an executive of a certain firm talks on the telephone is a random variable having the probability density

$$f(x) = \begin{cases} \dfrac{x}{4} & \text{for } 0 < x \leqq 2 \\ \dfrac{4}{x^3} & \text{for } x > 2 \\ 0 & \text{elsewhere} \end{cases}$$

With reference to part (b) of Exercise 60, find the expected length of one of these telephone conversations that has lasted at least 1 minute.

Answers to Odd-Numbered Exercises

1 (a) $g_1 = 0, g_2 = 1, g_3 = 4$, and $g_4 = 9$; **(b)** $f(0), f(-1) + f(1), f(-2) + f(2)$, and $f(3)$; **(c)** $0 \cdot f(0) + 1 \cdot \{f(-1) + f(1)\} + 4 \cdot \{f(-2) + f(2)\} + 9 \cdot f(3) = (-2)^2 \cdot f(-2) + (-1)^2 \cdot f(-1) + 0^2 \cdot f(0) + 1^2 \cdot f(1) + 2^2 \cdot f(2) + 3^2 \cdot f(3) = \sum_x g(x) \cdot f(x)$.

3 Replace $\int$ by $\sum$ in the proof of Theorem 3.

5 (a) $E(x) = \int_{-\infty}^{\infty} \int_{-\infty}^{\infty} xf(x, y)\,dy\,dx$;
(b) $E(x) = \int_{-\infty}^{\infty} xg(x)\,dx$.

7 $E(Y) = \frac{37}{12}$.

9 (a) 2.4 and 6.24; **(b)** 88.96.

11 $-\frac{11}{6}$.

13 $\frac{1}{2}$.

15 $\frac{1}{12}$.

19 $\mu = \frac{4}{3}, \mu_2' = 2$, and $\sigma^2 = \frac{2}{9}$.

25 $\mu_3 = \mu_3' - \mu\mu_2' + 2\mu^3$ and $\mu_4 = \mu_4' - 4\mu\mu_3' + 6\mu^2\mu_2' - 3\mu^4$.

27 (a) 3.2; **(b)** 2.6.

31 (a) $k = \sqrt{20}$; **(b)** $k = 10$.

33 $M_X(t) = \frac{2e^t}{3-e^t}$, $\mu_1' = \frac{3}{2}$, $\mu_2' = 3$, $\sigma^2 = \frac{3}{4}$.

35 $\mu = 4$, $\sigma^2 = 4$.

43 −0.14.

45 $\frac{1}{72}$.

49 (a) $\mu_Y = -7$, $\sigma_Y^2 = 155$; **(b)** $\mu_Z = 19$, $\sigma_Z^2 = 36$.

51 $\frac{805}{162}$.

53 $\text{var}(X) + \text{var}(Y) + 2\text{cov}(X, Y)$, $\text{var}(X) + \text{var}(Y) - 2\text{cov}(X, Y)$, $\text{var}(X) - \text{var}(Y)$.

55 −56.

57 3.

59 $\frac{5}{12}$.

61 (a) 98; **(b)** 29,997.

63 $5.

65 $3,000.

67 6 million liters.

69 $\frac{a}{a+b}$.

71 $\mu = 4$, $\sigma^2 = 16$.

73 $\mu = 1$, $\sigma^2 = 1$.

75 At least $\frac{63}{64}$.

77 0.9179.

79 $\mu = 3.6$, $\sigma = 0.0224$.

81 (a) 0.74, 0.68; **(b)** 1.91, 1.05.

83 0.8.

85 2.95 min.

Special Probability Distributions

1 Introduction

In this chapter we shall study some of the probability distributions that figure most prominently in statistical theory and applications. We shall also study their **parameters**, that is, the quantities that are constants for particular distributions but that can take on different values for different members of families of distributions of the same kind. The most common parameters are the lower moments, mainly μ and σ^2, and there are essentially two ways in which they can be obtained: We can evaluate the necessary sums directly or we can work with moment-generating functions. Although it would seem logical to use in each case whichever method is simplest, we shall sometimes use both. In some instances this will be done because the results are needed later; in others it will merely serve to provide the reader with experience in the application of the respective mathematical techniques. Also, to keep the size of this chapter within bounds, many of the details are left as exercises.

2 The Discrete Uniform Distribution

If a random variable can take on k different values with equal probability, we say that it has a **discrete uniform distribution**; symbolically, we have the following definition.

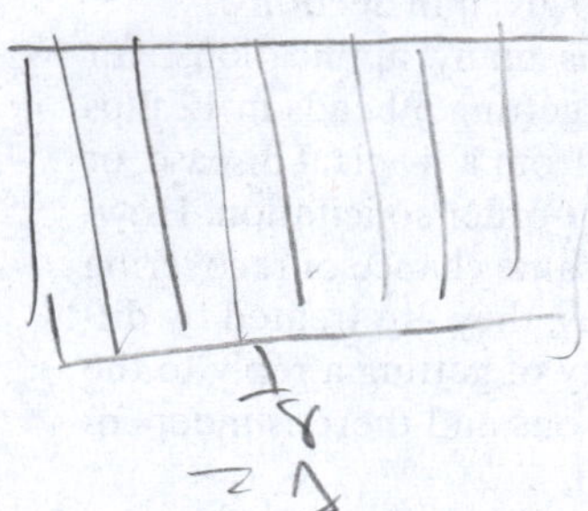

DEFINITION 1. DISCRETE UNIFORM DISTRIBUTION. *A random variable X has a* ***discrete uniform distribution*** *and it is referred to as a discrete uniform random variable if and only if its probability distribution is given by*

$$f(x) = \frac{1}{k} \quad \text{for } x = x_1, x_2, \ldots x_k$$

where $x_i \neq x_j$ *when* $i \neq j$.

In the special case where $x_i = i$, the discrete uniform distribution becomes $f(x) = \frac{1}{k}$ for $x = 1, 2, \ldots, k$, and in this form it applies, for example, to the number of points we roll with a balanced die. The mean and the variance of this discrete uniform distribution and its moment-generating function are treated in Exercises 1 and 2.

3 The Bernoulli Distribution

If an experiment has two possible outcomes, "success" and "failure," and their probabilities are, respectively, θ and $1-\theta$, then the number of successes, 0 or 1, has a **Bernoulli distribution**; symbolically, we have the following definition.

> **DEFINITION 2. BERNOULLI DISTRIBUTION.** *A random variable X has a **Bernoulli distribution** and it is referred to as a Bernoulli random variable if and only if its probability distribution is given by*
>
> $$f(x;\theta) = \theta^x(1-\theta)^{1-x} \quad \text{for } x = 0, 1$$

Thus, $f(0;\theta) = 1-\theta$ and $f(1;\theta) = \theta$ are combined into a single formula. Observe that we used the notation $f(x;\theta)$ to indicate explicitly that the Bernoulli distribution has the one parameter θ.

In connection with the Bernoulli distribution, a success may be getting heads with a balanced coin, it may be catching pneumonia, it may be passing (or failing) an examination, and it may be losing a race. This inconsistency is a carryover from the days when probability theory was applied only to games of chance (and one player's failure was the other's success). Also for this reason, we refer to an experiment to which the Bernoulli distribution applies as a **Bernoulli trial**, or simply a **trial**, and to sequences of such experiments as **repeated trials**.

4 The Binomial Distribution

Repeated trials play a very important role in probability and statistics, especially when the number of trials is fixed, the parameter θ (the probability of a success) is the same for each trial, and the trials are all independent. As we shall see, several random variables arise in connection with repeated trials. The one we shall study here concerns the total number of successes; others will be given in Section 5.

The theory that we shall discuss in this section has many applications; for instance, it applies if we want to know the probability of getting 5 heads in 12 flips of a coin, the probability that 7 of 10 persons will recover from a tropical disease, or the probability that 35 of 80 persons will respond to a mail-order solicitation. However, this is the case only if each of the 10 persons has the same chance of recovering from the disease and their recoveries are independent (say, they are treated by different doctors in different hospitals), and if the probability of getting a reply to the mail-order solicitation is the same for each of the 80 persons and there is independence (say, no two of them belong to the same household).

To derive a formula for the probability of getting "x successes in n trials" under the stated conditions, observe that the probability of getting x successes and $n-x$ failures *in a specific order* is $\theta^x(1-\theta)^{n-x}$. There is one factor θ for each success, one factor $1-\theta$ for each failure, and the x factors θ and $n-x$ factors $1-\theta$ are all multiplied together by virtue of the assumption of independence. Since this probability applies to any sequence of n trials in which there are x successes and $n-x$ failures, we have only to count how many sequences of this kind there are and then multiply $\theta^x(1-\theta)^{n-x}$ by that number. Clearly, the number of ways in which we can select the x trials on which there is to be a success is $\binom{n}{x}$, and it follows that the desired probability for "x successes in n trials" is $\binom{n}{x}\theta^x(1-\theta)^{n-x}$.

DEFINITION 3. BINOMIAL DISTRIBUTION. *A random variable* X *has a* ***binomial distribution*** *and it is referred to as a binomial random variable if and only if its probability distribution is given by*

$$b(x; n, \theta) = \binom{n}{x}\theta^x(1-\theta)^{n-x} \quad \text{for } x = 0, 1, 2, \ldots n$$

Thus, the number of successes in n trials is a random variable having a binomial distribution with the parameters n and θ. The name "binomial distribution" derives from the fact that the values of $b(x; n, \theta)$ for $x = 0, 1, 2, \ldots, n$ are the successive terms of the binomial expansion of $[(1-\theta)+\theta]^n$; this shows also that the sum of the probabilities equals 1, as it should.

EXAMPLE 1

Find the probability of getting five heads and seven tails in 12 flips of a balanced coin.

Solution

Substituting $x = 5$, $n = 12$, and $\theta = \frac{1}{2}$ into the formula for the binomial distribution, we get

$$b\left(5; 12, \frac{1}{2}\right) = \binom{12}{5}\left(\frac{1}{2}\right)^5\left(1-\frac{1}{2}\right)^{12-5}$$

and, looking up the value of $\binom{12}{5}$ in Table VII of "Statistical Tables", we find that the result is $792\left(\frac{1}{2}\right)^{12}$, or approximately 0.19.

EXAMPLE 2

Find the probability that 7 of 10 persons will recover from a tropical disease if we can assume independence and the probability is 0.80 that any one of them will recover from the disease.

Solution
Substituting $x = 7$, $n = 10$, and $\theta = 0.80$ into the formula for the binomial distribution, we get

$$b(7; 10, 0.80) = \binom{10}{7}(0.80)^7(1-0.80)^{10-7}$$

and, looking up the value of $\binom{10}{7}$ in Table VII of "Statistical Tables", we find that the result is $120(0.80)^7(0.20)^3$, or approximately 0.20.

If we tried to calculate the third probability asked for on the previous page, the one concerning the responses to the mail-order solicitation, by substituting $x = 35$, $n = 80$, and, say, $\theta = 0.15$, into the formula for the binomial distribution, we would find that this requires a prohibitive amount of work. In actual practice, binomial probabilities are rarely calculated directly, for they are tabulated extensively for various values of θ and n, and there exists an abundance of computer software yielding binomial probabilities as well as the corresponding cumulative probabilities

$$B(x; n, \theta) = \sum_{k=0}^{x} b(k; n, \theta)$$

upon simple commands. An example of such a printout (with somewhat different notation) is shown in Figure 1.

In the past, the National Bureau of Standards table and the book by H. G. Romig have been widely used; they are listed among the references at the end of this chapter. Also, Table I of "Statistical Tables" gives the values of $b(x; n, \theta)$ to four decimal places for $n = 1$ to $n = 20$ and $\theta = 0.05, 0.10, 0.15, \ldots, 0.45, 0.50$. To use this table when θ is greater than 0.50, we refer to the following identity.

```
MTB > BINOMIAL N=1Ø  P=Ø.63

 BINOMIAL PROBABILITIES FOR N = 1Ø AND P = .63ØØØØ

    K              P(X = K)              P(X LESS OR = K)
    Ø               .ØØØØ                     .ØØØØ
    1               .ØØØ8                     .ØØØ9
    2               .ØØ63                     .ØØ71
    3               .Ø285                     .Ø356
    4               .Ø849                     .12Ø5
    5               .1734                     .2939
    6               .2461                     .54ØØ
    7               .2394                     .7794
    8               .1529                     .9323
    9               .Ø578                     .99Ø2
   10               .ØØ98                    1.ØØØØ
```

Figure 1. Computer printout of binomial probabilities for $n = 10$ and $\theta = 0.63$.

THEOREM 1.

$$b(x; n, \theta) = b(n - x; n, 1 - \theta)$$

which the reader will be asked to prove in part (a) of Exercise 5. For instance, to find $b(11; 18, 0.70)$, we look up $b(7; 18, 0.30)$ and get 0.1376. Also, there are several ways in which binomial probabilities can be approximated when n is large; one of these will be mentioned in Section 7.

Let us now find formulas for the mean and the variance of the binomial distribution.

THEOREM 2. The mean and the variance of the binomial distribution are

$$\mu = n\theta \quad \text{and} \quad \sigma^2 = n\theta(1 - \theta)$$

Proof

$$\mu = \sum_{x=0}^{n} x \cdot \binom{n}{x} \theta^x (1-\theta)^{n-x}$$

$$= \sum_{x=1}^{n} \frac{n!}{(x-1)!(n-x)!} \theta^x (1-\theta)^{n-x}$$

where we omitted the term corresponding to $x = 0$, which is 0, and canceled the x against the first factor of $x! = x(x-1)!$ in the denominator of $\binom{n}{x}$. Then, factoring out the factor n in $n! = n(n-1)!$ and one factor θ, we get

$$\mu = n\theta \cdot \sum_{x=1}^{n} \binom{n-1}{x-1} \theta^{x-1} (1-\theta)^{n-x}$$

and, letting $y = x - 1$ and $m = n - 1$, this becomes

$$\mu = n\theta \cdot \sum_{y=0}^{m} \binom{m}{y} \theta^y (1-\theta)^{m-y} = n\theta$$

since the last summation is the sum of all the values of a binomial distribution with the parameters m and θ, and hence equal to 1.

To find expressions for μ_2' and σ^2, let us make use of the fact that $E(X^2) = E[X(X-1)] + E(X)$ and first evaluate $E[X(X-1)]$. Duplicating for all practical purposes the steps used before, we thus get

$$E[X(X-1)] = \sum_{x=0}^{n} x(x-1) \binom{n}{x} \theta^x (1-\theta)^{n-x}$$

$$= \sum_{x=2}^{n} \frac{n!}{(x-2)!(n-x)!} \theta^x (1-\theta)^{n-x}$$

$$= n(n-1)\theta^2 \cdot \sum_{x=2}^{n} \binom{n-2}{x-2} \theta^{x-2} (1-\theta)^{n-x}$$

and, letting $y = x - 2$ and $m = n - 2$, this becomes

$$E[X(X-1)] = n(n-1)\theta^2 \cdot \sum_{y=0}^{m} \binom{m}{y} \theta^y (1-\theta)^{m-y}$$
$$= n(n-1)\theta^2$$

Therefore,

$$\mu_2' = E[X(X-1)] + E(X) = n(n-1)\theta^2 + n\theta$$

and, finally,

$$\sigma^2 = \mu_2' - \mu^2$$
$$= n(n-1)\theta^2 + n\theta - n^2\theta^2$$
$$= n\theta(1-\theta)$$

An alternative proof of this theorem, requiring much less algebraic detail, is suggested in Exercise 6.

It should not have come as a surprise that the mean of the binomial distribution is given by the product $n\theta$. After all, if a balanced coin is flipped 200 times, we expect (in the sense of a mathematical expectation) $200 \cdot \frac{1}{2} = 100$ heads and 100 tails; similarly, if a balanced die is rolled 240 times, we expect $240 \cdot \frac{1}{6} = 40$ sixes, and if the probability is 0.80 that a person shopping at a department store will make a purchase, we would expect $400(0.80) = 320$ of 400 persons shopping at the department store to make a purchase.

The formula for the variance of the binomial distribution, being a measure of variation, has many important applications; but, to emphasize its significance, let us consider the random variable $Y = \frac{X}{n}$, where X is a random variable having a binomial distribution with the parameters n and θ. This random variable is the proportion of successes in n trials, and in Exercise 6 the reader will be asked to prove the following result.

THEOREM 3. If X has a binomial distribution with the parameters n and θ and $Y = \frac{X}{n}$, then

$$E(Y) = \theta \quad \text{and} \quad \sigma_Y^2 = \frac{\theta(1-\theta)}{n}$$

Now, if we apply Chebyshev's theorem with $k\sigma = c$, we can assert that *for any positive constant* c *the probability is at least*

$$1 - \frac{\theta(1-\theta)}{nc^2}$$

that the proportion of successes in n *trials falls between* $\theta - c$ *and* $\theta + c$. Hence, *when* $n \to \infty$, *the probability approaches* 1 *that the proportion of successes will differ from* θ *by less than any arbitrary constant* c. This result is called a **law of large numbers**, and it should be observed that it applies to the proportion of successes, not to their actual number. It is a fallacy to suppose that when n is large the number of successes must necessarily be close to $n\theta$.

```
MTB > BRANDOM 1ØØ N=1  P=.5 C1

  1ØØ BINOMIAL EXPERIMENTS WITH N = 1 AND P = .5ØØØ
     Ø.    Ø.    1.  1.  1.   1.   1.   Ø.   Ø.   1.
     1.    Ø.    Ø.  1.  Ø.   1.   1.   1.   Ø.   1.
     Ø.    Ø.    1.  Ø.  1.   1.   Ø.   1.   Ø.   Ø.
     1.    1.    Ø.  1.  Ø.   Ø.   1.   1.   1.   Ø.
     1.    Ø.    1.  Ø.  Ø.   Ø.   Ø.   1.   Ø.   Ø.
     1.    1.    Ø.  Ø.  Ø.   Ø.   Ø.   1.   Ø.   Ø.
     1.    1.    Ø.  Ø.  1.   1.   1.   Ø.   1.   1.
     1.    Ø.    1.  1.  Ø.   1.   1.   Ø.   Ø.   Ø.
     Ø.    Ø.    Ø.  1.  Ø.   Ø.   1.   Ø.   1.   1.
     1.    Ø.    1.  1.  1.   1.   Ø.   1.   Ø.   1.
  SUMMARY

  VALUE  FREQUENCY
    Ø       49
    1       51
```

Figure 2. Computer simulation of 100 flips of a balanced coin.

An easy illustration of this law of large numbers can be obtained through a **computer simulation** of the repeated flipping of a balanced coin. This is shown in Figure 2, where the 1's and 0's denote heads and tails.

Reading across successive rows, we find that among the first five simulated flips there are 3 heads, among the first ten there are 6 heads, among the first fifteen there are 8 heads, among the first twenty there are 12 heads, among the first twenty-five there are 14 heads, ..., and among all hundred there are 51 heads. The corresponding proportions, plotted in Figure 3, are $\frac{3}{5} = 0.60$, $\frac{6}{10} = 0.60$, $\frac{8}{15} = 0.53$, $\frac{12}{20} = 0.60$, $\frac{14}{25} = 0.56, \ldots$, and $\frac{51}{100} = 0.51$. Observe that the proportion of heads fluctuates but comes closer and closer to 0.50, the probability of heads for each flip of the coin.

Since the moment-generating function of the binomial distribution is easy to obtain, let us find it and use it to verify the results of Theorem 2.

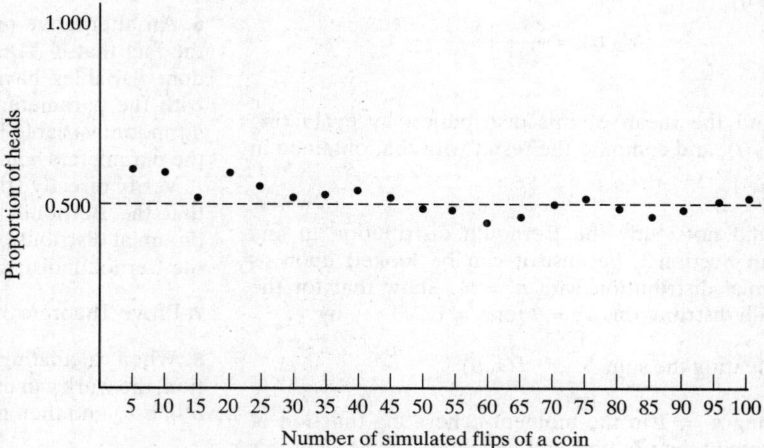

Figure 3. Graph illustrating the law of large numbers.

THEOREM 4. The moment-generating function of the binomial distribution is given by

$$M_X(t) = [1+\theta(e^t-1)]^n$$

If we differentiate $M_X(t)$ twice with respect to t, we get

$$\begin{aligned} M'_X(t) &= n\theta e^t[1+\theta(e^t-1)]^{n-1} \\ M''_X(t) &= n\theta e^t[1+\theta(e^t-1)]^{n-1} + n(n-1)\theta^2 e^{2t}[1+\theta(e^t-1)]^{n-2} \\ &= n\theta e^t(1-\theta+n\theta e^t)[1+\theta(e^t-1)]^{n-2} \end{aligned}$$

and, upon substituting $t = 0$, we get $\mu'_1 = n\theta$ and $\mu'_2 = n\theta(1-\theta+n\theta)$. Thus, $\mu = n\theta$ and $\sigma^2 = \mu'_2 - \mu^2 = n\theta(1-\theta+n\theta) - (n\theta)^2 = n\theta(1-\theta)$, which agrees with the formulas given in Theorem 2.

From the work of this section it may seem easier to find the moments of the binomial distribution with the moment-generating function than to evaluate them directly, but it should be apparent that the differentiation becomes fairly involved if we want to determine, say, μ'_3 or μ'_4. Actually, there exists yet an easier way of determining the moments of the binomial distribution; it is based on its **factorial moment-generating function**, which is explained in Exercise 12.

Exercises

1. If X has the discrete uniform distribution $f(x) = \frac{1}{k}$ for $x = 1, 2, \ldots, k$, show that
(a) its mean is $\mu = \frac{k+1}{2}$;
(b) its variance is $\sigma^2 = \frac{k^2-1}{12}$.

2. If X has the discrete uniform distribution $f(x) = \frac{1}{k}$ for $x = 1, 2, \ldots, k$, show that its moment-generating function is given by

$$M_X(t) = \frac{e^t(1-e^{kt})}{k(1-e^t)}$$

Also find the mean of this distribution by evaluating $\lim_{t\to 0} M'_X(t)$, and compare the result with that obtained in Exercise 1.

3. We did not study the Bernoulli distribution in any detail in Section 3, because it can be looked upon as a binomial distribution with $n = 1$. Show that for the Bernoulli distribution, $\mu'_r = \theta$ for $r = 1, 2, 3, \ldots$, by
(a) evaluating the sum $\sum_{x=0}^{1} x^r \cdot f(x;\theta)$;
(b) letting $n = 1$ in the moment-generating function of the binomial distribution and examining its Maclaurin's series.

4. This question has been intentionally omitted for this edition.

5. Verify that
(a) $b(x; n, \theta) = b(n-x; n, 1-\theta)$.
Also show that if $B(x; n, \theta) = \sum_{k=0}^{x} b(k; n, \theta)$ for $x = 0, 1, 2, \ldots, n$, then
(b) $b(x; n, \theta) = B(x; n, \theta) - B(x-1; n, \theta)$;
(c) $b(x; n, \theta) = B(n-x; n, 1-\theta) - B(n-x-1; n, 1-\theta)$;
(d) $B(x; n, \theta) = 1 - B(n-x-1; n, 1-\theta)$.

6. An alternative proof of Theorem 2 may be based on the fact that if $X_1, X_2, \ldots,$ and X_n are independent random variables having the same Bernoulli distribution with the parameter θ, then $Y = X_1 + X_2 + \cdots + X_n$ is a random variable having the binomial distribution with the parameters n and θ.

Verify directly (that is, without making use of the fact that the Bernoulli distribution is a special case of the binomial distribution) that the mean and the variance of the Bernoulli distribution are $\mu = \theta$ and $\sigma^2 = \theta(1-\theta)$.

7. Prove Theorem 3.

8. When calculating all the values of a binomial distribution, the work can usually be simplified by first calculating $b(0; n, \theta)$ and then using the recursion formula

$$b(x+1; n, \theta) = \frac{\theta(n-x)}{(x+1)(1-\theta)} \cdot b(x; n, \theta)$$

Verify this formula and use it to calculate the values of the binomial distribution with $n = 7$ and $\theta = 0.25$.

9. Use the recursion formula of Exercise 8 to show that for $\theta = \frac{1}{2}$ the binomial distribution has

(a) a maximum at $x = \dfrac{n}{2}$ when n is even;

(b) maxima at $x = \dfrac{n-1}{2}$ and $x = \dfrac{n+1}{2}$ when n is odd.

10. If X is a binomial random variable, for what value of θ is the probability $b(x; n, \theta)$ a maximum?

11. In the proof of Theorem 2 we determined the quantity $E[X(X-1)]$, called the second **factorial moment**. In general, the rth factorial moment of X is given by

$$\mu'_{(r)} = E[X(X-1)(X-2)\cdot \ldots \cdot (X-r+1)]$$

Express μ'_2, μ'_3, and μ'_4 in terms of factorial moments.

12. The **factorial moment-generating function** of a discrete random variable X is given by

$$F_X(t) = E(t^X) = \sum_x t^x \cdot f(x)$$

Show that the rth derivative of $F_X(t)$ with respect to t at $t = 1$ is $\mu'_{(r)}$, the rth factorial moment defined in Exercise 11.

13. With reference to Exercise 12, find the factorial moment-generating function of

(a) the Bernoulli distribution and show that $\mu'_{(1)} = \theta$ and $\mu'_{(r)} = 0$ for $r > 1$;

(b) the binomial distribution and use it to find μ and σ^2.

14. This question has been intentionally omitted for this edition.

15. This question has been intentionally omitted for this edition.

5 The Negative Binomial and Geometric Distributions

In connection with repeated Bernoulli trials, we are sometimes interested in the number of the trial on which the kth success occurs. For instance, we may be interested in the probability that the tenth child exposed to a contagious disease will be the third to catch it, the probability that the fifth person to hear a rumor will be the first one to believe it, or the probability that a burglar will be caught for the second time on his or her eighth job.

If the kth success is to occur on the xth trial, there must be $k-1$ successes on the first $x-1$ trials, and the probability for this is

$$b(k-1; x-1, \theta) = \binom{x-1}{k-1} \theta^{k-1}(1-\theta)^{x-k}$$

The probability of a success on the xth trial is θ, and the probability that the kth success occurs on the xth trial is, therefore,

$$\theta \cdot b(k-1; x-1, \theta) = \binom{x-1}{k-1} \theta^{k}(1-\theta)^{x-k}$$

DEFINITION 4. NEGATIVE BINOMIAL DISTRIBUTION. *A random variable X has a* ***negative binomial distribution*** *and it is referred to as a negative binomial random variable if and only if*

$$b^*(x; k, \theta) = \binom{x-1}{k-1} \theta^{k}(1-\theta)^{x-k} \quad \text{for } x = k, k+1, k+2, \ldots$$

Thus, the number of the trial on which the kth success occurs is a random variable having a negative binomial distribution with the parameters k and θ. The name "negative binomial distribution" derives from the fact that the values of $b^*(x; k, \theta)$ for $x = k, k+1, k+2, \ldots$ are the successive terms of the binomial expansion of

$\left(\frac{1}{\theta} - \frac{1-\theta}{\theta}\right)^{-k}$.[†] In the literature of statistics, negative binomial distributions are also referred to as **binomial waiting-time distributions** or as **Pascal distributions**.

EXAMPLE 3

If the probability is 0.40 that a child exposed to a certain contagious disease will catch it, what is the probability that the tenth child exposed to the disease will be the third to catch it?

Solution

Substituting $x = 10$, $k = 3$, and $\theta = 0.40$ into the formula for the negative binomial distribution, we get

$$\begin{aligned} b^*(10; 3, 0.40) &= \binom{9}{2}(0.40)^3(0.60)^7 \\ &= 0.0645 \end{aligned}$$

When a table of binomial probabilities is available, the determination of negative binomial probabilities can generally be simplified by making use of the following identity.

THEOREM 5.

$$b^*(x; k, \theta) = \frac{k}{x} \cdot b(k; x, \theta)$$

The reader will be asked to verify this theorem in Exercise 18.

EXAMPLE 4

Use Theorem 5 and Table I of "Statistical Tables" to rework Example 3.

Solution

Substituting $x = 10$, $k = 3$, and $\theta = 0.40$ into the formula of Theorem 5, we get

$$\begin{aligned} b^*(10; 3, 0.40) &= \frac{3}{10} \cdot b(3; 10, 0.40) \\ &= \frac{3}{10}(0.2150) \\ &= 0.0645 \end{aligned}$$

Moments of the negative binomial distribution may be obtained by proceeding as in the proof of Theorem 2; for the mean and the variance we obtain the following theorem.

THEOREM 6. The mean and the variance of the negative binomial distribution are

$$\mu = \frac{k}{\theta} \quad \text{and} \quad \sigma^2 = \frac{k}{\theta}\left(\frac{1}{\theta} - 1\right)$$

as the reader will be asked to verify in Exercise 19.

[†] Binomial expansions with negative exponents are explained in Feller, W., *An Introduction to Probability Theory and Its Applications*, Vol. I, 3rd ed. New York: John Wiley & Sons, Inc., 1968.

Since the negative binomial distribution with $k = 1$ has many important applications, it is given a special name; it is called the **geometric distribution**.

DEFINITION 5. GEOMETRIC DISTRIBUTION. *A random variable X has a **geometric distribution** and it is referred to as a geometric random variable if and only if its probability distribution is given by*

$$g(x; \theta) = \theta(1-\theta)^{x-1} \quad \text{for } x = 1, 2, 3, \ldots$$

EXAMPLE 5

If the probability is 0.75 that an applicant for a driver's license will pass the road test on any given try, what is the probability that an applicant will finally pass the test on the fourth try?

Solution

Substituting $x = 4$ and $\theta = 0.75$ into the formula for the geometric distribution, we get

$$\begin{aligned} g(4; 0.75) &= 0.75(1-0.75)^{4-1} \\ &= 0.75(0.25)^3 \\ &= 0.0117 \end{aligned}$$

Of course, this result is based on the assumption that the trials are all independent, and there may be some question here about its validity.

6 The Hypergeometric Distribution

finite math class

To obtain a formula analogous to that of the binomial distribution that applies to sampling without replacement, in which case the trials are not independent, let us consider a set of N elements of which M are looked upon as successes and the other $N - M$ as failures. As in connection with the binomial distribution, we are interested in the probability of getting x successes in n trials, but now we are choosing, without replacement, n of the N elements contained in the set.

There are $\binom{M}{x}$ ways of choosing x of the M successes and $\binom{N-M}{n-x}$ ways of choosing $n - x$ of the $N - M$ failures, and, hence, $\binom{M}{x}\binom{N-M}{n-x}$ ways of choosing x successes and $n - x$ failures. Since there are $\binom{N}{n}$ ways of choosing n of the N elements in the set, and we shall assume that they are all equally likely (which is what we mean when we say that the selection is random), the probability of "x successes in n trials" is $\binom{M}{x}\binom{N-M}{n-x} \Big/ \binom{N}{n}$.

DEFINITION 6. HYPERGEOMETRIC DISTRIBUTION. *A random variable X has a **hypergeometric distribution** and it is referred to as a hypergeometric random variable if and only if its probability distribution is given by*

$$h(x; n, N, M) = \frac{\binom{M}{x}\binom{N-M}{n-x}}{\binom{N}{n}} \quad \text{for } x = 0, 1, 2, \ldots, n$$
$$x \leq M \text{ and } n - x \leq N - M$$

Thus, for sampling without replacement, the number of successes in n trials is a random variable having a hypergeometric distribution with the parameters n, N, and M.

EXAMPLE 6

As part of an air-pollution survey, an inspector decides to examine the exhaust of 6 of a company's 24 trucks. If 4 of the company's trucks emit excessive amounts of pollutants, what is the probability that none of them will be included in the inspector's sample?

Solution

Substituting $x = 0$, $n = 6$, $N = 24$, and $M = 4$ into the formula for the hypergeometric distribution, we get

$$h(0; 6, 24, 4) = \frac{\binom{4}{0}\binom{20}{6}}{\binom{24}{6}}$$

$$= 0.2880$$

The method by which we find the mean and the variance of the hypergeometric distribution is very similar to that employed in the proof of Theorem 2.

THEOREM 7. The mean and the variance of the hypergeometric distribution are

$$\mu = \frac{nM}{N} \quad \text{and} \quad \sigma^2 = \frac{nM(N-M)(N-n)}{N^2(N-1)}$$

Proof To determine the mean, let us directly evaluate the sum

$$\mu = \sum_{x=0}^{n} x \cdot \frac{\binom{M}{x}(N - Mn - x)}{\binom{N}{n}}$$

$$= \sum_{x=1}^{n} \frac{M!}{(x-1)!(M-x)!} \cdot \frac{\binom{N-M}{n-x}}{\binom{N}{n}}$$

where we omitted the term corresponding to $x = 0$, which is 0, and canceled the x against the first factor of $x! = x(x-1)!$ in the denominator of $\binom{M}{x}$. Then, factoring out $M \Big/ \binom{N}{n}$, we get

$$\mu = \frac{M}{\binom{N}{n}} \cdot \sum_{x=1}^{n} \binom{M-1}{x-1} \binom{N-M}{n-x}$$

and, letting $y = x-1$ and $m = n-1$, this becomes

$$\mu = \frac{M}{\binom{N}{n}} \cdot \sum_{y=0}^{m} \binom{M-1}{y} \binom{N-M}{m-y}$$

Finally, using $\sum_{r=0}^{k} \binom{m}{r} \binom{n}{k-r} = \binom{m+n}{k}$, we get

$$\mu = \frac{M}{\binom{N}{n}} \cdot \binom{N-1}{m} = \frac{M}{\binom{N}{n}} \cdot \binom{N-1}{n-1} = \frac{nM}{N}$$

To obtain the formula for σ^2, we proceed as in the proof of Theorem 2 by first evaluating $E[X(X-1)]$ and then making use of the fact that $E(X^2) = E[X(X-1)] + E(X)$. Leaving it to the reader to show that

$$E[X(X-1)] = \frac{M(M-1)n(n-1)}{N(N-1)}$$

in Exercise 27, we thus get

$$\begin{aligned}\sigma^2 &= \frac{M(M-1)n(n-1)}{N(N-1)} + \frac{nM}{N} - \left(\frac{nM}{N}\right)^2 \\ &= \frac{nM(N-M)(N-n)}{N^2(N-1)}\end{aligned}$$

The moment-generating function of the hypergeometric distribution is fairly complicated. Details of this may be found in the book *The Advanced Theory of Statistics* by M. G. Kendall and A. Stuart.

When N is large and n is relatively small compared to N (the usual rule of thumb is that n should not exceed 5 percent of N), there is not much difference between sampling with replacement and sampling without replacement, and the formula for the binomial distribution with the parameters n and $\theta = \frac{M}{N}$ may be used to approximate hypergeometric probabilities.

EXAMPLE 7

Among the 120 applicants for a job, only 80 are actually qualified. If 5 of the applicants are randomly selected for an in-depth interview, find the probability that only 2 of the 5 will be qualified for the job by using

(a) the formula for the hypergeometric distribution;

(b) the formula for the binomial distribution with $\theta = \frac{80}{120}$ as an approximation.

Solution

(a) Substituting $x = 2$, $n = 5$, $N = 120$, and $M = 80$ into the formula for the hypergeometric distribution, we get

$$h(2; 5, 120, 80) = \frac{\binom{80}{2}\binom{40}{3}}{\binom{120}{5}}$$

$$= 0.164$$

rounded to three decimals;

(b) substituting $x = 2$, $n = 5$, and $\theta = \frac{80}{120} = \frac{2}{3}$ into the formula for the binomial distribution, we get

$$b\left(2; 5, \frac{2}{3}\right) = \binom{5}{2}\left(\frac{2}{3}\right)^2\left(1 - \frac{2}{3}\right)^3$$

$$= 0.165$$

rounded to three decimals. As can be seen from these results, the approximation is very close.

7 The Poisson Distribution

When n is large, the calculation of binomial probabilities with the formula of Definition 3 will usually involve a prohibitive amount of work. For instance, to calculate the probability that 18 of 3,000 persons watching a parade on a very hot summer day will suffer from heat exhaustion, we first have to determine $\binom{3{,}000}{18}$, and if the probability is 0.005 that any one of the 3,000 persons watching the parade will suffer from heat exhaustion, we also have to calculate the value of $(0.005)^{18}(0.995)^{2{,}982}$.

In this section we shall present a probability distribution that can be used to approximate binomial probabilities of this kind. Specifically, we shall investigate the limiting form of the binomial distribution when $n \to \infty$, $\theta \to 0$, while $n\theta$ remains constant. Letting this constant be λ, that is, $n\theta = \lambda$ and, hence, $\theta = \frac{\lambda}{n}$, we can write

$$b(x; n, \theta) = \binom{n}{x}\left(\frac{\lambda}{n}\right)^x\left(1-\frac{\lambda}{n}\right)^{n-x}$$

$$= \frac{n(n-1)(n-2)\cdot\ldots\cdot(n-x+1)}{x!}\left(\frac{\lambda}{n}\right)^x\left(1-\frac{\lambda}{n}\right)^{n-x}$$

Then, if we divide one of the x factors n in $\left(\frac{\lambda}{n}\right)^x$ into each factor of the product $n(n-1)(n-2)\cdot\ldots\cdot(n-x+1)$ and write

$$\left(1-\frac{\lambda}{n}\right)^{n-x} \quad \text{as} \quad \left[\left(1-\frac{\lambda}{n}\right)^{-n/\lambda}\right]^{-\lambda}\left(1-\frac{\lambda}{n}\right)^{-x}$$

we obtain

$$\frac{1\left(1-\frac{1}{n}\right)\left(1-\frac{2}{n}\right)\cdot\ldots\cdot\left(1-\frac{x-1}{n}\right)}{x!}(\lambda)^x\left[\left(1-\frac{\lambda}{n}\right)^{-n/\lambda}\right]^{-\lambda}\left(1-\frac{\lambda}{n}\right)^{-x}$$

Finally, if we let $n \to \infty$ while x and λ remain fixed, we find that

$$1\left(1-\frac{1}{n}\right)\left(1-\frac{2}{n}\right)\cdot\ldots\cdot\left(1-\frac{x-1}{n}\right) \to 1$$

$$\left(1-\frac{\lambda}{n}\right)^{-x} \to 1$$

$$\left(1-\frac{\lambda}{n}\right)^{-n/\lambda} \to e$$

and, hence, that the limiting distribution becomes

$$p(x; \lambda) = \frac{\lambda^x e^{-\lambda}}{x!} \quad \text{for } x = 0, 1, 2, \ldots$$

DEFINITION 7. POISSON DISTRIBUTION. *A random variable has a **Poisson distribution** and it is referred to as a Poisson random variable if and only if its probability distribution is given by*

$$p(x; \lambda) = \frac{\lambda^x e^{-\lambda}}{x!} \quad \text{for } x = 0, 1, 2, \ldots$$

Thus, in the limit when $n \to \infty$, $\theta \to 0$, and $n\theta = \lambda$ remains constant, the number of successes is a random variable having a Poisson distribution with the parameter λ. This distribution is named after the French mathematician Simeon Poisson (1781–1840). In general, the Poisson distribution will provide a good approximation to binomial probabilities when $n \geqq 20$ and $\theta \leqq 0.05$. When $n \geqq 100$ and $n\theta < 10$, the approximation will generally be excellent.

To get some idea about the closeness of the Poisson approximation to the binomial distribution, consider the computer printout of Figure 4, which shows, one above the other, the binomial distribution with $n = 150$ and $\theta = 0.05$ and the Poisson distribution with $\lambda = 150(0.05) = 7.5$.

EXAMPLE 8

Use Figure 4 to determine the value of x (from 5 to 15) for which the error is greatest when we use the Poisson distribution with $\lambda = 7.5$ to approximate the binomial distribution with $n = 150$ and $\theta = 0.05$.

Solution

Calculating the differences corresponding to $x = 5, x = 6, \ldots, x = 15$, we get 0.0006, −0.0017, −0.0034, −0.0037, −0.0027, −0.0011, 0.0003, 0.0011, 0.0013, 0.0011, and 0.0008. Thus, the maximum error (numerically) is −0.0037, and it corresponds to $x = 8$.

The examples that follow illustrate the Poisson approximation to the binomial distribution.

EXAMPLE 9

If 2 percent of the books bound at a certain bindery have defective bindings, use the Poisson approximation to the binomial distribution to determine the probability that 5 of 400 books bound by this bindery will have defective bindings.

Solution

Substituting $x = 5$, $\lambda = 400(0.02) = 8$, and $e^{-8} = 0.00034$ (from Table VIII of "Statistical Tables") into the formula of Definition 7, we get

$$p(5; 8) = \frac{8^5 \cdot e^{-8}}{5!} = \frac{(32{,}768)(0.00034)}{120} = 0.093$$

In actual practice, Poisson probabilities are seldom obtained by direct substitution into the formula of Definition 7. Sometimes we refer to tables of Poisson probabilities, such as Table II of "Statistical Tables", or more extensive tables in handbooks of statistical tables, but more often than not, nowadays, we refer to suitable computer software. The use of tables or computers is of special importance when we are concerned with probabilities relating to several values of x.

EXAMPLE 10

Records show that the probability is 0.00005 that a car will have a flat tire while crossing a certain bridge. Use the Poisson distribution to approximate the binomial probabilities that, among 10,000 cars crossing this bridge,

(a) exactly two will have a flat tire;

(b) at most two will have a flat tire.

Solution

(a) Referring to Table II of "Statistical Tables", we find that for $x = 2$ and $\lambda = 10,000(0.00005) = 0.5$, the Poisson probability is 0.0758.

(b) Referring to Table II of "Statistical Tables", we find that for $x = 0$, 1, and 2, and $\lambda = 0.5$, the Poisson probabilities are 0.6065, 0.3033, and 0.0758. Thus, the

```
MTB > BINOMIAL N=15Ø   P=Ø.Ø5
   BINOMIAL PROBABILITIES FOR N = 1.5Ø AND P = .Ø5ØØØØ

      K          P(X = K)      P(X LESS OR = K)
      Ø           .ØØØ5            .ØØØ5
      1           .ØØ36            .ØØ41
      2           .Ø141            .Ø182
      3           .Ø366            .Ø548
      4           .Ø7Ø8            .1256
      5           .1Ø88            .2344
      6           .1384            .3729
      7           .1499            .5228
      8           .141Ø            .6638
      9           .1171            .78Ø9
     10           .Ø869            .8678
     11           .Ø582            .926Ø
     12           .Ø355            .9615
     13           .Ø198            .9813
     14           .Ø1Ø2            .9915
     15           .ØØ49            .9964
     16           .ØØ22            .9986
     17           .ØØØ9            .9995
     18           .ØØØ3            .9998
     19           .ØØØ1            .9999

MTB > POISSON MU=7.5
   POISSON PROBABILITIES FOR MEAN =  7.5ØØ

      K          P(X = K)      P(X LESS OR = K)
      Ø           .ØØØ6            .ØØØ6
      1           .ØØ41            .ØØ47
      2           .Ø156            .Ø2Ø3
      3           .Ø389            .Ø591
      4           .Ø729            .1321
      5           .1Ø94            .2414
      6           .1367            .3782
      7           .1465            .5246
      8           .1373            .662Ø
      9           .1144            .7764
     10           .Ø858            .8622
     11           .Ø585            .92Ø8
     12           .Ø366            .9573
     13           .Ø211            .9784
     14           .Ø113            .9897
     15           .ØØ57            .9954
     16           .ØØ26            .998Ø
     17           .ØØ12            .9992
     18           .ØØØ5            .9997
     19           .ØØØ2            .9999
     20           .ØØØ1           1.ØØØØ
```

Figure 4. Computer printout of the binomial distribution with $n = 150$ and $\theta = 0.05$ and the Poisson distribution with $\lambda = 7.5$.

probability that at most 2 of 10,000 cars crossing the bridge will have a flat tire is

$$0.6065 + 0.3033 + 0.0758 = 0.9856$$

EXAMPLE 11

Use Figure 5 to rework the preceding example.

Solution

(a) Reading off the value for $K = 2$ in the P(X = K) column, we get 0.0758.

(b) Here we could add the values for $K = 0$, $K = 1$, and $K = 2$ in the P(X = K) column, or we could read the value for $K = 2$ in the P(X LESS OR = K) column, getting 0.9856.

Having derived the Poisson distribution as a limiting form of the binomial distribution, we can obtain formulas for its mean and its variance by applying the same limiting conditions ($n \to \infty$, $\theta \to 0$, and $n\theta = \lambda$ remains constant) to the mean and the variance of the binomial distribution. For the mean we get $\mu = n\theta = \lambda$ and for the variance we get $\sigma^2 = n\theta(1-\theta) = \lambda(1-\theta)$, which approaches λ when $\theta \to 0$.

THEOREM 8. The mean and the variance of the Poisson distribution are given by

$$\mu = \lambda \quad \text{and} \quad \sigma^2 = \lambda$$

These results can also be obtained by directly evaluating the necessary summations (see Exercise 33) or by working with the moment-generating function given in the following theorem.

THEOREM 9. The moment-generating function of the Poisson distribution is given by

$$M_X(t) = e^{\lambda(e^t - 1)}$$

Proof By Definition 7 and the definition of moment-generating function—*The **moment generating function** of a random variable X, where it exists, is given by* $M_X(t) = E(e^{tX}) = \sum_x e^{tX} \cdot f(x)$ *when X is discrete, and*

```
MTB > POISSON MU=.5

POISSON PROBABILITIES FOR MEAN = .5ØØ

K    P(X = K)  P(X LESS OR = K)
Ø    .6Ø65          .6Ø65
1    .3Ø33          .9Ø98
2    .Ø758          .9856
3    .Ø126          .9982
4    .ØØ16          .9998
5    .ØØØ2         1.ØØØØ
```

Figure 5. Computer printout of the Poisson distribution with $\lambda = 0.5$.

$M_X(t) = E(e^{tX}) = \int_{-\infty}^{\infty} e^{tx} \cdot f(x)dx$ *when X is continuous*—we get

$$M_X(t) = \sum_{x=0}^{\infty} e^{xt} \cdot \frac{\lambda^x e^{-\lambda}}{x!} = e^{-\lambda} \cdot \sum_{x=0}^{\infty} \frac{(\lambda e^t)^x}{x!}$$

where $\sum_{x=0}^{\infty} \frac{(\lambda e^t)^x}{x!}$ can be recognized as the Maclaurin's series of e^z with $z = \lambda e^t$. Thus,

$$M_X(t) = e^{-\lambda} \cdot e^{\lambda e^t} = e^{\lambda(e^t - 1)}$$

Then, if we differentiate $M_X(t)$ twice with respect to t, we get

$$M'_X(t) = \lambda e^t e^{\lambda(e^t-1)}$$
$$M''_X(t) = \lambda e^t e^{\lambda(e^t-1)} + \lambda^2 e^{2t} e^{\lambda(e^t-1)}$$

so that $\mu'_1 = M'_X(0) = \lambda$ and $\mu'_2 = M''_X(0) = \lambda + \lambda^2$. Thus, $\mu = \lambda$ and $\sigma^2 = \mu'_2 - \mu^2 = (\lambda + \lambda^2) - \lambda^2 = \lambda$, which agrees with Theorem 8.

Although the Poisson distribution has been derived as a limiting form of the binomial distribution, it has many applications that have no direct connection with binomial distributions. For example, the Poisson distribution can serve as a model for the number of successes that occur during a given time interval or in a specified region when (1) the numbers of successes occurring in nonoverlapping time intervals or regions are independent, (2) the probability of a single success occurring in a very short time interval or in a very small region is proportional to the length of the time interval or the size of the region, and (3) the probability of more than one success occurring in such a short time interval or falling in such a small region is negligible. Hence, a Poisson distribution might describe the number of telephone calls per hour received by an office, the number of typing errors per page, or the number of bacteria in a given culture when the average number of successes, λ, for the given time interval or specified region is known.

EXAMPLE 12

The average number of trucks arriving on any one day at a truck depot in a certain city is known to be 12. What is the probability that on a given day fewer than 9 trucks will arrive at this depot?

Solution

Let X be the number of trucks arriving on a given day. Then, using Table II of "Statistical Tables" with $\lambda = 12$, we get

$$P(X < 9) = \sum_{x=0}^{8} p(x; 12) = 0.1550$$

If, in a situation where the preceding conditions apply, successes occur at a mean rate of α per *unit* time or per *unit* region, then the number of successes in an interval of t units of time or t units of the specified region is a Poisson random variable with the mean $\lambda = \alpha t$ (see Exercise 31). Therefore, the number of successes, X, in a time interval of length t units or a region of size t units has the Poisson distribution

$$p(x; \alpha t) = \frac{e^{-\alpha t}(\alpha t)^x}{x!} \quad \text{for } x = 0, 1, 2, \ldots$$

EXAMPLE 13

A certain kind of sheet metal has, on the average, five defects per 10 square feet. If we assume a Poisson distribution, what is the probability that a 15-square-foot sheet of the metal will have at least six defects?

Solution

Let X denote the number of defects in a 15-square-foot sheet of the metal. Then, since the unit of area is 10 square feet, we have

$$\lambda = \alpha t = (5)(1.5) = 7.5$$

and

$$P(X \geqq 6) = 1 - P(X \leqq 5) = 1 - 0.2414 = 0.7586$$

according to the computer printout shown in Figure 4.

Exercises

16. The negative binomial distribution is sometimes defined in a different way as the distribution of the number of failures that precede the kth success. If the kth success occurs on the xth trial, it must be preceded by $x - k$ failures. Thus, find the distribution of $Y = X - k$, where X has the distribution of Definition 4.

17. With reference to Exercise 16, find expressions for μ_Y and σ_Y^2.

18. Prove Theorem 5.

19. Prove Theorem 6 by first determining $E(X)$ and $E[X(X+1)]$.

20. Show that the moment-generating function of the geometric distribution is given by

$$M_X(t) = \frac{\theta e^t}{1 - e^t(1-\theta)}$$

21. Use the moment-generating function derived in Exercise 20 to show that for the geometric distribution, $\mu = \frac{1}{\theta}$ and $\sigma^2 = \frac{1-\theta}{\theta^2}$.

22. Differentiating with respect to θ the expressions on both sides of the equation

$$\sum_{x=1}^{\infty} \theta(1-\theta)^{x-1} = 1$$

show that the mean of the geometric distribution is given by $\mu = \frac{1}{\theta}$. Then, differentiating again with respect to θ, show that $\mu_2' = \frac{2-\theta}{\theta^2}$ and hence that $\sigma^2 = \frac{1-\theta}{\theta^2}$.

23. If X is a random variable having a geometric distribution, show that

$$P(X = x + n | X > n) = P(X = x)$$

24. If the probability is $f(x)$ that a product fails the xth time it is being used, that is, on the xth trial, then its **failure rate** at the xth trial is the probability that it will fail on the xth trial given that it has not failed on the first $x - 1$ trials; symbolically, it is given by

$$Z(x) = \frac{f(x)}{1 - F(x-1)}$$

where $F(x)$ is the value of the corresponding distribution function at x. Show that if X is a geometric random variable, its failure rate is constant and equal to θ.

25. A variation of the binomial distribution arises when the n trials are all independent, but the probability of a success on the ith trial is θ_i, and these probabilities are not all equal. If X is the number of successes obtained under these conditions in n trials, show that

(a) $\mu_X = n\theta$, where $\theta = \frac{1}{n} \cdot \sum_{i=1}^{n} \theta_i$;

(b) $\sigma_X^2 = n\theta(1-\theta) - n\sigma_\theta^2$, where θ is as defined in part (a) and $\sigma_\theta^2 = \frac{1}{n} \cdot \sum_{i=1}^{n} (\theta_i - \theta)^2$.

26. When calculating all the values of a hypergeometric distribution, the work can often be simplified by first calculating $h(0; n, N, M)$ and then using the recursion formula

$$h(x+1; n, N, M) = \frac{(n-x)(M-x)}{(x+1)(N-M-n+x+1)} \cdot h(x; n, N, M)$$

Verify this formula and use it to calculate the values of the hypergeometric distribution with $n = 4$, $N = 9$, and $M = 5$.

27. Verify the expression given for $E[X(X-1)]$ in the proof of Theorem 7.

28. Show that if we let $\theta = \frac{M}{N}$ in Theorem 7, the mean and the variance of the hypergeometric distribution can be written as $\mu = n\theta$ and $\sigma^2 = n\theta(1-\theta) \cdot \frac{N-n}{N-1}$. How do these results tie in with the discussion in the theorem?

29. When calculating all the values of a Poisson distribution, the work can often be simplified by first calculating $p(0; \lambda)$ and then using the recursion formula

$$p(x+1; \lambda) = \frac{\lambda}{x+1} \cdot p(x; \lambda)$$

Verify this formula and use it and $e^{-2} = 0.1353$ to verify the values given in Table II of "Statistical Tables" for $\lambda = 2$.

30. Approximate the binomial probability $b(3; 100, 0.10)$ by using

(a) the formula for the binomial distribution and logarithms;

(b) Table II of "Statistical Tables."

31. Suppose that $f(x,t)$ is the probability of getting x successes during a time interval of length t when (i) the probability of a success during a very small time interval from t to $t + \Delta t$ is $\alpha \cdot \Delta t$, (ii) the probability of more than one success during such a time interval is negligible, and (iii) the probability of a success during such a time interval does not depend on what happened prior to time t.

(a) Show that under these conditions

$$f(x, t+\Delta t) = f(x,t)[1 - \alpha \cdot \Delta t] + f(x-1,t)\alpha \cdot \Delta t$$

and hence that

$$\frac{d[f(x,t)]}{dt} = \alpha[f(x-1,t) - f(x,t)]$$

(b) Show by direct substitution that a solution of this infinite system of differential equations (there is one for each value of x) is given by the Poisson distribution with $\lambda = \alpha t$.

32. Use repeated integration by parts to show that

$$\sum_{y=0}^{x} \frac{\lambda^y e^{-\lambda}}{y!} = \frac{1}{x!} \cdot \int_{\lambda}^{\infty} t^x e^{-t}\, dt$$

This result is important because values of the distribution function of a Poisson random variable may thus be obtained by referring to a table of incomplete gamma functions.

33. Derive the formulas for the mean and the variance of the Poisson distribution by first evaluating $E(X)$ and $E[X(X-1)]$.

34. Show that if the limiting conditions $n \to \infty, \theta \to 0$, while $n\theta$ remains constant, are applied to the moment-generating function of the binomial distribution, we get the moment-generating function of the Poisson distribution.

[*Hint*: Make use of the fact that $\lim_{n\to\infty} \left(1 + \frac{z}{n}\right)^n = e^z$.]

35. This question has been intentionally omitted for this edition.

36. Differentiating with respect to λ the expressions on both sides of the equation

$$\mu_r = \sum_{x=0}^{\infty} (x-\lambda)^r \cdot \frac{\lambda^x e^{-\lambda}}{x!}$$

derive the following recursion formula for the moments about the mean of the Poisson distribution:

$$\mu_{r+1} = \lambda \left[r\mu_{r-1} + \frac{d\mu_r}{d\lambda} \right]$$

for $r = 1, 2, 3, \ldots$. Also, use this recursion formula and the fact that $\mu_0 = 1$ and $\mu_1 = 0$ to find μ_2, μ_3, and μ_4, and verify the formula given for α_3 in Exercise 35.

37. Use Theorem 9 to find the moment-generating function of $Y = X - \lambda$, where X is a random variable having the Poisson distribution with the parameter λ, and use it to verify that $\sigma_Y^2 = \lambda$.

8 The Multinomial Distribution

An immediate generalization of the binomial distribution arises when each trial has more than two possible outcomes, the probabilities of the respective outcomes are the same for each trial, and the trials are all independent. This would be the case, for instance, when persons interviewed by an opinion poll are asked whether they are for a candidate, against her, or undecided or when samples of manufactured products are rated excellent, above average, average, or inferior.

To treat this kind of problem in general, let us consider the case where there are n independent trials permitting k mutually exclusive outcomes whose respective probabilities are $\theta_1, \theta_2, \ldots, \theta_k$ $\left(\text{with } \sum_{i=1}^{k} \theta_i = 1\right)$. Referring to the outcomes as being of the first kind, the second kind, ..., and the kth kind, we shall be interested in the probability of getting x_1 outcomes of the first kind, x_2 outcomes of the second kind, ..., and x_k outcomes of the kth kind $\left(\text{with } \sum_{i=1}^{k} x_i = n\right)$.

Proceeding as in the derivation of the formula for the binomial distribution, we first find that the probability of getting x_1 outcomes of the first kind, x_2 outcomes of the second kind, ..., and x_k outcomes of the kth kind *in a specific order* is $\theta_1^{x_1} \cdot \theta_2^{x_2} \cdot \ldots \cdot \theta_k^{x_k}$. To get the corresponding probability for that many outcomes of each kind *in any order*, we shall have to multiply the probability for any specific order by

$$\binom{n}{x_1, x_2, \ldots, x_k} = \frac{n!}{x_1! \cdot x_2! \cdot \ldots \cdot x_k!}$$

DEFINITION 8. MULTINOMIAL DISTRIBUTION. *The random variables* $X_1, X_2, \ldots, X_n$ *have a* ***multinomial distribution*** *and they are referred to as multinomial random variables if and only if their joint probability distribution is given by*

$$f(x_1, x_2, \ldots, x_k; n, \theta_1, \theta_2, \ldots, \theta_k) = \binom{n}{x_1, x_2, \ldots, x_k} \cdot \theta_1^{x_1} \cdot \theta_2^{x_2} \cdot \ldots \cdot \theta_k^{x_k}$$

for $x_i = 0, 1, \ldots n$ *for each* i, *where* $\sum_{i=1}^{k} x_i = n$ *and* $\sum_{i=1}^{k} \theta_i = 1$.

Thus, the numbers of outcomes of the different kinds are random variables having the multinomial distribution with the parameters $n, \theta_1, \theta_2, \ldots$, and θ_k. The name "multinomial" derives from the fact that for various values of the x_i, the probabilities equal corresponding terms of the multinomial expansion of $(\theta_1 + \theta_2 + \cdots + \theta_k)^n$.

EXAMPLE 14

A certain city has 3 newspapers, A, B, and C. Newspaper A has 50 percent of the readers in that city. Newspaper B, has 30 percent of the readers, and newspaper C has the remaining 20 percent. Find the probability that, among 8 randomly-chosen readers in that city, 5 will read newspaper A, 2 will read newspaper B, and 1 will read newspaper C. (For the purpose of this example, assume that no one reads more than one newspaper.)

Solution

Substituting $x_1 = 5, x_2 = 2, x_3 = 1, \theta_1 = 0.50, \theta_2 = 0.30, \theta_3 = 0.20$, and $n = 8$ into the formula of Definition 8, we get

$$f(5,2,1;8,0.50,0.30,0.20) = \frac{8!}{5!\cdot 2!\cdot 1!}(0.50)^5(0.30)^2(0.20) = 0.0945$$

9 The Multivariate Hypergeometric Distribution

Just as the hypergeometric distribution takes the place of the binomial distribution for sampling without replacement, there also exists a multivariate distribution analogous to the multinomial distribution that applies to sampling without replacement. To derive its formula, let us consider a set of N elements, of which M_1 are elements of the first kind, M_2 are elements of the second kind, ..., and M_k are elements of the kth kind, such that $\sum_{i=1}^{k} M_i = N$. As in connection with the multinomial distribution, we are interested in the probability of getting x_1 elements (outcomes) of the first kind, x_2 elements of the second kind, ..., and x_k elements of the kth kind, but now we are choosing, without replacement, n of the N elements of the set.

There are $\binom{M_1}{x_1}$ ways of choosing x_1 of the M_1 elements of the first kind, $\binom{M_2}{x_2}$ ways of choosing x_2 of the M_2 elements of the second kind, ..., and $\binom{M_k}{x_k}$ ways of choosing x_k of the M_k elements of the kth kind, and, hence, $\binom{M_1}{x_1}\binom{M_2}{x_2}\cdot\ldots\cdot\binom{M_k}{x_k}$ ways of choosing the required $\sum_{i=1}^{k} x_i = n$ elements. Since there are $\binom{N}{n}$ ways of choosing n of the N elements in the set and we assume that they are all equally likely (which is what we mean when we say that the selection is random), it follows that the desired probability is given by $\binom{M_1}{x_1}\binom{M_2}{x_2}\cdot\ldots\cdot\binom{M_k}{x_k}\Big/\binom{N}{n}$.

DEFINITION 9. MULTIVARIATE HYPERGEOMETRIC DISTRIBUTION. *The random variables* $X_1, X_2, \ldots, X_k$ *have a* ***multivariate hypergeometric distribution*** *and they are referred to as multivariate hypergeometric random variables if and only if their joint probability distribution is given by*

$$f(x_1, x_2, \ldots, x_k; n, M_1, M_2, \ldots, M_k) = \frac{\binom{M_1}{x_1}\binom{M_2}{x_2}\cdot\ldots\cdot\binom{M_k}{x_k}}{\binom{N}{n}}$$

for $x_i = 0, 1, \ldots n$ *and* $x_i \le M_i$ *for each* i, *where* $\sum_{i=1}^{k} x_i = n$ *and* $\sum_{i=1}^{k} M_i = N$.

Thus, the joint distribution of the random variables under consideration, that is, the distribution of the numbers of outcomes of the different kinds, is a multivariate hypergeometric distribution with the parameters $n, M_1, M_2, \ldots,$ and M_k.

EXAMPLE 15

A panel of prospective jurors includes six married men, three single men, seven married women, and four single women. If the selection is random, what is the probability that a jury will consist of four married men, one single man, five married women, and two single women?

Solution
Substituting $x_1 = 4, x_2 = 1, x_3 = 5, x_4 = 2, M_1 = 6, M_2 = 3, M_3 = 7, M_4 = 4,$ $N = 20$, and $n = 12$ into the formula of Definition 9, we get

$$f(4,1,5,2;12,6,3,7,4) = \frac{\binom{6}{4}\binom{3}{1}\binom{7}{5}\binom{4}{2}}{\binom{20}{12}}$$

$$= 0.0450$$

Exercises

38. If $X_1, X_2, \ldots, X_k$ have the multinomial distribution of Definition 8, show that the mean of the marginal distribution of X_i is $n\theta_i$ for $i = 1, 2, \ldots, k$.

39. If $X_1, X_2, \ldots, X_k$ have the multinomial distribution of Definition 8, show that the covariance of X_i and X_j is $-n\theta_i\theta_j$ for $i = 1, 2, \ldots, k, j = 1, 2, \ldots, k,$ and $i \neq j$.

10 The Theory in Practice

In this section we shall discuss an important application of the binomial distribution, namely **sampling inspection**.

In sampling inspection, a specified sample of a lot of manufactured product is inspected under controlled, supervised conditions. If the number of defectives found in the sample exceeds a given **acceptance number**, the lot is rejected. (A rejected lot may be subjected to closer inspection, but it is rarely scrapped.) A **sampling plan** consists of a specification of the number of items to be included in the sample taken from each lot, and a statement about the maximum number of defectives allowed before rejection takes place.

The probability that a lot will be accepted by a given sampling plan, of course, will depend upon p, the actual proportion of defectives in the lot. Since the value of p is unknown, we calculate the probability of accepting a lot for several different values of p. Suppose a sampling plan requires samples of size n from each lot, and that the lot size is large with respect to n. Suppose, further, that the acceptance number is c; that is, the lot will be accepted if c defectives or fewer are found in the sample. The probability of acceptance, the probability of finding c or fewer defectives in a sample of size n, is given by the binomial distribution to a close approximation. (Since sampling inspection is done without replacement, the assumption of equal probabilities from trial to trial, underlying the binomial distribution, is violated. But if the sample size is small relative to the lot size, this assumption is nearly satisfied.) Thus, for large lots, the probability of accepting a lot having the proportion of defectives p is closely approximated by the following definition.

DEFINITION 10. PROBABILITY OF ACCEPTANCE. *If* n *is the size of the sample taken from each large lot and* c *is the acceptance number, the* ***probability of acceptance*** *is closely approximated by*

$$L(p) = \sum_{k=0}^{c} b(k; n, p) = B(c; n, p)$$

where p *is the actual proportion of defectives in the lot.*

This equation simply states that the probability of c or fewer defectives in the sample is given by the probability of 0 defectives, plus the probability of 1 defective, ..., up to the probability of c defectives, with each probability being approximated by the binomial distribution having the parameters n and $\theta = p$. Definition 10 is closely related to the power function.

It can be seen from this definition that, for a given sampling plan (sample size, n, and acceptance number, c), the probability of acceptance depends upon p, the actual (unknown) proportion of defectives in the lot. Thus a curve can be drawn that gives the probability of accepting a lot as a function of the lot proportion defective, p. This curve, called the **operating characteristic curve**, or **OC curve**, defines the characteristics of the sampling plan.

To illustrate the construction of an OC curve, let us consider the sampling plan having $n = 20$ and $c = 3$. That is, samples of size 20 are drawn from each lot, and a lot is accepted if the sample contains 3 or fewer defectives. Referring to the line in Table I of "Statistical Tables" corresponding to $n = 20$ and $x = 3$, the probabilities that a random variable having the binomial distribution $b(x; 20, p)$ will assume a value less than or equal to 3 for various values of p are as follows:

p	0.05	0.10	0.15	0.20	0.25	0.30	0.35	0.40	0.45
$L(p)$	0.9841	0.8670	0.6477	0.4114	0.2252	0.1071	0.0444	0.0160	0.0049

A graph of $L(p)$ versus p is shown in Figure 6.

Inspection of the OC curve given in Figure 6 shows that the probability of acceptance is quite high (greater than 0.9) for small values of p, say values less than about 0.10. Also, the probability of acceptance is low (less than 0.10) for values of p greater than about 0.30. If the actual proportion of defectives in the lot lies between 0.10 and 0.30, however, it is somewhat of a tossup whether the lot will be accepted or rejected.

An "ideal" OC curve would be like the one shown in Figure 7. In this figure, there is no "gray area"; that is, it is certain that a lot with a given small value of p or less will be accepted, and it is certain that a lot with a value of p greater than the given value will be rejected. By comparison, the OC curve of Figure 6 seems to do a poor job of discriminating between "good" and "bad" lots. In such cases, a better OC curve can be obtained by increasing the sample size, n.

The OC curve of a sampling plan never can be like the ideal curve of Figure 7 with finite sample sizes, as there always will be some statistical error associated with sampling. However, sampling plans can be evaluated by choosing two values of p considered to be important and calculating the probabilities of lot acceptance at these values. First, a number, p_0, is chosen so that a lot containing a proportion of defectives less than or equal to p_0 is desired to be accepted. This value of p is

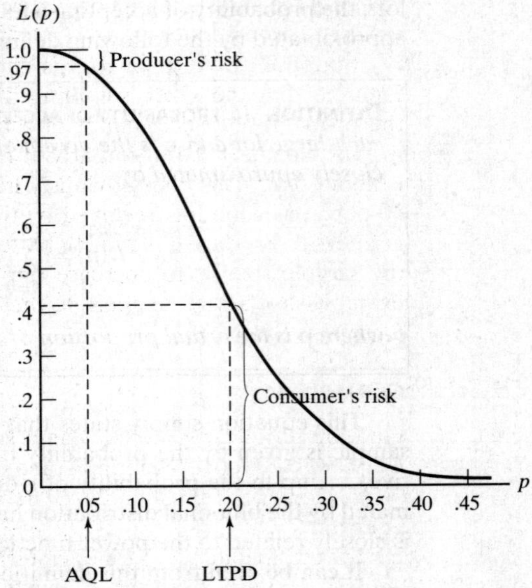

Figure 6. OC curve.

called the **acceptable quality level**, or **AQL**. Then, a second value of p, p_1, is chosen so that we wish to reject a lot containing a proportion of defectives greater than p_1. This value of p is called the **lot tolerance percentage defective**, or **LTPD**. We evaluate a sampling plan by finding the probability that a "good" lot (a lot with $p \leq p_0$) will be rejected and the probability that a "bad" lot (one with $p \geq p_1$) will be accepted.

The probability that a "good" lot will be rejected is called the **producer's risk**, and the probability that a "bad" lot will be accepted is called the **consumer's risk**. The producer's risk expresses the probability that a "good" lot (one with $p < p_0$) will erroneously be rejected by the sampling plan. It is the risk that the producer takes as a consequence of sampling variability. The consumer's risk is the probability that the consumer erroneously will receive a "bad" lot (one with $p > p_1$). These risks are analogous to the type I and type II errors, α and β (If the true value of the parameter θ is θ_0 and the statistician incorrectly concludes that $\theta = \theta_1$, he is committing an error referred to as a type I error. On the other hand, if the true value of the parameter θ

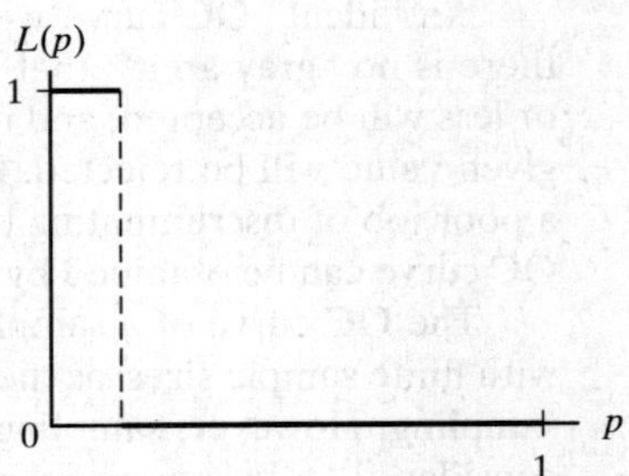

Figure 7. "Ideal" OC curve.

is θ_1 and the statistician incorrectly concludes that $\theta = \theta_0$, he is committing a type II error.)

Suppose an AQL of 0.05 is chosen ($p_0 = 0.05$). Then, it can be seen from Figure 6 that the given sampling plan has a producer's risk of about 0.03, since the probability of *acceptance* of a lot with an actual proportion defective of 0.05 is approximately 0.97. Similarly, if an LTPD of 0.20 is chosen, the consumer's risk is about 0.41. This plan obviously has an unacceptably high consumer's risk—over 40 percent of the lots received by the consumer will have 20 percent defectives or greater. To produce a plan with better characteristics, it will be necessary to increase the sample size, n, to decrease the acceptance number, c, or both. The following example shows what happens to these characteristics when c is decreased to 1, while n remains fixed at 20.

EXAMPLE 16

Find the producer's and consumer's risks corresponding to an AQL of 0.05 and an LTPD of 0.20 for the sampling plan defined by $n = 20$ and $c = 1$.

Solution

First, we calculate $L(p)$ for various values of p. Referring to Table I of "Statistical Tables" with $n = 20$ and $x = 1$, we obtain the following table:

p	0.05	0.10	0.15	0.20	0.25	0.30	0.35	0.40	0.45
$L(p)$	0.7358	0.3917	0.1756	0.0692	0.0243	0.0076	0.0021	0.0005	0.0001

A graph of this OC curve is shown in Figure 8. From this graph, we observe that the producer's risk is $1 - 0.7358 = 0.2642$, and the consumer's risk is 0.0692. Note that the work of constructing OC curves can be shortened considerably using computer software such as Excel or MINITAB.

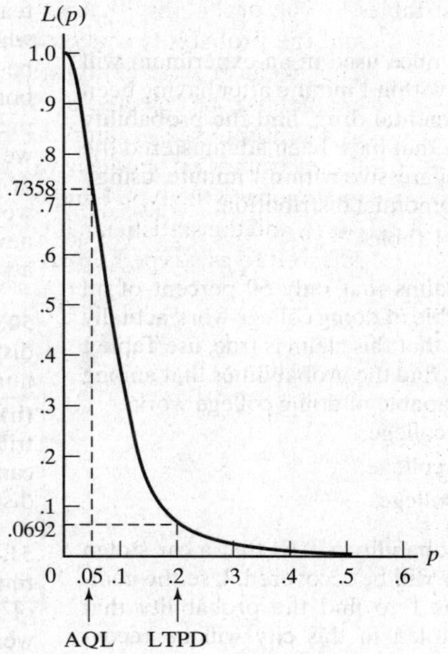

Figure 8. OC curve for Example 16.

Reduction of the acceptance number from 3 to 1 obviously has improved the consumer's risk, but now the producer's risk seems unacceptably high. Evidently, a larger sample size is needed.

The preceding example has been somewhat artificial. It would be quite unusual to specify an LTPD as high as 0.20 (20 percent defectives), and higher sample sizes than 20 usually are used for acceptance sampling. In practice, OC curves have been calculated for sampling plans having many different combinations of n and c. Choice then is made of the sampling plan whose OC curve has as nearly as possible the desired characteristics, AQL, LTPD, consumer's risk, and producer's risk for sample sizes in an acceptable range.

Applied Exercises

SECS. 1–4

40. A multiple-choice test consists of eight questions and three answers to each question (of which only one is correct). If a student answers each question by rolling a balanced die and checking the first answer if he gets a 1 or 2, the second answer if he gets a 3 or 4, and the third answer if he gets a 5 or 6, what is the probability that he will get exactly four correct answers?

41. An automobile safety engineer claims that 1 in 10 automobile accidents is due to driver fatigue. Using the formula for the binomial distribution and rounding to four decimals, what is the probability that at least 3 of 5 automobile accidents are due to driver fatigue?

42. In a certain city, incompatibility is given as the legal reason in 70 percent of all divorce cases. Find the probability that five of the next six divorce cases filed in this city will claim incompatibility as the reason, using
(a) the formula for the binomial distribution;
(b) Table I of "Statistical Tables."

43. If 40 percent of the mice used in an experiment will become very aggressive within 1 minute after having been administered an experimental drug, find the probability that exactly 6 of 15 mice that have been administered the drug will become very aggressive within 1 minute, using
(a) the formula for the binomial distribution;
(b) Table I of "Statistical Tables."

44. A social scientist claims that only 50 percent of all high school seniors capable of doing college work actually go to college. Assuming that this claim is true, use Table I of "Statistical Tables" to find the probabilities that among 18 high school seniors capable of doing college work
(a) exactly 10 will go to college;
(b) at least 10 will go to college;
(c) at most 8 will go to college.

45. Suppose that the probability is 0.63 that a car stolen in a certain Western city will be recovered. Use the computer printout of Figure 1 to find the probability that at least 8 of 10 cars stolen in this city will be recovered, using
(a) the values in the P(X = K) column;
(b) the values in the P(X LESS OR = K) column.

46. With reference to Exercise 45 and the computer printout of Figure 1, find the probability that among 10 cars stolen in the given city anywhere from 3 to 5 will be recovered, using
(a) the values in the P(X = K) column;
(b) the values in the P(X LESS OR = K) column.

47. With reference to Exercise 43, suppose that the percentage had been 42 instead of 40. Use a suitable table or a computer printout of the binomial distribution with $n = 15$ and $\theta = 0.42$ to rework both parts of that exercise.

48. With reference to Exercise 44, suppose that the percentage had been 51 instead of 50. Use a suitable table or a computer printout of the binomial distribution with $n = 18$ and $\theta = 0.51$ to rework the three parts of that exercise.

49. In planning the operation of a new school, one school board member claims that four out of five newly hired teachers will stay with the school for more than a year, while another school board member claims that it would be correct to say three out of five. In the past, the two board members have been about equally reliable in their predictions, so in the absence of any other information we would assign their judgments equal weight. If one or the other has to be right, what probabilities would we assign to their claims if it were found that 11 of 12 newly hired teachers stayed with the school for more than a year?

50. (a) To reduce the standard deviation of the binomial distribution by half, what change must be made in the number of trials?
(b) If n is multiplied by the factor k in the binomial distribution having the parameters n and θ, what statement can be made about the standard deviation of the resulting distribution?

51. A manufacturer claims that at most 5 percent of the time a given product will sustain fewer than 1,000 hours of operation before requiring service. Twenty products were selected at random from the production line and tested. It was found that three of them required service before 1,000 hours of operation. Comment on the manufacturer's claim.

52. (a) Use a computer program to calculate the probability of rolling between 14 and 18 "sevens" in 100 rolls of a pair of dice.

(b) Would it surprise you if more than 18 "sevens" were rolled? Why?

53. (a) Use a computer program to calculate the probability that more than 12 of 80 business telephone calls last longer than five minutes if it is assumed that 10 percent of such calls last that long.

(b) Can this result be used as evidence that the assumption is reasonable? Why?

54. Use Chebyshev's theorem and Theorem 3 to verify that the probability is at least $\frac{35}{36}$ that

(a) in 900 flips of a balanced coin the proportion of heads will be between 0.40 and 0.60;

(b) in 10,000 flips of a balanced coin the proportion of heads will be between 0.47 and 0.53;

(c) in 1,000,000 flips of a balanced coin the proportion of heads will be between 0.497 and 0.503.

Note that this serves to illustrate the law of large numbers.

55. You can get a feeling for the law of large numbers given Section 4 by flipping coins. Flip a coin 100 times and plot the accumulated proportion of heads after each five flips.

56. Record the first 200 numbers encountered in a newspaper, beginning with page 1 and proceeding in any convenient, systematic fashion. Include also numbers appearing in advertisements. For each of these numbers, note the leftmost digit, and record the proportions of 1's, 2's, 3's, ..., and 9's. (Note that 0 cannot be a leftmost digit. In the decimal number 0.0074, the leftmost digit is 7.) The results may seem quite surprising, but the law of large numbers tells you that you must be estimating correctly.

SECS. 5–7

57. If the probabilities of having a male or female child are both 0.50, find the probabilities that

(a) a family's fourth child is their first son;

(b) a family's seventh child is their second daughter;

(c) a family's tenth child is their fourth or fifth son.

58. If the probability is 0.75 that a person will believe a rumor about the transgressions of a certain politician, find the probabilities that

(a) the eighth person to hear the rumor will be the fifth to believe it;

(b) the fifteenth person to hear the rumor will be the tenth to believe it.

59. When taping a television commercial, the probability is 0.30 that a certain actor will get his lines straight on any one take. What is the probability that he will get his lines straight for the first time on the sixth take?

60. An expert sharpshooter misses a target 5 percent of the time. Find the probability that she will miss the target for the second time on the fifteenth shot using

(a) the formula for the negative binomial distribution;

(b) Theorem 5 and Table I of "Statistical Tables."

61. Adapt the formula of Theorem 5 so that it can be used to express geometric probabilities in terms of binomial probabilities, and use the formula and Table I of "Statistical Tables" to

(a) verify the result of Example 5;

(b) rework Exercise 59.

62. In a "torture test" a light switch is turned on and off until it fails. If the probability is 0.001 that the switch will fail any time it is turned on or off, what is the probability that the switch will not fail during the first 800 times that it is turned on or off? Assume that the conditions underlying the geometric distribution are met and use logarithms.

63. A quality control engineer inspects a random sample of two hand-held calculators from each incoming lot of size 18 and accepts the lot if they are both in good working condition; otherwise, the entire lot is inspected with the cost charged to the vendor. What are the probabilities that such a lot will be accepted without further inspection if it contains

(a) 4 calculators that are not in good working condition;

(b) 8 calculators that are not in good working condition;

(c) 12 calculators that are not in good working condition?

64. Among the 16 applicants for a job, 10 have college degrees. If 3 of the applicants are randomly chosen for interviews, what are the probabilities that

(a) none has a college degree;

(b) 1 has a college degree;

(c) 2 have college degrees;

(d) all 3 have college degrees?

65. Find the mean and the variance of the hypergeometric distribution with $n = 3, N = 16$, and $M = 10$, using

(a) the results of Exercise 64;

(b) the formulas of Theorem 7.

66. What is the probability that an IRS auditor will catch only 2 income tax returns with illegitimate deductions if she randomly selects 5 returns from among 15 returns, of which 9 contain illegitimate deductions?

67. Check in each case whether the condition for the binomial approximation to the hypergeometric distribution is satisfied:

(a) $N = 200$ and $n = 12$;

(b) $N = 500$ and $n = 20$;

(c) $N = 640$ and $n = 30$.

68. A shipment of 80 burglar alarms contains 4 that are defective. If 3 from the shipment are randomly selected

and shipped to a customer, find the probability that the customer will get exactly one bad unit using
(a) the formula of the hypergeometric distribution;
(b) the binomial distribution as an approximation.

69. Among the 300 employees of a company, 240 are union members, whereas the others are not. If 6 of the employees are chosen by lot to serve on a committee that administers the pension fund, find the probability that 4 of the 6 will be union members using
(a) the formula for the hypergeometric distribution;
(b) the binomial distribution as an approximation.

70. A panel of 300 persons chosen for jury duty includes 30 under 25 years of age. Since the jury of 12 persons chosen from this panel to judge a narcotics violation does not include anyone under 25 years of age, the youthful defendant's attorney complains that this jury is not really representative. Indeed, he argues, if the selection were random, the probability of having one of the 12 jurors under 25 years of age should be *many times* the probability of having none of them under 25 years of age. Actually, what is the ratio of these two probabilities?

71. Check in each case whether the values of n and θ satisfy the rule of thumb for a good approximation, an excellent approximation, or neither when we want to use the Poisson distribution to approximate binomial probabilities.
(a) $n = 125$ and $\theta = 0.10$;
(b) $n = 25$ and $\theta = 0.04$;
(c) $n = 120$ and $\theta = 0.05$;
(d) $n = 40$ and $\theta = 0.06$.

72. It is known from experience that 1.4 percent of the calls received by a switchboard are wrong numbers. Use the Poisson approximation to the binomial distribution to determine the probability that among 150 calls received by the switchboard 2 are wrong numbers.

73. With reference to Example 8, determine the value of x (from 5 to 15) for which the percentage error is greatest when we use the Poisson distribution with $\lambda = 7.5$ to approximate the binomial distribution with $n = 150$ and $\theta = 0.05$.

74. In a given city, 4 percent of all licensed drivers will be involved in at least one car accident in any given year. Use the Poisson approximation to the binomial distribution to determine the probability that among 150 licensed drivers randomly chosen in this city
(a) only 5 will be involved in at least one accident in any given year;
(b) at most 3 will be involved in at least one accident in any given year.

75. Records show that the probability is 0.0012 that a person will get food poisoning spending a day at a certain state fair. Use the Poisson approximation to the binomial distribution to find the probability that among 1,000 persons attending the fair at most 2 will get food poisoning.

76. With reference to Example 13 and the computer printout of Figure 4, find the probability that a 15-square-foot sheet of the metal will have anywhere from 8 to 12 defects, using
(a) the values in the P(X = K) column;
(b) the values in the P(X LESS OR = K) column.

77. The number of complaints that a dry-cleaning establishment receives per day is a random variable having a Poisson distribution with $\lambda = 3.3$. Use the formula for the Poisson distribution to find the probability that it will receive only two complaints on any given day.

78. The number of monthly breakdowns of a super computer is a random variable having a Poisson distribution with $\lambda = 1.8$. Use the formula for the Poisson distribution to find the probabilities that this computer will function
(a) without a breakdown;
(b) with only one breakdown.

79. Use Table II of "Statistical Tables" to verify the results of Exercise 78.

80. In the inspection of a fabric produced in continuous rolls, the number of imperfections per yard is a random variable having the Poisson distribution with $\lambda = 0.25$. Find the probability that 2 yards of the fabric will have at most one imperfection using
(a) Table II of "Statistical Tables";
(b) the computer printout of Figure 5.

81. In a certain desert region the number of persons who become seriously ill each year from eating a certain poisonous plant is a random variable having a Poisson distribution with $\lambda = 5.2$. Use Table II of "Statistical Tables" to find the probabilities of
(a) 3 such illnesses in a given year;
(b) at least 10 such illnesses in a given year;
(c) anywhere from 4 to 6 such illnesses in a given year.

82. (a) Use a computer program to calculate the *exact* probability of obtaining one or more defectives in a sample of size 100 taken from a lot of 1,000 manufactured products assumed to contain six defectives.
(b) Approximate this probability using the appropriate binomial distribution.
(c) Approximate this probability using the appropriate Poisson distribution and compare the results of parts (a), (b), and (c).

SECS. 8–9

83. The probabilities are 0.40, 0.50, and 0.10 that, in city driving, a certain kind of compact car will average less than 28 miles per gallon, from 28 to 32 miles per gallon, or more than 32 miles per gallon. Find the probability that among 10 such cars tested, 3 will average less than

28 miles per gallon, 6 will average from 28 to 32 miles per gallon, and 1 will average more than 32 miles per gallon.

84. Suppose that the probabilities are 0.60, 0.20, 0.10, and 0.10 that a state income tax return will be filled out correctly, that it will contain only errors favoring the taxpayer, that it will contain only errors favoring the state, or that it will contain both kinds of errors. What is the probability that among 12 such income tax returns randomly chosen for audit, 5 will be filled out correctly, 4 will contain only errors favoring the taxpayer, 2 will contain only errors favoring the state, and 1 will contain both kinds of errors?

85. According to the Mendelian theory of heredity, if plants with round yellow seeds are crossbred with plants with wrinkled green seeds, the probabilities of getting a plant that produces round yellow seeds, wrinkled yellow seeds, round green seeds, or wrinkled green seeds are, respectively, $\frac{9}{16}$, $\frac{3}{16}$, $\frac{3}{16}$, and $\frac{1}{16}$. What is the probability that among nine plants thus obtained there will be four that produce round yellow seeds, two that produce wrinkled yellow seeds, three that produce round green seeds, and none that produce wrinkled green seeds?

86. Among 25 silver dollars struck in 1903 there are 15 from the Philadelphia mint, 7 from the New Orleans mint, and 3 from the San Francisco mint. If 5 of these silver dollars are picked at random, find the probabilities of getting

(a) 4 from the Philadelphia mint and 1 from the New Orleans mint;

(b) 3 from the Philadelphia mint and 1 from each of the other 2 mints.

87. If 18 defective glass bricks include 10 that have cracks but no discoloration, 5 that have discoloration but no cracks, and 3 that have cracks and discoloration, what is the probability that among 6 of the bricks (chosen at random for further checks) 3 will have cracks but no discoloration, 1 will have discoloration but no cracks, and 2 will have cracks and discoloration?

SEC. 10

88. A sampling inspection program has a 0.10 probability of rejecting a lot when the true proportion of defectivesis 0.01, and a 0.95 probability of rejecting the lot when the true proportion of defectives is 0.03. If 0.01 is the AQL and 0.03 is the LTPD, what are the producer's and consumer's risks?

89. The producer's risk in a sampling program is 0.05 and the consumer's risk is 0.10; the AQL is 0.03 and the LTPD is 0.07.

(a) What is the probability of accepting a lot whose true proportion of defectives is 0.03?

(b) What is the probability of accepting a lot whose true proportion of defectives is 0.07?

90. Suppose the acceptance number in Example 16 is changed from 1 to 2. Keeping the producer's risk at 0.05 and the consumer's risk at 0.10, what are the new values of the AQL and the LTPD?

91. From Figure 6,

(a) find the producer's risk if the AQL is 0.10;

(b) find the LTPD corresponding to a consumer's risk of 0.05.

92. Sketch the OC curve for a sampling plan having a sample size of 15 and an acceptance number of 1.

93. Sketch the OC curve for a sampling plan having a sample size of 25 and an acceptance number of 2.

94. Sketch the OC curve for a sampling plan having a sample size of 10 and an acceptance number of 0.

95. Find the AQL and the LTPD of the sampling plan in Exercise 93 if both the producer's and consumer's risks are 0.10.

96. If the AQL is 0.1 and the LTPD is 0.25 in the sampling plan given in Exercise 92, find the producer's and consumer's risks.

97. (a) In Exercise 92 change the acceptance number from 1 to 0 and sketch the OC curve.

(b) How do the producer's and consumer's risks change if the AQL is 0.05 and the LTPD is 0.3 in both sampling plans?

References

Useful information about various special probability distributions may be found in

DERMAN, C., GLESER, L., and OLKIN, I., *Probability Models and Applications*. New York: Macmillan Publishing Co., Inc., 1980,

HASTINGS, N. A. J., and PEACOCK, J. B., *Statistical Distributions*. London: Butterworth & Co. Ltd., 1975,

and

JOHNSON, N. L., and KOTZ, S., *Discrete Distributions*, Boston: Houghton Mifflin Company, 1969.

Binomial probabilities for $n = 2$ to $n = 49$ may be found in

Tables of the Binomial Probability Distribution, National Bureau of Standards Applied Mathematics Series No. 6, Washington, D.C.: U.S. Government Printing Office, 1950,

and for $n = 50$ to $n = 100$ in
ROMIG, H. G., 50–100 *Binomial Tables*. New York: John Wiley & Sons, Inc., 1953.

The most widely used table of Poisson probabilities is

MOLINA, E. C., *Poisson's Exponential Binomial Limit*. Melbourne, Fla.: Robert E. Krieger Publishing Company, 1973 Reprint.

Answers to Odd-Numbered Exercises

11 $\mu_2' = \mu_{(2)}' + \mu_{(1)}'$, $\mu_3' = \mu_{(3)}' + 3\mu_{(2)}' + \mu_{(1)}'$, and $\mu_4' = \mu_{(4)}' + 6\mu_{(3)}' + 7\mu_{(2)}' + \mu_{(1)}'$.

13 **(a)** $F_X(t) = 1 - \theta + \theta t$; **(b)** $F_X(t) = [1 + \theta(t-1)]^n$.

15 **(a)** $\alpha_3 = 0$ when $\theta = \frac{1}{2}$; **(b)** $\alpha_3 \to 0$ when $n \to \infty$.

17 $\mu_Y = k\left(\frac{1}{\theta} - 1\right)$; $\sigma_Y^2 = \frac{k}{\theta}\left(\frac{1}{\theta} - 1\right)$.

37 $M_Y(t) = e^{\lambda(e^t - t - 1)}$; $\sigma_Y^2 = M_Y''(0) = \lambda$.

41 0.0086.

43 **(a)** 0.2066; **(b)** 0.2066.

45 **(a)** 0.2205; **(b)** 0.2206.

47 0.2041.

49 0.9222.

51 0.0754.

53 **(a)** 0.0538.

57 **(a)** 0.0625; **(b)** 0.0469; **(c)** 0.2051.

59 0.0504.

61 **(a)** 0.0117; **(b)** 0.0504.

63 **(a)** 0.5948; **(b)** 0.2941; **(c)** 0.0980.

65 **(a)** $\mu = \frac{15}{8}$ and $\sigma^2 = \frac{39}{64}$; **(b)** $\mu = \frac{15}{8}$ and $\sigma^2 = \frac{39}{64}$.

67 **(a)** The condition is not satisfied. **(b)** The condition is satisfied. **(c)** The condition is satisfied.

69 **(a)** 0.2478; **(b)** 0.2458.

71 **(a)** Neither rule of thumb is satisfied. **(b)** The rule of thumb for good approximation is satisfied. **(c)** The rule of thumb for excellent approximation is satisfied. **(d)** Neither rule of thumb is satisfied.

73 $x = 15$.

75 0.8795.

77 0.2008.

79 **(a)** 0.1653; **(b)** 0.2975.

81 **(a)** 0.1293; **(b)** 0.0397; **(c)** 0.4944.

83 0.0841.

85 0.0292.

87 0.0970.

89 **(a)** 0.95; **(b)** 0.10.

91 **(a)** 0.17; **(b)** 0.35.

95 AQL = 0.07, LTPD = 0.33.

97 **(b)** Plan 1 ($c = 0$): producer's risk = 0.0861 and consumer's risk = 0.1493; Plan 2 ($c = 1$): producer's risk = 0.4013 and consumer's risk = 0.0282.

Special Probability Densities

1 Introduction

In this chapter we shall study some of the probability densities that figure most prominently in statistical theory and in applications. In addition to the ones given in the text, several others are introduced in the exercises following Section 4. We shall derive parameters and moment-generating functions, again leaving some of the details as exercises.

2 The Uniform Distribution

DEFINITION 1. UNIFORM DISTRIBUTION. *A random variable X has a* ***uniform distribution*** *and it is referred to as a continuous uniform random variable if and only if its probability density is given by*

$$u(x; \alpha, \beta) = \begin{cases} \dfrac{1}{\beta - \alpha} & \text{for } \alpha < x < \beta \\ 0 & \text{elsewhere} \end{cases}$$

The parameters α and β of this probability density are real constants, with $\alpha < \beta$, and may be pictured as in Figure 1. In Exercise 2 the reader will be asked to verify the following theorem.

THEOREM 1. The mean and the variance of the uniform distribution are given by

$$\mu = \frac{\alpha + \beta}{2} \quad \text{and} \quad \sigma^2 = \frac{1}{12}(\beta - \alpha)^2$$

From Chapter 6 of *John E. Freund's Mathematical Statistics with Applications*, Eighth Edition. Irwin Miller, Marylees Miller.

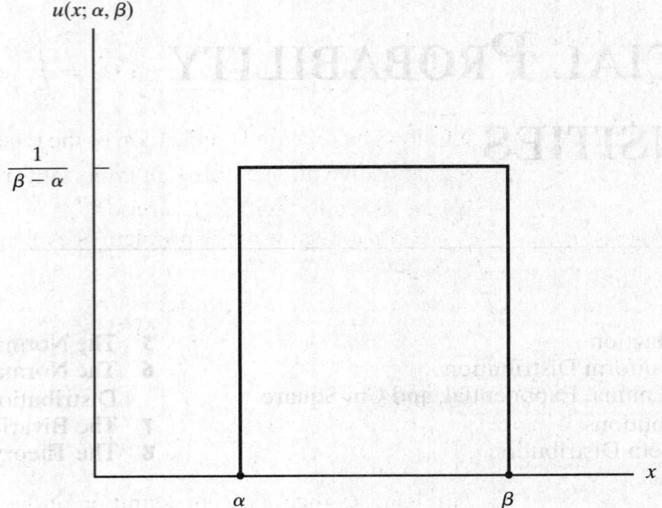

Figure 1. The uniform distribution.

Although the uniform distribution has some direct applications, its main value is that, due to its simplicity, it lends itself readily to the task of illustrating various aspects of statistical theory.

3 The Gamma, Exponential, and Chi-Square Distributions

Let's start with random variables having probability densities of the form

$$f(x) = \begin{cases} kx^{\alpha-1}e^{-x/\beta} & \text{for } x > 0 \\ 0 & \text{elsewhere} \end{cases}$$

where $\alpha > 0$, $\beta > 0$, and k must be such that the total area under the curve is equal to 1. To evaluate k, we first make the substitution $y = \frac{x}{\beta}$, which yields

$$\int_0^{\infty} kx^{\alpha-1}e^{-x/\beta}\,dx = k\beta^{\alpha}\int_0^{\infty} y^{\alpha-1}e^{-y}\,dy$$

The integral thus obtained depends on α alone, and it defines the well-known **gamma function**

$$\Gamma(\alpha) = \int_0^{\infty} y^{\alpha-1}e^{-y}\,dy \qquad \text{for } \alpha > 0$$

which is treated in detail in most advanced calculus texts. Integrating by parts, which is left to the reader in Exercise 7, we find that the gamma function satisfies the recursion formula

$$\Gamma(\alpha) = (\alpha - 1) \cdot \Gamma(\alpha - 1)$$

for $\alpha > 1$, and since

$$\Gamma(1) = \int_0^\infty e^{-y} dy = 1$$

it follows by repeated application of the recursion formula that $\Gamma(\alpha) = (\alpha - 1)!$ when α is a positive integer. Also, an important special value is $\Gamma\left(\frac{1}{2}\right) = \sqrt{\pi}$, as the reader will be asked to verify in Exercise 9.

Returning now to the problem of evaluating k, we equate the integral we obtained to 1, getting

$$\int_0^\infty kx^{\alpha-1}e^{-x/\beta} dx = k\beta^\alpha \Gamma(\alpha) = 1$$

and hence

$$k = \frac{1}{\beta^\alpha \Gamma(\alpha)}$$

This leads to the following definition of the **gamma distribution**.

DEFINITION 2. GAMMA DISTRIBUTION. *A random variable X has a **gamma distribution** and it is referred to as a gamma random variable if and only if its probability density is given by*

$$g(x; \alpha, \beta) = \begin{cases} \dfrac{1}{\beta^\alpha \Gamma(\alpha)} x^{\alpha-1} e^{-x/\beta} & \text{for } x > 0 \\ 0 & \text{elsewhere} \end{cases}$$

where $\alpha > 0$ and $\beta > 0$.

When α is not a positive integer, the value of $\Gamma(\alpha)$ will have to be looked up in a special table. To give the reader some idea about the shape of the graphs of gamma densities, those for several special values of α and β are shown in Figure 2.

Some special cases of the gamma distribution play important roles in statistics; for instance, for $\alpha = 1$ and $\beta = \theta$, we obtain the following definition.

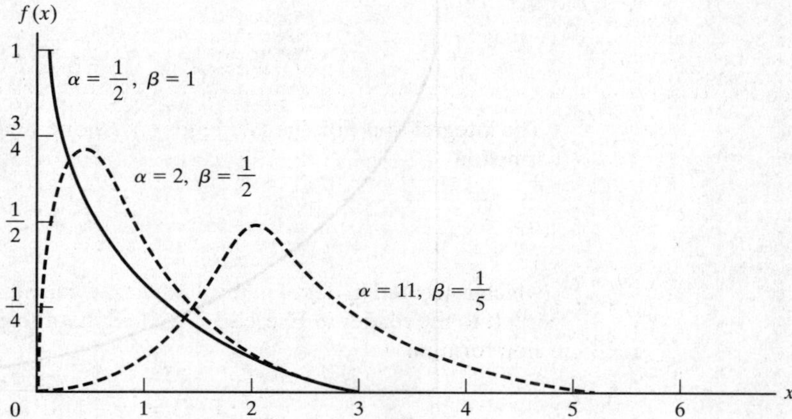

Figure 2. Graphs of gamma distributions.

DEFINITION 3. EXPONENTIAL DISTRIBUTION. *A random variable X has an* ***exponential distribution*** *and it is referred to as an exponential random variable if and only if its probability density is given by*

$$g(x;\theta) = \begin{cases} \frac{1}{\theta}e^{-x/\theta} & \text{for } x > 0 \\ 0 & \text{elsewhere} \end{cases}$$

where $\theta > 0$.

This density is pictured in Figure 3.

Let us consider there is the probability of getting x successes during a time interval of length t when (i) the probability of a success during a very small time interval from t to $t + \Delta t$ is $\alpha \cdot \Delta t$, (ii) the probability of more than one success during such a time interval is negligible, and (iii) the probability of a success during such a time interval does not depend on what happened prior to time t. The number of successes is a value of the discrete random variable X having the Poisson distribution with $\lambda = \alpha t$. Let us determine the probability density of the continuous random variable Y, the **waiting time** until the first success. Clearly,

$$\begin{aligned} F(y) = P(Y \leqq y) &= 1 - P(Y > y) \\ &= 1 - P(0 \text{ successes in a time interval of length } y) \\ &= 1 - p(0; \alpha y) \\ &= 1 - \frac{e^{-\alpha y}(\alpha y)^0}{0!} \\ &= 1 - e^{-\alpha y} \quad \text{for } y > 0 \end{aligned}$$

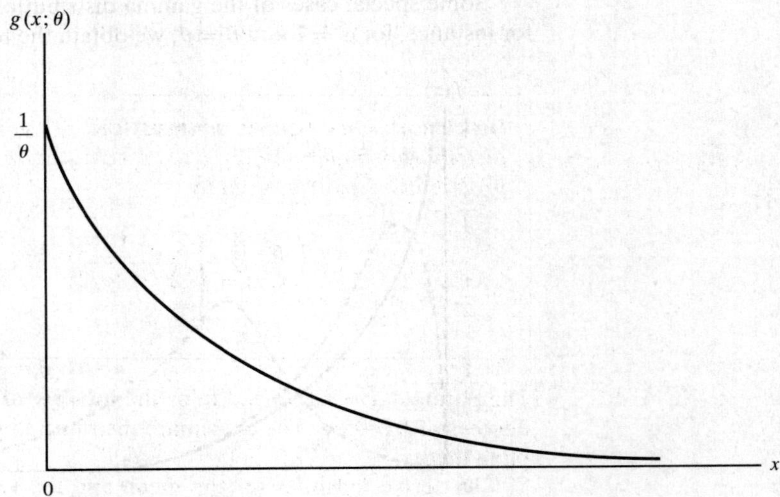

Figure 3. Exponential distribution.

and $F(y) = 0$ for $y \leqq 0$. Having thus found the distribution function of Y, we find that differentiation with respect to y yields

$$f(y) = \begin{cases} \alpha e^{-\alpha y} & \text{for } y > 0 \\ 0 & \text{elsewhere} \end{cases}$$

which is the exponential distribution with $\theta = \dfrac{1}{\alpha}$.

The exponential distribution applies not only to the occurrence of the first success in a **Poisson process** but, by virtue of condition (see Exercise 16), it applies also to the waiting times between successes.

EXAMPLE 1

At a certain location on highway I-10, the number of cars exceeding the speed limit by more than 10 miles per hour in half an hour is a random variable having a Poisson distribution with $\lambda = 8.4$. What is the probability of a waiting time of less than 5 minutes between cars exceeding the speed limit by more than 10 miles per hour?

Solution

Using half an hour as the unit of time, we have $\alpha = \lambda = 8.4$. Therefore, the waiting time is a random variable having an exponential distribution with $\theta = \frac{1}{8.4}$, and since 5 minutes is $\frac{1}{6}$ of the unit of time, we find that the desired probability is

$$\int_0^{1/6} 8.4e^{-8.4x}\,dx = -e^{-8.4x}\Big|_0^{1/6} = -e^{-1.4} + 1$$

which is approximately 0.75.

Another special case of the gamma distribution arises when $\alpha = \dfrac{\nu}{2}$ and $\beta = 2$, where ν is the lowercase Greek letter *nu*.

DEFINITION 4. CHI-SQUARE DISTRIBUTION. *A random variable X has a* ***chi-square distribution*** *and it is referred to as a chi-square random variable if and only if its probability density is given by*

$$f(x, \nu) = \begin{cases} \dfrac{1}{2^{\nu/2}\Gamma(\nu/2)} x^{\frac{\nu-2}{2}} e^{-\frac{x}{2}} & \text{for } x > 0 \\ 0 & \text{elsewhere} \end{cases}$$

The parameter ν is referred to as the **number of degrees of freedom**, or simply the **degrees of freedom**. The chi-square distribution plays a very important role in sampling theory.

To derive formulas for the mean and the variance of the gamma distribution, and hence the exponential and chi-square distributions, let us first prove the following theorem.

THEOREM 2. The rth moment about the origin of the gamma distribution is given by

$$\mu'_r = \frac{\beta^r \Gamma(\alpha + r)}{\Gamma(\alpha)}$$

Proof By using the definition of the rth moment about the origin,

$$\mu'_r = \int_0^\infty x^r \cdot \frac{1}{\beta^\alpha \Gamma(\alpha)} x^{\alpha-1} e^{-x/\beta} dx = \frac{\beta^r}{\Gamma(\alpha)} \cdot \int_0^\infty y^{\alpha+r-1} e^{-y} dy$$

where we let $y = \frac{x}{\beta}$. Since the integral on the right is $\Gamma(r + \alpha)$ according to the definition of gamma function, this completes the proof.

Using this theorem, let us now derive the following results about the gamma distribution.

THEOREM 3. The mean and the variance of the gamma distribution are given by

$$\mu = \alpha\beta \quad \text{and} \quad \sigma^2 = \alpha\beta^2$$

Proof From Theorem 2 with $r = 1$ and $r = 2$, we get

$$\mu'_1 = \frac{\beta\Gamma(\alpha + 1)}{\Gamma(\alpha)} = \alpha\beta$$

and

$$\mu'_2 = \frac{\beta^2\Gamma(\alpha + 2)}{\Gamma(\alpha)} = \alpha(\alpha + 1)\beta^2$$

so $\mu = \alpha\beta$ and $\sigma^2 = \alpha(\alpha + 1)\beta^2 - (\alpha\beta)^2 = \alpha\beta^2$.

Substituting into these formulas $\alpha = 1$ and $\beta = \theta$ for the exponential distribution and $\alpha = \frac{\nu}{2}$ and $\beta = 2$ for the chi-square distribution, we obtain the following corollaries.

COROLLARY 1. The mean and the variance of the exponential distribution are given by

$$\mu = \theta \quad \text{and} \quad \sigma^2 = \theta^2$$

COROLLARY 2. The mean and the variance of the chi-square distribution are given by

$$\mu = \nu \quad \text{and} \quad \sigma^2 = 2\nu$$

For future reference, let us give here also the moment-generating function of the gamma distribution.

THEOREM 4. The moment-generating function of the gamma distribution is given by

$$M_X(t) = (1 - \beta t)^{-\alpha}$$

The reader will be asked to prove this result and use it to find some of the lower moments in Exercises 12 and 13.

4 The Beta Distribution

The uniform density $f(x) = 1$ for $0 < x < 1$ and $f(x) = 0$ elsewhere is a special case of the **beta distribution**, which is defined in the following way.

DEFINITION 5. BETA DISTRIBUTION. *A random variable X has a* **beta distribution** *and it is referred to as a beta random variable if and only if its probability density is given by*

$$f(x; \alpha, \beta) = \begin{cases} \dfrac{\Gamma(\alpha+\beta)}{\Gamma(\alpha)\cdot\Gamma(\beta)} x^{\alpha-1}(1-x)^{\beta-1} & \text{for } 0 < x < 1 \\ 0 & \text{elsewhere} \end{cases}$$

where $\alpha > 0$ *and* $\beta > 0$.

In recent years, the beta distribution has found important applications in **Bayesian inference**, where parameters are looked upon as random variables, and there is a need for a fairly "flexible" probability density for the parameter θ of the binomial distribution, which takes on nonzero values only on the interval from 0 to 1. By "flexible" we mean that the probability density can take on a great variety of different shapes, as the reader will be asked to verify for the beta distribution in Exercise 27.

We shall not prove here that the total area under the curve of the beta distribution, like that of any probability density, is equal to 1, but in the proof of the theorem that follows, we shall make use of the fact that

$$\int_0^1 \frac{\Gamma(\alpha+\beta)}{\Gamma(\alpha)\cdot\Gamma(\beta)} x^{\alpha-1}(1-x)^{\beta-1}\,dx = 1$$

and hence that

$$\int_0^1 x^{\alpha-1}(1-x)^{\beta-1}\,dx = \frac{\Gamma(\alpha)\cdot\Gamma(\beta)}{\Gamma(\alpha+\beta)}$$

This integral defines the **beta function**, whose values are denoted $\mathrm{B}(\alpha, \beta)$; in other words, $\mathrm{B}(\alpha, \beta) = \dfrac{\Gamma(\alpha)\cdot\Gamma(\beta)}{\Gamma(\alpha+\beta)}$. Detailed discussion of the beta function may be found in any textbook on advanced calculus.

THEOREM 5. The mean and the variance of the beta distribution are given by

$$\mu = \frac{\alpha}{\alpha+\beta} \quad \text{and} \quad \sigma^2 = \frac{\alpha\beta}{(\alpha+\beta)^2(\alpha+\beta+1)}$$

Proof By definition,

$$\mu = \frac{\Gamma(\alpha+\beta)}{\Gamma(\alpha)\cdot\Gamma(\beta)} \cdot \int_0^1 x \cdot x^{\alpha-1}(1-x)^{\beta-1} dx$$
$$= \frac{\Gamma(\alpha+\beta)}{\Gamma(\alpha)\cdot\Gamma(\beta)} \cdot \frac{\Gamma(\alpha+1)\cdot\Gamma(\beta)}{\Gamma(\alpha+\beta+1)}$$
$$= \frac{\alpha}{\alpha+\beta}$$

where we recognized the integral as $B(\alpha+1, \beta)$ and made use of the fact that $\Gamma(\alpha+1) = \alpha\cdot\Gamma(\alpha)$ and $\Gamma(\alpha+\beta+1) = (\alpha+\beta)\cdot\Gamma(\alpha+\beta)$. Similar steps, which will be left to the reader in Exercise 28, yield

$$\mu_2' = \frac{(\alpha+1)\alpha}{(\alpha+\beta+1)(\alpha+\beta)}$$

and it follows that

$$\sigma^2 = \frac{(\alpha+1)\alpha}{(\alpha+\beta+1)(\alpha+\beta)} - \left(\frac{\alpha}{\alpha+\beta}\right)^2$$
$$= \frac{\alpha\beta}{(\alpha+\beta)^2(\alpha+\beta+1)}$$

Exercises

1. Show that if a random variable has a uniform density with the parameters α and β, the probability that it will take on a value less than $\alpha + p(\beta-\alpha)$ is equal to p.

2. Prove Theorem 1.

3. If a random variable X has a uniform density with the parameters α and β, find its distribution function.

4. Show that if a random variable has a uniform density with the parameters α and β, the rth moment about the mean equals
(a) 0 when r is odd;
(b) $\frac{1}{r+1}\left(\frac{\beta-\alpha}{2}\right)^r$ when r is even.

5. Use the results of Exercise 4 to find α_3 and α_4 for the uniform density with the parameters α and β.

6. A random variable is said to have a **Cauchy distribution** if its density is given by

$$f(x) = \frac{\frac{\beta}{\pi}}{(x-\alpha)^2+\beta^2} \quad \text{for } -\infty < x < \infty$$

Show that for this distribution μ_1' and μ_2' do not exist.

7. Use integration by parts to show that $\Gamma(\alpha) = (\alpha-1)\cdot\Gamma(\alpha-1)$ for $\alpha > 1$.

8. Perform a suitable change of variable to show that the integral defining the gamma function can be written as

$$\Gamma(\alpha) = 2^{1-\alpha}\cdot\int_0^\infty z^{2\alpha-1}e^{-\frac{1}{2}z^2}dz \quad \text{for } \alpha > 0$$

9. Using the form of the gamma function of Exercise 8, we can write

$$\Gamma\left(\frac{1}{2}\right) = \sqrt{2}\int_0^\infty e^{-\frac{1}{2}z^2}dz$$

and hence

$$\left[\Gamma\left(\frac{1}{2}\right)\right]^2 = 2\left\{\int_0^\infty e^{-\frac{1}{2}x^2}dx\right\}\left\{\int_0^\infty e^{-\frac{1}{2}y^2}dy\right\}$$
$$= 2\int_0^\infty\int_0^\infty e^{-\frac{1}{2}(x^2+y^2)}\,dx\,dy$$

Change to polar coordinates to evaluate this double integral, and thus show that $\Gamma(\frac{1}{2}) = \sqrt{\pi}$.

10. Find the probabilities that the value of a random variable will exceed 4 if it has a gamma distribution with
(a) $\alpha = 2$ and $\beta = 3$;
(b) $\alpha = 3$ and $\beta = 4$.

11. Show that a gamma distribution with $\alpha > 1$ has a relative maximum at $x = \beta(\alpha - 1)$. What happens when $0 < \alpha < 1$ and when $\alpha = 1$?

12. Prove Theorem 4, making the substitution $y = x\left(\frac{1}{\beta} - t\right)$ in the integral defining $M_X(t)$.

13. Expand the moment-generating function of the gamma distribution as a binomial series, and read off the values of μ_1', μ_2', μ_3', and μ_4'.

14. Use the results of Exercise 13 to find α_3 and α_4 for the gamma distribution.

15. Show that if a random variable has an exponential density with the parameter θ, the probability that it will take on a value less than $-\theta \cdot \ln(1-p)$ is equal to p for $0 \leqq p < 1$.

16. If X has an exponential distribution, show that

$$P[(X \geq t+T)|(x \geq T)] = P(X \geq t)$$

17. This question has been intentionally omitted for this edition.

18. With reference to Exercise 17, using the fact that the moments of Y about the origin are the corresponding moments of X about the mean, find α_3 and α_4 for the exponential distribution with the parameter θ.

19. Show that if $\nu > 2$, the chi-square distribution has a relative maximum at $x = \nu - 2$. What happens when $\nu = 2$ or $0 < \nu < 2$?

20. A random variable X has a **Rayleigh distribution** if and only if its probability density is given by

$$f(x) = \begin{cases} 2\alpha x e^{-\alpha x^2} & \text{for } x > 0 \\ 0 & \text{elsewhere} \end{cases}$$

where $\alpha > 0$. Show that for this distribution

(a) $\mu = \frac{1}{2}\sqrt{\frac{\pi}{\alpha}}$;

(b) $\sigma^2 = \frac{1}{\alpha}\left(1 - \frac{\pi}{4}\right)$.

21. A random variable X has a **Pareto distribution** if and only if its probability density is given by

$$f(x) = \begin{cases} \dfrac{\alpha}{x^{\alpha+1}} & \text{for } x > 1 \\ 0 & \text{elsewhere} \end{cases}$$

where $\alpha > 0$. Show that μ_r' exists only if $r < \alpha$.

22. With reference to Exercise 21, show that for the Pareto distribution

$$\mu = \frac{\alpha}{\alpha - 1} \text{ provided } \alpha > 1.$$

23. A random variable X has a **Weibull distribution** if and only if its probability density is given by

$$f(x) = \begin{cases} kx^{\beta-1}e^{-\alpha x^{\beta}} & \text{for } x > 0 \\ 0 & \text{elsewhere} \end{cases}$$

where $\alpha > 0$ and $\beta > 0$.

(a) Express k in terms of α and β.

(b) Show that $\mu = \alpha^{-1/\beta}\Gamma\left(1 + \frac{1}{\beta}\right)$.

Note that Weibull distributions with $\beta = 1$ are exponential distributions.

24. If the random variable T is the time to failure of a commercial product and the values of its probability density and distribution function at time t are $f(t)$ and $F(t)$, then its failure rate at time t is given by $\frac{f(t)}{1 - F(t)}$. Thus, the failure rate at time t is the probability density of failure at time t given that failure does not occur prior to time t.

(a) Show that if T has an exponential distribution, the failure rate is constant.

(b) Show that if T has a Weibull distribution (see Exercise 23), the failure rate is given by $\alpha\beta t^{\beta-1}$.

25. Verify that the integral of the beta density from $-\infty$ to ∞ equals 1 for

(a) $\alpha = 2$ and $\beta = 4$;

(b) $\alpha = 3$ and $\beta = 3$.

26. Show that if $\alpha > 1$ and $\beta > 1$, the beta density has a relative maximum at

$$x = \frac{\alpha - 1}{\alpha + \beta - 2}.$$

27. Sketch the graphs of the beta densities having

(a) $\alpha = 2$ and $\beta = 2$;

(b) $\alpha = \frac{1}{2}$ and $\beta = 1$;

(c) $\alpha = 2$ and $\beta = \frac{1}{2}$;

(d) $\alpha = 2$ and $\beta = 5$.

[*Hint*: To evaluate $\Gamma(\frac{3}{2})$ and $\Gamma(\frac{5}{2})$, make use of the recursion formula $\Gamma(\alpha) = (\alpha - 1) \cdot \Gamma(\alpha - 1)$ and the result of Exercise 9.]

28. Verify the expression given for μ_2' in the proof of Theorem 5.

29. Show that the parameters of the beta distribution can be expressed as follows in terms of the mean and the variance of this distribution:

(a) $\alpha = \mu\left[\frac{\mu(1-\mu)}{\sigma^2} - 1\right]$;

(b) $\beta = (1 - \mu)\left[\frac{\mu(1-\mu)}{\sigma^2} - 1\right]$.

30. Karl Pearson, one of the founders of modern statistics, showed that the differential equation

$$\frac{1}{f(x)} \cdot \frac{d[f(x)]}{dx} = \frac{d - x}{a + bx + cx^2}$$

yields (for appropriate values of the constants a, b, c, and d) most of the important distributions of statistics. Verify that the differential equation gives
(a) the gamma distribution when $a = c = 0$, $b > 0$, and $d > -b$;
(b) the exponential distribution when $a = c = d = 0$ and $b > 0$;
(c) the beta distribution when $a = 0$, $b = -c$, $\frac{d-1}{b} < 1$, and $\frac{d}{b} > -1$.

5 The Normal Distribution

The **normal distribution**, which we shall study in this section, is in many ways the cornerstone of modern statistical theory. It was investigated first in the eighteenth century when scientists observed an astonishing degree of regularity in errors of measurement. They found that the patterns (distributions) that they observed could be closely approximated by continuous curves, which they referred to as "normal curves of errors" and attributed to the laws of chance. The mathematical properties of such normal curves were first studied by Abraham de Moivre (1667–1745), Pierre Laplace (1749–1827), and Karl Gauss (1777–1855).

DEFINITION 6. NORMAL DISTRIBUTION. *A random variable X has a* ***normal distribution*** *and it is referred to as a normal random variable if and only if its probability density is given by*

$$n(x; \mu, \sigma) = \frac{1}{\sigma\sqrt{2\pi}} e^{-\frac{1}{2}\left(\frac{x-\mu}{\sigma}\right)^2} \quad \text{for } -\infty < x < \infty$$

where $\sigma > 0$.

The graph of a normal distribution, shaped like the cross section of a bell, is shown in Figure 4.

The notation used here shows explicitly that the two parameters of the normal distribution are μ and σ. It remains to be shown, however, that the parameter μ is, in fact, $E(X)$ and that the parameter σ is, in fact, the square root of $\text{var}(X)$, where X is a random variable having the normal distribution with these two parameters.

First, though, let us show that the formula of Definition 6 can serve as a probability density. Since the values of $n(x; \mu, \sigma)$ are evidently positive as long as $\sigma > 0$,

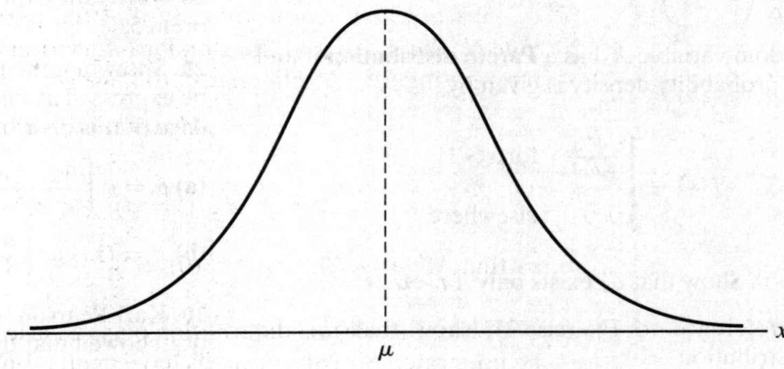

Figure 4. Graph of normal distribution.

we must show that the total area under the curve is equal to 1. Integrating from $-\infty$ to ∞ and making the substitution $z = \dfrac{x-\mu}{\sigma}$, we get

$$\int_{-\infty}^{\infty} \frac{1}{\sigma\sqrt{2\pi}} e^{-\frac{1}{2}\left(\frac{x-\mu}{\sigma}\right)^2} dx = \frac{1}{\sqrt{2\pi}} \int_{-\infty}^{\infty} e^{-\frac{1}{2}z^2} dz = \frac{2}{\sqrt{2\pi}} \int_{0}^{\infty} e^{-\frac{1}{2}z^2} dz$$

Then, since the integral on the right equals $\dfrac{\Gamma\left(\frac{1}{2}\right)}{\sqrt{2}} = \dfrac{\sqrt{\pi}}{\sqrt{2}}$ according to Exercise 9, it follows that the total area under the curve is equal to $\dfrac{2}{\sqrt{2\pi}} \cdot \dfrac{\sqrt{\pi}}{\sqrt{2}} = 1$.

Next let us prove the following theorem.

THEOREM 6. The moment-generating function of the normal distribution is given by

$$M_X(t) = e^{\mu t + \frac{1}{2}\sigma^2 t^2}$$

Proof By definition,

$$M_X(t) = \int_{-\infty}^{\infty} e^{xt} \cdot \frac{1}{\sigma\sqrt{2\pi}} e^{-\frac{1}{2}\left(\frac{x-\mu}{\sigma}\right)^2} dx$$

$$= \frac{1}{\sigma\sqrt{2\pi}} \cdot \int_{-\infty}^{\infty} e^{-\frac{1}{2\sigma^2}[-2xt\sigma^2 + (x-\mu)^2]} dx$$

and if we complete the square, that is, use the identity

$$-2xt\sigma^2 + (x-\mu)^2 = [x - (\mu + t\sigma^2)]^2 - 2\mu t\sigma^2 - t^2\sigma^4$$

we get

$$M_X(t) = e^{\mu t + \frac{1}{2}t^2\sigma^2} \left\{ \frac{1}{\sigma\sqrt{2\pi}} \cdot \int_{-\infty}^{\infty} e^{-\frac{1}{2}\left[\frac{x-(\mu+t\sigma^2)}{\sigma}\right]^2} dx \right\}$$

Since the quantity inside the braces is the integral from $-\infty$ to ∞ of a normal density with the parameters $\mu + t\sigma^2$ and σ, and hence is equal to 1, it follows that

$$M_X(t) = e^{\mu t + \frac{1}{2}\sigma^2 t^2}$$

We are now ready to verify that the parameters μ and σ in Definition 6 are, indeed, the mean and the standard deviation of the normal distribution. Twice differentiating $M_X(t)$ with respect to t, we get

$$M'_X(t) = (\mu + \sigma^2 t) \cdot M_X(t)$$

and

$$M''_X(t) = [(\mu + \sigma^2 t)^2 + \sigma^2] \cdot M_X(t)$$

so that $M'_X(0) = \mu$ and $M''_X(0) = \mu^2 + \sigma^2$. Thus, $E(X) = \mu$ and $\text{var}(X) = (\mu^2 + \sigma^2) - \mu^2 = \sigma^2$.

Since the normal distribution plays a basic role in statistics and its density cannot be integrated directly, its areas have been tabulated for the special case where $\mu = 0$ and $\sigma = 1$.

DEFINITION 7. STANDARD NORMAL DISTRIBUTION. *The normal distribution with $\mu = 0$ and $\sigma = 1$ is referred to as the* ***standard normal distribution***.

The entries in standard normal distribution table, represented by the shaded area of Figure 5, are the values of

$$\int_0^z \frac{1}{\sqrt{2\pi}} e^{-\frac{1}{2}x^2}\, dx$$

that is, the probabilities that a random variable having the standard normal distribution will take on a value on the interval from 0 to z, for $z = 0.00, 0.01, 0.02, \ldots, 3.08$, and 3.09 and also $z = 4.0$, $z = 5.0$, and $z = 6.0$. By virtue of the symmetry of the normal distribution about its mean, it is unnecessary to extend the table to negative values of z.

EXAMPLE 2

Find the probabilities that a random variable having the standard normal distribution will take on a value

(a) less than 1.72;
(b) less than −0.88;
(c) between 1.30 and 1.75;
(d) between −0.25 and 0.45.

Solution

(a) We look up the entry corresponding to $z = 1.72$ in the standard normal distribution table, add 0.5000 (see Figure 6), and get $0.4573 + 0.5000 = 0.9573$.

(b) We look up the entry corresponding to $z = 0.88$ in the table, subtract it from 0.5000 (see Figure 6), and get $0.5000 - 0.3106 = 0.1894$.

(c) We look up the entries corresponding to $z = 1.75$ and $z = 1.30$ in the table, subtract the second from the first (see Figure 6), and get $0.4599 - 0.4032 = 0.0567$.

(d) We look up the entries corresponding to $z = 0.25$ and $z = 0.45$ in the table, add them (see Figure 6), and get $0.0987 + 0.1736 = 0.2723$.

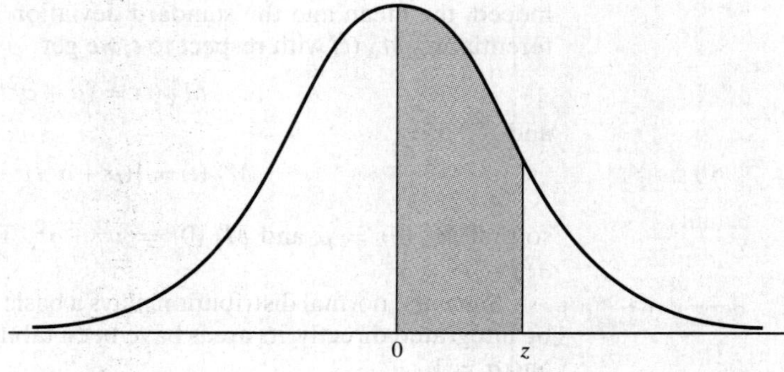

Figure 5. Tabulated areas under the standard normal distribution.

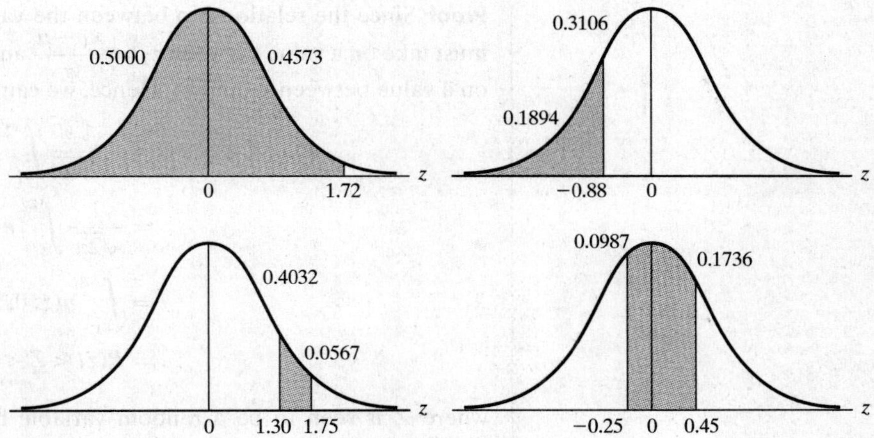

Figure 6. Diagrams for Example 2.

Occasionally, we are required to find a value of z corresponding to a specified probability that falls between values listed in the table. In that case, for convenience, we always choose the z value corresponding to the tabular value that comes closest to the specified probability. However, if the given probability falls midway between tabular values, we shall choose for z the value falling midway between the corresponding values of z.

EXAMPLE 3

With reference to the standard normal distribution table, find the values of z that correspond to entries of

(a) 0.3512;
(b) 0.2533.

Solution

(a) Since 0.3512 falls between 0.3508 and 0.3531, corresponding to $z = 1.04$ and $z = 1.05$, and since 0.3512 is closer to 0.3508 than 0.3531, we choose $z = 1.04$.

(b) Since 0.2533 falls midway between 0.2517 and 0.2549, corresponding to $z = 0.68$ and $z = 0.69$, we choose $z = 0.685$.

To determine probabilities relating to random variables having normal distributions other than the standard normal distribution, we make use of the following theorem.

THEOREM 7. If X has a normal distribution with the mean μ and the standard deviation σ, then

$$Z = \frac{X - \mu}{\sigma}$$

has the standard normal distribution.

Proof Since the relationship between the values of X and Z is linear, Z must take on a value between $z_1 = \frac{x_1 - \mu}{\sigma}$ and $z_2 = \frac{x_2 - \mu}{\sigma}$ when X takes on a value between x_1 and x_2. Hence, we can write

$$\begin{aligned}
P(x_1 < X < x_2) &= \frac{1}{\sqrt{2\pi}\sigma} \int_{x_1}^{x_2} e^{-\frac{1}{2}\left(\frac{x-\mu}{\sigma}\right)^2} dx \\
&= \frac{1}{\sqrt{2\pi}} \int_{z_1}^{z_2} e^{-\frac{1}{2}z^2} dz \\
&= \int_{z_1}^{z_2} n(z; 0, 1)\, dz \\
&= P(z_1 < Z < z_2)
\end{aligned}$$

where Z is seen to be a random variable having the standard normal distribution.

Thus, to use the standard normal distribution table in connection with any random variable having a normal distribution, we simply perform the change of scale $z = \frac{x - \mu}{\sigma}$.

EXAMPLE 4

Suppose that the amount of cosmic radiation to which a person is exposed when flying by jet across the United States is a random variable having a normal distribution with a mean of 4.35 mrem and a standard deviation of 0.59 mrem. What is the probability that a person will be exposed to more than 5.20 mrem of cosmic radiation on such a flight?

Solution

Looking up the entry corresponding to $z = \frac{5.20 - 4.35}{0.59} = 1.44$ in the table and subtracting it from 0.5000 (see Figure 7), we get $0.5000 - 0.4251 = 0.0749$.

Probabilities relating to random variables having the normal distribution and several other continuous distributions can be found directly with the aid of computer

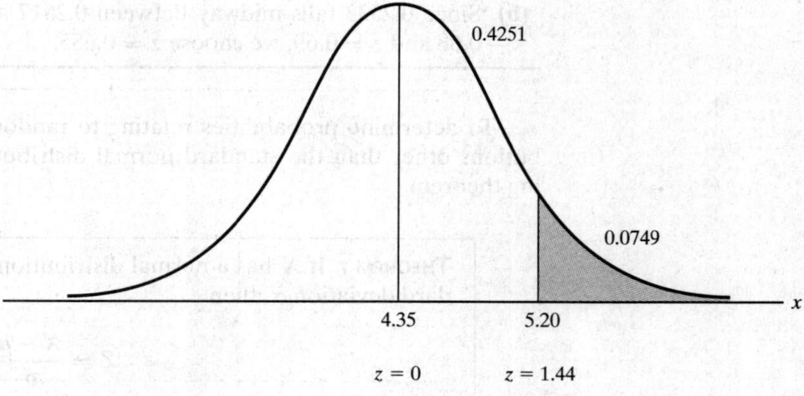

Figure 7. Diagram for Example 4.

programs especially written for statistical applications. The following example illustrates such calculations using MINITAB statistical software.

EXAMPLE 5

Use a computer program to find the probability that a random variable having

(a) the chi-square distribution with 25 degrees of freedom will assume a value greater than 30;

(b) the normal distribution with the mean 18.7 and the standard deviation 9.1 will assume a value between 10.6 and 24.8.

Solution

Using MINITAB software, we select the option "cumulative distribution" to obtain the following:

(a)
```
MTB>CDF C1;
SUBC>Chisquare 25
     30.0000 0.7757
```

Thus, the required probability is $1.0000 - 0.7757 = 0.2243$.

(b)
```
MTB>CDF C2;                  and    MTB>CDF C3;
SUBC>Normal 18.7 9.1.               SUBC>Normal 18.7 9.1.
     10.6000 0.1867                      24.800 0.7487
```

Thus, the required probability is $0.7487 - 0.1867 = 0.5620$.

6 The Normal Approximation to the Binomial Distribution

The normal distribution is sometimes introduced as a continuous distribution that provides a close approximation to the binomial distribution when n, the number of trials, is very large and θ, the probability of a success on an individual trial, is close to $\frac{1}{2}$. Figure 8 shows the histograms of binomial distributions with $\theta = \frac{1}{2}$ and $n = 2$,

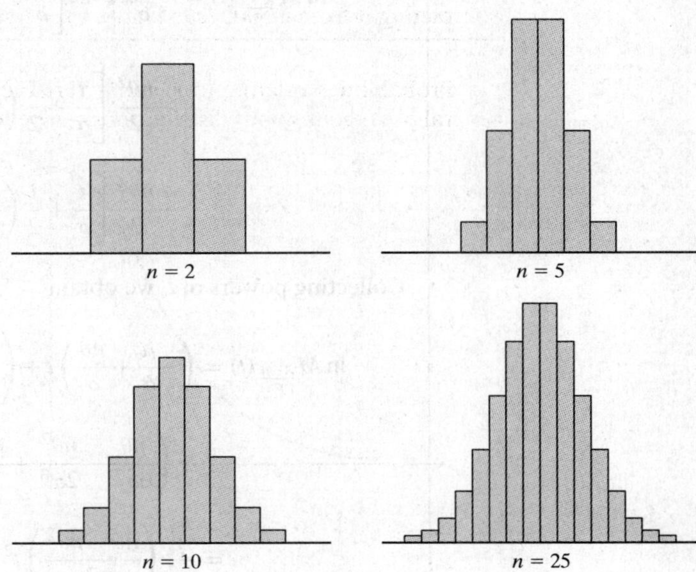

Figure 8. Binomial distributions with $\theta = \frac{1}{2}$.

5, 10, and 25, and it can be seen that with increasing n these distributions approach the symmetrical bell-shaped pattern of the normal distribution.

To provide a theoretical foundation for this argument, let us first prove the following theorem.

THEOREM 8. If X is a random variable having a binomial distribution with the parameters n and θ, then the moment-generating function of

$$Z = \frac{X - n\theta}{\sqrt{n\theta(1-\theta)}}$$

approaches that of the standard normal distribution when $n \to \infty$.

Proof Making use of theorems relating to moment-generating functions of the binomial distribution, we can write

$$M_Z(t) = M_{\frac{X-\mu}{\sigma}}(t) = e^{-\mu t/\sigma} \cdot [1 + \theta(e^{t/\sigma} - 1)]^n$$

where $\mu = n\theta$ and $\sigma = \sqrt{n\theta(1-\theta)}$. Then, taking logarithms and substituting the Maclaurin's series of $e^{t/\sigma}$, we get

$$\ln M_{\frac{X-\mu}{\sigma}}(t) = -\frac{\mu t}{\sigma} + n \cdot \ln[1 + \theta(e^{t/\sigma} - 1)]$$

$$= -\frac{\mu t}{\sigma} + n \cdot \ln\left[1 + \theta\left\{\frac{t}{\sigma} + \frac{1}{2}\left(\frac{t}{\sigma}\right)^2 + \frac{1}{6}\left(\frac{t}{\sigma}\right)^3 + \cdots\right\}\right]$$

and, using the infinite series $\ln(1+x) = x - \frac{1}{2}x^2 + \frac{1}{3}x^3 - \cdots$, which converges for $|x| < 1$, to expand this logarithm, it follows that

$$\ln M_{\frac{X-\mu}{\sigma}}(t) = -\frac{\mu t}{\sigma} + n\theta\left[\frac{t}{\sigma} + \frac{1}{2}\left(\frac{t}{\sigma}\right)^2 + \frac{1}{6}\left(\frac{t}{\sigma}\right)^3 + \cdots\right]$$

$$-\frac{n\theta^2}{2}\left[\frac{t}{\sigma} + \frac{1}{2}\left(\frac{t}{\sigma}\right)^2 + \frac{1}{6}\left(\frac{t}{\sigma}\right)^3 + \cdots\right]^2$$

$$+\frac{n\theta^3}{3}\left[\frac{t}{\sigma} + \frac{1}{2}\left(\frac{t}{\sigma}\right)^2 + \frac{1}{6}\left(\frac{t}{\sigma}\right)^3 + \cdots\right]^3 - \cdots$$

Collecting powers of t, we obtain

$$\ln M_{\frac{X-\mu}{\sigma}}(t) = \left(-\frac{\mu}{\sigma} + \frac{n\theta}{\sigma}\right)t + \left(\frac{n\theta}{2\sigma^2} - \frac{n\theta^2}{2\sigma^2}\right)t^2$$

$$+\left(\frac{n\theta}{6\sigma^3} - \frac{n\theta^2}{2\sigma^3} + \frac{n\theta^3}{3\sigma^3}\right)t^3 + \cdots$$

$$= \frac{1}{\sigma^2}\left(\frac{n\theta - n\theta^2}{2}\right)t^2 + \frac{n}{\sigma^3}\left(\frac{\theta - 3\theta^2 + 2\theta^3}{6}\right)t^3 + \cdots$$

since $\mu = n\theta$. Then, substituting $\sigma = \sqrt{n\theta(1-\theta)}$, we find that

$$\ln M_{\frac{X-\mu}{\sigma}}(t) = \frac{1}{2}t^2 + \frac{n}{\sigma^3}\left(\frac{\theta - 3\theta^2 + 2\theta^3}{6}\right)t^3 + \cdots$$

For $r > 2$ the coefficient of t^r is a constant times $\frac{n}{\sigma^r}$, which approaches 0 when $n \to \infty$. It follows that

$$\lim_{n\to\infty} \ln M_{\frac{X-\mu}{\sigma}}(t) = \frac{1}{2}t^2$$

and since the limit of a logarithm equals the logarithm of the limit (provided the two limits exist), we conclude that

$$\lim_{n\to\infty} M_{\frac{X-\mu}{\sigma}}(t) = e^{\frac{1}{2}t^2}$$

which is the moment-generating function of Theorem 6 with $\mu = 0$ and $\sigma = 1$.

This completes the proof of Theorem 8, but have we shown that when $n \to \infty$ the distribution of Z, the **standardized** binomial random variable, approaches the standard normal distribution? Not quite. To this end, we must refer to two theorems that we shall state here without proof:

1. *There is a one-to-one correspondence between moment-generating functions and probability distributions (densities) when the former exist.*
2. *If the moment-generating function of one random variable approaches that of another random variable, then the distribution (density) of the first random variable approaches that of the second random variable under the same limiting conditions.*

Strictly speaking, our results apply only when $n \to \infty$, but the normal distribution is often used to approximate binomial probabilities even when n is fairly small. A good rule of thumb is to use this approximation only when $n\theta$ and $n(1-\theta)$ are both greater than 5.

EXAMPLE 6

Use the normal approximation to the binomial distribution to determine the probability of getting 6 heads and 10 tails in 16 flips of a balanced coin.

Solution

To find this approximation, we must use the **continuity correction** according to which each nonnegative integer k is represented by the interval from $k - \frac{1}{2}$ to $k + \frac{1}{2}$. With reference to Figure 9, we must thus determine the area under the curve between 5.5 and 6.5, and since $\mu = 16 \cdot \frac{1}{2} = 8$ and $\sigma = \sqrt{16 \cdot \frac{1}{2} \cdot \frac{1}{2}} = 2$, we must find the area between

$$z = \frac{5.5-8}{2} = -1.25 \quad \text{and} \quad z = \frac{6.5-8}{2} = -0.75$$

The entries in the standard normal distribution table corresponding to $z = 1.25$ and $z = 0.75$ are 0.3944 and 0.2734, and we find that the normal approximation to the

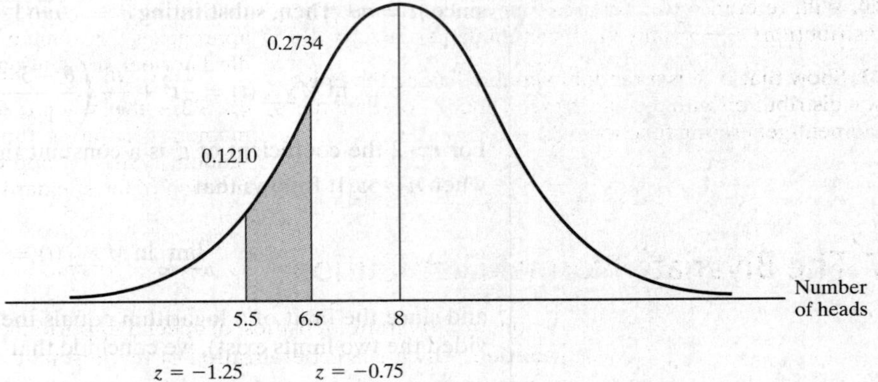

Figure 9. Diagram for Example 6.

probability of "6 heads and 10 tails" is $0.3944 - 0.2734 = 0.1210$. Since the corresponding value in the binomial probabilities table of "Statistical Tables" is 0.1222, we find that the error of the approximation is -0.0012 and that the percentage error is $\frac{0.0012}{0.1222} \cdot 100 = 0.98\%$ in absolute value.

The normal approximation to the binomial distribution used to be applied quite extensively, particularly in approximating probabilities associated with large sets of values of binomial random variables. Nowadays, most of this work is done with computers, as illustrated in Example 5, and we have mentioned the relationship between the binomial and normal distributions primarily because of its theoretical applications.

Exercises

31. Show that the normal distribution has
(a) a relative maximum at $x = \mu$;
(b) inflection points at $x = \mu - \sigma$ and $x = \mu + \sigma$.

32. Show that the differential equation of Exercise 30 with $b = c = 0$ and $a > 0$ yields a normal distribution.

33. This question has been intentionally omitted for this edition.

34. If X is a random variable having a normal distribution with the mean μ and the standard deviation σ, find the moment-generating function of $Y = X - c$, where c is a constant, and use it to rework Exercise 33.

35. This question has been intentionally omitted for this edition.

36. This question has been intentionally omitted for this edition.

37. If X is a random variable having the standard normal distribution and $Y = X^2$, show that $\text{cov}(X, Y) = 0$ even though X and Y are evidently not independent.

38. Use the Maclaurin's series expansion of the moment-generating function of the standard normal distribution to show that
(a) $\mu_r = 0$ when r is odd;
(b) $\mu_r = \dfrac{r!}{2^{r/2}\left(\frac{r}{2}\right)!}$ when r is even.

39. If we let $K_X(t) = \ln M_{X-\mu}(t)$, the coefficient of $\frac{t^r}{r!}$ in the Maclaurin's series of $K_X(t)$ is called the **rth cumulant**, and it is denoted by κ_r. Equating coefficients of like powers, show that
(a) $\kappa_2 = \mu_2$;
(b) $\kappa_3 = \mu_3$;
(c) $\kappa_4 = \mu_4 - 3\mu_2^2$.

40. With reference to Exercise 39, show that for normal distributions $\kappa_2 = \sigma^2$ and all other cumulants are zero.

41. Show that if X is a random variable having the Poisson distribution with the parameter λ and $\lambda \to \infty$, then the moment-generating function of

$$Z = \frac{X - \lambda}{\sqrt{\lambda}}$$

that is, that of a standardized Poisson random variable, approaches the moment-generating function of the standard normal distribution.

42. Show that when $\alpha \to \infty$ and β remains constant, the moment-generating function of a standardized gamma random variable approaches the moment-generating function of the standard normal distribution.

7 The Bivariate Normal Distribution

Among multivariate densities, of special importance is the **multivariate normal distribution**, which is a generalization of the normal distribution in one variable. As it is best (indeed, virtually necessary) to present this distribution in matrix notation, we shall give here only the **bivariate** case; discussions of the general case are listed among the references at the end of this chapter.

DEFINITION 8. BIVARIATE NORMAL DISTRIBUTION. *A pair of random variables X and Y have a* ***bivariate normal distribution*** *and they are referred to as jointly normally distributed random variables if and only if their joint probability density is given by*

$$f(x,y) = \frac{e^{-\frac{1}{2(1-\rho)^2}\left[\left(\frac{x-\mu_1}{\sigma_1}\right)^2 - 2\rho\left(\frac{x-\mu_1}{\sigma_1}\right)\left(\frac{y-\mu_2}{\sigma_2}\right) + \left(\frac{y-\mu_2}{\sigma_2}\right)^2\right]}}{2\pi\sigma_1\sigma_2\sqrt{1-\rho^2}}$$

for $-\infty < x < \infty$ *and* $-\infty < y < \infty$, *where* $\sigma_1 > 0$, $\sigma_2 > 0$, *and* $-1 < \rho < 1$.

To study this joint distribution, let us first show that the parameters μ_1, μ_2, σ_1, and σ_2 are, respectively, the means and the standard deviations of the two random variables X and Y. To begin with, we integrate on y from $-\infty$ to ∞, getting

$$g(x) = \frac{e^{-\frac{1}{2(1-\rho^2)}\left(\frac{x-\mu_1}{\sigma_1}\right)^2}}{2\pi\sigma_1\sigma_2\sqrt{1-\rho^2}} \int_{-\infty}^{\infty} e^{-\frac{1}{2(1-\rho^2)}\left[\left(\frac{y-\mu_2}{\sigma_2}\right)^2 - 2\rho\left(\frac{x-\mu_1}{\sigma_1}\right)\left(\frac{y-\mu_2}{\sigma_2}\right)\right]} dy$$

for the marginal density of X. Then, temporarily making the substitution $u = \dfrac{x-\mu_1}{\sigma_1}$ to simplify the notation and changing the variable of integration by letting $v = \dfrac{y-\mu_2}{\sigma_2}$, we obtain

$$g(x) = \frac{e^{-\frac{1}{2(1-\rho^2)}u^2}}{2\pi\sigma_1\sqrt{1-\rho^2}} \int_{-\infty}^{\infty} e^{-\frac{1}{2(1-\rho^2)}(v^2-2\rho uv)} dv$$

After completing the square by letting

$$v^2 - 2\rho uv = (v - \rho u)^2 - \rho^2 u^2$$

and collecting terms, this becomes

$$g(x) = \frac{e^{-\frac{1}{2}u^2}}{\sigma_1\sqrt{2\pi}} \left\{ \frac{1}{\sqrt{2\pi}\sqrt{1-\rho^2}} \int_{-\infty}^{\infty} e^{-\frac{1}{2}\left(\frac{v-\rho u}{\sqrt{1-\rho^2}}\right)^2} dv \right\}$$

Finally, identifying the quantity in parentheses as the integral of a normal density from $-\infty$ to ∞, and hence equaling 1, we get

$$g(x) = \frac{e^{-\frac{1}{2}u^2}}{\sigma_1\sqrt{2\pi}} = \frac{1}{\sigma_1\sqrt{2\pi}} e^{-\frac{1}{2}\left(\frac{x-\mu_1}{\sigma_1}\right)^2}$$

for $-\infty < x < \infty$. It follows by inspection that the marginal density of X is a normal distribution with the mean μ_1 and the standard deviation σ_1 and, by symmetry, that the marginal density of Y is a normal distribution with the mean μ_2 and the standard deviation σ_2.

As far as the parameter ρ is concerned, where ρ is the lowercase Greek letter *rho*, it is called the **correlation coefficient**, and the necessary integration will show that $\text{cov}(X, Y) = \rho\sigma_1\sigma_2$. Thus, the parameter ρ measures how the two random variables X and Y vary together.

When we deal with a pair of random variables having a bivariate normal distribution, their conditional densities are also of importance; let us prove the following theorem.

THEOREM 9. If X and Y have a bivariate normal distribution, the conditional density of Y given $X = x$ is a normal distribution with the mean

$$\mu_{Y|x} = \mu_2 + \rho\frac{\sigma_2}{\sigma_1}(x - \mu_1)$$

and the variance

$$\sigma^2_{Y|x} = \sigma^2_2(1 - \rho^2)$$

and the conditional density of X given $Y = y$ is a normal distribution with the mean

$$\mu_{X|y} = \mu_1 + \rho\frac{\sigma_1}{\sigma_2}(y - \mu_2)$$

and the variance

$$\sigma^2_{X|y} = \sigma^2_1(1 - \rho^2)$$

Proof Writing $w(y|x) = \dfrac{f(x,y)}{g(x)}$ in accordance with the definition of conditional density and letting $u = \dfrac{x - \mu_1}{\sigma_1}$ and $v = \dfrac{y - \mu_2}{\sigma_2}$ to simplify the notation, we get

$$w(y|x) = \frac{\dfrac{1}{2\pi\sigma_1\sigma_2\sqrt{1-\rho^2}} e^{-\frac{1}{2(1-\rho^2)}[u^2 - 2\rho uv + v^2]}}{\dfrac{1}{\sqrt{2\pi}\sigma_1} e^{-\frac{1}{2}u^2}}$$

$$= \frac{1}{\sqrt{2\pi}\sigma_2\sqrt{1-\rho^2}} e^{-\frac{1}{2(1-\rho^2)}[v^2 - 2\rho uv + \rho^2 u^2]}$$

$$= \frac{1}{\sqrt{2\pi}\sigma_2\sqrt{1-\rho^2}} e^{-\frac{1}{2}\left[\frac{v - \rho u}{\sqrt{1-\rho^2}}\right]^2}$$

Then, expressing this result in terms of the original variables, we obtain

$$w(y|x) = \frac{1}{\sigma_2\sqrt{2\pi}\sqrt{1-\rho^2}} e^{-\frac{1}{2}\left[\frac{y-\left\{\mu_2+\rho\frac{\sigma_2}{\sigma_1}(x-\mu_1)\right\}}{\sigma_2\sqrt{1-\rho^2}}\right]^2}$$

for $-\infty < y < \infty$, and it can be seen by inspection that this is a normal density with the mean $\mu_{Y|x} = \mu_2 + \rho\frac{\sigma_2}{\sigma_1}(x - \mu_1)$ and the variance $\sigma^2_{Y|x} = \sigma^2_2(1 - \rho^2)$. The corresponding results for the conditional density of X given $Y = y$ follow by symmetry.

The bivariate normal distribution has many important properties, some statistical and some purely mathematical. Among the former, there is the following property, which the reader will be asked to prove in Exercise 43.

THEOREM 10. If two random variables have a bivariate normal distribution, they are independent if and only if $\rho = 0$.

In this connection, if $\rho = 0$, the random variables are said to be **uncorrelated**.

Also, we have shown that for two random variables having a bivariate normal distribution the two marginal densities are normal, but the converse is not necessarily true. In other words, the marginal distributions may both be normal without the joint distribution being a bivariate normal distribution. For instance, if the bivariate density of X and Y is given by

$$f^*(x,y) = \begin{cases} 2f(x,y) & \text{inside squares 2 and 4 of Figure 10} \\ 0 & \text{inside squares 1 and 3 of Figure 10} \\ f(x,y) & \text{elsewhere} \end{cases}$$

where $f(x, y)$ is the value of the bivariate normal density with $\mu_1 = 0, \mu_2 = 0$, and $\rho = 0$ at (x, y), it is easy to see that the marginal densities of X and Y are normal even though their joint density is not a bivariate normal distribution.

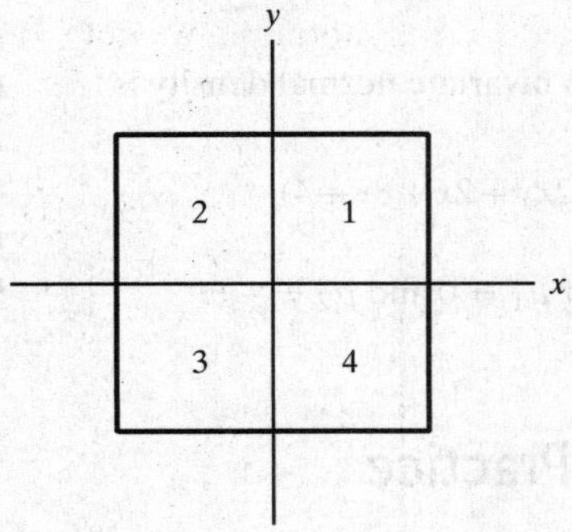

Figure 10. Sample space for the bivariate density given by $f^*(x, y)$.

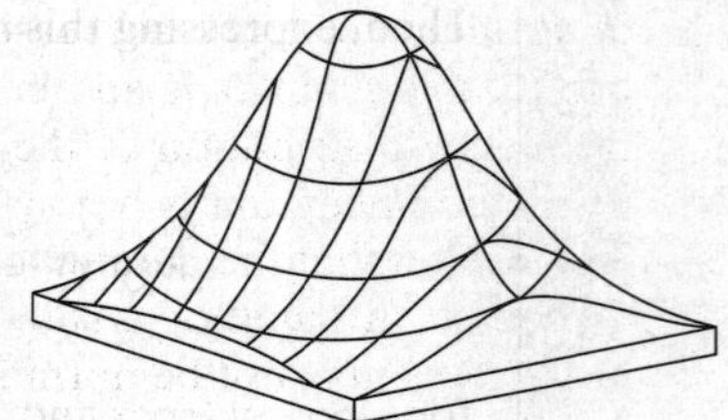

Figure 11. Bivariate normal surface.

Many interesting properties of the bivariate normal density are obtained by studying the **bivariate normal surface**, pictured in Figure 11, whose equation is $z = f(x, y)$, where $f(x, y)$ is the value of the bivariate normal density at (x, y). As the reader will be asked to verify in some of the exercises that follow, the bivariate normal surface has a maximum at (μ_1, μ_2), any plane parallel to the z-axis intersects the surface in a curve having the shape of a normal distribution, and any plane parallel to the xy-plane that intersects the surface intersects it in an ellipse called a **contour of constant probability density**. When $\rho = 0$ and $\sigma_1 = \sigma_2$, the contours of constant probability density are circles, and it is customary to refer to the corresponding joint density as a **circular normal distribution**.

Exercises

43. To prove Theorem 10, show that if X and Y have a bivariate normal distribution, then
(a) their independence implies that $\rho = 0$;
(b) $\rho = 0$ implies that they are independent.

44. Show that any plane perpendicular to the xy-plane intersects the bivariate normal surface in a curve having the shape of a normal distribution.

45. If the exponent of e of a bivariate normal density is

$$\frac{-1}{102}[(x+2)^2 - 2.8(x+2)(y-1) + 4(y-1)^2]$$

find
(a) $\mu_1, \mu_2, \sigma_1, \sigma_2$, and ρ;
(b) $\mu_{Y|x}$ and $\sigma^2_{Y|x}$.

46. If the exponent of e of a bivariate normal density is

$$\frac{-1}{54}(x^2 + 4y^2 + 2xy + 2x + 8y + 4)$$

find σ_1, σ_2, and ρ, given that $\mu_1 = 0$ and $\mu_2 = -1$.

47. If X and Y have the bivariate normal distribution with $\mu_1 = 2, \mu_2 = 5, \sigma_1 = 3, \sigma_2 = 6$, and $\rho = \frac{2}{3}$, find $\mu_{Y|1}$ and $\sigma_{Y|1}$.

48. If X and Y have a bivariate normal distribution and $U = X + Y$ and $V = X - Y$, find an expression for the correlation coefficient of U and V.

49. If X and Y have a bivariate normal distribution, it can be shown that their joint moment-generating function is given by

$$\begin{aligned} M_{X,Y}(t_1, t_2) &= E(e^{t_1 X + t_2 Y}) \\ &= e^{t_1\mu_1 + t_2\mu_2 + \frac{1}{2}(\sigma_1^2 t_1^2 + 2\rho\sigma_1\sigma_2 t_1 t_2 + \sigma_2^2 t_2^2)} \end{aligned}$$

Verify that
(a) the first partial derivative of this function with respect to t_1 at $t_1 = 0$ and $t_2 = 0$ is μ_1;
(b) the second partial derivative with respect to t_1 at $t_1 = 0$ and $t_2 = 0$ is $\sigma_1^2 + \mu_1^2$;
(c) the second partial derivative with respect to t_1 and t_2 at $t_1 = 0$ and $t_2 = 0$ is $\rho\sigma_1\sigma_2 + \mu_1\mu_2$.

8 The Theory in Practice

In many of the applications of statistics it is assumed that the data are approximately normally distributed. Thus, it is important to make sure that the assumption of normality can, at least reasonably, be supported by the data. Since the normal distribution is symmetric and bell-shaped, examination of the histogram picturing the frequency distribution of the data is useful in checking the assumption of normality. If the histogram is not symmetric, or if it is symmetric but not bell-shaped, the assumption that the data set comes from a normal distribution cannot be supported. Of course, this method is subjective; data that appear to have symmetric, bell-shaped histograms may not be normally distributed.

Another somewhat less subjective method for checking data is the **normal-scores plot**. This plot makes use of ordinary graph paper. It is based on the calculation of **normal scores**, z_p. If n observations are ordered from smallest to largest, they divide the area under the normal curve into $n+1$ equal parts, each having the area $1/(n+1)$. The normal score for the first of these areas is the value of z such that the area under the standard normal curve to the left of z is $1/(n+1)$, or $-z_{1/(n+1)}$. Thus, the normal scores for $n = 4$ observations are $-z_{0.20} = -0.84, -z_{0.40} = -0.25, z_{0.40} = 0.25$, and $z_{20} = 0.84$. The ordered observations then are plotted against the corresponding normal scores on ordinary graph paper.

EXAMPLE 7

Find the normal scores and the coordinates for making a normal-scores plot of the following six observations:

$$3,\ 2,\ 7,\ 4,\ 3,\ 5$$

Solution

Since $n = 6$, there are 6 normal scores, as follows: $-z_{0.14} = -1.08, -z_{0.29} = -0.55, -z_{0.43} = -0.18, z_{0.43} = 0.18, z_{0.29} = 0.55$, and $z_{0.14} = 1.08$. When the observations are ordered and tabulated together with the normal scores, the following table results:

Observation:	2	3	3	4	5	7
Normal score:	−1.08	−0.55	−0.18	0.18	0.55	1.08

The coordinates for a normal-scores plot make use of a cumulative percentage distribution of the data. The cumulative percentage distribution is as follows:

Class Boundary	*Cumulative Percentage*	*Normal Score*
4395	5	−1.64
4595	17	−0.95
4795	37	−0.33
4995	69	0.50
5195	87	1.13
5395	97	1.88

A graph of the class boundaries versus the normal scores is shown in Figure 12. It can be seen from this graph that the points lie in an almost perfect straight line, strongly suggesting that the underlying data are very close to being normally distributed.

In modern practice, use of MINITAB or other statistical software eases the computation considerably. In addition, MINITAB offers three tests for normality that are less subjective than mere examination of a normal-scores plot.

Sometimes a normal-scores plot showing a curve can be changed to a straight line by means of an appropriate transformation. The procedure involves identifying

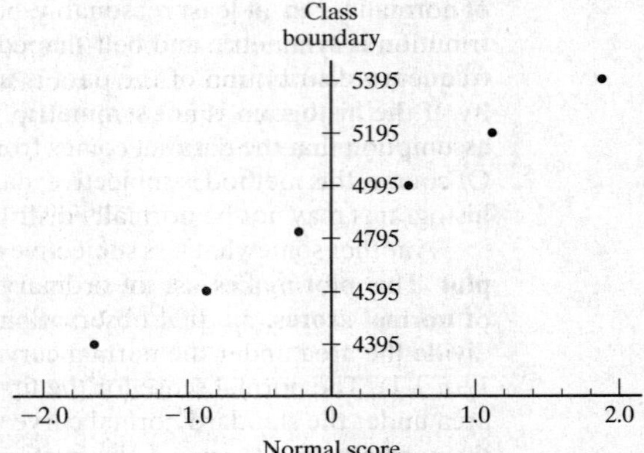

Figure 12. Normal-scores plot.

the type of transformation needed, making the transformation, and then checking the transformed data by means of a normal-scores plot to see if they can be assumed to have a normal distribution.

When data appear not to be normally distributed because of *too many large values*, the following transformations are good candidates to try:

$$\text{logarithmic transformation } u = \log(x)$$

$$\text{square-root transformation } u = \sqrt{x}$$

$$\text{reciprocal transformation } u = \frac{1}{x}$$

When data exhibit *too many small values*, the following transformations may produce approximately normal data:

$$\text{power transformation} \qquad u = x^a, \text{ where } a > 1$$

$$\text{exponential transformation} \qquad u = a^x, \text{ where } a > 1$$

On rare occasions, it helps to make a linear transformation of the form $u = a + bx$ first, and then to use one of the indicated transformations. This strategy becomes necessary when some of the data have negative values and logarithmic, square-root, or certain power transformations are to be tried. However, making a linear transformation alone cannot be effective. If x is a value of a normally distributed random variable, then the random variable having the values $a + bx$ also has the normal distribution. Thus, a linear transformation alone cannot transform nonnormally distributed data into normality.

EXAMPLE 8

Make a normal-scores plot of the following data. If the plot does not appear to show normality, make an appropriate transformation, and check the transformed data for normality.

54.9 8.3 5.2 32.4 15.5

Solution

The normal scores are −0.95, −0.44, 0, 0.44, and 0.95. A normal-scores plot of these data (Figure 13[a]) shows sharp curvature. Since two of the five values are very large compared with the other three values, a logarithmic transformation (base 10) was used to transform the data to

1.74 0.92 0.72 1.51 1.19

A normal-scores plot of these transformed data (Figure 13[b]) shows a nearly straight line, indicating that the transformed data are approximately normally distributed.

If lack of normality seems to result from one or a small number of aberrant observations called outliers, a single large observation, a single small observation, or both, it is not likely that the data can be transformed to normality. It is difficult to give a hard-and-fast rule for identifying outliers. For example, it may be inappropriate to define an **outlier** as an observation whose value is more than three standard deviations from the mean, since such an observation can occur with a reasonable probability in a large number of observations taken from a normal distribution. Ordinarily, an observation that clearly does not lie on the straight line defined by the other observations in a normal-scores plot can be considered an outlier. In the presence of suspected outliers, it is customary to examine normal-scores plots of the data after the outlier or outliers have been omitted.

Outlying observations may result from several causes, such as an error in recording data, an error of observation, or an unusual event such as a particle of dust settling on a material during thin-film deposition. There is always a great temptation to drop outliers from a data set entirely on the basis that they do not seem to belong to the main body of data. But an outlier can be as informative about the process from which the data were taken as the remainder of the data. Outliers which occur infrequently, but regularly in successive data sets, give evidence that should not be ignored. For example, a hole with an unusually large diameter might result from a drill not having been inserted properly into the chuck. Perhaps the condition was corrected after one or two holes were drilled, and the operator failed to discard the parts with the "bad" hole, thus producing one or two outliers. While outliers sometimes are separated from the other data for the purpose of performing a preliminary

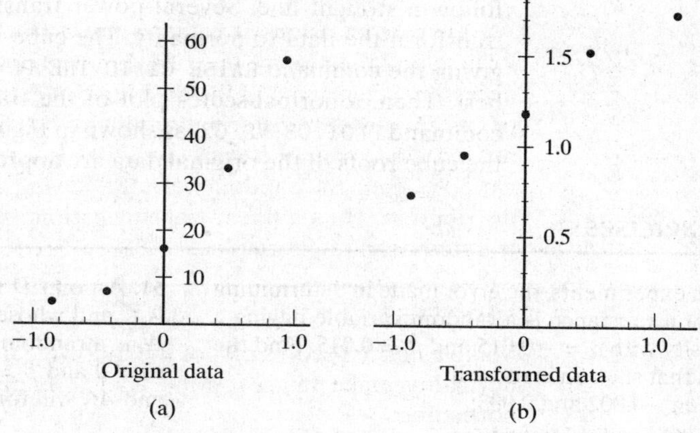

Figure 13. Normal-scores plot for Example 8.

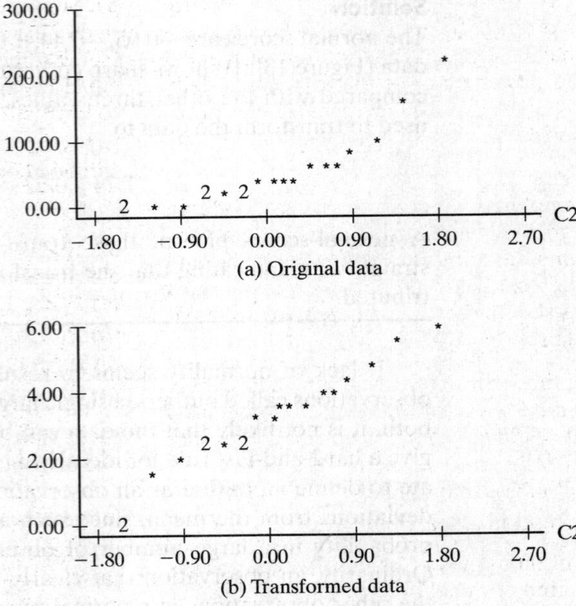

Figure 14. Normal-scores plots.

analysis, they should be discarded only after a good reason for their existence has been found.

Normal scores and normal-score plots can be obtained with a variety of statistical software. To illustrate the procedure using MINITAB, 20 numbers are entered with the following command and data-entry instructions

```
SET C1:
0 215 31 7 15 80 17 41 51 3 58 158 0 11 42 11 17 32 64 100
END
```

Then the command `NSCORES C1 PUT IN C2` is given to find the normal scores and place them in the second column. A normal-scores plot, generated by the command `PLOT C1 VS C2`, is shown in Figure 14(a). The points in this graph clearly do not follow a straight line. Several power transformations were tried in an attempt to transform the data to normality. The cube-root transformation $u = X^{1/3}$, made by giving the command `RAISE C1 TO THE POWER .3333 PUT IN C3`, seemed to work best. Then, a normal-scores plot of the transformed data was generated with the command `PLOT C3 VS C2`, as shown in Figure 14(b). It appears from this graph that the cube roots of the original data are approximately normally distributed.

Applied Exercises — SECS. 1–4

50. In certain experiments, the error made in determining the density of a substance is a random variable having a uniform density with $\alpha = -0.015$ and $\beta = 0.015$. Find the probabilities that such an error will
(a) be between −0.002 and 0.003;
(b) exceed 0.005 in absolute value.

51. A point D is chosen on the line AB, whose midpoint is C and whose length is a. If X, the distance from D to A, is a random variable having the uniform density with $\alpha = 0$ and $\beta = a$, what is the probability that AD, BD, and AC will form a triangle?

52. In a certain city, the daily consumption of electric power in millions of kilowatt-hours can be treated as a random variable having a gamma distribution with $\alpha = 3$ and $\beta = 2$. If the power plant of this city has a daily capacity of 12 million kilowatt-hours, what is the probability that this power supply will be inadequate on any given day?

53. If a company employs n salespersons, its gross sales in thousands of dollars may be regarded as a random variable having a gamma distribution with $\alpha = 80\sqrt{n}$ and $\beta = 2$. If the sales cost is \$8,000 per salesperson, how many salespersons should the company employ to maximize the expected profit?

54. The amount of time that a watch will run without having to be reset is a random variable having an exponential distribution with $\theta = 120$ days. Find the probabilities that such a watch will
(a) have to be reset in less than 24 days;
(b) not have to be reset in at least 180 days.

55. The mileage (in thousands of miles) that car owners get with a certain kind of radial tire is a random variable having an exponential distribution with $\theta = 40$. Find the probabilities that one of these tires will last
(a) at least 20,000 miles;
(b) at most 30,000 miles.

56. The number of bad checks that a bank receives during a 5-hour business day is a Poisson random variable with $\lambda = 2$. What is the probability that it will not receive a bad check on any one day during the first 2 hours of business?

57. The number of planes arriving per day at a small private airport is a random variable having a Poisson distribution with $\lambda = 28.8$. What is the probability that the time between two such arrivals is at least 1 hour?

58. If the annual proportion of erroneous income tax returns filed with the IRS can be looked upon as a random variable having a beta distribution with $\alpha = 2$ and $\beta = 9$, what is the probability that in any given year there will be fewer than 10 percent erroneous returns?

59. A certain kind of appliance requires repairs on the average once every 2 years. Assuming that the times between repairs are exponentially distributed, what is the probability that such an appliance will work at least 3 years without requiring repairs?

60. If the annual proportion of new restaurants that fail in a given city may be looked upon as a random variable having a beta distribution with $\alpha = 1$ and $\beta = 4$, find
(a) the mean of this distribution, that is, the annual proportion of new restaurants that can be expected to fail in the given city;
(b) the probability that at least 25 percent of all new restaurants will fail in the given city in any one year.

61. Suppose that the service life in hours of a semiconductor is a random variable having a Weibull distribution (see Exercise 23) with $\alpha = 0.025$ and $\beta = 0.500$.
(a) How long can such a semiconductor be expected to last?
(b) What is the probability that such a semiconductor will still be in operating condition after 4,000 hours?

SECS. 5–7

62. If Z is a random variable having the standard normal distribution, find
(a) $P(Z < 1.33)$;
(b) $P(Z \geqq -0.79)$;
(c) $P(0.55 < Z < 1.22)$;
(d) $P(-1.90 \leq Z \leq 0.44)$.

63. If Z is a random variable having the standard normal distribution, find the probabilities that it will take on a value
(a) greater than 1.14;
(b) greater than −0.36;
(c) between −0.46 and −0.09;
(d) between −0.58 and 1.12.

64. If Z is a random variable having the standard normal distribution, find the respective values z_1, z_2, z_3, and z_4 such that
(a) $P(0 < Z < z_1) = 0.4306$;
(b) $P(Z \geqq z_2) = 0.7704$;
(c) $P(Z > z_3) = 0.2912$;
(d) $P(-z_4 \leqq Z < z_4) = 0.9700$.

65. Find z if the standard-normal-curve area
(a) between 0 and z is 0.4726;
(b) to the left of z is 0.9868;
(c) to the right of z is 0.1314;
(d) between $-z$ and z is 0.8502.

66. If X is a random variable having a normal distribution, what are the probabilities of getting a value
(a) within one standard deviation of the mean;
(b) within two standard deviations of the mean;
(c) within three standard deviations of the mean;
(d) within four standard deviations of the mean?

67. If z_α is defined by

$$\int_{z_\alpha}^{\infty} n(z; 0, 1)\, dz = \alpha$$

find its values for
(a) $\alpha = 0.05$;
(b) $\alpha = 0.025$;
(c) $\alpha = 0.01$;
(d) $\alpha = 0.005$.

68. (a) Use a computer program to find the probability that a random variable having the normal distribution with the mean -1.786 and the standard deviation 1.0416 will assume a value between -2.159 and 0.5670.

(b) Interpolate in the standard normal distribution table to find this probability and compare your result with the more exact value found in part (a).

69. (a) Use a computer program to find the probability that a random variable having the normal distribution with mean 5.853 and the standard deviation 1.361 will assume a value greater than 8.625.

(b) Interpolate in the standard normal distribution table to find this probability and compare your result with the more exact value found in part (a).

70. Suppose that during periods of meditation the reduction of a person's oxygen consumption is a random variable having a normal distribution with $\mu = 37.6$ cc per minute and $\sigma = 4.6$ cc per minute. Find the probabilities that during a period of meditation a person's oxygen consumption will be reduced by

(a) at least 44.5 cc per minute;

(b) at most 35.0 cc per minute;

(c) anywhere from 30.0 to 40.0 cc per minute.

71. In a photographic process, the developing time of prints may be looked upon as a random variable having the normal distribution with $\mu = 15.40$ seconds and $\sigma = 0.48$ second. Find the probabilities that the time it takes to develop one of the prints will be

(a) at least 16.00 seconds;

(b) at most 14.20 seconds;

(c) anywhere from 15.00 to 15.80 seconds.

72. A random variable has a normal distribution with $\sigma = 10$. If the probability that the random variable will take on a value less than 82.5 is 0.8212, what is the probability that it will take on a value greater than 58.3?

73. Suppose that the actual amount of instant coffee that a filling machine puts into "6-ounce" jars is a random variable having a normal distribution with $\sigma = 0.05$ ounce. If only 3 percent of the jars are to contain less than 6 ounces of coffee, what must be the mean fill of these jars?

74. Check in each case whether the normal approximation to the binomial distribution may be used according to the rule of thumb in Section 6.

(a) $n = 16$ and $\theta = 0.20$;

(b) $n = 65$ and $\theta = 0.10$;

(c) $n = 120$ and $\theta = 0.98$.

75. Suppose that we want to use the normal approximation to the binomial distribution to determine $b(1; 150, 0.05)$.

(a) Based on the rule of thumb in Section 6, would we be justified in using the approximation?

(b) Make the approximation and round to four decimals.

(c) If a computer printout shows that $b(1; 150, 0.05) = 0.0036$ rounded to four decimals, what is the percentage error of the approximation obtained in part (b)?

This serves to illustrate that the rule of thumb is just that and no more; making approximations like this also requires a good deal of professional judgment.

76. Use the normal approximation to the binomial distribution to determine (to four decimals) the probability of getting 7 heads and 7 tails in 14 flips of a balanced coin. Also refer to the binomial probabilities table of "Statistical Tables" to find the error of this approximation.

77. With reference to Exercise 75, show that the Poisson distribution would have yielded a better approximation.

78. If 23 percent of all patients with high blood pressure have bad side effects from a certain kind of medicine, use the normal approximation to find the probability that among 120 patients with high blood pressure treated with this medicine more than 32 will have bad side effects.

79. If the probability is 0.20 that a certain bank will refuse a loan application, use the normal approximation to determine (to three decimals) the probability that the bank will refuse at most 40 of 225 loan applications.

80. To illustrate the law of large numbers, use the normal approximation to the binomial distribution to determine the probabilities that the proportion of heads will be anywhere from 0.49 to 0.51 when a balanced coin is flipped

(a) 100 times;

(b) 1,000 times;

(c) 10,000 times.

SEC. 8

81. Check the following data for normality by finding normal scores and making a normal-scores plot:

3.9 4.6 4.5 1.6 4.2

82. Check the following data for normality by finding normal scores and making a normal-scores plot:

36 22 3 13 31 45

83. This question has been intentionally omitted for this edition.

84. The weights (in pounds) of seven shipments of bolts are

37 45 11 51 13 48 61

Make a normal-scores plot of these weights. Can they be regarded as having come from a normal distribution?

85. This question has been intentionally omitted for this edition.

86. Use a computer program to make a normal-scores plot for the data on the time to make coke in successive runs of a coke oven (given in hours).

7.8	9.2	6.4	8.2	7.6	5.9	7.4	7.1	6.7	8.5
10.1	8.6	7.7	5.9	9.3	6.4	6.8	7.9	7.2	10.2
6.9	7.4	7.8	6.6	8.1	9.5	6.4	7.6	8.4	9.2

Also test these data for normality using the three tests given by MINITAB.

87. Eighty pilots were tested in a flight simulator and the time for each to take corrective action for a given emergency was measured in seconds, with the following results:

11.1	5.2	3.6	7.6	12.4	6.8	3.8	5.7	9.0	6.0	4.9	12.6
7.4	5.3	14.2	8.0	12.6	13.7	3.8	10.6	6.8	5.4	9.7	6.7
14.1	5.3	11.1	13.4	7.0	8.9	6.2	8.3	7.7	4.5	7.6	5.0
9.4	3.5	7.9	11.0	8.6	10.5	5.7	7.0	5.6	9.1	5.1	4.5
6.2	6.8	4.3	8.5	3.6	6.1	5.8	10.0	6.4	4.0	5.4	7.0
4.1	8.1	5.8	11.8	6.1	9.1	3.3	12.5	8.5	10.8	6.5	7.9
6.8	10.1	4.9	5.4	9.6	8.2	4.2	3.4				

Use a computer to make a normal-scores plot of these data, and test for normality.

References

Useful information about various special probability densities, in outline form, may be found in

DERMAN, C., GLESER, L., and OLKIN, I., *Probability Models and Applications*. New York: Macmillan Publishing Co., Inc., 1980,

HASTINGS, N. A. J., and PEACOCK, J. B., *Statistical Distributions*. London: Butterworth and Co. Ltd., 1975,

and

JOHNSON, N. L., and KOTZ, S., *Continuous Univariate Distributions*, Vols. 1 and 2. Boston: Houghton Mifflin Company, 1970.

A direct proof that the standardized binomial distribution approaches the standard normal distribution when $n \to \infty$ is given in

KEEPING, E. S., *Introduction to Statistical Inference*. Princeton, N.J.: D. Van Nostrand Co., Inc., 1962.

A detailed treatment of the mathematical and statistical properties of the bivariate normal surface may be found in

YULE, G. U., and KENDALL, M. G., *An Introduction to the Theory of Statistics*, 14th ed. New York: Hafner Publishing Co., Inc., 1950.

The multivariate normal distribution is treated in matrix notation in

BICKEL, P. J., and DOKSUM, K. A., *Mathematical Statistics: Basic Ideas and Selected Topics*, San Francisco: Holden-Day, Inc., 1977,

HOGG, R. V., and CRAIG, A. T., *Introduction to Mathematical Statistics*, 4th ed. New York: Macmillan Publishing Co., Inc., 1978,

LINDGREN, B. W., *Statistical Theory*, 3rd ed. New York: Macmillan Publishing Co., Inc., 1976.

Answers to Odd-Numbered Exercises

3 $F(x) = \begin{cases} 0 & \text{for } x \leq \alpha \\ \dfrac{x-\alpha}{\beta-\alpha} & \text{for } \alpha < x < \beta \\ 1 & \text{for } x \geq \beta \end{cases}$

5 $\alpha_3 = 0$ and $\alpha_4 = \frac{9}{5}$.

11 For $0 < \alpha < 1$ the function $\to \infty$ when $x \to 0$; for $\alpha = 1$ the function has an absolute maximum at $x = 0$.

13 $\mu_1' = \alpha\beta, \mu_2' = \alpha(\alpha+1)\beta^2, \mu_3' = \alpha(\alpha+1)(\alpha+2)\beta^3$, and $\mu_4' = \alpha(\alpha+1)(\alpha+2)(\alpha+3)\beta^4$.

17 $M_Y(t) = \dfrac{e^{-\theta t}}{1-\theta t}$.

19 For $0 < \nu < 2$ the function $\to \infty$ when $x \to 0$, for $\nu = 2$ the function has an absolute maximum at $x = 0$.

23 **(a)** $k = \alpha\beta$.

33 $\mu_3 = 0$ and $\mu_4 = 3\sigma^4$.

45 **(a)** $\mu_1 = -2$, $\mu_2 = 1$, $\sigma_1 = 10$, $\sigma_2 = 5$, and $\rho = 0.7$.

47 $\mu_{Y|1} = \frac{11}{3}$, $\sigma_{Y|1} = \sqrt{20} = 4.47$.

51 $\frac{1}{2}$.

53 n = 100.

55 **(a)** 0.6065; **(b)** 0.5276.

57 0.1827.

59 0.2231.

61 **(a)** 3200 hours; **(b)** 0.2060.

63 **(a)** 0.1271; **(b)** 0.6406; **(c)** 0.1413; **(d)** 0.5876.

65 **(a)** 1.92; **(b)** 2.22; **(c)** 1.12; **(d)** ±1.44.

67 **(a)** 1.645; **(b)** 1.96; **(c)** 2.33; **(d)** 2.575.

69 **(a)** 0.0208.

71 **(a)** 0.1056; **(b)** 0.0062; **(c)** 0.5934.

73 6.094 ounces.

75 **(a)** yes; **(b)** 0.0078; **(c)** 117%.

77 0.0041.

79 0.227.

Functions of Random Variables

1 Introduction

In this chapter we shall concern ourselves with the problem of finding the probability distributions or densities of **functions of one or more random variables**. That is, given a set of random variables $X_1, X_2, \ldots, X_n$ and their joint probability distribution or density, we shall be interested in finding the probability distribution or density of some random variable $Y = u(X_1, X_2, \ldots, X_n)$. This means that the values of Y are related to those of the X's by means of the equation

$$y = u(x_1, x_2, \ldots, x_n)$$

Several methods are available for solving this kind of problem. The ones we shall discuss in the next four sections are called the **distribution function technique**, the **transformation technique**, and the **moment-generating function technique**. Although all three methods can be used in some situations, in most problems one technique will be preferable (easier to use than the others). This is true, for example, in some instances where the function in question is linear in the random variables $X_1, X_2, \ldots, X_n$, and the moment-generating function technique yields the simplest derivations.

2 Distribution Function Technique

A straightforward method of obtaining the probability density of a function of continuous random variables consists of first finding its distribution function and then its probability density by differentiation. Thus, if $X_1, X_2, \ldots, X_n$ are continuous random variables with a given joint probability density, the probability density of $Y = u(X_1, X_2, \ldots, X_n)$ is obtained by first determining an expression for the probability

$$F(y) = P(Y \leqq y) = P[u(X_1, X_2, \ldots, X_n) \leqq y]$$

and then differentiating to get

$$f(y) = \frac{dF(y)}{dy}$$

From Chapter 7 of *John E. Freund's Mathematical Statistics with Applications*, Eighth Edition. Irwin Miller, Marylees Miller.

EXAMPLE 1

If the probability density of X is given by

$$f(x) = \begin{cases} 6x(1-x) & \text{for } 0 < x < 1 \\ 0 & \text{elsewhere} \end{cases}$$

find the probability density of $Y = X^3$.

Solution

Letting $G(y)$ denote the value of the distribution function of Y at y, we can write

$$\begin{aligned} G(y) &= P(Y \leqq y) \\ &= P(X^3 \leqq y) \\ &= P(X \leqq y^{1/3}) \\ &= \int_0^{y^{1/3}} 6x(1-x)\, dx \\ &= 3y^{2/3} - 2y \end{aligned}$$

and hence

$$g(y) = 2(y^{-1/3} - 1)$$

for $0 < y < 1$; elsewhere, $g(y) = 0$. In Exercise 15 the reader will be asked to verify this result by a different technique.

EXAMPLE 2

If $Y = |X|$, show that

$$g(y) = \begin{cases} f(y) + f(-y) & \text{for } y > 0 \\ 0 & \text{elsewhere} \end{cases}$$

where $f(x)$ is the value of the probability density of X at x and $g(y)$ is the value of the probability density of Y at y. Also, use this result to find the probability density of $Y = |X|$ when X has the standard normal distribution.

Solution

For $y > 0$ we have

$$\begin{aligned} G(y) &= P(Y \leqq y) \\ &= P(|X| \leqq y) \\ &= P(-y \leqq X \leqq y) \\ &= F(y) - F(-y) \end{aligned}$$

and, upon differentiation,

$$g(y) = f(y) + f(-y)$$

Also, since $|x|$ cannot be negative, $g(y) = 0$ for $y < 0$. Arbitrarily letting $g(0) = 0$, we can thus write

$$g(y) = \begin{cases} f(y) + f(-y) & \text{for } y > 0 \\ 0 & \text{elsewhere} \end{cases}$$

If X has the standard normal distribution and $Y = |X|$, it follows that

$$g(y) = n(y; 0, 1) + n(-y; 0, 1)$$
$$= 2n(y; 0, 1)$$

for $y > 0$ and $g(y) = 0$ elsewhere. An important application of this result may be found in Example 9.

EXAMPLE 3

If the joint density of X_1 and X_2 is given by

$$f(x_1, x_2) = \begin{cases} 6e^{-3x_1-2x_2} & \text{for } x_1 > 0, x_2 > 0 \\ 0 & \text{elsewhere} \end{cases}$$

find the probability density of $Y = X_1 + X_2$.

Solution

Integrating the joint density over the shaded region of Figure 1, we get

$$F(y) = \int_0^y \int_0^{y-x_2} 6e^{-3x_1-2x_2}\,dx_1\,dx_2$$
$$= 1 + 2e^{-3y} - 3e^{-2y}$$

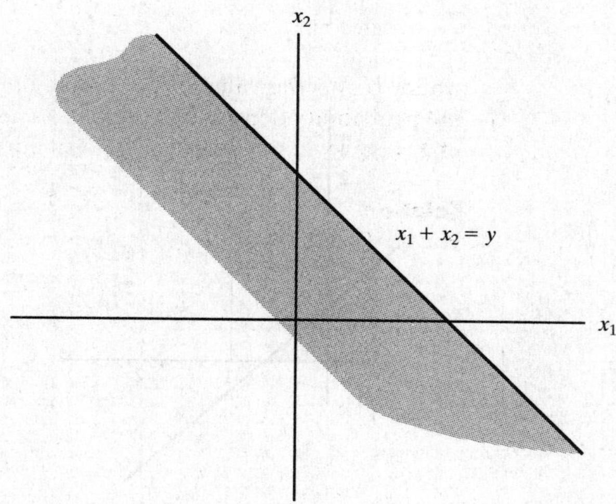

Figure 1. Diagram for Example 3.

and, differentiating with respect to y, we obtain

$$f(y) = 6(e^{-2y} - e^{-3y})$$

for $y > 0$; elsewhere, $f(y) = 0$.

Exercises

1. If X has an exponential distribution with the parameter θ, use the distribution function technique to find the probability density of the random variable $Y = \ln X$.

2. If the probability density of X is given by

$$f(x) = \begin{cases} 2xe^{-x^2} & \text{for } x > 0 \\ 0 & \text{elsewhere} \end{cases}$$

and $Y = X^2$, find
(a) the distribution function of Y;
(b) the probability density of Y.

3. If X has the uniform density with the parameters $\alpha = 0$ and $\beta = 1$, use the distribution function technique to find the probability density of the random variable $Y = \sqrt{X}$.

4. If the joint probability density of X and Y is given by

$$f(x, y) = \begin{cases} 4xye^{-(x^2+y^2)} & \text{for } x > 0, y > 0 \\ 0 & \text{elsewhere} \end{cases}$$

and $Z = \sqrt{X^2 + Y^2}$, find
(a) the distribution function of Z;
(b) the probability density of Z.

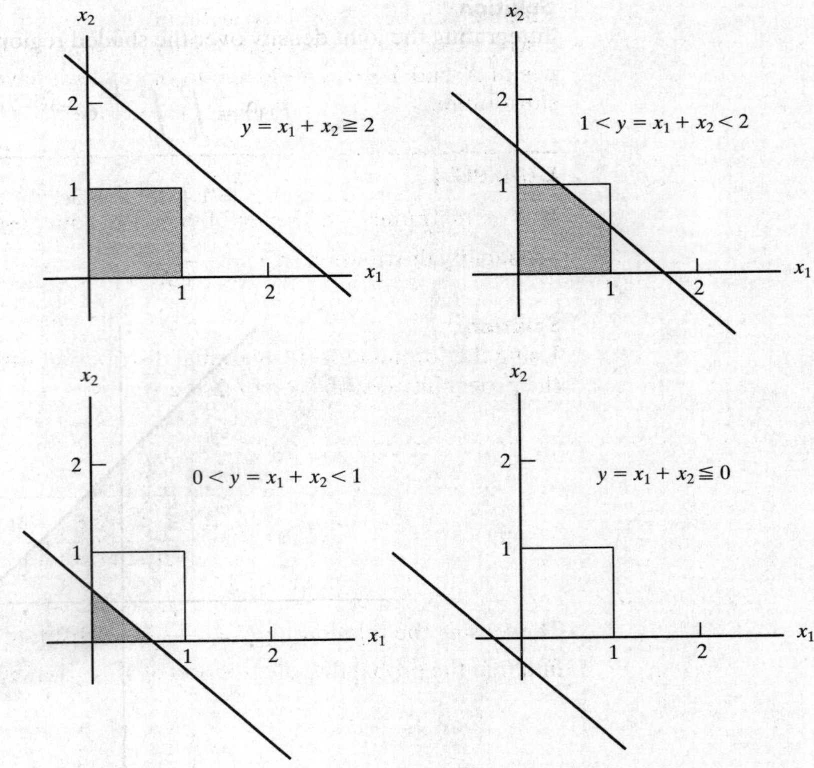

Figure 2. Diagram for Exercise 6.

5. If X_1 and X_2 are independent random variables having exponential densities with the parameters θ_1 and θ_2, use the distribution function technique to find the probability density of $Y = X_1 + X_2$ when

(a) $\theta_1 \neq \theta_2$;
(b) $\theta_1 = \theta_2$.

(Example 3 is a special case of this with $\theta_1 = \frac{1}{3}$ and $\theta_2 = \frac{1}{2}$.)

6. Let X_1 and X_2 be independent random variables having the uniform density with $\alpha = 0$ and $\beta = 1$. Referring to Figure 2, find expressions for the distribution function of $Y = X_1 + X_2$ for

(a) $y \leqq 0$;
(b) $0 < y < 1$;
(c) $1 < y < 2$;
(d) $y \geqq 2$.

Also find the probability density of Y.

7. With reference to the two random variables of Exercise 5, show that if $\theta_1 = \theta_2 = 1$, the random variable

$$Z = \frac{X_1}{X_1 + X_2}$$

has the uniform density with $\alpha = 0$ and $\beta = 1$.

8. If the joint density of X and Y is given by

$$f(x, y) = \begin{cases} e^{-(x+y)} & \text{for } x > 0, y > 0 \\ 0 & \text{elsewhere} \end{cases}$$

and $Z = \dfrac{X+Y}{2}$, find the probability density of Z by the distribution function technique.

3 Transformation Technique: One Variable

Let us show how the probability distribution or density of a function of a random variable can be determined without first getting its distribution function. In the discrete case there is no real problem as long as the relationship between the values of X and $Y = u(X)$ is one-to-one; all we have to do is make the appropriate substitution.

EXAMPLE 4

If X is the number of heads obtained in four tosses of a balanced coin, find the probability distribution of $Y = \dfrac{1}{1+X}$.

Solution

Using the formula for the binomial distribution with $n = 4$ and $\theta = \frac{1}{2}$, we find that the probability distribution of X is given by

x	0	1	2	3	4
$f(x)$	$\frac{1}{16}$	$\frac{4}{16}$	$\frac{6}{16}$	$\frac{4}{16}$	$\frac{1}{16}$

Then, using the relationship $y = \dfrac{1}{1+x}$ to substitute values of Y for values of X, we find that the probability distribution of Y is given by

y	1	$\frac{1}{2}$	$\frac{1}{3}$	$\frac{1}{4}$	$\frac{1}{5}$
$g(y)$	$\frac{1}{16}$	$\frac{4}{16}$	$\frac{6}{16}$	$\frac{4}{16}$	$\frac{1}{16}$

If we had wanted to make the substitution directly in the formula for the binomial distribution with $n = 4$ and $\theta = \frac{1}{2}$, we could have substituted $x = \frac{1}{y} - 1$ for x in

$$f(x) = \binom{4}{x}\left(\frac{1}{2}\right)^4 \quad \text{for } x = 0, 1, 2, 3, 4$$

getting

$$g(y) = f\left(\frac{1}{y} - 1\right) = \binom{4}{\frac{1}{y} - 1}\left(\frac{1}{2}\right)^4 \quad \text{for } y = 1, \frac{1}{2}, \frac{1}{3}, \frac{1}{4}, \frac{1}{5}$$

Note that in the preceding example the probabilities remained unchanged; the only difference is that in the result they are associated with the various values of Y instead of the corresponding values of X. That is all there is to the **transformation** (or **change-of-variable**) **technique** in the discrete case as long as the relationship is one-to-one. If the relationship is not one-to-one, we may proceed as in the following example.

EXAMPLE 5

With reference to Example 4, find the probability distribution of the random variable $Z = (X - 2)^2$.

Solution

Calculating the probabilities $h(z)$ associated with the various values of Z, we get

$$h(0) = f(2) = \frac{6}{16}$$

$$h(1) = f(1) + f(3) = \frac{4}{16} + \frac{4}{16} = \frac{8}{16}$$

$$h(4) = f(0) + f(4) = \frac{1}{16} + \frac{1}{16} = \frac{2}{16}$$

and hence

z	0	1	4
$h(z)$	$\frac{3}{8}$	$\frac{4}{8}$	$\frac{1}{8}$

To perform a transformation of variable in the continuous case, we shall assume that the function given by $y = u(x)$ is differentiable and either increasing or decreasing for all values within the range of X for which $f(x) \neq 0$, so the inverse function, given by $x = w(y)$, exists for all the corresponding values of y and is differentiable except where $u'(x) = 0$.† Under these conditions, we can prove the following theorem.

†To avoid points where $u'(x)$ might be 0, we generally do not include the endpoints of the intervals for which probability densities are nonzero. This is the practice that we follow throughout this chapter.

THEOREM 1. Let $f(x)$ be the value of the probability density of the continuous random variable X at x. If the function given by $y = u(x)$ is differentiable and either increasing or decreasing for all values within the range of X for which $f(x) \neq 0$, then, for these values of x, the equation $y = u(x)$ can be uniquely solved for x to give $x = w(y)$, and for the corresponding values of y the probability density of $Y = u(X)$ is given by

$$g(y) = f[w(y)] \cdot |w'(y)| \quad \text{provided } u'(x) \neq 0$$

Elsewhere, $g(y) = 0$.

Proof First, let us prove the case where the function given by $y = u(x)$ is increasing. As can be seen from Figure 3, X must take on a value between $w(a)$ and $w(b)$ when Y takes on a value between a and b. Hence,

$$\begin{aligned} P(a < Y < b) &= P[w(a) < X < w(b)] \\ &= \int_{w(a)}^{w(b)} f(x)\,dx \\ &= \int_{a}^{b} f[w(y)]w'(y)\,dy \end{aligned}$$

where we performed the change of variable $y = u(x)$, or equivalently $x = w(y)$, in the integral. The integrand gives the probability density of Y as long as $w'(y)$ exists, and we can write

$$g(y) = f[w(y)]w'(y)$$

When the function given by $y = u(x)$ is decreasing, it can be seen from Figure 3 that X must take on a value between $w(b)$ and $w(a)$ when Y takes on a value between a and b. Hence,

$$\begin{aligned} P(a < Y < b) &= P[w(b) < X < w(a)] \\ &= \int_{w(b)}^{w(a)} f(x)\,dx \\ &= \int_{b}^{a} f[w(y)]w'(y)\,dy \\ &= -\int_{a}^{b} f[w(y)]w'(y)\,dy \end{aligned}$$

where we performed the same change of variable as before, and it follows that

$$g(y) = -f[w(y)]w'(y)$$

Since $w'(y) = \dfrac{dx}{dy} = \dfrac{1}{\frac{dy}{dx}}$ is positive when the function given by $y = u(x)$ is increasing, and $-w'(y)$ is positive when the function given by $y = u(x)$ is decreasing, we can combine the two cases by writing

$$g(y) = f[w(y)] \cdot |w'(y)|$$

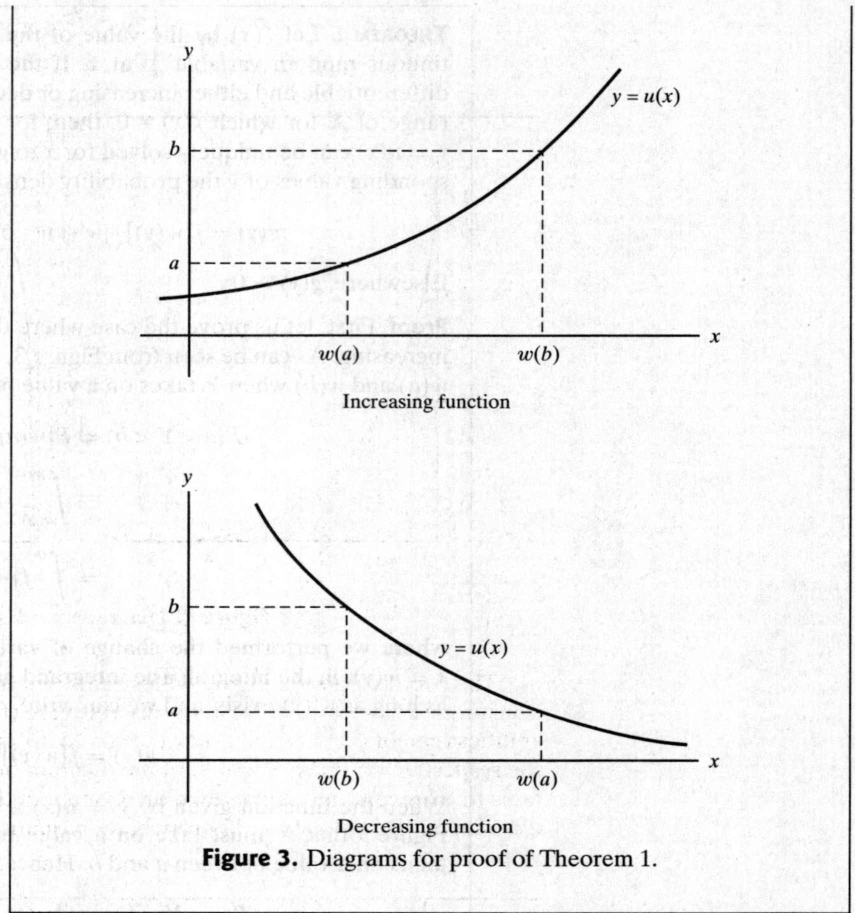

Figure 3. Diagrams for proof of Theorem 1.

EXAMPLE 6

If X has the exponential distribution given by

$$f(x) = \begin{cases} e^{-x} & \text{for } x > 0 \\ 0 & \text{elsewhere} \end{cases}$$

find the probability density of the random variable $Y = \sqrt{X}$.

Solution

The equation $y = \sqrt{x}$, relating the values of X and Y, has the unique inverse $x = y^2$, which yields $w'(y) = \frac{dx}{dy} = 2y$. Therefore,

$$g(y) = e^{-y^2}|2y| = 2ye^{-y^2}$$

for $y > 0$ in accordance with Theorem 1. Since the probability of getting a value of Y less than or equal to 0, like the probability of getting a value of X less than or equal to 0, is zero, it follows that the probability density of Y is given by

$$g(y) = \begin{cases} 2ye^{-y^2} & \text{for } y > 0 \\ 0 & \text{elsewhere} \end{cases}$$

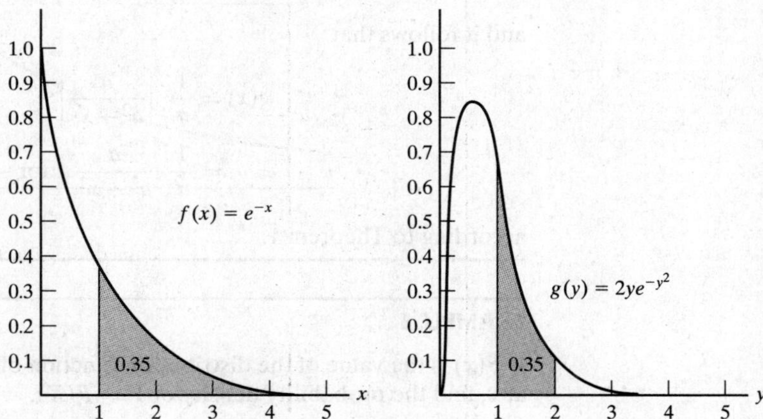

Figure 4. Diagrams for Example 6.

The two diagrams of Figure 4 illustrate what happened in this example when we transformed from X to Y. As in the discrete case (for instance, Example 4), the probabilities remain the same, but they pertain to different values (intervals of values) of the respective random variables. In the diagram on the left, the 0.35 probability pertains to the event that X will take on a value on the interval from 1 to 4, and in the diagram on the right, the 0.35 probability pertains to the event that Y will take on a value on the interval from 1 to 2.

EXAMPLE 7

If the double arrow of Figure 5 is spun so that the random variable Θ has the uniform density

$$f(\theta) = \begin{cases} \frac{1}{\pi} & \text{for } -\frac{\pi}{2} < \theta < \frac{\pi}{2} \\ 0 & \text{elsewhere} \end{cases}$$

determine the probability density of X, the abscissa of the point on the x-axis to which the arrow will point.

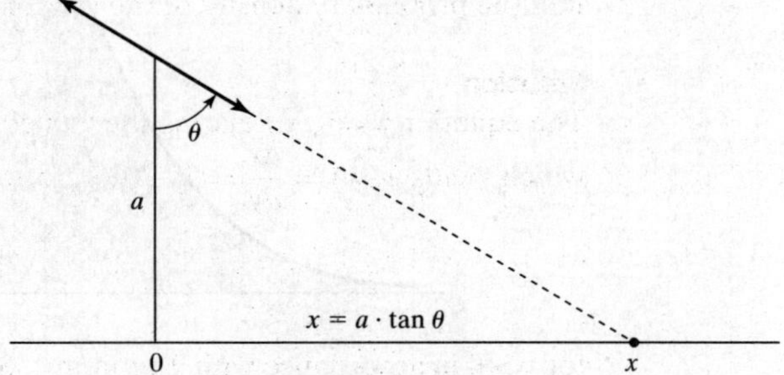

Figure 5. Diagram for Example 7.

Solution

As is apparent from the diagram, the relationship between x and θ is given by $x = a \cdot \tan\theta$, so that

$$\frac{d\theta}{dx} = \frac{a}{a^2 + x^2}$$

and it follows that

$$g(x) = \frac{1}{\pi} \cdot \left| \frac{a}{a^2 + x^2} \right|$$
$$= \frac{1}{\pi} \cdot \frac{a}{a^2 + x^2} \quad \text{for} -\infty < x < \infty$$

according to Theorem 1.

EXAMPLE 8

If $F(x)$ is the value of the distribution function of the continuous random variable X at x, find the probability density of $Y = F(X)$.

Solution

As can be seen from Figure 6, the value of Y corresponding to any particular value of X is given by the area under the curve, that is, the area under the graph of the density of X to the left of x. Differentiating $y = F(x)$ with respect to x, we get

$$\frac{dy}{dx} = F'(x) = f(x)$$

and hence

$$\frac{dx}{dy} = \frac{1}{\frac{dy}{dx}} = \frac{1}{f(x)}$$

provided $f(x) \neq 0$. It follows from Theorem 1 that

$$g(y) = f(x) \cdot \left| \frac{1}{f(x)} \right| = 1$$

for $0 < y < 1$, and we can say that y has the uniform density with $\alpha = 0$ and $\beta = 1$.

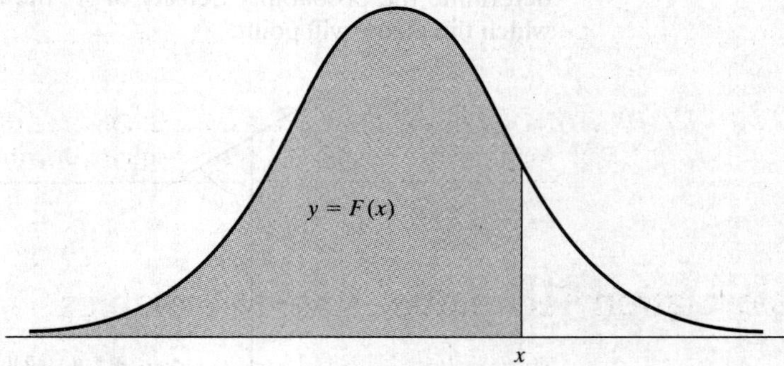

Figure 6. Diagram for Example 8.

The transformation that we performed in this example is called the **probability integral transformation**. Not only is the result of theoretical importance, but it facilitates the **simulation** of observed values of continuous random variables. A reference to how this is done, especially in connection with the normal distribution, is given in the end of the chapter.

When the conditions underlying Theorem 1 are not met, we can be in serious difficulties, and we may have to use the method of Section 2 or a generalization of Theorem 1 referred to among the references at the end of the chapter; sometimes, there is an easy way out, as in the following example.

EXAMPLE 9

If X has the standard normal distribution, find the probability density of $Z = X^2$.

Solution

Since the function given by $z = x^2$ is decreasing for negative values of x and increasing for positive values of x, the conditions of Theorem 1 are not met. However, the transformation from X to Z can be made in two steps: First, we find the probability density of $Y = |X|$, and then we find the probability density of $Z = Y^2 (= X^2)$.

As far as the first step is concerned, we already studied the transformation $Y = |X|$ in Example 2; in fact, we showed that if X has the standard normal distribution, then $Y = |X|$ has the probability density

$$g(y) = 2n(y; 0, 1) = \frac{2}{\sqrt{2\pi}} e^{-\frac{1}{2}y^2}$$

for $y > 0$, and $g(y) = 0$ elsewhere. For the second step, the function given by $z = y^2$ is increasing for $y > 0$, that is, for all values of Y for which $g(y) \neq 0$. Thus, we can use Theorem 1, and since

$$\frac{dy}{dz} = \frac{1}{2} z^{-\frac{1}{2}}$$

we get

$$h(z) = \frac{2}{\sqrt{2\pi}} e^{-\frac{1}{2}z} \left| \frac{1}{2} z^{-\frac{1}{2}} \right|$$

$$= \frac{1}{\sqrt{2\pi}} z^{-\frac{1}{2}} e^{-\frac{1}{2}z}$$

for $z > 0$, and $h(z) = 0$ elsewhere. Observe that since $\Gamma(\frac{1}{2}) = \sqrt{\pi}$, the distribution we have arrived at for Z is a chi-square distribution with $\nu = 1$.

---

4 Transformation Technique: Several Variables

The method of the preceding section can also be used to find the distribution of a random variable that is a function of two or more random variables. Suppose, for instance, that we are given the joint distribution of two random variables X_1 and X_2 and that we want to determine the probability distribution or the probability density of the random variable $Y = u(X_1, X_2)$. If the relationship between y and x_1 with x_2 held constant or the relationship between y and x_2 with x_1 held constant permits, we can proceed in the discrete case as in Example 4 to find the joint distribution of Y and X_2 or that of X_1 and Y and then sum on the values of the other random variable to get the marginal distribution of Y. In the continuous case, we first use Theorem 1 with the transformation formula written as

$$g(y, x_2) = f(x_1, x_2) \cdot \left| \frac{\partial x_1}{\partial y} \right|$$

or as

$$g(x_1, y) = f(x_1, x_2) \cdot \left| \frac{\partial x_2}{\partial y} \right|$$

where $f(x_1, x_2)$ and the partial derivative must be expressed in terms of y and x_2 or x_1 and y. Then we integrate out the other variable to get the marginal density of Y.

EXAMPLE 10

If X_1 and X_2 are independent random variables having Poisson distributions with the parameters λ_1 and λ_2, find the probability distribution of the random variable $Y = X_1 + X_2$.

Solution

Since X_1 and X_2 are independent, their joint distribution is given by

$$f(x_1, x_2) = \frac{e^{-\lambda_1}(\lambda_1)^{x_1}}{x_1!} \cdot \frac{e^{-\lambda_2}(\lambda_2)^{x_2}}{x_2!}$$

$$= \frac{e^{-(\lambda_1+\lambda_2)}(\lambda_1)^{x_1}(\lambda_2)^{x_2}}{x_1!x_2!}$$

for $x_1 = 0, 1, 2, \ldots$ and $x_2 = 0, 1, 2, \ldots$. Since $y = x_1 + x_2$ and hence $x_1 = y - x_2$, we can substitute $y - x_2$ for x_1, getting

$$g(y, x_2) = \frac{e^{-(\lambda_1+\lambda_2)}(\lambda_2)^{x_2}(\lambda_1)^{y-x_2}}{x_2!(y-x_2)!}$$

for $y = 0, 1, 2, \ldots$ and $x_2 = 0, 1, \ldots, y$, for the joint distribution of Y and X_2. Then, summing on x_2 from 0 to y, we get

$$h(y) = \sum_{x_2=0}^{y} \frac{e^{-(\lambda_1+\lambda_2)}(\lambda_2)^{x_2}(\lambda_1)^{y-x_2}}{x_2!(y-x_2)!}$$

$$= \frac{e^{-(\lambda_1+\lambda_2)}}{y!} \sum_{x_2=0}^{y} \frac{y!}{x_2!(y-x_2)!}(\lambda_2)^{x_2}(\lambda_1)^{y-x_2}$$

after factoring out $e^{-(\lambda_1+\lambda_2)}$ and multiplying and dividing by $y!$. Identifying the summation at which we arrived as the binomial expansion of $(\lambda_1+\lambda_2)^y$, we finally get

$$h(y) = \frac{e^{-(\lambda_1+\lambda_2)}(\lambda_1+\lambda_2)^y}{y!} \quad \text{for } y = 0, 1, 2, \ldots$$

and we have thus shown that the sum of two independent random variables having Poisson distributions with the parameters λ_1 and λ_2 has a Poisson distribution with the parameter $\lambda = \lambda_1 + \lambda_2$.

EXAMPLE 11

If the joint probability density of X_1 and X_2 is given by

$$f(x_1, x_2) = \begin{cases} e^{-(x_1+x_2)} & \text{for } x_1 > 0, x_2 > 0 \\ 0 & \text{elsewhere} \end{cases}$$

find the probability density of $Y = \dfrac{X_1}{X_1 + X_2}$.

Solution

Since y decreases when x_2 increases and x_1 is held constant, we can use Theorem 1 to find the joint density of X_1 and Y. Since $y = \dfrac{x_1}{x_1 + x_2}$ yields $x_2 = x_1 \cdot \dfrac{1-y}{y}$ and hence

$$\frac{\partial x_2}{\partial y} = -\frac{x_1}{y^2}$$

it follows that

$$g(x_1, y) = e^{-x_1/y}\left|-\frac{x_1}{y^2}\right| = \frac{x_1}{y^2} \cdot e^{-x_1/y}$$

for $x_1 > 0$ and $0 < y < 1$. Finally, integrating out x_1 and changing the variable of integration to $u = x_1/y$, we get

$$\begin{aligned} h(y) &= \int_0^{\infty} \frac{x_1}{y^2} \cdot e^{-x_1/y} dx_1 \\ &= \int_0^{\infty} u \cdot e^{-u} du \\ &= \Gamma(2) \\ &= 1 \end{aligned}$$

for $0 < y < 1$, and $h(y) = 0$ elsewhere. Thus, the random variable Y has the uniform density with $\alpha = 0$ and $\beta = 1$. (Note that in Exercise 7 the reader was asked to show this by the distribution function technique.)

The preceding example could also have been worked by a general method where we begin with the joint distribution of two random variables X_1 and X_2 and determine the joint distribution of two new random variables $Y_1 = u_1(X_1, X_2)$ and $Y_2 = u_2(X_1, X_2)$. Then we can find the marginal distribution of Y_1 or Y_2 by summation or integration.

This method is used mainly in the continuous case, where we need the following theorem, which is a direct generalization of Theorem 1.

THEOREM 2. Let $f(x_1, x_2)$ be the value of the joint probability density of the continuous random variables X_1 and X_2 at (x_1, x_2). If the functions given by $y_1 = u_1(x_1, x_2)$ and $y_2 = u_2(x_1, x_2)$ are partially differentiable with respect to both x_1 and x_2 and represent a one-to-one transformation for all values within the range of X_1 and X_2 for which $f(x_1, x_2) \neq 0$, then, for these values of x_1 and x_2, the equations $y_1 = u_1(x_1, x_2)$ and $y_2 = u_2(x_1, x_2)$ can be uniquely solved for x_1 and x_2 to give $x_1 = w_1(y_1, y_2)$ and $x_2 = w_2(y_1, y_2)$, and for the corresponding values of y_1 and y_2, the joint probability density of $Y_1 = u_1(X_1, X_2)$ and $Y_2 = u_2(X_1, X_2)$ is given by

$$g(y_1, y_2) = f[w_1(y_1, y_2), w_2(y_1, y_2)] \cdot |J|$$

Here, J, called the **Jacobian** of the transformation, is the determinant

$$J = \begin{vmatrix} \dfrac{\partial x_1}{\partial y_1} & \dfrac{\partial x_1}{\partial y_2} \\ \dfrac{\partial x_2}{\partial y_1} & \dfrac{\partial x_2}{\partial y_2} \end{vmatrix}$$

Elsewhere, $g(y_1, y_2) = 0$.

We shall not prove this theorem, but information about Jacobians and their applications can be found in most textbooks on advanced calculus. There they are used mainly in connection with multiple integrals, say, when we want to change from rectangular coordinates to polar coordinates or from rectangular coordinates to spherical coordinates.

EXAMPLE 12

With reference to the random variables X_1 and X_2 of Example 11, find

(a) the joint density of $Y_1 = X_1 + X_2$ and $Y_2 = \dfrac{X_1}{X_1 + X_2}$;

(b) the marginal density of Y_2.

Solution

(a) Solving $y_1 = x_1 + x_2$ and $y_2 = \dfrac{x_1}{x_1 + x_2}$ for x_1 and x_2, we get $x_1 = y_1 y_2$ and $x_2 = y_1(1 - y_2)$, and it follows that

$$J = \begin{vmatrix} y_2 & y_1 \\ 1 - y_2 & -y_1 \end{vmatrix} = -y_1$$

Since the transformation is one-to-one, mapping the region $x_1 > 0$ and $x_2 > 0$ in the x_1x_2-plane into the region $y_1 > 0$ and $0 < y_2 < 1$ in the y_1y_2-plane, we can use Theorem 2 and it follows that

$$g(y_1, y_2) = e^{-y_1}|-y_1| = y_1 e^{-y_1}$$

for $y_1 > 0$ and $0 < y_2 < 1$; elsewhere, $g(y_1, y_2) = 0$.

(b) Using the joint density obtained in part (a) and integrating out y_1, we get

$$\begin{aligned} h(y_2) &= \int_0^\infty g(y_1, y_2)\, dy_1 \\ &= \int_0^\infty y_1 e^{-y_1}\, dy_1 \\ &= \Gamma(2) \\ &= 1 \end{aligned}$$

for $0 < y_2 < 1$; elsewhere, $h(y_2) = 0$.

EXAMPLE 13

If the joint density of X_1 and X_2 is given by

$$f(x_1, x_2) = \begin{cases} 1 & \text{for } 0 < x_1 < 1, 0 < x_2 < 1 \\ 0 & \text{elsewhere} \end{cases}$$

find

(a) the joint density of $Y = X_1 + X_2$ and $Z = X_2$;

(b) the marginal density of Y.

Note that in Exercise 6 the reader was asked to work the same problem by the distribution function technique.

Solution

(a) Solving $y = x_1 + x_2$ and $z = x_2$ for x_1 and x_2, we get $x_1 = y - z$ and $x_2 = z$, so that

$$J = \begin{vmatrix} 1 & -1 \\ 0 & 1 \end{vmatrix} = 1$$

Because this transformation is one-to-one, mapping the region $0 < x_1 < 1$ and $0 < x_2 < 1$ in the x_1x_2-plane into the region $z < y < z+1$ and $0 < z < 1$ in the yz-plane (see Figure 7), we can use Theorem 2 and we get

$$g(y, z) = 1 \cdot |1| = 1$$

for $z < y < z+1$ and $0 < z < 1$; elsewhere, $g(y, z) = 0$.

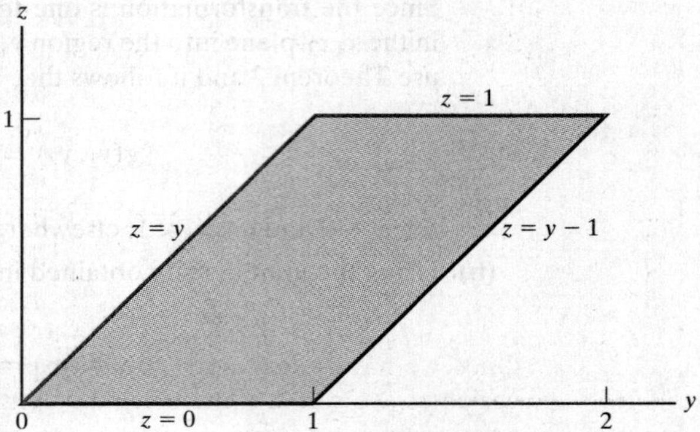

Figure 7. Transformed sample space for Example 13.

(b) Integrating out z separately for $y \leqq 0$, $0 < y < 1$, $1 < y < 2$, and $y \geqq 2$, we get

$$h(y) = \begin{cases} 0 & \text{for } y \leqq 0 \\ \int_0^y 1 \cdot dz = y & \text{for } 0 < y < 1 \\ \int_{y-1}^1 1 \cdot dz = 2 - y & \text{for } 1 < y < 2 \\ 0 & \text{for } y \geqq 2 \end{cases}$$

and to make the density function continuous, we let $h(1) = 1$. We have thus shown that the sum of the given random variables has the **triangular probability density** whose graph is shown in Figure 8.

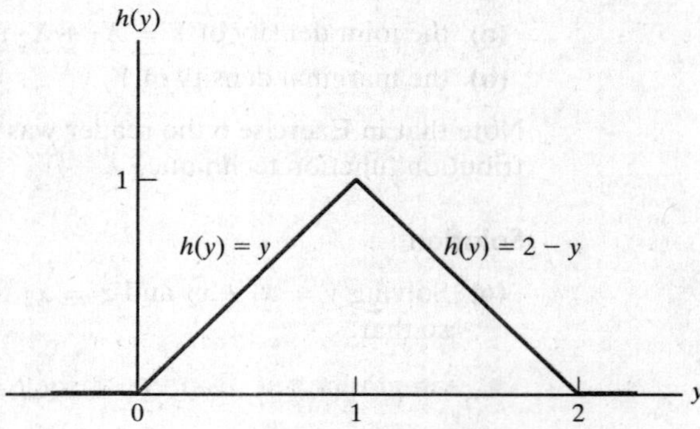

Figure 8. Triangular probability density.

So far we have considered here only functions of two random variables, but the method based on Theorem 2 can easily be generalized to functions of three or more random variables. For instance, if we are given the joint probability density of three random variables X_1, X_2, and X_3 and we want to find the joint probability density of the random variables $Y_1 = u_1(X_1, X_2, X_3)$, $Y_2 = u_2(X_1, X_2, X_3)$, and

$Y_3 = u_3(X_1, X_2, X_3)$, the general approach is the same, but the Jacobian is now the 3×3 determinant

$$J = \begin{vmatrix} \frac{\partial x_1}{\partial y_1} & \frac{\partial x_1}{\partial y_2} & \frac{\partial x_1}{\partial y_3} \\ \frac{\partial x_2}{\partial y_1} & \frac{\partial x_2}{\partial y_2} & \frac{\partial x_2}{\partial y_3} \\ \frac{\partial x_3}{\partial y_1} & \frac{\partial x_3}{\partial y_2} & \frac{\partial x_3}{\partial y_3} \end{vmatrix}$$

Once we have determined the joint probability density of the three new random variables, we can find the marginal density of any two of the random variables, or any one, by integration.

EXAMPLE 14

If the joint probability density of X_1, X_2, and X_3 is given by

$$f(x_1, x_2, x_3) = \begin{cases} e^{-(x_1+x_2+x_3)} & \text{for } x_1 > 0, x_2 > 0, x_3 > 0 \\ 0 & \text{elsewhere} \end{cases}$$

find

(a) the joint density of $Y_1 = X_1 + X_2 + X_3$, $Y_2 = X_2$, and $Y_3 = X_3$;

(b) the marginal density of Y_1.

Solution

(a) Solving the system of equations $y_1 = x_1 + x_2 + x_3$, $y_2 = x_2$, and $y_3 = x_3$ for x_1, x_2, and x_3, we get $x_1 = y_1 - y_2 - y_3$, $x_2 = y_2$, and $x_3 = y_3$. It follows that

$$J = \begin{vmatrix} 1 & -1 & -1 \\ 0 & 1 & 0 \\ 0 & 0 & 1 \end{vmatrix} = 1$$

and, since the transformation is one-to-one, that

$$\begin{aligned} g(y_1, y_2, y_3) &= e^{-y_1} \cdot |1| \\ &= e^{-y_1} \end{aligned}$$

for $y_2 > 0$, $y_3 > 0$, and $y_1 > y_2 + y_3$; elsewhere, $g(y_1, y_2, y_3) = 0$.

(b) Integrating out y_2 and y_3, we get

$$\begin{aligned} h(y_1) &= \int_0^{y_1} \int_0^{y_1 - y_3} e^{-y_1}\, dy_2\, dy_3 \\ &= \frac{1}{2} y_1^2 \cdot e^{-y_1} \end{aligned}$$

for $y_1 > 0$; $h(y_1) = 0$ elsewhere. Observe that we have shown that the sum of three independent random variables having the gamma distribution with $\alpha = 1$ and $\beta = 1$ is a random variable having the gamma distribution with $\alpha = 3$ and $\beta = 1$.

As the reader will find in Exercise 39, it would have been easier to obtain the result of part (b) of Example 14 by using the method based on Theorem 1.

Exercises

9. If X has a hypergeometric distribution with $M = 3$, $N = 6$, and $n = 2$, find the probability distribution of Y, the number of successes minus the number of failures.

10. With reference to Exercise 9, find the probability distribution of the random variable $Z = (X - 1)^2$.

11. If X has a binomial distribution with $n = 3$ and $\theta = \frac{1}{3}$, find the probability distributions of

(a) $Y = \dfrac{X}{1+X}$;

(b) $U = (X - 1)^4$.

12. If X has a geometric distribution with $\theta = \frac{1}{3}$, find the formula for the probability distribution of the random variable $Y = 4 - 5X$.

13. This question has been intentionally omitted for this edition.

14. This question has been intentionally omitted for this edition.

15. Use the transformation technique to rework Exercise 2.

16. If the probability density of X is given by

$$f(x) = \begin{cases} \dfrac{kx^3}{(1+2x)^6} & \text{for } x > 0 \\ 0 & \text{elsewhere} \end{cases}$$

where k is an appropriate constant, find the probability density of the random variable $Y = \dfrac{2X}{1+2X}$. Identify the distribution of Y, and thus determine the value of k.

17. If the probability density of X is given by

$$f(x) = \begin{cases} \dfrac{x}{2} & \text{for } 0 < x < 2 \\ 0 & \text{elsewhere} \end{cases}$$

find the probability density of $Y = X^3$. Also, plot the graphs of the probability densities of X and Y and indicate the respective areas under the curves that represent $P(\frac{1}{2} < X < 1)$ and $P(\frac{1}{8} < Y < 1)$.

18. If X has a uniform density with $\alpha = 0$ and $\beta = 1$, show that the random variable $Y = -2$. In X has a gamma distribution. What are its parameters?

19. This question has been intentionally omitted for this edition.

20. Consider the random variable X with the probability density

$$f(x) = \begin{cases} \dfrac{3x^2}{2} & \text{for } -1 < x < 1 \\ 0 & \text{elsewhere} \end{cases}$$

(a) Use the result of Example 2 to find the probability density of $Y = |X|$.

(b) Find the probability density of $Z = X^2 (= Y^2)$.

21. Consider the random variable X with the uniform density having $\alpha = 1$ and $\beta = 3$.

(a) Use the result of Example 2 to find the probability density of $Y = |X|$.

(b) Find the probability density of $Z = X^4 (= Y^4)$.

22. If the joint probability distribution of X_1 and X_2 is given by

$$f(x_1, x_2) = \frac{x_1 x_2}{36}$$

for $x_1 = 1, 2, 3$ and $x_2 = 1, 2, 3$, find

(a) the probability distribution of $X_1 X_2$;

(b) the probability distribution of X_1/X_2.

23. With reference to Exercise 22, find

(a) the joint distribution of $Y_1 = X_1 + X_2$ and $Y_2 = X_1 - X_2$;

(b) the marginal distribution of Y_1.

24. If the joint probability distribution of X and Y is given by

$$f(x, y) = \frac{(x - y)^2}{7}$$

for $x = 1, 2$ and $y = 1, 2, 3$, find

(a) the joint distribution of $U = X + Y$ and $V = X - Y$;

(b) the marginal distribution of U.

25. If X_1, X_2, and X_3 have the multinomial distribution with $n = 2$, $\theta_1 = \frac{1}{4}$, $\theta_2 = \frac{1}{3}$, and $\theta_3 = \frac{5}{12}$, find the joint probability distribution of $Y_1 = X_1 + X_2$, $Y_2 = X_1 - X_2$, and $Y_3 = X_3$.

26. This question has been intentionally omitted for this edition.

27. If X_1 and X_2 are independent random variables having binomial distributions with the respective parameters n_1 and θ and n_2 and θ, show that $Y = X_1 + X_2$ has the binomial distribution with the parameters $n_1 + n_2$ and θ.

28. If X_1 and X_2 are independent random variables having the geometric distribution with the parameter θ, show that $Y = X_1 + X_2$ is a random variable having the negative binomial distribution with the parameters θ and $k = 2$.

29. If X and Y are independent random variables having the standard normal distribution, show that the random variable $Z = X + Y$ is also normally distributed. (*Hint*: Complete the square in the exponent.) What are the mean and the variance of this normal distribution?

30. Consider two random variables X and Y with the joint probability density

$$f(x,y) = \begin{cases} 12xy(1-y) & \text{for } 0 < x < 1, 0 < y < 1 \\ 0 & \text{elsewhere} \end{cases}$$

Find the probability density of $Z = XY^2$ by using Theorem 1 to determine the joint probability density of Y and Z and then integrating out y.

31. Rework Exercise 30 by using Theorem 2 to determine the joint probability density of $Z = XY^2$ and $U = Y$ and then finding the marginal density of Z.

32. Consider two independent random variables X_1 and X_2 having the same Cauchy distribution

$$f(x) = \frac{1}{\pi(1+x^2)} \quad \text{for } -\infty < x < \infty$$

Find the probability density of $Y_1 = X_1 + X_2$ by using Theorem 1 to determine the joint probability density of X_1 and Y_1 and then integrating out x_1. Also, identify the distribution of Y_1.

33. Rework Exercise 32 by using Theorem 2 to determine the joint probability density of $Y_1 = X_1 + X_2$ and $Y_2 = X_1 - X_2$ and then finding the marginal density of Y_1.

34. Consider two random variables X and Y whose joint probability density is given by

$$f(x,y) = \begin{cases} \frac{1}{2} & \text{for } x > 0, y > 0, x + y < 2 \\ 0 & \text{elsewhere} \end{cases}$$

Find the probability density of $U = Y - X$ by using Theorem 1.

35. Rework Exercise 34 by using Theorem 2 to determine the joint probability density of $U = Y - X$ and $V = X$ and then finding the marginal density of U.

36. Let X_1 and X_2 be two continuous random variables having the joint probability density

$$f(x_1,x_2) = \begin{cases} 4x_1x_2 & \text{for } 0 < x_1 < 1, 0 < x_2 < 1 \\ 0 & \text{elsewhere} \end{cases}$$

Find the joint probability density of $Y_1 = X_1^2$ and $Y_2 = X_1X_2$.

37. Let X and Y be two continuous random variables having the joint probability density

$$f(x,y) = \begin{cases} 24xy & \text{for } 0 < x < 1, 0 < y < 1, x + y < 1 \\ 0 & \text{elsewhere} \end{cases}$$

Find the joint probability density of $Z = X + Y$ and $W = X$.

38. Let X and Y be two independent random variables having identical gamma distributions.

(a) Find the joint probability density of the random variables $U = \dfrac{X}{X+Y}$ and $V = X + Y$.

(b) Find and identify the marginal density of U.

39. The method of transformation based on Theorem 1 can be generalized so that it applies also to random variables that are functions of two or more random variables. So far we have used this method only for functions of two random variables, but when there are three, for example, we introduce the new random variable in place of one of the original random variables, and then we eliminate (by summation or integration) the other two random variables with which we began. Use this method to rework Example 14.

40. In Example 13 we found the probability density of the sum of two independent random variables having the uniform density with $\alpha = 0$ and $\beta = 1$. Given a third random variable X_3, which has the same uniform density and is independent of both X_1 and X_2, show that if $U = Y + X_3 = X_1 + X_2 + X_3$, then

(a) the joint probability density of U and Y is given by

$$g(u,y) = \begin{cases} y & \text{for Regions I and II of Figure 9} \\ 2 - y & \text{for Regions III and IV of Figure 9} \\ 0 & \text{elsewhere} \end{cases}$$

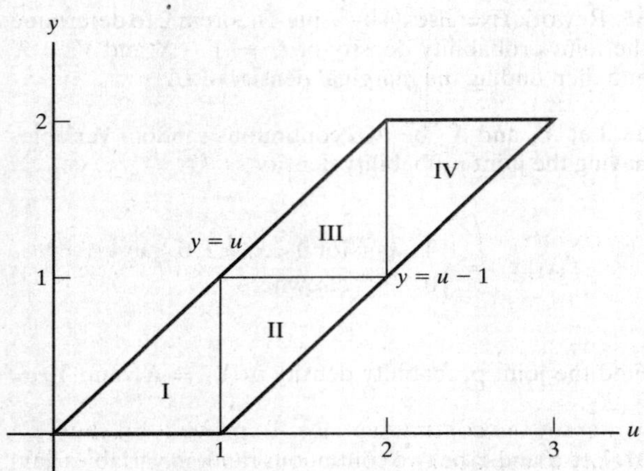

Figure 9. Diagram for Exercise 40.

(b) the probability density of U is given by

$$h(u) = \begin{cases} 0 & \text{for } u \leqq 0 \\ \frac{1}{2}u^2 & \text{for } 0 < u < 1 \\ \frac{1}{2}u^2 - \frac{3}{2}(u-1)^2 & \text{for } 1 < u < 2 \\ \frac{1}{2}u^2 - \frac{3}{2}(u-1)^2 + \frac{3}{2}(u-2)^2 & \text{for } 2 < u < 3 \\ 0 & \text{for } u \geqq 3 \end{cases}$$

Note that if we let $h(1) = h(2) = \frac{1}{2}$, this will make the probability density of U continuous.

5 Moment-Generating Function Technique

Moment-generating functions can play an important role in determining the probability distribution or density of a function of random variables when the function is a linear combination of n *independent* random variables. We shall illustrate this technique here when such a linear combination is, in fact, the sum of n independent random variables, leaving it to the reader to generalize it in Exercises 45 and 46.

The method is based on the following theorem that the moment-generating function of the sum of n independent random variables equals the product of their moment-generating functions.

THEOREM 3. If $X_1, X_2, \ldots,$ and X_n are independent random variables and $Y = X_1 + X_2 + \cdots + X_n$, then

$$M_Y(t) = \prod_{i=1}^{n} M_{X_i}(t)$$

where $M_{X_i}(t)$ is the value of the moment-generating function of X_i at t.

Proof Making use of the fact that the random variables are independent and hence

$$f(x_1, x_2, \ldots, x_n) = f_1(x_1) \cdot f_2(x_2) \cdot \ldots \cdot f_n(x_n)$$

according to the following definition "**INDEPENDENCE OF DISCRETE RANDOM VARIABLES**. *If* $f(x_1, x_2, \ldots, x_n)$ *is the value of the joint probability distribution of the discrete random variables* $X_1, X_2, \ldots, X_n$ *at* $(x_1, x_2, \ldots, x_n)$ *and* $f_i(x_i)$ *is the value of the marginal distribution of* X_i *at* x_i *for* $i = 1, 2, \ldots, n$, *then the* n *random variables are* ***independent*** *if and only if* $f(x_1, x_2, \ldots, x_n) = f_1(x_1) \cdot f_2(x_2) \cdot \ldots \cdot f_n(x_n)$ *for all* $(x_1, x_2, \ldots, x_n)$ *within their range*", we can write

$$M_Y(t) = E(e^{Yt})$$
$$= E\left[e^{(X_1+X_2+\cdots+X_n)t}\right]$$
$$= \int_{-\infty}^{\infty} \cdots \int_{-\infty}^{\infty} e^{(x_1+x_2+\cdots+x_n)t} f(x_1, x_2, \ldots, x_n)\, dx_1\, dx_2 \cdots dx_n$$
$$= \int_{-\infty}^{\infty} e^{x_1 t} f_1(x_1)\, dx_1 \cdot \int_{-\infty}^{\infty} e^{x_2 t} f_2(x_2)\, dx_2 \cdots \int_{-\infty}^{\infty} e^{x_n t} f_n(x_n)\, dx_n$$
$$= \prod_{i=1}^{n} M_{X_i}(t)$$

which proves the theorem for the continuous case. To prove it for the discrete case, we have only to replace all the integrals by sums.

Note that if we want to use Theorem 3 to find the probability distribution or the probability density of the random variable $Y = X_1 + X_2 + \cdots + X_n$, we must be able to identify whatever probability distribution or density corresponds to $M_Y(t)$.

EXAMPLE 15

Find the probability distribution of the sum of n independent random variables $X_1, X_2, \ldots, X_n$ having Poisson distributions with the respective parameters $\lambda_1, \lambda_2, \ldots, \lambda_n$.

Solution

By the theorem "The moment-generating function of the Poisson distribution is given by $M_X(t) = e^{\lambda(e^t-1)}$" we have

$$M_{X_i}(t) = e^{\lambda_i(e^t-1)}$$

hence, for $Y = X_1 + X_2 + \cdots + X_n$, we obtain

$$M_Y(t) = \prod_{i=1}^{n} e^{\lambda_i(e^t-1)} = e^{(\lambda_1+\lambda_2+\cdots+\lambda_n)(e^t-1)}$$

which can readily be identified as the moment-generating function of the Poisson distribution with the parameter $\lambda = \lambda_1 + \lambda_2 + \cdots + \lambda_n$. Thus, the distribution of the sum of n independent random variables having Poisson distributions with the parameters λ_i is a Poisson distribution with the parameter $\lambda = \lambda_1 + \lambda_2 + \cdots + \lambda_n$. Note that in Example 10 we proved this for $n = 2$.

EXAMPLE 16

If $X_1, X_2, \ldots, X_n$ are independent random variables having exponential distributions with the same parameter θ, find the probability density of the random variable $Y = X_1 + X_2 + \cdots + X_n$.

Solution

Since the exponential distribution is a gamma distribution with $\alpha = 1$ and $\beta = \theta$, we have

$$M_{X_i}(t) = (1 - \theta t)^{-1}$$

and hence

$$M_Y(t) = \prod_{i=1}^{n}(1-\theta t)^{-1} = (1-\theta t)^{-n}$$

Identifying the moment-generating function of Y as that of a gamma distribution with $\alpha = n$ and $\beta = \theta$, we conclude that the distribution of the sum of n independent random variables having exponential distributions with the same parameter θ is a gamma distribution with the parameters $\alpha = n$ and $\beta = \theta$. Note that this agrees with the result of Example 14, where we showed that the sum of three independent random variables having exponential distributions with the parameter $\theta = 1$ has a gamma distribution with $\alpha = 3$ and $\beta = 1$.

Theorem 3 also provides an easy and elegant way of deriving the moment-generating function of the binomial distribution. Suppose that $X_1, X_2, \ldots, X_n$ are independent random variables having the same Bernoulli distribution $f(x;\theta) = \theta^x(1-\theta)^{1-x}$ for $x = 0, 1$. We have

$$M_{X_i}(t) = e^{0 \cdot t}(1-\theta) + e^{1 \cdot t}\theta = 1 + \theta(e^t - 1)$$

so that Theorem 3 yields

$$M_Y(t) = \prod_{i=1}^{n}[1+\theta(e^t-1)] = [1+\theta(e^t-1)]^n$$

This moment-generating function is readily identified as that of the binomial distribution with the parameters n and θ. Of course, $Y = X_1 + X_2 + \cdots + X_n$ is the total number of successes in n trials, since X_1 is the number of successes on the first trial, X_2 is the number of successes on the second trial, ..., and X_n is the number of successes on the nth trial. This is a fruitful way of looking at the binomial distribution.

Exercises

41. Use the moment-generating function technique to rework Exercise 27.

42. Find the moment-generating function of the negative binomial distribution by making use of the fact that if k independent random variables have geometric distributions with the same parameter θ, their sum is a random variable having the negative binomial distribution with the parameters θ and k.

43. If n independent random variables have the same gamma distribution with the parameters α and β, find the moment-generating function of their sum and, if possible, identify its distribution.

44. If n independent random variables X_i have normal distributions with the means μ_i and the standard deviations σ_i, find the moment-generating function of their sum and identify the corresponding distribution, its mean, and its variance.

45. Prove the following generalization of Theorem 3: If $X_1, X_2, \ldots$, and X_n are independent random variables and $Y = a_1X_1 + a_2X_2 + \cdots + a_nX_n$, then

$$M_Y(t) = \prod_{i=1}^{n} M_{X_i}(a_i t)$$

where $M_{X_i}(t)$ is the value of the moment-generating function of X_i at t.

46. Use the result of Exercise 45 to show that, if n independent random variables X_i have normal distributions with the means μ_i and the standard deviations σ_i, then $Y = a_1X_1 + a_2X_2 + \cdots + a_nX_n$ has a normal distribution. What are the mean and the variance of this distribution?

6 The Theory in Application

Examples of the need for transformations in solving practical problems abound. To illustrate these applications, we give three examples. The first example illustrates an application of the transformation technique to a simple problem in electrical engineering.

EXAMPLE 17

Suppose the resistance in a simple circuit varies randomly in response to environmental conditions. To determine the effect of this variation on the current flowing through the circuit, an experiment was performed in which the resistance (R) was varied with equal probabilities on the interval $0 < R \leq A$ and the ensuing voltage (E) was measured. Find the distribution of the random variable I, the current flowing through the circuit.

Solution

Using the well-known relation $E = IR$, we have $I = u(R) = \frac{E}{R}$. The probability distribution of R is given by $f(R) = \frac{1}{A}$ for $0 < R \leq A$. Thus, $w(I) = \frac{E}{I}$, and the probability density of I is given by

$$g(I) = f(R) \cdot |w'(I)| = \frac{1}{A}\left|-\frac{E}{R^2}\right| = \frac{E}{AR^2} \qquad R > 0$$

It should be noted, with respect to this example, that this is a designed experiment in as much as the distribution of R was preselected as a uniform distribution. If the nominal value of R is to be the mean of this distribution, some other distribution might have been selected to impart better properties to this estimate.

The next example illustrates transformations of data to normality.

EXAMPLE 18

What underlying distribution of the data is assumed when the square-root transformation is used to obtain approximately normally distributed data? (Assume the data are nonnegative, that is, the probability of a negative observation is zero.)

Solution

A simple alternate way to use the distribution-function technique is to write down the differential element of the density function, $f(x)\,dx$, of the transformed observations, y, and to substitute x^2 for y. (When we do this, we must remember that the differential element, dy, must be changed to $dx = 2x\,dx$.) We obtain

$$f(x)\,dx = \frac{1}{\sqrt{2\pi}\sigma} \cdot 2x \cdot e^{-\frac{1}{2}(x^2-\mu)^2/\sigma^2}\,dx$$

The required density function is given by

$$f(x) = \sqrt{\frac{2}{\pi\sigma^2}}\,xe^{-\frac{1}{2}(x^2-\mu)^2/\sigma^2}$$

This distribution is not immediately recognizable, but it can be graphed quickly using appropriate computer software.

The final example illustrates an application to waiting-time problems.

EXAMPLE 19

Let us assume that the decay of a radioactive element is exponentially distributed, so that $f(x) = \lambda e^{-\lambda x}$ for $\lambda > 0$ and $x > 0$; that is, the time for the nucleus to emit the first α particle is x (in seconds). It can be shown that such a process has no memory; that is, the time *between successive emissions* also can be described by this distribution. Thus, it follows that successive emissions of α particles are independent. If the parameter λ equals 5, find the probability that a given substance will emit 2 particles in less than or equal to 3 seconds.

Solution

Let x_i be the waiting time between emissions i and $i+1$, for $i = 0, 1, 2, \ldots, n-1$. Then the total time for n emissions to take place is the sum $T = x_0 + x_1 + \cdots + x_{n-1}$. The moment-generating function of this sum is given in Example 16 to be

$$M_T(t) = (1 - t/\lambda)^{-n}$$

This can be recognized as the moment-generating function of the gamma distribution with parameters $\alpha = n = 2$ and $\beta = 1/\lambda = 1/5$. The required probability is given by

$$P\left(T \le 3; \alpha = 10, \beta = \frac{1}{5}\right) = \frac{1}{\frac{1}{5}\Gamma(2)} \int_0^3 x\, e^{-5x}\, dx$$

Integrating by parts, the integral becomes

$$P(T \le 3) = -\frac{1}{5} x e^{-5x}\Big|_0^3 - \int_0^3 -\frac{1}{5} e^{-5x}\, dx = 1 - 1.6e^{-15}$$

Without further evaluation, it is clear that this event is virtually certain to occur.

Applied Exercises — SECS. 1–2

47. This question has been intentionally omitted for this edition.

48. This question has been intentionally omitted for this edition.

49. This question has been intentionally omitted for this edition.

50. Let X be the amount of premium gasoline (in 1,000 gallons) that a service station has in its tanks at the beginning of a day, and Y the amount that the service station sells during that day. If the joint density of X and Y is given by

$$f(x, y) = \begin{cases} \dfrac{1}{200} & \text{for } 0 < y < x < 20 \\ 0 & \text{elsewhere} \end{cases}$$

use the distribution function technique to find the probability density of the amount that the service station has left in its tanks at the end of the day.

51. The percentages of copper and iron in a certain kind of ore are, respectively, X_1 and X_2. If the joint density of these two random variables is given by

$$f(x_1, x_2) = \begin{cases} \dfrac{3}{11}(5x_1 + x_2) & \text{for } x_1 > 0, x_2 > 0, \\ & \text{and } x_1 + 2x_2 < 2 \\ 0 & \text{elsewhere} \end{cases}$$

use the distribution function technique to find the probability density of $Y = X_1 + X_2$. Also find $E(Y)$, the expected total percentage of copper and iron in the ore.

SECS. 3–4

52. According to the Maxwell–Boltzmann law of theoretical physics, the probability density of V, the velocity of a gas molecule, is

$$f(v) = \begin{cases} kv^2 e^{-\beta v^2} & \text{for } v > 0 \\ 0 & \text{elsewhere} \end{cases}$$

where β depends on its mass and the absolute temperature and k is an appropriate constant. Show that the kinetic energy $E = \frac{1}{2}mV^2$, where m the mass of the molecule is a random variable having a gamma distribution.

53. This question has been intentionally omitted for this edition.

54. This question has been intentionally omitted for this edition.

55. This question has been intentionally omitted for this edition.

56. Use a computer program to generate 10 "pseudorandom" numbers having the standard normal distribution.

57. Describe how the probability integral transformation might have been used by the writers of the software that you used to produce the result of Exercise 56.

SEC. 5

58. A lawyer has an unlisted number on which she receives on the average 2.1 calls every half-hour and a listed number on which she receives on the average 10.9 calls every half-hour. If it can be assumed that the numbers of calls that she receives on these phones are independent random variables having Poisson distributions, what are the probabilities that in half an hour she will receive altogether

(a) 14 calls;

(b) at most 6 calls?

59. In a newspaper ad, a car dealer lists a 2001 Chrysler, a 2010 Ford, and a 2008 Buick. If the numbers of inquiries he will get about these cars may be regarded as independent random variables having Poisson distributions with the parameters $\lambda_1 = 3.6$, $\lambda_2 = 5.8$, and $\lambda_3 = 4.6$, what are the probabilities that altogether he will receive

(a) fewer than 10 inquiries about these cars;

(b) anywhere from 15 to 20 inquiries about these cars;

(c) at least 18 inquiries about these cars?

60. With reference to Exercise 59, what is the probability that the car dealer will receive six inquiries about the 2010 Ford and eight inquiries about the other two cars?

61. If the number of complaints a dry-cleaning establishment receives per day is a random variable having the Poisson distribution with $\lambda = 3.3$, what are the probabilities that it will receive

(a) 2 complaints on any given day;

(b) 5 complaints altogether on any two given days;

(c) at least 12 complaints altogether on any three given days?

62. The number of fish that a person catches per hour at Woods Canyon Lake is a random variable having the Poisson distribution with $\lambda = 1.6$. What are the probabilities that a person fishing there will catch

(a) four fish in 2 hours;

(b) at least two fish in 3 hours;

(c) at most three fish in 4 hours?

63. If the number of minutes it takes a service station attendant to balance a tire is a random variable having an exponential distribution with the parameter $\theta = 5$, what are the probabilities that the attendant will take

(a) less than 8 minutes to balance two tires;

(b) at least 12 minutes to balance three tires?

64. If the number of minutes that a doctor spends with a patient is a random variable having an exponential distribution with the parameter $\theta = 9$, what are the probabilities that it will take the doctor at least 20 minutes to treat

(a) one patient; **(b)** two patients; **(c)** three patients?

65. If X is the number of 7's obtained when rolling a pair of dice three times, find the probability that $Y = X^2$ will exceed 2.

66. If X has the exponential distribution given by $f(x) = 0.5\,e^{-0.5x}, x > 0$, find the probability that $x > 1$.

SEC. 6

67. If, d, the diameter of a circle is selected at random from the density function

$$f(d) = k\left(1 - \frac{d}{5}\right), 0 < d < 5,$$

(a) find the value of k so that $f(d)$ is a probability density;

(b) find the density function of the areas of the circles so selected.

68. Show that the underlying distribution function of Example 18 is, indeed, a probability distribution, and use a computer program to graph the density function.

69. If $X = \ln Y$ has a normal distribution with the mean μ and the standard deviation σ, find the probability density of Y which is said to have the **log-normal** distribution.

70. The logarithm of the ratio of the output to the input current of a transistor is called its current gain. If current gain measurements made on a certain transistor are normally distributed with $\mu = 1.8$ and $\sigma = 0.05$, find the probability that the current gain will exceed the required minimum value of 6.0.

References

The use of the probability integral transformation in problems of simulation is discussed in

JOHNSON, R. A., *Miller and Freund's Probability and Statistics for Engineers*, 6th ed. Upper Saddle River, N.J.: Prentice Hall, 2000.

A generalization of Theorem 1, which applies when the interval within the range of X for which $f(x) \neq 0$ can be partitioned into k subintervals so that the conditions of Theorem 1 apply separately for each of the subintervals, may be found in

WALPOLE, R. E., and MYERS, R. H., *Probability and Statistics for Engineers and Scientists*, 4th ed. New York: Macmillan Publishing Company, Inc., 1989.

More detailed and more advanced treatments of the material in this chapter are given in many advanced texts on mathematical statistics; for instance, in

HOGG, R. V., and CRAIG, A. T., *Introduction to Mathematical Statistics*, 4th ed. New York: Macmillan Publishing Company, Inc., 1978,

ROUSSAS, G. G., *A First Course in Mathematical Statistics*. Reading, Mass.: Addison-Wesley Publishing Company, Inc., 1973,

WILKS, S. S., *Mathematical Statistics*. New York: John Wiley & Sons, Inc., 1962.

Answers to Odd-Numbered Exercises

1 $g(y) = \frac{1}{\theta} e^{y} e^{-(1/\theta)e^{y}}$ for $-\infty < y < \infty$.

3 $g(y) = 2y$ for $0 < y < 1$ and $g(y) = 0$ elsewhere.

5 (a) $f(y) = \frac{1}{\theta_1 - \theta_2} \cdot (e^{-y/\theta_1} - e^{-y/\theta_2})$ for $y > 0$ and $f(y) = 0$ elsewhere; **(b)** $f(y) = \frac{1}{\theta^2} \cdot ye^{-y/\theta}$ for $y > 0$ and $f(y) = 0$ elsewhere.

9 $h(-2) = \frac{1}{5}, h(0) = \frac{3}{5}$, and $h(2) = \frac{1}{5}$.

11 (a) $g(0) = \frac{8}{27}, g(\frac{1}{2}) = \frac{12}{27}, g(\frac{2}{3}) = \frac{6}{27}, g(\frac{3}{4}) = \frac{1}{27}$;
(b) $g(0) = \frac{12}{27}, g(1) = \frac{14}{27}, g(16) = \frac{1}{27}$.

13 $g(0) = \frac{1}{3}, g(1) = \frac{1}{3}, g(2) = \frac{1}{3}$.

17 $g(y) = \frac{1}{6} y^{\frac{-1}{3}}$.

21 (a) $g(y) = \frac{1}{8} y^{-3/4}$ for $0 < y < 1$ and $g(y) = \frac{1}{4}$ for $1 < y < 3$;
(b) $h(z) = \frac{1}{16} \cdot z^{-3/4}$ for $1 < z < 81$ and $h(z) = 0$ elsewhere.

23 (a) $f(2, 0) = \frac{1}{36}, f(3, -1) = \frac{2}{36}, f(3, 1) = \frac{2}{36}, f(4, -2) = \frac{3}{36}, f(4, 0) = \frac{4}{36}, f(4, 2) = \frac{3}{36}, f(5, -1) = \frac{6}{36}, f(5, 1) = \frac{6}{36}$, and $f(6, 0) = \frac{9}{36}$;
(b) $g(2) = \frac{1}{36}, g(3) = \frac{4}{36}, g(4) = \frac{10}{36}, g(5) = \frac{12}{36}$, and $g(6) = \frac{9}{36}$.

25 (b) $g(0, 0, 2) = \frac{25}{144}, g(1, -1, 1) = \frac{5}{18}, g(1, 1, 1) = \frac{5}{24}, g(2, -2, 0) = \frac{1}{9}, g(2, 0, 0) = \frac{1}{6}$, and $g(2, 2, 0) = \frac{1}{16}$.

29 $\mu = 0$ and $\sigma^2 = 2$.

31 $g(z, u) = 12z(u^{-3} - u^{-2})$ over the region bounded by $z = 0, u = 1$, and $z = u^2$, and $g(z, u) = 0$ elsewhere; $h(z) = 6z + 6 - 12\sqrt{z}$ for $0 < z < 1$ and $h(z) = 0$ elsewhere.

33 The marginal distribution is the Cauchy distribution $g(y) = \frac{1}{\pi} \cdot \frac{2}{4 + y^2}$ for $-\infty < y < \infty$.

35 $f(u, v) = \frac{1}{2}$ over the region bounded by $v = 0, u = -v$, and $2v + u = 2$, and $f(u, v) = 0$ elsewhere; $g(u) = \frac{1}{4}(2 + u)$ for $-2 < u \leq 0, g(u) = \frac{1}{4}(2 - u)$ for $0 < u < 2$ and $g(u) = 0$ elsewhere.

37 $g(w, z) = 24w(z - w)$ over the region bounded by $w = 0, z = 1$, and $z = w$; $g(w, z) = 0$ elsewhere.

43 It is a gamma distribution with the parameters αn and β.

51 $g(y) = \frac{9}{11} \cdot y^2$ for $0 < y \leq 1, g(y) = \frac{3(2 - y)(7y - 4)}{11}$ for $1 < y < 2$, and $g(y) = 0$ elsewhere.

53 $h(r) = 2r$ for $0 < r < 1$ and $h(r) = 0$ elsewhere.

55 $g(v, w) = 5e^{-v}$ for $0.2 < w < 0.4$ and $v > 0$; $h(v) = e^{-v}$ for $v > 0$ and $h(v) = 0$ elsewhere.

59 (a) 0.1093; **(b)** 0.3817; **(c)** 0.1728.

61 (a) 0.2008; **(b)** 0.1420; **(c)** 0.2919.

63 (a) 0.475; **(b)** 0.570.

65 $\frac{2}{27}$.

67 (a) $\frac{2}{5}$; **(b)** $g(A) = \frac{2}{5}\left(\frac{1}{\sqrt{\pi}} A^{-1/2} - 1\right)$ for $0 < A < \frac{25}{4}\pi$ and $g(A) = 0$ elsewhere.

69 $g(y) = \frac{1}{\sqrt{2\pi}\sigma} \cdot \frac{1}{y} \cdot e^{-\frac{1}{2}\left(\frac{\ln y - \mu}{\sigma}\right)^2}$ for $y > 0$ and $g(y) = 0$ elsewhere.

Sampling Distributions

1 Introduction

Statistics concerns itself mainly with conclusions and predictions resulting from chance outcomes that occur in carefully planned experiments or investigations. Drawing such conclusions usually involves taking sample observations from a given population and using the results of the sample to make inferences about the population itself, its mean, its variance, and so forth. To do this requires that we first find the distributions of certain functions of the random variables whose values make up the sample, called **statistics**. (An example of such a statistic is the sample mean.) The properties of these distributions then allow us to make probability statements about the resulting inferences drawn from the sample about the population. The theory to be given in this chapter forms an important foundation for the theory of statistical inference.

Inasmuch as statistical inference can be loosely defined as a process of drawing conclusions from a sample about the population from which it is drawn, it is useful to have the following definition.

> **DEFINITION 1. POPULATION.** *A set of numbers from which a sample is drawn is referred to as a* ***population****. The distribution of the numbers constituting a population is called the* ***population distribution****.*

To illustrate, suppose a scientist must choose and then weigh 5 of 40 guinea pigs as part of an experiment, a layman might say that the ones she selects constitute the sample. This is how the term "sample" is used in everyday language. In statistics, it is preferable to look upon the weights of the 5 guinea pigs as a sample from the population, which consists of the weights of all 40 guinea pigs. In this way, the population as well as the sample consists of numbers. Also, suppose that, to estimate the average useful life of a certain kind of transistor, an engineer selects 10 of these transistors, tests them over a period of time, and records for each one the time to failure. If these times to failure are values of independent random variables having an exponential distribution with the parameter θ, we say that they constitute a sample from this exponential population.

As can well be imagined, not all samples lend themselves to valid generalizations about the populations from which they came. In fact, most of the methods of inference discussed in this chapter are based on the assumption that we are dealing with

random samples. In practice, we often deal with random samples from populations that are finite, but large enough to be treated as if they were infinite. Thus, most statistical theory and most of the methods we shall discuss apply to samples from infinite populations, and we shall begin here with a definition of random samples from infinite populations. Random samples from finite populations will be treated later in Section 3.

> **DEFINITION 2. RANDOM SAMPLE.** *If $X_1, X_2, \ldots, X_n$ are independent and identically distributed random variables, we say that they constitute a* ***random sample*** *from the infinite population given by their common distribution.*

If $f(x_1, x_2, \ldots, x_n)$ is the value of the joint distribution of such a set of random variables at $(x_1, x_2, \ldots, x_n)$, by virtue of independence we can write

$$f(x_1, x_2, \ldots, x_n) = \prod_{i=1}^{n} f(x_i)$$

where $f(x_i)$ is the value of the population distribution at x_i. Observe that Definition 2 and the subsequent discussion apply also to sampling with replacement from finite populations; sampling without replacement from finite populations is discussed in section 3.

Statistical inferences are usually based on **statistics**, that is, on random variables that are functions of a set of random variables $X_1, X_2, \ldots, X_n$, constituting a random sample. Typical of what we mean by "statistic" are the **sample mean** and the **sample variance**.

> **DEFINITION 3. SAMPLE MEAN AND SAMPLE VARIANCE.** *If $X_1, X_2, \ldots, X_n$ constitute a random sample, then the* ***sample mean*** *is given by*
>
> $$\overline{X} = \frac{\sum_{i=1}^{n} X_i}{n}$$
>
> *and the* ***sample variance*** *is given by*
>
> $$S^2 = \frac{\sum_{i=1}^{n} (X_i - \overline{X})^2}{n-1}$$

As they are given here, these definitions apply only to random samples, but the sample mean and the sample variance can, similarly, be defined for any set of random variables $X_1, X_2, \ldots, X_n$.

It is common practice also to apply the terms "random sample," "statistic," "sample mean," and "sample variance" to the values of the random variables instead of the random variables themselves. Intuitively, this makes more sense and it conforms with colloquial usage. Thus, we might calculate

$$\overline{x} = \frac{\sum_{i=1}^{n} x_i}{n} \quad \text{and} \quad s^2 = \frac{\sum_{i=1}^{n} (x_i - \overline{x})^2}{n-1}$$

for observed sample data and refer to these statistics as the sample mean and the sample variance. Here, the x_i, $\overline{x}$, and s^2 are values of the corresponding random

---

† The note has been intentionally omitted for this edition.

variables X_i, $\overline{X}$, and S^2. Indeed, the formulas for $\bar{x}$ and s^2 are used even when we deal with any kind of data, not necessarily sample data, in which case we refer to $\bar{x}$ and s^2 simply as the mean and the variance.

These, and other statistics that will be introduced in this chapter, are those mainly used in statistical inference. Sample statistics such as the sample mean and sample variance play an important role in estimating the parameters of the population from which the corresponding random samples were drawn.

2 The Sampling Distribution of the Mean

Inasmuch as the values of sampling statistics can be expected to vary from sample to sample, it is necessary that we find the distribution of such statistics. We call these distributions **sampling distributions**, and we make important use of them in determining the properties of the inferences we draw from the sample about the parameters of the population from which it is drawn.

First let us study some theory about the **sampling distribution of the mean**, making only some very general assumptions about the nature of the populations sampled.

THEOREM 1. If $X_1, X_2, \ldots, X_n$ constitute a random sample from an infinite population with the mean μ and the variance σ^2, then

$$E(\overline{X}) = \mu \quad \text{and} \quad \text{var}(\overline{X}) = \frac{\sigma^2}{n}$$

Proof Letting $Y = \overline{X}$ and hence setting $a_i = \frac{1}{n}$, we get

$$E(\overline{X}) = \sum_{i=1}^{n} \frac{1}{n} \cdot \mu = n\left(\frac{1}{n} \cdot \mu\right) = \mu$$

since $E(X_i) = \mu$. Then, by the corollary of a theorem "If the random variables $X_1, X_2, \ldots, X_n$ are independent and $Y = \sum_{i=1}^{n} a_i X_i$, then $\text{var}(Y) = \sum_{i=1}^{n} a_i^2 \cdot \text{var}(X_i)$", we conclude that

$$\text{var}(\overline{X}) = \sum_{i=1}^{n} \frac{1}{n^2} \cdot \sigma^2 = n\left(\frac{1}{n^2} \cdot \sigma^2\right) = \frac{\sigma^2}{n}$$

It is customary to write $E(\overline{X})$ as $\mu_{\overline{X}}$ and $\text{var}(\overline{X})$ as $\sigma^2_{\overline{X}}$ and refer to $\sigma_{\overline{X}}$ as the **standard error of the mean**. The formula for the standard error of the mean, $\sigma_{\overline{X}} = \frac{\sigma}{\sqrt{n}}$, shows that the standard deviation of the distribution of $\overline{X}$ decreases when n, the **sample size**, is increased. This means that when n becomes larger and we actually have more information (the values of more random variables), we can expect values of $\overline{X}$ to be closer to μ, the quantity that they are intended to estimate.

THEOREM 2. For any positive constant c, the probability that $\overline{X}$ will take on a value between $\mu - c$ and $\mu + c$ is at least

$$1 - \frac{\sigma^2}{nc^2}$$

When $n \to \infty$, this probability approaches 1.

This result, called a **law of large numbers**, is primarily of theoretical interest. Of much more practical value is the **central limit theorem**, one of the most important theorems of statistics, which concerns the limiting distribution of the **standardized mean** of n random variables when $n \to \infty$. We shall prove this theorem here only for the case where the n random variables are a random sample from a population whose moment-generating function exists. More general conditions under which the theorem holds are given in Exercises 7 and 9, and the most general conditions under which it holds are referred to at the end of this chapter.

THEOREM 3. CENTRAL LIMIT THEOREM. If $X_1, X_2, \ldots, X_n$ constitute a random sample from an infinite population with the mean μ, the variance σ^2, and the moment-generating function $M_X(t)$, then the limiting distribution of

$$Z = \frac{\overline{X} - \mu}{\sigma/\sqrt{n}}$$

as $n \to \infty$ is the standard normal distribution.

Proof First using the third part and then the second of the given theorem "If a and b are constants, then **1.** $M_{X+a}(t) = E[e^{(X+a)t}] = e^{at} \cdot M_X(t)$; **2.** $M_{bX}(t) = E(e^{bXt}) = M_X(bt)$; **3.** $M_{\frac{X+a}{b}}(t) = E[e^{\left(\frac{X+a}{b}\right)t}] = e^{\frac{a}{b}t} \cdot M_X\left(\frac{t}{b}\right)$", we get

$$M_Z(t) = M_{\frac{\overline{X}-\mu}{\sigma/\sqrt{n}}}(t) = e^{-\sqrt{n}\,\mu t/\sigma} \cdot M_{\overline{X}}\left(\frac{\sqrt{n}t}{\sigma}\right)$$

$$= e^{-\sqrt{n}\,\mu t/\sigma} \cdot M_{n\overline{X}}\left(\frac{t}{\sigma\sqrt{n}}\right)$$

Since $n\overline{X} = X_1 + X_2 + \cdots + X_n$,

$$M_Z(t) = e^{-\sqrt{n}\,\mu t/\sigma} \cdot \left[M_X\left(\frac{t}{\sigma\sqrt{n}}\right)\right]^n$$

and hence that

$$\ln M_Z(t) = -\frac{\sqrt{n}\,\mu t}{\sigma} + n \cdot \ln M_X\left(\frac{t}{\sigma\sqrt{n}}\right)$$

Expanding $M_X\left(\frac{t}{\sigma\sqrt{n}}\right)$ as a power series in t, we obtain

$$\ln M_Z(t) = -\frac{\sqrt{n}\,\mu t}{\sigma} + n \cdot \ln\left[1 + \mu_1' \frac{t}{\sigma\sqrt{n}} + \mu_2' \frac{t^2}{2\sigma^2 n} + \mu_3' \frac{t^3}{6\sigma^3 n\sqrt{n}} + \cdots\right]$$

where μ_1', μ_2', and μ_3' are the moments about the origin of the population distribution, that is, those of the original random variables X_i.

If n is sufficiently large, we can use the expansion of $\ln(1+x)$ as a power series in x, getting

$$\ln M_Z(t) = -\frac{\sqrt{n}\,\mu t}{\sigma} + n\left\{\left[\mu_1'\frac{t}{\sigma\sqrt{n}} + \mu_2'\frac{t^2}{2\sigma^2 n} + \mu_3'\frac{t^3}{6\sigma^3 n\sqrt{n}} + \cdots\right]\right.$$

$$-\frac{1}{2}\left[\mu_1'\frac{t}{\sigma\sqrt{n}} + \mu_2'\frac{t^2}{2\sigma^2 n} + \mu_3'\frac{t^3}{6\sigma^3 n\sqrt{n}} + \cdots\right]^2$$

$$\left.+\frac{1}{3}\left[\mu_1'\frac{t}{\sigma\sqrt{n}} + \mu_2'\frac{t^2}{2\sigma^2 n} + \mu_3'\frac{t^3}{6\sigma^3 n\sqrt{n}} + \cdots\right]^3 - \cdots\right\}$$

Then, collecting powers of t, we obtain

$$\ln M_Z(t) = \left(-\frac{\sqrt{n}\,\mu}{\sigma} + \frac{\sqrt{n}\,\mu_1'}{\sigma}\right)t + \left(\frac{\mu_2'}{2\sigma^2} - \frac{\mu_1'^2}{2\sigma^2}\right)t^2$$

$$+\left(\frac{\mu_3'}{6\sigma^3\sqrt{n}} - \frac{\mu_1'\cdot\mu_2'}{2\sigma^3\sqrt{n}} + \frac{\mu_1'^3}{3\sigma^3\sqrt{n}}\right)t^3 + \cdots$$

and since $\mu_1' = \mu$ and $\mu_2' - (\mu_1')^2 = \sigma^2$, this reduces to

$$\ln M_Z(t) = \frac{1}{2}t^2 + \left(\frac{\mu_3'}{6} - \frac{\mu_1'\mu_2'}{2} + \frac{\mu_1'^3}{6}\right)\frac{t^3}{\sigma^3\sqrt{n}} + \cdots$$

Finally, observing that the coefficient of t^3 is a constant times $\frac{1}{\sqrt{n}}$ and in general, for $r \geqq 2$, the coefficient of t^r is a constant times $\frac{1}{\sqrt{n^{r-2}}}$, we get

$$\lim_{n\to\infty} \ln M_Z(t) = \frac{1}{2}t^2$$

and hence

$$\lim_{n\to\infty} M_Z(t) = e^{\frac{1}{2}t^2}$$

since the limit of a logarithm equals the logarithm of the limit (provided these limits exist). An illustration of this theorem is given in Exercise 13 and 14.

Sometimes, the central limit theorem is interpreted incorrectly as implying that the distribution of $\overline{X}$ approaches a normal distribution when $n \to \infty$. This is incorrect because $\text{var}(\overline{X}) \to 0$ when $n \to \infty$; on the other hand, the central limit theorem does justify approximating the distribution of $\overline{X}$ with a normal distribution having the mean μ and the variance $\frac{\sigma^2}{n}$ when n is large. In practice, this approximation is used when $n \geqq 30$ regardless of the actual shape of the population sampled. For smaller values of n the approximation is questionable, but see Theorem 4.

EXAMPLE 1

A soft-drink vending machine is set so that the amount of drink dispensed is a random variable with a mean of 200 milliliters and a standard deviation of 15 milliliters. What is the probability that the average (mean) amount dispensed in a random sample of size 36 is at least 204 milliliters?

Solution

According to Theorem 1, the distribution of $\overline{X}$ has the mean $\mu_{\overline{X}} = 200$ and the standard deviation $\sigma_{\overline{X}} = \dfrac{15}{\sqrt{36}} = 2.5$, and according to the central limit theorem, this distribution is approximately normal. Since $z = \dfrac{204-200}{2.5} = 1.6$, it follows from Table III of "Statistical Tables" that $P(\overline{X} \geqq 204) = P(Z \geqq 1.6) = 0.5000 - 0.4452 = 0.0548$.

It is of interest to note that when the population we are sampling is normal, the distribution of $\overline{X}$ is a normal distribution regardless of the size of n.

THEOREM 4. If $\overline{X}$ is the mean of a random sample of size n from a normal population with the mean μ and the variance σ^2, its sampling distribution is a normal distribution with the mean μ and the variance σ^2/n.

Proof According to Theorems "If a and b are constants, then **1.** $M_{X+a}(t) = E[e^{(X+a)t}] = e^{at} \cdot M_X(t)$; **2.** $M_{bX}(t) = E(e^{bXt}) = M_X(bt)$; **3.** $M_{\frac{X+a}{b}}(t) = E[e^{\left(\frac{X+a}{b}\right)t}] = e^{\frac{a}{b}t} \cdot M_X\left(\frac{t}{b}\right)$. If $X_1, X_2, \ldots,$ and X_n are independent random variables and $Y = X_1 + X_2 + \cdots + X_n$, then $M_Y(t) = \prod_{i=1}^{n} M_{X_i}(t)$ where $M_{X_i}(t)$ is the value of the moment-generating function of X_i at t", we can write

$$M_{\overline{X}}(t) = \left[M_X\left(\frac{t}{n}\right)\right]^n$$

and since the moment-generating function of a normal distribution with the mean μ and the variance σ^2 is given by

$$M_X(t) = e^{\mu t + \frac{1}{2}\sigma^2 t^2}$$

according to the theorem $M_X(t) = e^{\mu t + \frac{1}{2}\sigma^2 t^2}$, we get

$$M_{\overline{X}}(t) = \left[e^{\mu \cdot \frac{t}{n} + \frac{1}{2}\left(\frac{t}{n}\right)^2 \sigma^2}\right]^n$$

$$= e^{\mu t + \frac{1}{2} t^2 \left(\frac{\sigma^2}{n}\right)}$$

This moment-generating function is readily seen to be that of a normal distribution with the mean μ and the variance σ^2/n.

3 The Sampling Distribution of the Mean: Finite Populations

If an experiment consists of selecting one or more values from a finite set of numbers $\{c_1, c_2, \ldots, c_N\}$, this set is referred to as a **finite population of size N**. In the definition that follows, it will be assumed that we are sampling *without replacement* from a finite population of size N.

DEFINITION 4. RANDOM SAMPLE—FINITE POPULATION. *If* X_1 *is the first value drawn from a finite population of size* N, X_2 *is the second value drawn,* ..., X_n *is the* n*th value drawn, and the joint probability distribution of these* n *random variables is given by*

$$f(x_1, x_2, \ldots, x_n) = \frac{1}{N(N-1)\cdot \ldots \cdot (N-n+1)}$$

for each ordered n*-tuple of values of these random variables, then* $X_1, X_2, \ldots, X_n$ *are said to constitute a* ***random sample*** *from the given finite population.*

As in Definition 2, the random sample is a set of random variables, but here again it is common practice also to apply the term "random sample" to the values of the random variables, that is, to the actual numbers drawn.

From the joint probability distribution of Definition 4, it follows that the probability for each subset of n of the N elements of the finite population (regardless of the order in which the values are drawn) is

$$\frac{n!}{N(N-1)\cdot \ldots \cdot (N-n+1)} = \frac{1}{\binom{N}{n}}$$

This is often given as an alternative definition or as a criterion for the selection of a random sample of size n from a finite population of size N: *Each of the* $\binom{N}{n}$ *possible samples must have the same probability.*

It also follows from the joint probability distribution of Definition 4 that the marginal distribution of X_r is given by

$$f(x_r) = \frac{1}{N} \qquad \text{for } x_r = c_1, c_2, \ldots, c_N$$

for $r = 1, 2, \ldots, n$, and we refer to the mean and the variance of this discrete uniform distribution as the mean and the variance of the finite population. Therefore,

DEFINITION 5. SAMPLE MEAN AND VARIANCE—FINITE POPULATION. *The* ***sample mean*** *and the* ***sample variance*** *of the finite population* $\{c_1, c_2, \ldots, c_N\}$ *are*

$$\mu = \sum_{i=1}^{N} c_i \cdot \frac{1}{N} \quad \textit{and} \quad \sigma^2 = \sum_{i=1}^{N} (c_i - \mu)^2 \cdot \frac{1}{N}$$

Finally, it follows from the joint probability distribution of Definition 4 that the joint marginal distribution of any two of the random variables $X_1, X_2, \ldots, X_n$ is given by

$$g(x_r, x_s) = \frac{1}{N(N-1)}$$

for each ordered pair of elements of the finite population. Thus, we can prove the following theorem.

THEOREM 5. If X_r and X_s are the rth and sth random variables of a random sample of size n drawn from the finite population $\{c_1, c_2, \ldots, c_N\}$, then

$$\text{cov}(X_r, X_s) = -\frac{\sigma^2}{N-1}$$

Proof According to the definition given here "**COVARIANCE.** $\mu_{1,1}$ *is called **the covariance** of* X *and* Y, *and it is denoted by* σ_{XY}, *cov*(X, Y), or C(X, Y)",

$$\text{cov}(X_r, X_s) = \sum_{i=1}^{N} \sum_{\substack{j=1 \\ i \neq j}}^{N} \frac{1}{N(N-1)} (c_i - \mu)(c_j - \mu)$$

$$= \frac{1}{N(N-1)} \cdot \sum_{i=1}^{N} (c_i - \mu) \left[\sum_{\substack{j=1 \\ j \neq i}}^{N} (c_j - \mu) \right]$$

and since $\sum_{\substack{j=1 \\ j \neq i}}^{N} (c_j - \mu) = \sum_{j=1}^{N} (c_j - \mu) - (c_i - \mu) = -(c_i - \mu)$, we get

$$\text{cov}(X_r, X_s) = -\frac{1}{N(N-1)} \cdot \sum_{i=1}^{N} (c_i - \mu)^2$$

$$= -\frac{1}{N-1} \cdot \sigma^2$$

Making use of all these results, let us now prove the following theorem, which, for random samples from finite populations, corresponds to Theorem 1.

THEOREM 6. If $\overline{X}$ is the mean of a random sample of size n taken without replacement from a finite population of size N with the mean μ and the variance σ^2, then

$$E(\overline{X}) = \mu \quad \text{and} \quad \text{var}(\overline{X}) = \frac{\sigma^2}{n} \cdot \frac{N-n}{N-1}$$

Proof Substituting $a_i = \frac{1}{N}$, $\text{var}(X_i) = \sigma^2$, and $\text{cov}(X_i, X_j) = -\frac{\sigma^2}{N-1}$ into the formula $E(Y) = \sum_{i=1}^{n} a_i E(X_i)$, we get

$$E(\overline{X}) = \sum_{i=1}^{n} \frac{1}{n} \cdot \mu = \mu$$

and

$$\text{var}(\overline{X}) = \sum_{i=1}^{n} \frac{1}{n^2} \cdot \sigma^2 + 2 \cdot \sum_{i<j}\sum \frac{1}{n^2} \left(-\frac{\sigma^2}{N-1} \right)$$

$$= \frac{\sigma^2}{n} + 2 \cdot \frac{n(n-1)}{2} \cdot \frac{1}{n^2} \left(-\frac{\sigma^2}{N-1} \right)$$

$$= \frac{\sigma^2}{n} \cdot \frac{N-n}{N-1}$$

It is of interest to note that the formulas we obtained for $\text{var}(\overline{X})$ in Theorems 1 and 6 differ only by the **finite population correction factor** $\frac{N-n}{N-1}$.† Indeed, when N is large compared to n, the difference between the two formulas for $\text{var}(\overline{X})$ is usually negligible, and the formula $\sigma_{\overline{X}} = \frac{\sigma}{\sqrt{n}}$ is often used as an approximation when we are sampling from a large finite population. A general rule of thumb is to use this approximation when the sample does not constitute more than 5 percent of the population.

Exercises

1. This question has been intentionally omitted for this edition.

2. This question has been intentionally omitted for this edition.

3. With reference to Exercise 2, show that if the two samples come from normal populations, then $\overline{X}_1 - \overline{X}_2$ is a random variable having a normal distribution with the mean $\mu_1 - \mu_2$ and the variance $\frac{\sigma_1^2}{n_1} + \frac{\sigma_2^2}{n_2}$. (*Hint*: Proceed as in the proof of Theorem 4.)

4. If $X_1, X_2, \ldots, X_n$ are independent random variables having identical Bernoulli distributions with the parameter θ, then $\overline{X}$ is the proportion of successes in n trials, which we denote by $\hat{\Theta}$. Verify that

(a) $E(\hat{\Theta}) = \theta$;

(b) $\text{var}(\hat{\Theta}) = \frac{\theta(1-\theta)}{n}$.

5. If the first n_1 random variables of Exercise 2 have Bernoulli distributions with the parameter θ_1 and the other n_2 random variables have Bernoulli distributions with the parameter θ_2, show that, in the notation of Exercise 4,

(a) $E(\hat{\Theta}_1 - \hat{\Theta}_2) = \theta_1 - \theta_2$;

(b) $\text{var}(\hat{\Theta}_1 - \hat{\Theta}_2) = \frac{\theta_1(1-\theta_1)}{n_1} + \frac{\theta_2(1-\theta_2)}{n_2}$.

6. This question has been intentionally omitted for this edition.

7. The following is a sufficient condition for the central limit theorem: If the random variables $X_1, X_2, \ldots, X_n$ are independent and uniformly bounded (that is, there exists a positive constant k such that the probability is zero that any one of the random variables X_i will take on a value greater than k or less than $-k$), then if the variance of

$$Y_n = X_1 + X_2 + \cdots + X_n$$

becomes infinite when $n \to \infty$, the distribution of the standardized mean of the X_i approaches the standard normal distribution. Show that this sufficient condition holds for a sequence of independent random variables X_i having the respective probability distributions

$$f_i(x_i) = \begin{cases} \frac{1}{2} & \text{for } x_i = 1 - (\frac{1}{2})^i \\ \frac{1}{2} & \text{for } x_i = (\frac{1}{2})^i - 1 \end{cases}$$

8. Consider the sequence of independent random variables $X_1, X_2, X_3, \ldots$ having the uniform densities

$$f_i(x_i) = \begin{cases} \frac{1}{2 - \frac{1}{i}} & \text{for } 0 < x_i < 2 - \frac{1}{i} \\ 0 & \text{elsewhere} \end{cases}$$

Use the sufficient condition of Exercise 7 to show that the central limit theorem holds.

9. The following is a sufficient condition, the *Laplace–Liapounoff condition*, for the central limit theorem: If $X_1, X_2, X_3, \ldots$ is a sequence of independent random variables, each having an absolute third moment

$$c_i = E(|X_i - \mu_i|^3)$$

and if

$$\lim_{n \to \infty} [\text{var}(Y_n)]^{-\frac{3}{2}} \cdot \sum_{i=1}^{n} c_i = 0$$

where $Y_n = X_1 + X_2 + \cdots + X_n$, then the distribution of the standardized mean of the X_i approaches the standard normal distribution when $n \to \infty$. Use this condition to show that the central limit theorem holds for the sequence of random variables of Exercise 7.

10. Use the condition of Exercise 9 to show that the central limit theorem holds for the sequence of random variables of Exercise 8.

†Since there are many problems in which we are interested in the standard deviation rather than the variance, the term "finite population correction factor" often refers to $\sqrt{\frac{N-n}{N-1}}$ instead of $\frac{N-n}{N-1}$. This does not matter, of course, as long as the usage is clearly understood.

11. Explain why, when we sample with replacement from a finite population, the results of Theorem 1 apply rather than those of Theorem 6.

12. This question has been intentionally omitted for this edition.

13. Use MINITAB or some other statistical computer program to generate 20 samples of size 10 each from the uniform density function $f(x) = 1, 0 \leq x \leq 1$.
(a) Find the mean of each sample and construct a histogram of these sample means.
(b) Calculate the mean and the variance of the 20 sample means.

14. Referring to Exercise 13, now change the sample size to 30.
(a) Does this histogram more closely resemble that of a normal distribution than that of Exercise 13? Why?
(b) Which resembles it more closely?
(c) Calculate the mean and the variance of the 20 sample means.

15. If a random sample of size n is selected without replacement from the finite population that consists of the integers $1, 2, \ldots, N$, show that
(a) the mean of $\overline{X}$ is $\dfrac{N+1}{2}$;
(b) the variance of $\overline{X}$ is $\dfrac{(N+1)(N-n)}{12n}$;
(c) the mean and the variance of $Y = n \cdot \overline{X}$ are

$$E(Y) = \frac{n(N+1)}{2} \quad \text{and} \quad \text{var}(Y) = \frac{n(N+1)(N-n)}{12}$$

16. Find the mean and the variance of the finite population that consists of the 10 numbers 15, 13, 18, 10, 6, 21, 7, 11, 20, and 9.

17. Show that the variance of the finite population $\{c_1, c_2, \ldots, c_N\}$ can be written as

$$\sigma^2 = \frac{\sum_{i=1}^{N} c_i^2}{N} - \mu^2$$

Also, use this formula to recalculate the variance of the finite population of Exercise 16.

18. Show that, analogous to the formula of Exercise 17, the formula for the sample variance can be written as

$$S^2 = \frac{\sum_{i=1}^{n} X_i^2}{n-1} - \frac{n\overline{X}^2}{n-1}$$

Also, use this formula to calculate the variance of the following sample data on the number of service calls received by a tow truck operator on eight consecutive working days: 13, 14, 13, 11, 15, 14, 17, and 11.

19. Show that the formula for the sample variance can be written as

$$S^2 = \frac{n\left(\sum_{i=1}^{n} X_i^2\right) - \left(\sum_{i=1}^{n} X_i\right)^2}{n(n-1)}$$

Also, use this formula to recalculate the variance of the sample data of Exercise 18.

4 The Chi-Square Distribution

If X has the standard normal distribution, then X^2 has the special gamma distribution, which is referred to as the **chi-square distribution**, and this accounts for the important role that the chi-square distribution plays in problems of sampling from normal populations. Theorem 11 will show the importance of this distribution in making inferences about sample variances.

The chi-square distribution is often denoted by "χ^2 distribution," where χ is the lowercase Greek letter *chi*. We also use χ^2 for values of random variables having chi-square distributions, but we shall refrain from denoting the corresponding random variables by X^2, where X is the capital Greek letter *chi*. This avoids having to reiterate in each case whether X is a random variable with values x or a random variable with values χ.

If a random variable X has the chi-square distribution with ν degrees of freedom if its probability density is given by

$$f(x) = \begin{cases} \dfrac{1}{2^{\nu/2}\Gamma(\nu/2)} x^{\frac{\nu-2}{2}} e^{-x/2} & \text{for } x > 0 \\ 0 & \text{elsewhere} \end{cases}$$

The mean and the variance of the chi-square distribution with ν degrees of freedom are ν and 2ν, respectively, and its moment-generating function is given by

$$M_X(t) = (1-2t)^{-\nu/2}$$

The chi-square distribution has several important mathematical properties, which are given in Theorems 7 through 10.

THEOREM 7. If X has the standard normal distribution, then X^2 has the chi-square distribution with $\nu = 1$ degree of freedom.

More generally, let us prove the following theorem.

THEOREM 8. If $X_1, X_2, \ldots, X_n$ are independent random variables having standard normal distributions, then

$$Y = \sum_{i=1}^{n} X_i^2$$

has the chi-square distribution with $\nu = n$ degrees of freedom.

Proof Using the moment-generating function given previously with $\nu = 1$ and Theorem 7, we find that

$$M_{X_i^2}(t) = (1-2t)^{-\frac{1}{2}}$$

and it follows the theorem "$M_Y(t) = \prod_{i=1}^{n} M_{X_i}(t)$" that

$$M_Y(t) = \prod_{i=1}^{n}(1-2t)^{-\frac{1}{2}} = (1-2t)^{-\frac{n}{2}}$$

This moment-generating function is readily identified as that of the chi-square distribution with $\nu = n$ degrees of freedom.

Two further properties of the chi-square distribution are given in the two theorems that follow; the reader will be asked to prove them in Exercises 20 and 21.

THEOREM 9. If $X_1, X_2, \ldots, X_n$ are independent random variables having chi-square distributions with $\nu_1, \nu_2, \ldots, \nu_n$ degrees of freedom, then

$$Y = \sum_{i=1}^{n} X_i$$

has the chi-square distribution with $\nu_1 + \nu_2 + \cdots + \nu_n$ degrees of freedom.

THEOREM 10. If X_1 and X_2 are independent random variables, X_1 has a chi-square distribution with ν_1 degrees of freedom, and $X_1 + X_2$ has a chi-square distribution with $\nu > \nu_1$ degrees of freedom, then X_2 has a chi-square distribution with $\nu - \nu_1$ degrees of freedom.

The chi-square distribution has many important applications. Foremost are those based, directly or indirectly, on the following theorem.

THEOREM 11. If $\overline{X}$ and S^2 are the mean and the variance of a random sample of size n from a normal population with the mean μ and the standard deviation σ, then

1. $\overline{X}$ and S^2 are independent;
2. the random variable $\frac{(n-1)S^2}{\sigma^2}$ has a chi-square distribution with $n-1$ degrees of freedom.

Proof Since a detailed proof of part 1 would go beyond the scope of this chapter we shall assume the independence of $\overline{X}$ and S^2 in our proof of part 2. In addition to the references to proofs of part 1 at the end of this chapter, Exercise 31 outlines the major steps of a somewhat simpler proof based on the idea of a conditional moment-generating function, and in Exercise 30 the reader will be asked to prove the independence of $\overline{X}$ and S^2 for the special case where $n = 2$.

To prove part 2, we begin with the identity

$$\sum_{i=1}^{n}(X_i-\mu)^2 = \sum_{i=1}^{n}(X_i-\overline{X})^2 + n(\overline{X}-\mu)^2$$

which the reader will be asked to verify in Exercise 22. Now, if we divide each term by σ^2 and substitute $(n-1)S^2$ for $\sum_{i=1}^{n}(X_i-\overline{X})^2$, it follows that

$$\sum_{i=1}^{n}\left(\frac{X_i-\mu}{\sigma}\right)^2 = \frac{(n-1)S^2}{\sigma^2} + \left(\frac{\overline{X}-\mu}{\sigma/\sqrt{n}}\right)^2$$

With regard to the three terms of this identity, we know from Theorem 8 that the one on the left-hand side of the equation is a random variable having a chi-square distribution with n degrees of freedom. Also, according to Theorems 4 and 7, the second term on the right-hand side of the equation is a random variable having a chi-square distribution with 1 degree of freedom. Now, since $\overline{X}$ and S^2 are assumed to be independent, it follows that the two terms on the right-hand side of the equation are independent, and we conclude that $\frac{(n-1)S^2}{\sigma^2}$ is a random variable having a chi-square distribution with $n-1$ degrees of freedom.

Since the chi-square distribution arises in many important applications, integrals of its density have been extensively tabulated. Table V of "Statistical Tables" contains values of $\chi^2_{\alpha,\nu}$ for $\alpha = 0.995, 0.99, 0.975, 0.95, 0.05, 0.025, 0.01, 0.005$, and $\nu = 1, 2, \ldots, 30$, where $\chi^2_{\alpha,\nu}$ is such that the area to its right under the chi-square curve with ν degrees of freedom (see Figure 1) is equal to α. That is, $\chi^2_{\alpha,\nu}$ is such that if X is a random variable having a chi-square distribution with ν degrees of freedom, then

$$P(X \geqq \chi^2_{\alpha,\nu}) = \alpha$$

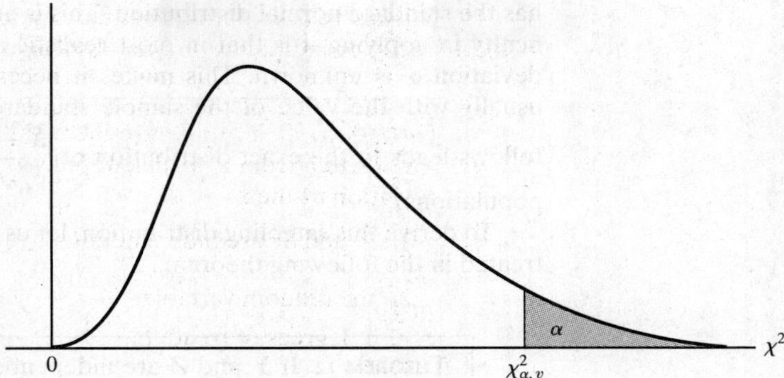

Figure 1. Chi-square distribution.

When ν is greater than 30, Table V of "Statistical Tables" cannot be used and probabilities related to chi-square distributions are usually approximated with normal distributions, as in Exercise 25 or 28.

EXAMPLE 2

Suppose that the thickness of a part used in a semiconductor is its critical dimension and that the process of manufacturing these parts is considered to be under control if the true variation among the thicknesses of the parts is given by a standard deviation not greater than $\sigma = 0.60$ thousandth of an inch. To keep a check on the process, random samples of size $n = 20$ are taken periodically, and it is regarded to be "out of control" if the probability that S^2 will take on a value greater than or equal to the observed sample value is 0.01 or less (even though $\sigma = 0.60$). What can one conclude about the process if the standard deviation of such a periodic random sample is $s = 0.84$ thousandth of an inch?

Solution

The process will be declared "out of control" if $\dfrac{(n-1)s^2}{\sigma^2}$ with $n = 20$ and $\sigma = 0.60$ exceeds $\chi^2_{0.01,19} = 36.191$. Since

$$\frac{(n-1)s^2}{\sigma^2} = \frac{19(0.84)^2}{(0.60)^2} = 37.24$$

exceeds 36.191, the process is declared out of control. Of course, it is assumed here that the sample may be regarded as a random sample from a normal population.

5 The *t* Distribution

In Theorem 4 we showed that for random samples from a normal population with the mean μ and the variance σ^2, the random variable $\overline{X}$ has a normal distribution with the mean μ and the variance $\dfrac{\sigma^2}{n}$; in other words,

$$\frac{\overline{X} - \mu}{\sigma/\sqrt{n}}$$

has the standard normal distribution. This is an important result, but the major difficulty in applying it is that in most realistic applications the population standard deviation σ is unknown. This makes it necessary to replace σ with an estimate, usually with the value of the sample standard deviation S. Thus, the theory that follows leads to the exact distribution of $\dfrac{\overline{X}-\mu}{S/\sqrt{n}}$ for random samples from normal populations.

To derive this sampling distribution, let us first study the more general situation treated in the following theorem.

THEOREM 12. If Y and Z are independent random variables, Y has a chi-square distribution with ν degrees of freedom, and Z has the standard normal distribution, then the distribution of

$$T = \frac{Z}{\sqrt{Y/\nu}}$$

is given by

$$f(t) = \frac{\Gamma\left(\frac{\nu+1}{2}\right)}{\sqrt{\pi\nu}\,\Gamma\left(\frac{\nu}{2}\right)} \cdot \left(1+\frac{t^2}{\nu}\right)^{-\frac{\nu+1}{2}} \qquad \text{for } -\infty < t < \infty$$

and it is called the ***t* distribution with ν degrees of freedom**.

Proof Since Y and Z are independent, their joint probability density is given by

$$f(y,z) = \frac{1}{\sqrt{2\pi}} e^{-\frac{1}{2}z^2} \cdot \frac{1}{\Gamma\left(\frac{\nu}{2}\right) 2^{\frac{\nu}{2}}} y^{\frac{\nu}{2}-1} e^{-\frac{y}{2}}$$

for $y > 0$ and $-\infty < z < \infty$, and $f(y,z) = 0$ elsewhere. Then, to use the change-of-variable technique, we solve $t = \dfrac{z}{\sqrt{y/\nu}}$ for z, getting $z = t\sqrt{y/\nu}$ and hence $\dfrac{\partial z}{\partial t} = \sqrt{y/\nu}$. Thus, the joint density of Y and T is given by

$$g(y,t) = \begin{cases} \dfrac{1}{\sqrt{2\pi\nu}\,\Gamma\left(\frac{\nu}{2}\right) 2^{\frac{\nu}{2}}} y^{\frac{\nu-1}{2}} e^{-\frac{y}{2}\left(1+\frac{t^2}{\nu}\right)} & \text{for } y > 0 \text{ and } -\infty < t < \infty \\ 0 & \text{elsewhere} \end{cases}$$

and, integrating out y with the aid of the substitution $w = \dfrac{y}{2}\left(1+\dfrac{t^2}{\nu}\right)$, we finally get

$$f(t) = \frac{\Gamma\left(\frac{\nu+1}{2}\right)}{\sqrt{\pi\nu}\,\Gamma\left(\frac{\nu}{2}\right)} \cdot \left(1+\frac{t^2}{\nu}\right)^{-\frac{\nu+1}{2}} \qquad \text{for } -\infty < t < \infty$$

The t distribution was introduced originally by W. S. Gosset, who published his scientific writings under the pen name "Student," since the company for which he worked, a brewery, did not permit publication by employees. Thus, the t distribution is also known as the **Student t distribution**, or **Student's t distribution**. As shown in Figure 2, graphs of t distributions having different numbers of degrees of freedom resemble that of the standard normal distribution, but have larger variances. In fact, for large values of υ, the t distribution approaches the standard normal distribution.

In view of its importance, the t distribution has been tabulated extensively. Table IV of "Statistical Tables", for example, contains values of $t_{\alpha,\nu}$ for $\alpha = 0.10, 0.05, 0.025, 0.01, 0.005$ and $\nu = 1, 2, \ldots, 29$, where $t_{\alpha,\nu}$ is such that the area to its right under the curve of the t distribution with ν degrees of freedom (see Figure 3) is equal to α. That is, $t_{\alpha,\nu}$ is such that if T is a random variable having a t distribution with ν degrees of freedom, then

$$P(T \geqq t_{\alpha,\nu}) = \alpha$$

The table does not contain values of $t_{\alpha,\nu}$ for $\alpha > 0.50$, since the density is symmetrical about $t = 0$ and hence $t_{1-\alpha,\nu} = -t_{\alpha,\nu}$. When ν is 30 or more, probabilities related to the t distribution are usually approximated with the use of normal distributions (see Exercise 35).

Among the many applications of the t distribution, its major application (for which it was originally developed) is based on the following theorem.

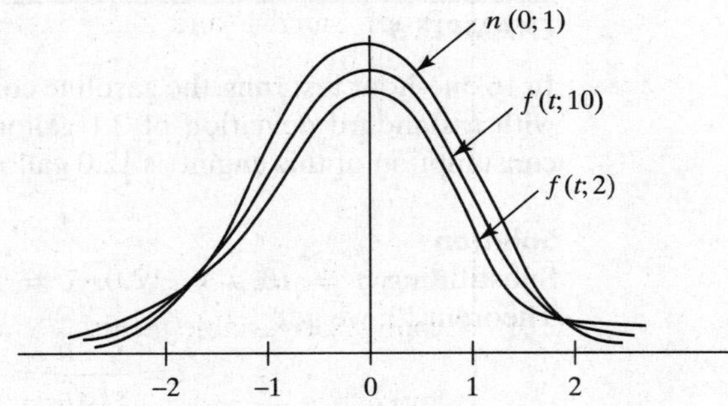

Figure 2. Comparison of t distributions and standard normal distribution.

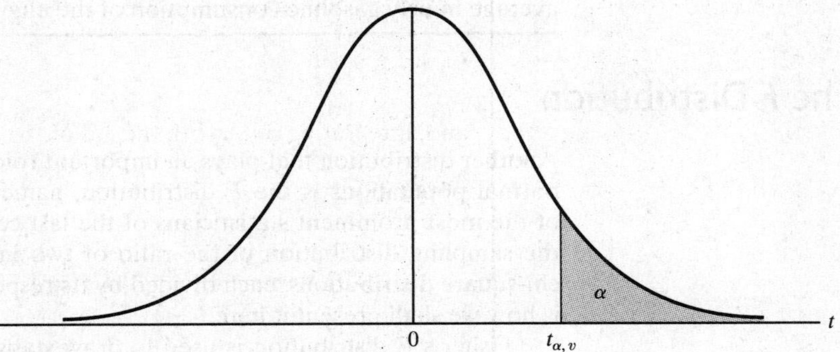

Figure 3. t distribution.

THEOREM 13. If $\overline{X}$ and S^2 are the mean and the variance of a random sample of size n from a normal population with the mean μ and the variance σ^2, then

$$T = \frac{\overline{X} - \mu}{S/\sqrt{n}}$$

has the t distribution with $n - 1$ degrees of freedom.

Proof By Theorems 11 and 4, the random variables

$$Y = \frac{(n-1)S^2}{\sigma^2} \quad \text{and} \quad Z = \frac{\overline{X} - \mu}{\sigma/\sqrt{n}}$$

have, respectively, a chi-square distribution with $n - 1$ degrees of freedom and the standard normal distribution. Since they are also independent by part 1 of Theorem 11, substitution into the formula for T of Theorem 12 yields

$$T = \frac{\dfrac{\overline{X} - \mu}{\sigma/\sqrt{n}}}{\sqrt{S^2/\sigma^2}} = \frac{\overline{X} - \mu}{S/\sqrt{n}}$$

and this completes the proof.

EXAMPLE 3

In 16 one-hour test runs, the gasoline consumption of an engine averaged 16.4 gallons with a standard deviation of 2.1 gallons. Test the claim that the average gasoline consumption of this engine is 12.0 gallons per hour.

Solution

Substituting $n = 16, \mu = 12.0, \bar{x} = 16.4$, and $s = 2.1$ into the formula for t in Theorem 13, we get

$$t = \frac{\bar{x} - \mu}{s/\sqrt{n}} = \frac{16.4 - 12.0}{2.1/\sqrt{16}} = 8.38$$

Since Table IV of "Statistical Tables" shows that for $\nu = 15$ the probability of getting a value of T greater than 2.947 is 0.005, the probability of getting a value greater than 8 must be negligible. Thus, it would seem reasonable to conclude that the true average hourly gasoline consumption of the engine exceeds 12.0 gallons.

6 The *F* Distribution

Another distribution that plays an important role in connection with sampling from normal populations is the F distribution, named after Sir Ronald A. Fisher, one of the most prominent statisticians of the last century. Originally, it was studied as the sampling distribution of the ratio of two independent random variables with chi-square distributions, each divided by its respective degrees of freedom, and this is how we shall present it here.

Fisher's F distribution is used to draw statistical inferences about the ratio of two sample variances. As such, it plays a key role in the analysis of variance, used in conjunction with experimental designs.

THEOREM 14. If U and V are independent random variables having chi-square distributions with ν_1 and ν_2 degrees of freedom, then

$$F = \frac{U/\nu_1}{V/\nu_2}$$

is a random variable having an ***F* distribution**, that is, a random variable whose probability density is given by

$$g(f) = \frac{\Gamma\left(\frac{\nu_1+\nu_2}{2}\right)}{\Gamma\left(\frac{\nu_1}{2}\right)\Gamma\left(\frac{\nu_2}{2}\right)}\left(\frac{\nu_1}{\nu_2}\right)^{\frac{\nu_1}{2}} \cdot f^{\frac{\nu_1}{2}-1}\left(1+\frac{\nu_1}{\nu_2}f\right)^{-\frac{1}{2}(\nu_1+\nu_2)}$$

for $f > 0$ and $g(f) = 0$ elsewhere.

Proof By virtue of independence, the joint density of U and V is given by

$$f(u, v) = \frac{1}{2^{\nu_1/2}\Gamma\left(\frac{\nu_1}{2}\right)} \cdot u^{\frac{\nu_1}{2}-1}e^{-\frac{u}{2}} \cdot \frac{1}{2^{\nu_2/2}\Gamma\left(\frac{\nu_2}{2}\right)} \cdot v^{\frac{\nu_2}{2}-1}e^{-\frac{v}{2}}$$

$$= \frac{1}{2^{(\nu_1+\nu_2)/2}\Gamma\left(\frac{\nu_1}{2}\right)\Gamma\left(\frac{\nu_2}{2}\right)} \cdot u^{\frac{\nu_1}{2}-1}v^{\frac{\nu_2}{2}-1}e^{-\frac{u+v}{2}}$$

for $u > 0$ and $v > 0$, and $f(u, v) = 0$ elsewhere. Then, to use the change-of-variable technique, we solve

$$f = \frac{u/\nu_1}{v/\nu_2}$$

for u, getting $u = \frac{\nu_1}{\nu_2} \cdot vf$ and hence $\frac{\partial u}{\partial f} = \frac{\nu_1}{\nu_2} \cdot v$. Thus, the joint density of F and V is given by

$$g(f, v) = \frac{\left(\frac{\nu_1}{\nu_2}\right)^{\nu_1/2}}{2^{(\nu_1+\nu_2)/2}\Gamma\left(\frac{\nu_1}{2}\right)\Gamma\left(\frac{\nu_2}{2}\right)} \cdot f^{\frac{\nu_1}{2}-1}v^{\frac{\nu_1+\nu_2}{2}-1}e^{-\frac{v}{2}\left(\frac{\nu_1 f}{\nu_2}+1\right)}$$

for $f > 0$ and $v > 0$, and $g(f, v) = 0$ elsewhere. Now, integrating out v by making the substitution $w = \frac{v}{2}\left(\frac{\nu_1 f}{\nu_2}+1\right)$, we finally get

$$g(f) = \frac{\Gamma\left(\frac{\nu_1+\nu_2}{2}\right)}{\Gamma\left(\frac{\nu_1}{2}\right)\Gamma\left(\frac{\nu_2}{2}\right)}\left(\frac{\nu_1}{\nu_2}\right)^{\frac{\nu_1}{2}} \cdot f^{\frac{\nu_1}{2}-1}\left(1+\frac{\nu_1}{\nu_2}f\right)^{-\frac{1}{2}(\nu_1+\nu_2)}$$

for $f > 0$, and $g(f) = 0$ elsewhere.

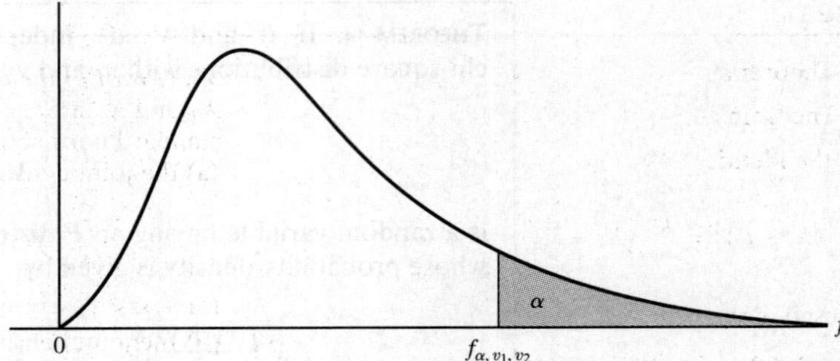

Figure 4. *F* distribution.

In view of its importance, the F distribution has been tabulated extensively. Table VI of "Statistical Tables", for example, contains values of f_{α,ν_1,ν_2} for $\alpha = 0.05$ and 0.01 and for various values of ν_1 and ν_2, where f_{α,ν_1,ν_2} is such that the area to its right under the curve of the F distribution with ν_1 and ν_2 degrees of freedom (see Figure 4) is equal to α. That is, f_{α,ν_1,ν_2} is such that

$$P(F \geqq f_{\alpha,\nu_1,\nu_2}) = \alpha$$

Applications of Theorem 14 arise in problems in which we are interested in comparing the variances σ_1^2 and σ_2^2 of two normal populations; for instance, in problems in which we want to estimate the ratio $\frac{\sigma_1^2}{\sigma_2^2}$ or perhaps to test whether $\sigma_1^2 = \sigma_2^2$. We base such inferences on **independent random samples** of sizes n_1 and n_2 from the two populations and Theorem 11, according to which

$$\chi_1^2 = \frac{(n_1 - 1)s_1^2}{\sigma_1^2} \quad \text{and} \quad \chi_2^2 = \frac{(n_2 - 1)s_2^2}{\sigma_2^2}$$

are values of random variables having chi-square distributions with $n_1 - 1$ and $n_2 - 1$ degrees of freedom. By "independent random samples," we mean that the $n_1 + n_2$ random variables constituting the two random samples are all independent, so that the two chi-square random variables are independent and the substitution of their values for U and V in Theorem 14 yields the following result.

THEOREM 15. If S_1^2 and S_2^2 are the variances of independent random samples of sizes n_1 and n_2 from normal populations with the variances σ_1^2 and σ_2^2, then

$$F = \frac{S_1^2/\sigma_1^2}{S_2^2/\sigma_2^2} = \frac{\sigma_2^2 S_1^2}{\sigma_1^2 S_2^2}$$

is a random variable having an F distribution with $n_1 - 1$ and $n_2 - 1$ degrees of freedom.

The F distribution is also known as the **variance-ratio distribution**.

Exercises

20. Prove Theorem 9.

21. Prove Theorem 10.

22. Verify the identity

$$\sum_{i=1}^{n}(X_i-\mu)^2=\sum_{i=1}^{n}(X_i-\overline{X})^2+n(\overline{X}-\mu)^2$$

which we used in the proof of Theorem 11.

23. Use Theorem 11 to show that, for random samples of size n from a normal population with the variance σ^2, the sampling distribution of S^2 has the mean σ^2 and the variance $\frac{2\sigma^4}{n-1}$. (A general formula for the variance of S^2 for random samples from any population with finite second and fourth moments may be found in the book by H. Cramér listed among the references at the end of this chapter.)

24. Show that if $X_1, X_2, \ldots, X_n$ are independent random variables having the chi-square distribution with $\nu = 1$ and $Y_n = X_1 + X_2 + \cdots + X_n$, then the limiting distribution of

$$Z=\frac{\frac{Y_n}{n}-1}{\sqrt{2/n}}$$

as $n \to \infty$ is the standard normal distribution.

25. Based on the result of Exercise 24, show that if X is a random variable having a chi-square distribution with ν degrees of freedom and ν is large, the distribution of $\frac{X-\nu}{\sqrt{2\nu}}$ can be approximated with the standard normal distribution.

26. Use the method of Exercise 25 to find the approximate value of the probability that a random variable having a chi-square distribution with $\nu = 50$ will take on a value greater than 68.0.

27. If the range of X is the set of all positive real numbers, show that for $k > 0$ the probability that $\sqrt{2X}-\sqrt{2\nu}$ will take on a value less than k equals the probability that $\frac{X-\nu}{\sqrt{2\nu}}$ will take on a value less than $k+\frac{k^2}{2\sqrt{2\nu}}$.

28. Use the results of Exercises 25 and 27 to show that if X has a chi-square distribution with ν degrees of freedom, then for large ν the distribution of $\sqrt{2X}-\sqrt{2\nu}$ can be approximated with the standard normal distribution. Also, use this method of approximation to rework Exercise 26.

29. Find the percentage errors of the approximations of Exercises 26 and 28, given that the actual value of the probability (rounded to five decimals) is 0.04596.

30. (*Proof of the independence of* $\overline{X}$ *and* S^2 *for* $n = 2$) If X_1 and X_2 are independent random variables having the standard normal distribution, show that

(a) the joint density of X_1 and $\overline{X}$ is given by

$$f(x_1,\bar{x})=\frac{1}{\pi}\cdot e^{-\bar{x}^2}e^{-(x_1-\bar{x})^2}$$

for $-\infty < x_1 < \infty$ and $-\infty < \bar{x} < \infty$;

(b) the joint density of $U = |X_1 - \overline{X}|$ and $\overline{X}$ is given by

$$g(u,\bar{x})=\frac{2}{\pi}\cdot e^{-(\bar{x}^2+u^2)}$$

for $u > 0$ and $-\infty < \bar{x} < \infty$, since $f(x_1, \bar{x})$ is symmetrical about $\bar{x}$ for fixed $\bar{x}$;

(c) $S^2 = 2(X_1 - \overline{X})^2 = 2U^2$;

(d) the joint density of $\overline{X}$ and S^2 is given by

$$h(s^2,\bar{x})=\frac{1}{\sqrt{\pi}}e^{-\bar{x}^2}\cdot\frac{1}{\sqrt{2\pi}}(s^2)^{-\frac{1}{2}}e^{-\frac{1}{2}s^2}$$

for $s^2 > 0$ and $-\infty < \bar{x} < \infty$, demonstrating that $\overline{X}$ and S^2 are independent.

31. (*Proof of the independence of* $\overline{X}$ *and* S^2) If $X_1, X_2, \ldots, X_n$ constitute a random sample from a normal population with the mean μ and the variance σ^2,

(a) find the conditional density of X_1 given $X_2 = x_2, X_3 = x_3, \ldots, X_n = x_n$, and then set $X_1 = n\overline{X} - X_2 - \cdots - X_n$ and use the transformation technique to find the conditional density of $\overline{X}$ given $X_2 = x_2, X_3 = x_3, \ldots, X_n = x_n$;

(b) find the joint density of $\overline{X}, X_2, X_3, \ldots, X_n$ by multiplying the conditional density of $\overline{X}$ obtained in part (a) by the joint density of $X_2, X_3, \ldots, X_n$, and show that

$$g(x_2,x_3,\ldots,x_n|\bar{x})=\sqrt{n}\left(\frac{1}{\sigma\sqrt{2\pi}}\right)^{n-1}e^{-\frac{(n-1)s^2}{2\sigma^2}}$$

for $-\infty < x_i < \infty, i = 2, 3, \ldots, n$;

(c) show that the conditional moment-generating function of $\frac{(n-1)S^2}{\sigma^2}$ given $\overline{X} = \bar{x}$ is

$$E\left[e^{\frac{(n-1)S^2}{\sigma^2}\cdot t}\middle|\bar{x}\right]=(1-2t)^{-\frac{n-1}{2}}\qquad\text{for } t<\frac{1}{2}$$

Since this result is free of $\bar{x}$, it follows that $\overline{X}$ and S^2 are independent; it also shows that $\frac{(n-1)S^2}{\sigma^2}$ has a chi-square distribution with $n - 1$ degrees of freedom.

This proof, due to J. Shuster, is listed among the references at the end of this chapter.

32. This question has been intentionally omitted for this edition.

33. Show that for $\nu > 2$ the variance of the t distribution with ν degrees of freedom is $\frac{\nu}{\nu - 2}$. (*Hint*: Make the substitution $1 + \frac{t^2}{\nu} = \frac{1}{u}$.)

34. Show that for the t distribution with $\nu > 4$ degrees of freedom

(a) $\mu_4 = \frac{3\nu^2}{(\nu-2)(\nu-4)}$;

(b) $\alpha_4 = 3 + \frac{6}{\nu - 4}$.

(*Hint*: Make the substitution $1 + \frac{t^2}{\nu} = \frac{1}{u}$.)

35. This question has been intentionally omitted for this edition.

36. By what name did we refer to the t distribution with $\nu = 1$ degree of freedom?

37. This question has been intentionally omitted for this edition.

38. Show that for $\nu_2 > 2$ the mean of the F distribution is $\frac{\nu_2}{\nu_2 - 2}$, making use of the definition of F in Theorem 14 and the fact that for a random variable V having the chi-square distribution with ν_2 degrees of freedom, $E\left(\frac{1}{V}\right) = \frac{1}{\nu_2 - 2}$.

39. Verify that if X has an F distribution with ν_1 and ν_2 degrees of freedom and $\nu_2 \to \infty$, the distribution of $Y = \nu_1 X$ approaches the chi-square distribution with ν_1 degrees of freedom.

40. Verify that if T has a t distribution with ν degrees of freedom, then $X = T^2$ has an F distribution with $\nu_1 = 1$ and $\nu_2 = \nu$ degrees of freedom.

41. If X has an F distribution with ν_1 and ν_2 degrees of freedom, show that $Y = \frac{1}{X}$ has an F distribution with ν_2 and ν_1 degrees of freedom.

42. Use the result of Exercise 41 to show that

$$f_{1-\alpha,\nu_1,\nu_2} = \frac{1}{f_{\alpha,\nu_2,\nu_1}}$$

43. Verify that if Y has a beta distribution with $\alpha = \frac{\nu_1}{2}$ and $\beta = \frac{\nu_2}{2}$, then

$$X = \frac{\nu_2 Y}{\nu_1(1-Y)}$$

has an F distribution with ν_1 and ν_2 degrees of freedom.

44. Show that the F distribution with 4 and 4 degrees of freedom is given by

$$g(f) = \begin{cases} 6f(1+f)^{-4} & \text{for } f > 0 \\ 0 & \text{elsewhere} \end{cases}$$

and use this density to find the probability that for independent random samples of size $n = 5$ from normal populations with the same variance, S_1^2/S_2^2 will take on a value less than $\frac{1}{2}$ or greater than 2.

7 Order Statistics

The sampling distributions presented so far in this chapter depend on the assumption that the population from which the sample was taken has the normal distribution. This assumption often is satisfied, at least approximately for large samples, as illustrated by the central limit theorem. However, small samples sometimes must be used in practice, for example in statistical quality control or where taking and measuring a sample is very expensive. In an effort to deal with the problem of small samples in cases where it may be unreasonable to assume a normal population, statisticians have developed **nonparametric statistics**, whose sampling distributions do not depend upon any assumptions about the population from which the sample is taken. Statistical inferences based upon such statistics are called **nonparametric** inference. We will identify a class of nonparametric statistics called **order statistics** and discuss their statistical properties.

Consider a random sample of size n from an infinite population with a continuous density, and suppose that we arrange the values of $X_1, X_2, \ldots,$ and X_n according to size. If we look upon the smallest of the x's as a value of the random variable Y_1, the next largest as a value of the random variable Y_2, the next largest after that as a value of the random variable $Y_3, \ldots$, and the largest as a value of the random variable Y_n, we refer to these Y's as **order statistics**. In particular, Y_1 is the first order statistic, Y_2 is the second order statistic, Y_3 is the third order statistic, and so on. (We are limiting this discussion to infinite populations with continuous densities so that there is zero probability that any two of the x's will be alike.)

To be more explicit, consider the case where $n = 2$ and the relationship between the values of the X's and the Y's is

$$y_1 = x_1 \quad \text{and} \quad y_2 = x_2 \quad \text{when} \quad x_1 < x_2$$
$$y_1 = x_2 \quad \text{and} \quad y_2 = x_1 \quad \text{when} \quad x_2 < x_1$$

Similarly, for $n = 3$ the relationship between the values of the respective random variables is

$$y_1 = x_1, \quad y_2 = x_2, \quad \text{and} \quad y_3 = x_3, \quad \text{when} \quad x_1 < x_2 < x_3$$
$$y_1 = x_1, \quad y_2 = x_3, \quad \text{and} \quad y_3 = x_2, \quad \text{when} \quad x_1 < x_3 < x_2$$
$$\cdots\cdots\cdots\cdots\cdots\cdots\cdots\cdots\cdots\cdots$$
$$y_1 = x_3, \quad y_2 = x_2, \quad \text{and} \quad y_3 = x_1, \quad \text{when} \quad x_3 < x_2 < x_1$$

Let us now derive a formula for the probability density of the rth order statistic for $r = 1, 2, \ldots, n$.

THEOREM 16. For random samples of size n from an infinite population that has the value $f(x)$ at x, the probability density of the rth order statistic Y_r is given by

$$g_r(y_r) = \frac{n!}{(r-1)!(n-r)!}\left[\int_{-\infty}^{y_r} f(x)\,dx\right]^{r-1} f(y_r)\left[\int_{y_r}^{\infty} f(x)\,dx\right]^{n-r}$$

for $-\infty < y_r < \infty$.

Proof Suppose that the real axis is divided into three intervals, one from $-\infty$ to y_r, a second from y_r to $y_r + h$ (where h is a positive constant), and the third from $y_r + h$ to ∞. Since the population we are sampling has the value $f(x)$ at x, the probability that $r - 1$ of the sample values fall into the first interval, 1 falls into the second interval, and $n - r$ fall into the third interval is

$$\frac{n!}{(r-1)!1!(n-r)!}\left[\int_{-\infty}^{y_r} f(x)\,dx\right]^{r-1}\left[\int_{y_r}^{y_r+h} f(x)\,dx\right]\left[\int_{y_r+h}^{\infty} f(x)\,dx\right]^{n-r}$$

according to the formula for the multinomial distribution. Using the mean-value theorem for integrals from calculus, we have

$$\int_{y_r}^{y_r+h} f(x)\,dx = f(\xi)\cdot h \qquad \text{where } y_r \leqq \xi \leqq y_r + h$$

and if we let $h \to 0$, we finally get

$$g_r(y_r) = \frac{n!}{(r-1)!(n-r)!}\left[\int_{-\infty}^{y_r} f(x)\,dx\right]^{r-1} f(y_r)\left[\int_{y_r}^{\infty} f(x)\,dx\right]^{n-r}$$

for $-\infty < y_r < \infty$ for the probability density of the rth order statistic.

In particular, the sampling distribution of Y_1, the smallest value in a random sample of size n, is given by

$$g_1(y_1) = n \cdot f(y_1)\left[\int_{y_1}^{\infty} f(x)\,dx\right]^{n-1} \qquad \text{for } -\infty < y_1 < \infty$$

while the sampling distribution of Y_n, the largest value in a random sample of size n, is given by

$$g_n(y_n) = n \cdot f(y_n)\left[\int_{-\infty}^{y_n} f(x)\,dx\right]^{n-1} \qquad \text{for } -\infty < y_n < \infty$$

Also, in a random sample of size $n = 2m+1$ the **sample median** $\tilde{X}$ is Y_{m+1}, whose sampling distribution is given by

$$h(\tilde{x}) = \frac{(2m+1)!}{m!m!}\left[\int_{-\infty}^{\tilde{x}} f(x)\,dx\right]^{m} f(\tilde{x})\left[\int_{\tilde{x}}^{\infty} f(x)\,dx\right]^{m} \qquad \text{for } -\infty < \tilde{x} < \infty$$

[For random samples of size $n = 2m$, the median is defined as $\frac{1}{2}(Y_m + Y_{m+1})$.]

In some instances it is possible to perform the integrations required to obtain the densities of the various order statistics; for other populations there may be no choice but to approximate these integrals by using numerical methods.

EXAMPLE 4

Show that for random samples of size n from an exponential population with the parameter θ, the sampling distributions of Y_1 and Y_n are given by

$$g_1(y_1) = \begin{cases} \dfrac{n}{\theta} \cdot e^{-ny_1/\theta} & \text{for } y_1 > 0 \\ 0 & \text{elsewhere} \end{cases}$$

and

$$g_n(y_n) = \begin{cases} \dfrac{n}{\theta} \cdot e^{-y_n/\theta}[1 - e^{-y_n/\theta}]^{n-1} & \text{for } y_n > 0 \\ 0 & \text{elsewhere} \end{cases}$$

and that, for random samples of size $n = 2m+1$ from this kind of population, the sampling distribution of the median is given by

$$h(\tilde{x}) = \begin{cases} \dfrac{(2m+1)!}{m!m!\theta} \cdot e^{-\tilde{x}(m+1)/\theta}[1 - e^{-\tilde{x}/\theta}]^{m} & \text{for } \tilde{x} > 0 \\ 0 & \text{elsewhere} \end{cases}$$

Solution

The integrations required to obtain these results are straightforward, and they will be left to the reader in Exercise 45.

The following is an interesting result about the sampling distribution of the median, which holds when the population density is continuous and nonzero at the **population median** $\tilde{\mu}$, which is such that $\int_{-\infty}^{\tilde{\mu}} f(x)\,dx = \frac{1}{2}$.

> **THEOREM 17.** For large n, the sampling distribution of the median for random samples of size $2n+1$ is approximately normal with the mean $\tilde{\mu}$ and the variance $\dfrac{1}{8[f(\tilde{\mu})]^2 n}$.

Note that for random samples of size $2n+1$ from a normal population we have $\mu = \tilde{\mu}$, so

$$f(\tilde{\mu}) = f(\mu) = \frac{1}{\sigma\sqrt{2\pi}}$$

and the variance of the median is approximately $\dfrac{\pi\sigma^2}{4n}$. If we compare this with the variance of the mean, which for random samples of size $2n+1$ from an infinite population is $\dfrac{\sigma^2}{2n+1}$, we find that for large samples from normal populations the mean is **more reliable** than the median; that is, the mean is subject to smaller chance fluctuations than the median.

Exercises

45. Verify the results of Example 4, that is, the sampling distributions of Y_1, Y_n, and $\tilde{X}$ shown there for random samples from an exponential population.

46. Find the sampling distributions of Y_1 and Y_n for random samples of size n from a continuous uniform population with $\alpha = 0$ and $\beta = 1$.

47. Find the sampling distribution of the median for random samples of size $2m+1$ from the population of Exercise 46.

48. Find the mean and the variance of the sampling distribution of Y_1 for random samples of size n from the population of Exercise 46.

49. Find the sampling distributions of Y_1 and Y_n for random samples of size n from a population having the beta distribution with $\alpha = 3$ and $\beta = 2$.

50. Find the sampling distribution of the median for random samples of size $2m+1$ from the population of Exercise 49.

51. Find the sampling distribution of Y_1 for random samples of size $n = 2$ taken
(a) without replacement from the finite population that consists of the first five positive integers;
(b) with replacement from the same population.
(*Hint*: Enumerate all possibilities.)

52. Duplicate the method used in the proof of Theorem 16 to show that the joint density of Y_1 and Y_n is given by

$$g(y_1, y_n) = n(n-1)f(y_1)f(y_n)\left[\int_{y_1}^{y_n} f(x)\,dx\right]^{n-2}$$
$$\text{for } -\infty < y_1 < y_n < \infty$$

and $g(y_1, y_n) = 0$ elsewhere.
(a) Use this result to find the joint density of Y_1 and Y_n for random samples of size n from an exponential population.
(b) Use this result to find the joint density of Y_1 and Y_n for the population of Exercise 46.

53. With reference to part (b) of Exercise 52, find the covariance of Y_1 and Y_n.

54. Use the formula for the joint density of Y_1 and Y_n shown in Exercise 52 and the transformation technique of several variables to find an expression for the joint density of Y_1 and the **sample range** $R = Y_n - Y_1$.

55. Use the result of Exercise 54 and that of part (a) of Exercise 52 to find the sampling distribution of R for random samples of size n from an exponential population.

56. Use the result of Exercise 54 to find the sampling distribution of R for random samples of size n from the continuous uniform population of Exercise 46.

57. Use the result of Exercise 56 to find the mean and the variance of the sampling distribution of R for random samples of size n from the continuous uniform population of Exercise 46.

58. There are many problems, particularly in industrial applications, in which we are interested in the proportion of a population that lies between certain limits. Such limits are called **tolerance limits**. The following steps lead to the sampling distribution of the statistic P, which is the proportion of a population (having a continuous density) that lies between the smallest and the largest values of a random sample of size n.

(a) Use the formula for the joint density of Y_1 and Y_n shown in Exercise 52 and the transformation technique of several variables to show that the joint density of Y_1 and P, whose values are given by

$$p = \int_{y_1}^{y_n} f(x)\, dx$$

is

$$h(y_1, p) = n(n-1)f(y_1)p^{n-2}$$

(b) Use the result of part (a) and the transformation technique of several variables to show that the joint density of P and W, whose values are given by

$$w = \int_{-\infty}^{y_1} f(x)\, dx$$

is

$$\varphi(w, p) = n(n-1)p^{n-2}$$

for $w > 0, p > 0, w + p < 1$, and $\varphi(w, p) = 0$ elsewhere.

(c) Use the result of part (b) to show that the marginal density of P is given by

$$g(p) = \begin{cases} n(n-1)p^{n-2}(1-p) & \text{for } 0 < p < 1 \\ 0 & \text{elsewhere} \end{cases}$$

This is the desired density of the proportion of the population that lies between the smallest and the largest values of a random sample of size n, and it is of interest to note that it does not depend on the form of the population distribution.

59. Use the result of Exercise 58 to show that, for the random variable P defined there,

$$E(P) = \frac{n-1}{n+1} \quad \text{and} \quad \text{var}(P) = \frac{2(n-1)}{(n+1)^2(n+2)}$$

What can we conclude from this about the distribution of P when n is large?

8 The Theory in Practice

More on Random Samples

While it is practically impossible to take a purely random sample, there are several methods that can be employed to assure that a sample is close enough to randomness to be useful in representing the distribution from which it came. In selecting a sample from a production line, *systematic sampling* can be used to select units at evenly spaced periods of time or having evenly spaced run numbers. In selecting a random sample from products in a warehouse, a *two-stage sampling process* can be used, numbering the containers and using a random device, such as a set of random numbers generated by a computer, to choose the containers. Then, a second set of random numbers can be used to select the unit or units in each container to be included in the sample. There are many other methods, employing mechanical devices or computer-generated random numbers, that can be used to aid in selecting a random sample.

Selection of a sample that reasonably can be regarded as random sometimes requires ingenuity, but it always requires care. Care should be taken to assure that only the specified distribution is represented. Thus, if a sample of product is meant to represent an entire production line, it should not be taken from the first shift only. Care should be taken to assure independence of the observations. Thus, the production-line sample should not be taken from a "chunk" of products produced at about the same time; they represent the same set of conditions and settings, and the resulting observations are closely related to each other. Human judgment in selecting samples usually includes personal bias, often unconscious, and such judgments should be avoided. Whenever possible, the use of mechanical devices or random numbers is preferable to methods involving personal choice.

The Assumption of Normality

It is not unusual to expect that errors are made in taking and recording observations. This phenomenon was described by early nineteenth-century astronomers who noted that different observers obtained somewhat different results when determining the location of a star.

Observational error can arise from one or both of two sources, **random error**, or statistical error, and **bias**. Random errors occur as the result of many imperfections of measurement; among these imperfections are imprecise markings on measurement scales, parallax (not viewing readings straight on) errors in setting up apparatus, slight differences in materials, expansion and contraction, minor changes in ambient conditions, and so forth. Bias occurs when there is a relatively consistent error, such as not obtaining a representative sample in a survey, using a measuring instrument that is not properly calibrated, and recording errors.

Errors involving bias can be corrected by discerning the source of the error and making appropriate “fixes” to eliminate the bias. Random error, however, is something we must live with, as no human endeavor can be made perfect in applications. Let us assume, however, that the many individual sources of random error, known or unknown, are additive. In fact this is usually the case, at least to a good approximation. Then we can write

$$X = \mu + E_1 + E_2 + \cdots + E_n$$

where the random variable X is an observed value, μ is the “true” value of the observation, and the E_i are the n random errors that affect the value of the observation. We shall assume that

$$E(X) = \mu + E(E_1) + E(E_2) + \cdots + E(E_n) = \mu$$

In other words, we are assuming that the random errors have a mean of zero, at least in the long run. We also can write

$$\text{var}(X) = (\mu + E_1 + E_2 + \cdots + E_n) = n\sigma^2$$

In other words, the variance of the *sum* of the random errors is $n\sigma^2$.

It follows that $\overline{X} = \mu + \overline{E}$, where $\overline{E}$ is the sample mean of the errors $E_1, E_2, \ldots, E_n$, and $\sigma^2_{\overline{X}} = \sigma^2/n$. The central limit theorem given by Theorem 3 allows us to conclude that

$$Z = \frac{\overline{X} - \mu}{\sigma\sqrt{n}}$$

is a random variable whose distribution as $n \to \infty$ is the standard normal distribution.

It is not difficult to see from this argument that most repeated measurements of the same phenomenon are, at least approximately, normally distributed. It is this conclusion that underscores the importance of the chi-square, t, and F distributions in applications that are based on the assumption of data from normally distributed populations. It also demonstrates why the normal distribution is of major importance in statistics.

Applied Exercises SECS. 1–3

In the following exercises it is assumed that all samples are drawn without replacement unless otherwise specified.

60. How many different samples of size $n = 3$ can be drawn from a finite population of size
(a) $N = 12$; **(b)** $N = 20$; **(c)** $N = 50$?

61. What is the probability of each possible sample if
(a) a random sample of size $n = 4$ is to be drawn from a finite population of size $N = 12$;
(b) a random sample of size $n = 5$ is to be drawn from a finite population of size $N = 22$?

62. If a random sample of size $n = 3$ is drawn from a finite population of size $N = 50$, what is the probability that a particular element of the population will be included in the sample?

63. For random samples from an infinite population, what happens to the standard error of the mean if the sample size is
(a) increased from 30 to 120;
(b) increased from 80 to 180;
(c) decreased from 450 to 50;
(d) decreased from 250 to 40?

64. Find the value of the finite population correction factor $\frac{N-n}{N-1}$ for
(a) $n = 5$ and $N = 200$;
(b) $n = 50$ and $N = 300$;
(c) $n = 200$ and $N = 800$.

65. A random sample of size $n = 100$ is taken from an infinite population with the mean $\mu = 75$ and the variance $\sigma^2 = 256$.
(a) Based on Chebyshev's theorem, with what probability can we assert that the value we obtain for $\overline{X}$ will fall between 67 and 83?
(b) Based on the central limit theorem, with what probability can we assert that the value we obtain for $\overline{X}$ will fall between 67 and 83?

66. A random sample of size $n = 81$ is taken from an infinite population with the mean $\mu = 128$ and the standard deviation $\sigma = 6.3$. With what probability can we assert that the value we obtain for $\overline{X}$ will not fall between 126.6 and 129.4 if we use
(a) Chebyshev's theorem;
(b) the central limit theorem?

67. Rework part (b) of Exercise 66, assuming that the population is not infinite but finite and of size $N = 400$.

68. A random sample of size $n = 225$ is to be taken from an exponential population with $\theta = 4$. Based on the central limit theorem, what is the probability that the mean of the sample will exceed 4.5?

69. A random sample of size $n = 200$ is to be taken from a uniform population with $\alpha = 24$ and $\beta = 48$. Based on the central limit theorem, what is the probability that the mean of the sample will be less than 35?

70. A random sample of size 64 is taken from a normal population with $\mu = 51.4$ and $\sigma = 6.8$. What is the probability that the mean of the sample will
(a) exceed 52.9;
(b) fall between 50.5 and 52.3;
(c) be less than 50.6?

71. A random sample of size 100 is taken from a normal population with $\sigma = 25$. What is the probability that the mean of the sample will differ from the mean of the population by 3 or more either way?

72. Independent random samples of sizes 400 are taken from each of two populations having equal means and the standard deviations $\sigma_1 = 20$ and $\sigma_2 = 30$. Using Chebyshev's theorem and the result of Exercise 2, what can we assert with a probability of at least 0.99 about the value we will get for $\overline{X}_1 - \overline{X}_2$? (By "independent" we mean that the samples satisfy the conditions of Exercise 2.)

73. Assume that the two populations of Exercise 72 are normal and use the result of Exercise 3 to find k such that

$$P(-k < \overline{X}_1 - \overline{X}_2 < k) = 0.99$$

74. Independent random samples of sizes $n_1 = 30$ and $n_2 = 50$ are taken from two normal populations having the means $\mu_1 = 78$ and $\mu_2 = 75$ and the variances $\sigma_1^2 = 150$ and $\sigma_2^2 = 200$. Use the results of Exercise 3 to find the probability that the mean of the first sample will exceed that of the second sample by at least 4.8.

75. The actual proportion of families in a certain city who own, rather than rent, their home is 0.70. If 84 families in this city are interviewed at random and their responses to the question of whether they own their home are looked upon as values of independent random variables having identical Bernoulli distributions with the parameter $\theta = 0.70$, with what probability can we assert that the value we obtain for the sample proportion $\hat{\Theta}$ will fall between 0.64 and 0.76, using the result of Exercise 4 and
(a) Chebyshev's theorem;
(b) the central limit theorem?

76. The actual proportion of men who favor a certain tax proposal is 0.40 and the corresponding proportion for women is 0.25; $n_1 = 500$ men and $n_2 = 400$

women are interviewed at random, and their individual responses are looked upon as the values of independent random variables having Bernoulli distributions with the respective parameters $\theta_1 = 0.40$ and $\theta_2 = 0.25$. What can we assert, according to Chebyshev's theorem, with a probability of at least 0.9375 about the value we will get for $\hat{\Theta}_1 - \hat{\Theta}_2$, the difference between the two sample proportions of favorable responses? Use the result of Exercise 5.

SECS. 4–6

(In Exercises 78 through 83, refer to Tables IV, V, and VI of "Statistical Tables.")

77. Integrate the appropriate chi-square density to find the probability that the variance of a random sample of size 5 from a normal population with $\sigma^2 = 25$ will fall between 20 and 30.

78. The claim that the variance of a normal population is $\sigma^2 = 25$ is to be rejected if the variance of a random sample of size 16 exceeds 54.668 or is less than 12.102. What is the probability that this claim will be rejected even though $\sigma^2 = 25$?

79. The claim that the variance of a normal population is $\sigma^2 = 4$ is to be rejected if the variance of a random sample of size 9 exceeds 7.7535. What is the probability that this claim will be rejected even though $\sigma^2 = 4$?

80. A random sample of size $n = 25$ from a normal population has the mean $\bar{x} = 47$ and the standard deviation $s = 7$. If we base our decision on the statistic of Theorem 13, can we say that the given information supports the conjecture that the mean of the population is $\mu = 42$?

81. A random sample of size $n = 12$ from a normal population has the mean $\bar{x} = 27.8$ and the variance $s^2 = 3.24$. If we base our decision on the statistic of Theorem 13, can we say that the given information supports the claim that the mean of the population is $\mu = 28.5$?

82. If S_1 and S_2 are the standard deviations of independent random samples of sizes $n_1 = 61$ and $n_2 = 31$ from normal populations with $\sigma_1^2 = 12$ and $\sigma_2^2 = 18$, find $P(S_1^2/S_2^2 > 1.16)$.

83. If S_1^2 and S_2^2 are the variances of independent random samples of sizes $n_1 = 10$ and $n_2 = 15$ from normal populations with equal variances, find $P(S_1^2/S_2^2 < 4.03)$.

84. Use a computer program to verify the five entries in Table IV of "Statistical Tables" corresponding to 11 degrees of freedom.

85. Use a computer program to verify the eight entries in Table V of "Statistical Tables" corresponding to 21 degrees of freedom.

86. Use a computer program to verify the five values of $f_{0.05}$ in Table VI of "Statistical Tables" corresponding to 7 and 6 to 10 degrees of freedom.

87. Use a computer program to verify the six values of $f_{0.01}$ in Table VI of "Statistical Tables" corresponding to $\nu_1 = 15$ and $\nu_2 = 12, 13, \ldots, 17$.

SEC. 7

88. Find the probability that in a random sample of size $n = 4$ from the continuous uniform population of Exercise 46, the smallest value will be at least 0.20.

89. Find the probability that in a random sample of size $n = 3$ from the beta population of Exercise 77, the largest value will be less than 0.90.

90. Use the result of Exercise 56 to find the probability that the range of a random sample of size $n = 5$ from the given uniform population will be at least 0.75.

91. Use the result of part (c) of Exercise 58 to find the probability that in a random sample of size $n = 10$ at least 80 percent of the population will lie between the smallest and largest values.

92. Use the result of part (c) of Exercise 58 to set up an equation in n whose solution will give the sample size that is required to be able to assert with probability $1 - \alpha$ that the proportion of the population contained between the smallest and largest sample values is at least p. Show that for $p = 0.90$ and $\alpha = 0.05$ this equation can be written as

$$(0.90)^{n-1} = \frac{1}{2n+18}$$

This kind of equation is difficult to solve, but it can be shown that an approximate solution for n is given by

$$\frac{1}{2} + \frac{1}{4} \cdot \frac{1+p}{1-p} \cdot \chi^2_{\alpha,4}$$

where $\chi^2_{\alpha,4}$ must be looked up in Table V of "Statistical Tables". Use this method to find an approximate solution of the equation for $p = 0.90$ and $\alpha = 0.05$.

SEC. 8

93. Cans of food, stacked in a warehouse, are sampled to determine the proportion of damaged cans. Explain why a sample that includes only the top can in each stack would not be a random sample.

94. An inspector chooses a sample of parts coming from an automated lathe by visually inspecting all parts, and then including 10 percent of the "good" parts in the sample with the use of a table of random digits.

(a) Why does this method not produce a random sample of the production of the lathe?

(b) Of what population can this be considered to be a random sample?

95. Sections of aluminum sheet metal of various lengths, used for construction of airplane fuselages, are lined up on a conveyor belt that moves at a constant speed. A sample is selected by taking whatever section is passing in front of a station at five-minute intervals. Explain why this sample may not be random; that is, it is not an accurate representation of the population of all aluminum sections.

96. A process error may cause the oxide thicknesses on the surface of a silicon wafer to be "wavy," with a constant difference between the wave heights. What precautions are necessary in taking a random sample of oxide thicknesses at various positions on the wafer to assure that the observations are independent?

References

Necessary and sufficient conditions for the strongest form of the central limit theorem for independent random variables, the *Lindeberg–Feller* conditions, are given in

FELLER, W., *An Introduction to Probability Theory and Its Applications*, Vol. I, 3rd ed. New York: John Wiley & Sons, Inc., 1968,

as well as in other advanced texts on probability theory.

Extensive tables of the normal, chi-square, F, and t distributions may be found in

PEARSON, E. S., and HARTLEY, H. O., *Biometrika Tables for Statisticians*, Vol. I. New York: John Wiley & Sons, Inc., 1968.

A general formula for the variance of the sampling distribution of the second sample moment M_2 (which differs from S^2 only insofar as we divide by n instead of $n-1$) is derived in

CRAMÉR, H., *Mathematical Methods of Statistics*. Princeton, N.J.: Princeton University Press, 1950,

and a proof of Theorem 17 is given in

WILKS, S. S., *Mathematical Statistics*. New York: John Wiley & Sons, Inc., 1962.

Proofs of the independence of $\overline{X}$ and S^2 for random samples from normal populations are given in many advanced texts on mathematical statistics. For instance, a proof based on moment-generating functions may be found in the above-mentioned book by S. S. Wilks, and a somewhat more elementary proof, illustrated for $n=3$, may be found in

KEEPING, E. S., *Introduction to Statistical Inference*. Princeton, N.J.: D. Van Nostrand Co., Inc., 1962.

The proof outlined in Exercise 48 is given in

SHUSTER, J., "A Simple Method of Teaching the Independence of $\overline{X}$ and S^2," *The American Statistician*, Vol. 27, No. 1, 1973.

Answers to Odd-Numbered Exercises

11 When we sample with replacement from a finite population, we satisfy the conditions for random sampling from an infinite population; that is, the random variables are independent and identically distributed.

17 $\mu = 13.0$; $\sigma^2 = 25.6$.

19 $s^2 = 4$.

29 21.9% and 5.53%.

47 $h(\tilde{x}) = \dfrac{(2m+1)!}{m!m!}\tilde{x}(1-\tilde{x})^m$ for $0 < x < 1$; $h(\tilde{x}) = 0$ elsewhere.

49 $g_1(y_1) = 12ny_1^2(1-y_1)(1-4y_1)^3$.

51 (a)

y_1	1	2	3	4
$g_1(y_1)$	$\frac{4}{10}$	$\frac{3}{10}$	$\frac{2}{10}$	$\frac{1}{10}$

(b)

y_1	1	2	3	4	5
$g_1(y_1)$	$\frac{9}{25}$	$\frac{7}{25}$	$\frac{5}{25}$	$\frac{3}{25}$	$\frac{1}{25}$

53 $\dfrac{1}{(n+1)^2(n+2)}$.

55 $f(R) = \dfrac{n-1}{\theta}e^{-R/\theta}[1-e^{-R/\theta}]^{n-2}$ for $R > 0$; $f(R) = 0$ elsewhere.

57 $E(R) = \dfrac{n-1}{n+1}$; $\sigma^2 = \dfrac{2(n-1)}{(n+1)^2(n+2)}$.

61 (a) $\frac{1}{495}$; **(b)** $\frac{1}{77}$.

63 (a) It is divided by 2. **(b)** It is divided by 1.5. **(c)** It is multiplied by 3. **(d)** It is multiplied by 2.5.

65 (a) 0.96; **(b)** 0.9999994.

67 0.0250.

69 0.0207.

71 0.2302.

73 4.63.

75 (a) 0.3056; **(b)** 0.7698.

77 0.216.

79 0.5.

81 $t = -1.347$; the data support the claim.

83 0.99.

89 0.851.

91 0.6242.

Decision Theory[†]

1 Introduction

In applied situations, mathematical expectations are often used as a guide in choosing among alternatives, that is, in making decisions, because it is generally considered rational to select alternatives with the "most promising" mathematical expectations—the ones that maximize expected profits, minimize expected losses, maximize expected sales, minimize expected costs, and so on.

Although this approach to decision making has great intuitive appeal, it is not without complications, for there are many problems in which it is difficult, if not impossible, to assign numerical values to the consequences of one's actions and to the probabilities of all eventualities.

EXAMPLE 1

A manufacturer of leather goods must decide whether to expand his plant capacity now or wait at least another year. His advisors tell him that if he expands now and economic conditions remain good, there will be a profit of $164,000 during the next fiscal year; if he expands now and there is a recession, there will be a loss of $40,000; if he waits at least another year and economic conditions remain good, there will be a profit of $80,000; and if he waits at least another year and there is a recession, there will be a small profit of $8,000. What should the manufacturer decide to do if he wants to minimize the expected loss during the next fiscal year and he feels that the odds are 2 to 1 that there will be a recession?

Solution
Schematically, all these "payoffs" can be represented as in the following table, where the entries are the losses that correspond to the various possibilities and, hence, gains are represented by negative numbers:

[†] Although the material in this chapter is basic to an understanding of the foundations of statistics, it is often omitted in a first course in mathematical statistics.

From Chapter 9 of *John E. Freund's Mathematical Statistics with Applications*, Eighth Edition. Irwin Miller, Marylees Miller.

	Expand now	*Delay expansion*
Economic conditions remain good	−164,000	−80,000
There is a recession	40,000	−8,000

We are working with losses here rather than profits to make this example fit the general scheme that we shall present in Sections 2 and 3.

Since the probabilities that economic conditions will remain good and that there will be a recession are, respectively, $\frac{1}{3}$ and $\frac{2}{3}$, the manufacturer's expected loss for the next fiscal year is

$$-164{,}000 \cdot \frac{1}{3} + 40{,}000 \cdot \frac{2}{3} = -28{,}000$$

if he expands his plant capacity now, and

$$-80{,}000 \cdot \frac{1}{3} + (-8{,}000) \cdot \frac{2}{3} = -32{,}000$$

if he waits at least another year. Since an expected profit (negative expected loss) of \$32,000 is preferable to an expected profit (negative expected loss) of \$28,000, it follows that the manufacturer should delay expanding the capacity of his plant.

The result at which we arrived in this example assumes that the values given in the table and also the odds for a recession are properly assessed. As the reader will be asked to show in Exercises 10 and 11, changes in these quantities can easily lead to different results.

EXAMPLE 2

With reference to Example 1, suppose that the manufacturer has no idea about the odds that there will be a recession. What should he decide to do if he is a confirmed pessimist?

Solution

Being the kind of person who always expects the worst to happen, he might argue that if he expands his plant capacity now he could lose \$40,000, if he delays expansion there would be a profit of at least \$8,000 and, hence, that he will minimize the maximum loss (or maximize the minimum profit) if he waits at least another year.

The criterion used in this example is called the **minimax criterion**, and it is only one of many different criteria that can be used in this kind of situation. One such criterion, based on optimism rather than pessimism, is referred to in Exercise 15, and another, based on the fear of "losing out on a good deal," is referred to in Exercise 16.

2 The Theory of Games

The examples of the preceding section may well have given the impression that the manufacturer is playing a game—a game between him and Nature (or call it fate or whatever "controls" whether there will be a recession). Each of the "players" has the choice of two moves: The manufacturer has the choice between actions a_1 and

a_2 (to expand his plant capacity now or to delay expansion for at least a year), and Nature controls the choice between θ_1 and θ_2 (whether economic conditions are to remain good or whether there is to be a recession). Depending on the choice of their moves, there are the "payoffs" shown in the following table:

		Player A (*The Manufacturer*)	
		a_1	a_2
Player B (*Nature*)	θ_1	$L(a_1, \theta_1)$	$L(a_2, \theta_1)$
	θ_2	$L(a_1, \theta_2)$	$L(a_2, \theta_2)$

The amounts $L(a_1, \theta_1), L(a_2, \theta_1), \ldots$, are referred to as the values of the **loss function** that characterizes the particular "game"; in other words, $L(a_i, \theta_j)$ is the loss of Player A (the amount he has to pay Player B) when he chooses alternative a_i and Player B chooses alternative θ_j. Although it does not really matter, we shall assume here that these amounts are in dollars. In actual practice, they can also be expressed in terms of any goods or services, in units of utility (desirability or satisfaction), and even in terms of life or death (as in Russian roulette or in the conduct of a war).

The analogy we have drawn here is not really farfetched; the problem of Example 2 is typical of the kind of situation treated in the **theory of games**, a relatively new branch of mathematics that has stimulated considerable interest in recent years. This theory is not limited to parlor games, as its name might suggest, but it applies to any kind of competitive situation and, as we shall see, it has led to a unified approach to solving problems of statistical inference.

To introduce some of the basic concepts of the theory of games, let us begin by explaining what we mean by a **zero-sum two-person game**. In this term, "two-person" means that there are two players (or, more generally, two parties with conflicting interests), and "zero-sum" means that whatever one player loses the other player wins. Thus, in a zero-sum game there is no "cut for the house" as in professional gambling, and no capital is created or destroyed during the course of play. Of course, the theory of games also includes games that are neither zero-sum nor limited to two players, but, as can well be imagined, such games are generally much more complicated. Exercise 27 is an example of a game that is not zero-sum.

Games are also classified according to the number of **strategies** (moves, choices, or alternatives) that each player has at his disposal. For instance, if each player has to choose one of two alternatives (as in Example 1), we say that it is a 2×2 game; if one player has 3 possible moves while the other has 4, the game is 3×4 or 4×3, as the case may be. In this section we shall consider only **finite** games, that is, games in which each player has only a finite, or fixed, number of possible moves, but later we shall consider also games where each player has infinitely many moves.

It is customary in the theory of games to refer to the two players as Player A and Player B as we did in the preceding table, but the moves (choices or alternatives) of Player A are usually labeled I, II, III, ..., instead of $a_1, a_2, a_3, \ldots$, and those of Player B are usually labeled 1, 2, 3, ..., instead of $\theta_1, \theta_2, \theta_3, \ldots$. The **payoffs**, the amounts of money or other considerations that change hands when the players choose their respective strategies, are usually shown in a table like that on this page, which is referred to as a **payoff matrix** in the theory of games. (As before, positive payoffs represent losses of Player A and negative payoffs represent losses of Player B.) Let us also add that it is always assumed in the theory of games that each player must choose a strategy without knowing what the opponent is going to do and that once a player has made a choice it cannot be changed.

> **DEFINITION 1. PAYOFF MATRIX.** *A* ***payoff*** *in game theory is the amount of money (or other numerical consideration) that changes hands when the players choose their respective strategies. Positive payoffs represent losses of Player* A *and negative payoffs represent losses of player* B. *A* ***strategy*** *is a choice of actions by either player. The matrix giving the payoff to a given player for each choice of strategy by both players is called the* ***payoff matrix***.

The objectives of the theory of games are to determine **optimum strategies** (that is, strategies that are most profitable to the respective players) and the corresponding payoff, which is called the **value** of the game.

EXAMPLE 3

Given the 2×2 zero-sum two-person game

		Player A	
		I	II
Player B	1	7	−4
	2	8	10

find the optimum strategies of Players *A* and *B* and the value of the game.

Solution

As can be seen by inspection, it would be foolish for Player *B* to choose Strategy 1, since Strategy 2 will yield more than Strategy 1 regardless of the choice made by Player *A*. In a situation like this we say that Strategy 1 is **dominated** by Strategy 2 (or that Strategy 2 **dominates** Strategy 1), and it stands to reason that any strategy that is dominated by another should be discarded. If we do this here, we find that Player *B*'s optimum strategy is Strategy 2, the only one left, and the Player *A*'s optimum strategy is Strategy I, since a loss of 8 units is obviously preferable to a loss of 10 units. Also, the value of the game, the payoff corresponding to Strategies I and 2, is 8 units.

EXAMPLE 4

Given the 3×2 zero-sum two-person game

		Player A		
		I	II	III
Player B	1	−4	1	7
	2	4	3	5

find the optimum strategies of Players *A* and *B* and the value of the game.

Solution

In this game neither strategy of Player *B* dominates the other, but the third strategy of Player *A* is dominated by each of the other two. Expressing the units as dollars, a profit of \$4 or a loss of \$1 is preferable to a loss of \$7, and a loss of \$4 or a loss of

$3 is preferable to a loss of $5. Thus, we can discard the third column of the payoff matrix and study the 2×2 game

		Player A	
		I	II
Player B	1	−4	1
	2	4	3

where now Strategy 2 of Player B dominates Strategy 1. Thus, the optimum choice of Player B is Strategy 2, the optimum choice of Player A is Strategy II (since a loss of $3 is preferable to a loss of $4), and the value of the game is $3.

The process of discarding dominated strategies can be of great help in the solution of a game (that is, in finding optimum strategies and the value of the game), but it is the exception rather than the rule that it will lead to a complete solution. Dominances may not even exist, as is illustrated by the following 3×3 zero-sum two-person game:

		Player A		
		I	II	III
Player B	1	−1	6	−2
	2	2	4	6
	3	−2	−6	12

So, we must look for other ways of arriving at optimum strategies. From the point of view of Player A, we might argue as follows: If he chooses Strategy I, the worst that can happen is that he loses $2; if he chooses Strategy II, the worst that can happen is that he loses $6; and if he chooses Strategy III, the worst that can happen is that he loses $12. Thus, he could minimize the maximum loss by choosing Strategy I.

Applying the same kind of argument to select a strategy for Player B, we find that if she chooses Strategy 1, the worst that can happen is that she loses $2; if she chooses Strategy 2, the worst that can happen is that she wins $2; and if she chooses Strategy 3, the worst that can happen is that she loses $6. Thus, she could minimize the maximum loss (or maximize the minimum gain, which is the same) by choosing Strategy 2.

DEFINITION 2. MINIMAX STRATEGY. *A strategy that minimizes the maximum loss of a player is called a* ***minimax strategy****. The choice of a minimax strategy to make a decision is called the* ***minimax criterion****.*

The selection of Strategies I and 2, the **minimax strategies**, is really quite reasonable. By choosing Strategy I, Player A makes sure that his opponent can win at most $2, and by choosing Strategy 2, Player B makes sure that she will actually win this amount. Thus $2 is the value of the game, which means that the game favors Player B, but we could make it **equitable** by charging Player B $2 for the privilege of playing the game and giving the $2 to Player A.

A very important aspect of the minimax strategies I and 2 of this example is that they are completely "spyproof" in the sense that neither player can profit from knowing the other's choice. In our example, even if Player A announced publicly that he will choose Strategy I, it would still be best for Player B to choose Strategy 2, and if Player B announced publicly that she will choose Strategy 2, it would still be best for Player A to choose Strategy I. Unfortunately, not all games are spyproof.

EXAMPLE 5

Show that the minimax strategies of Players A and B are not spyproof in the following game:

		Player A	
		I	II
Player B	1	8	−5
	2	2	6

Solution

Player A can minimize his maximum loss by choosing Strategy II, and Player B can minimize her maximum loss by choosing Strategy 2. However, if Player A knew that Player B was going to base her choice on the minimax criterion, he could switch to Strategy I and thus reduce his loss from \$6 to \$2. Of course, if Player B discovered that Player A would try to outsmart her in this way, she could in turn switch to Strategy 1 and increase her gain to \$8. In any case, the minimax strategies of the two players are not spyproof, thus leaving room for all sorts of trickery or deception.

There exists an easy way of determining for any given game whether minimax strategies are spyproof. What we have to look for are **saddle points**, that is, pairs of strategies for which the corresponding entry in the payoff matrix is the smallest value of its row and the greatest value of its column.

DEFINITION 3. SADDLE POINT. *A* ***saddle point*** *of a game is a pair of strategies for which the corresponding entry in the payoff matrix is the smallest value of its row and the greatest value of its column. A game that has a saddle point is said to be* ***strictly determined.***

In Example 5 there is no saddle point, since the smallest value of each row is also the smallest value of its column. On the other hand, in the game of Example 3 there is a saddle point corresponding to Strategies I and 2 since 8, the smallest value of the second row, is the greatest value of the first column. Also, the 3×2 game of Example 4 has a saddle point corresponding to Strategies II and 2 since 3, the smallest value of the second row, is the greatest value of the second column, and the 3×3 game on the previous page has a saddle point corresponding to Strategies I and 2 since 2, the smallest value of the second row, is the greatest value of the first column. In general, if a game has a saddle point, it is said to be **strictly determined**, and the strategies corresponding to the saddle point are spyproof (and hence optimum) minimax strategies. The fact that there can be more than one saddle point in a game is illustrated in Exercise 2; it also follows from this exercise that it does not matter in that case which of the saddle points is used to determine the optimum strategies of the two players.

If a game does not have a saddle point, minimax strategies are not spyproof, and each player can outsmart the other if he or she knows how the opponent will react in a given situation. To avoid this possibility, each player should somehow mix up his or her behavior patterns intentionally, and the best way of doing this is by introducing an element of chance into the selection of strategies.

EXAMPLE 6

With reference to the game of Example 5, suppose that Player A uses a gambling device (dice, cards, numbered slips of paper, a table of random numbers) that leads to the choice of Strategy I with probability x and to the choice of Strategy II with probability $1-x$. Find the value of x that will minimize Player A's maximum expected loss.

Solution

If Player B chooses Strategy 1, Player A can expect to lose

$$E = 8x - 5(1-x)$$

dollars, and if Player B chooses Strategy 2, Player A can expect to lose

$$E = 2x + 6(1-x)$$

dollars. Graphically, this situation is described in Figure 1, where we have plotted the lines whose equations are $E = 8x - 5(1-x)$ and $E = 2x + 6(1-x)$ for values of x from 0 to 1.

Applying the minimax criterion to the expected losses of Player A, we find from Figure 1 that the greater of the two values of E for any given value of x is smallest where the two lines intersect, and to find the corresponding value of x, we have only to solve the equation

$$8x - 5(1-x) = 2x + 6(1-x)$$

which yields $x = \frac{11}{17}$. Thus, if Player A uses 11 slips of paper numbered I and 6 slips of paper numbered II, shuffles them thoroughly, and then acts according to which kind he randomly draws, he will be holding his maximum expected loss down to $8 \cdot \frac{11}{17} - 5 \cdot \frac{6}{17} = 3\frac{7}{17}$, or \$3.41 to the nearest cent.

As far as Player B of the preceding example is concerned, in Exercise 22 the reader will be asked to use a similar argument to show that Player B will maximize her minimum gain (which is the same as minimizing her maximum loss) by choosing between Strategies 1 and 2 with respective probabilities of $\frac{4}{17}$ and $\frac{13}{17}$ and that she will thus assure for herself an expected gain of $3\frac{7}{17}$, or \$3.41 to the nearest cent. Incidentally, the \$3.41 to which Player A can hold down his expected loss and Player B can raise her expected gain is called the value of this game.

DEFINITION 4. RANDOMIZED STRATEGY. *If a player's choice of strategy is left to chance, the overall strategy is called a* ***randomized strategy****, or a* ***mixed strategy****. By contrast, in a game where each player makes a definite choice of a given strategy, each strategy is called a* ***pure strategy****.*

The examples of this section were all given without any "physical" interpretation because we were interested only in introducing some of the basic concepts of the theory of games. If we apply these methods to Example 1, we find that the "game" has a saddle point and that the manufacturer's minimax strategy is to delay expanding the capacity of his plant. Of course, this assumes, questionably so, that Nature (which controls whether there is going to be a recession) is a malevolent opponent.

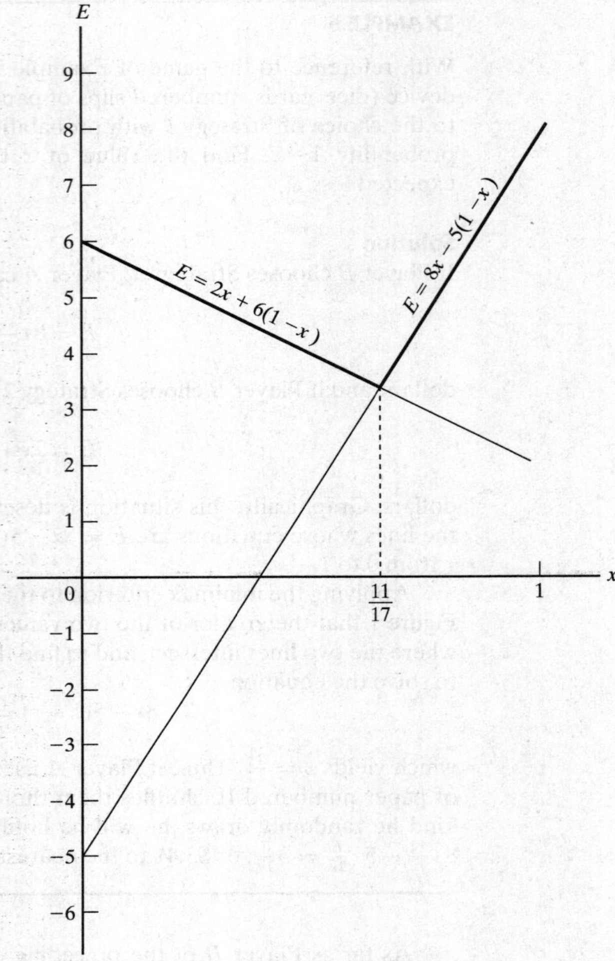

Figure 1. Diagram for Example 6.

Also, it would seem that in a situation like this the manufacturer ought to have some idea about the chances for a recession and hence that the problem should be solved by the first method of Section 1.

Exercises

1. An $n \times n$ matrix is called a *Latin square* if each row and each column contains the integers $1, 2, \ldots, n$. The following is an example of a 3×3 Latin square.

$$\begin{vmatrix} 1 & 2 & 3 \\ 2 & 3 & 1 \\ 3 & 1 & 2 \end{vmatrix}$$

Show that any strategy is the minimax strategy for either player in a game whose payoff matrix is an $n \times n$ Latin square. What is the value of the game?

2. If a zero-sum two-person game has a saddle point corresponding to the ith row and the jth column of the payoff matrix and another corresponding to the kth row and the lth column, show that

(a) there are also saddle points corresponding to the ith row and the lth column of the payoff matrix and the kth row and the jth column;

(b) the payoff must be the same for all four saddle points.

3 Statistical Games

In statistical inference we base decisions about populations on sample data, and it is by no means farfetched to look upon such an inference as a game between Nature, which controls the relevant feature (or features) of the population, and the person (scientist or statistician) who must arrive at some decision about Nature's choice. For instance, if we want to estimate the mean μ of a normal population on the basis of a random sample of size n, we could say that Nature has control over the "true" value of μ. On the other hand, we might estimate μ in terms of the value of the sample mean or that of the sample median, and presumably there is some penalty or reward that depends on the size of our error.

In spite of the obvious similarity between this problem and the ones of the preceding section, there are essentially two features in which **statistical games** are different. First, there is the question that we already met when we tried to apply the theory of games to the decision problem of Example 1, that is, the question of whether it is reasonable to treat Nature as a malevolent opponent. Obviously not, but this does not simplify matters; if we could treat Nature as a rational opponent, we would know, at least, what to expect.

The other distinction is that in the games of Section 2 each player had to choose his strategy without any knowledge of what his opponent had done or was planning to do, whereas in a statistical game the statistician is supplied with sample data that provide him with some information about Nature's choice. This also complicates matters, but it merely amounts to the fact that we are dealing with more complicated kinds of games. To illustrate, let us consider the following decision problem: *We are told that a coin is either balanced with heads on one side and tails on the other or two-headed. We cannot inspect the coin, but we can flip it once and observe whether it comes up heads or tails. Then we must decide whether or not it is two-headed, keeping in mind that there is a penalty of* \$1 *if our decision is wrong and no penalty (or reward) if our decision is right.* If we ignored the fact that we can observe one flip of the coin, we could treat the problem as the following game:

		Player A (The Statistician) a_1	a_2
Player B (Nature)	θ_1	$L(a_1, \theta_1) = 0$	$L(a_2, \theta_1) = 1$
	θ_2	$L(a_1, \theta_2) = 1$	$L(a_2, \theta_2) = 0$

which should remind the reader of the scheme in Section 2. Now, θ_1 is the "state of Nature" that the coin is two-headed, θ_2 is the "state of Nature" that the coin is balanced with heads on one side and tails on the other, a_1 is the statistician's decision that the coin is two-headed, and a_2 is the statistician's decision that the coin is balanced with heads on one side and tails on the other. The entries in the table are the corresponding values of the given loss function.

Now let us consider also the fact that we (Player A, or the statistician) know what happened in the flip of the coin; that is, we know whether a random variable X has taken on the value $x = 0$ (heads) or $x = 1$ (tails). Since we shall want to make use of this information in choosing between a_1 and a_2, we need a function, a **decision function**, that tells us what action to take when $x = 0$ and what action to take when $x = 1$.

> **DEFINITION 5. DECISION FUNCTION.** *The function that tells the statistician which decision to make for each action of Nature is called the* ***decision function*** *of a statistical game. The values of this function are given by* $d_i(x)$, *where* d_i *refers to the ith decision made by the statistician and x is a value of the random variable X whose values give the actions that can be taken by Nature.*

One possibility is to choose a_1 when $x = 0$ and a_2 when $x = 1$, and we can express this symbolically by writing

$$d_1(x) = \begin{cases} a_1 & \text{when } x = 0 \\ a_2 & \text{when } x = 1 \end{cases}$$

or, more simply, $d_1(0) = a_1$ and $d_1(1) = a_2$. The purpose of the subscript is to distinguish this decision function from others, for instance, from

$$d_2(0) = a_1 \quad \text{and} \quad d_2(1) = a_1$$

which tells us to choose a_1 regardless of the outcome of the experiment, from

$$d_3(0) = a_2 \quad \text{and} \quad d_3(1) = a_2$$

which tells us to choose a_2 regardless of the outcome of the experiment, and from

$$d_4(0) = a_2 \quad \text{and} \quad d_4(1) = a_1$$

which tells us to choose a_2 when $x = 0$ and a_1 when $x = 1$.

To compare the merits of all these decision functions, let us first determine the expected losses to which they lead for the various strategies of Nature.

> **DEFINITION 6. RISK FUNCTION.** *The function that gives the expected loss to which each value of the decision function leads for each action of Nature is called the* ***risk function****. This function is given by*
>
> $$R(d_i, \theta_j) = E\{L[d_i(X), \theta_j]\}$$
>
> *where the expectation is taken with respect to the random variable X.*

Since the probabilities for $x = 0$ and $x = 1$ are, respectively, 1 and 0 for θ_1, and $\frac{1}{2}$ and $\frac{1}{2}$ for θ_2, we get

$$R(d_1, \theta_1) = 1 \cdot L(a_1, \theta_1) + 0 \cdot L(a_2, \theta_1) = 1 \cdot 0 + 0 \cdot 1 = 0$$

$$R(d_1, \theta_2) = \frac{1}{2} \cdot L(a_1, \theta_2) + \frac{1}{2} \cdot L(a_2, \theta_2) = \frac{1}{2} \cdot 1 + \frac{1}{2} \cdot 0 = \frac{1}{2}$$

$$R(d_2, \theta_1) = 1 \cdot L(a_1, \theta_1) + 0 \cdot L(a_1, \theta_1) = 1 \cdot 0 + 0 \cdot 0 = 0$$

$$R(d_2, \theta_2) = \frac{1}{2} \cdot L(a_1, \theta_2) + \frac{1}{2} \cdot L(a_1, \theta_2) = \frac{1}{2} \cdot 1 + \frac{1}{2} \cdot 1 = 1$$

$$R(d_3, \theta_1) = 1 \cdot L(a_2, \theta_1) + 0 \cdot L(a_2, \theta_1) = 1 \cdot 1 + 0 \cdot 1 = 1$$

$$R(d_3, \theta_2) = \frac{1}{2} \cdot L(a_2, \theta_2) + \frac{1}{2} \cdot L(a_2, \theta_2) = \frac{1}{2} \cdot 0 + \frac{1}{2} \cdot 0 = 0$$

$$R(d_4, \theta_1) = 1 \cdot L(a_2, \theta_1) + 0 \cdot L(a_1, \theta_1) = 1 \cdot 1 + 0 \cdot 0 = 1$$

$$R(d_4, \theta_2) = \frac{1}{2} \cdot L(a_2, \theta_2) + \frac{1}{2} \cdot L(a_1, \theta_2) = \frac{1}{2} \cdot 0 + \frac{1}{2} \cdot 1 = \frac{1}{2}$$

where the values of the loss function were obtained from the table under Section 3.

We have thus arrived at the following 4×2 zero-sum two-person game, in which the payoffs are the corresponding values of the risk function:

		Player A (*The Statistician*)			
		d_1	d_2	d_3	d_4
Player B (*Nature*)	θ_1	0	0	1	1
	θ_2	$\frac{1}{2}$	1	0	$\frac{1}{2}$

As can be seen by inspection, d_2 is dominated by d_1 and d_4 is dominated by d_3, so that d_2 and d_4 can be discarded; in decision theory we say that they are **inadmissible**. Actually, this should not come as a surprise, since in d_2 as well as d_4 we accept alternative a_1 (that the coin is two-headed) even though it came up tails.

This leaves us with the 2×2 zero-sum two-person game in which Player A has to choose between d_1 and d_3. It can easily be verified that if Nature is looked upon as a malevolent opponent, the optimum strategy is to randomize between d_1 and d_3 with respective probabilities of $\frac{2}{3}$ and $\frac{1}{3}$, and the value of the game (the expected risk) is $\frac{1}{3}$ of a dollar. If Nature is not looked upon as a malevolent opponent, some other criterion will have to be used for choosing between d_1 and d_3, and this will be discussed in the sections that follow. Incidentally, we formulated this problem with reference to a two-headed coin and an ordinary coin, but we could just as well have formulated it more abstractly as a decision problem in which we must decide on the basis of a single observation whether a random variable has the Bernoulli distribution with the parameter $\theta = 0$ or the parameter $\theta = \frac{1}{2}$.

To illustrate further the concepts of a loss function and a risk function, let us consider the following example, in which Nature as well as the statistician has a continuum of strategies.

EXAMPLE 7

A random variable has the uniform density

$$f(x) = \begin{cases} \frac{1}{\theta} & \text{for } 0 < x < \theta \\ 0 & \text{elsewhere} \end{cases}$$

and we want to estimate the parameter θ (the "move" of Nature) on the basis of a single observation. If the decision function is to be of the form $d(x) = kx$, where $k \geqq 1$, and the losses are proportional to the absolute value of the errors, that is,

$$L(kx, \theta) = c|kx - \theta|$$

where c is a positive constant, find the value of k that will minimize the risk.

Solution
For the risk function we get

$$R(d,\theta) = \int_0^{\theta/k} c(\theta - kx)\cdot\frac{1}{\theta}\,dx + \int_{\theta/k}^{\theta} c(kx-\theta)\cdot\frac{1}{\theta}\,dx$$
$$= c\theta\left(\frac{k}{2} - 1 + \frac{1}{k}\right)$$

and there is nothing we can do about the factor θ; but it can easily be verified that $k = \sqrt{2}$ will minimize $\frac{k}{2} - 1 + \frac{1}{k}$. Thus, if we actually took the observation and got $x = 5$, our estimate of θ would be $5\sqrt{2}$, or approximately 7.07.

4 Decision Criteria

In Example 7 we were able to find a decision function that minimized the risk regardless of the true state of Nature (that is, regardless of the true value of the parameter θ), but this is the exception rather than the rule. Had we not limited ourselves to decision functions of the form $d(x) = kx$, then the decision function given by $d(x) = \theta_1$ would be best when θ happens to equal θ_1, the one given by $d(x) = \theta_2$ would be best when θ happens to equal $\theta_2, \ldots$, and it is obvious that there can be no decision function that is best for all values of θ.

In general, we thus have to be satisfied with decision functions that are best only with respect to some criterion, and the two criteria that we shall study in this chapter are (1) the **minimax criterion**, according to which we choose the decision function d for which $R(d,\theta)$, maximized with respect to θ, is a minimum; and (2) the **Bayes criterion**.

DEFINITION 7. BAYES RISK. *If Θ is assumed to be a random variable having a given distribution, the quantity*

$$E[R(d,\Theta)]$$

*where the expectation is taken with respect to Θ, is called the **Bayes risk**. Choosing the decision function* d *for which the Bayes risk is a minimum is called the **Bayes criterion**.*

It is of interest to note that in the example of Section 1 we used both of these criteria. When we quoted odds for a recession, we assigned probabilities to the two states of Nature, θ_1 and θ_2, and when we suggested that the manufacturer minimize his expected loss, we suggested, in fact, that he use the Bayes criterion. Also, when we asked in Section 2 what the manufacturer might do if he were a confirmed pessimist, we suggested that he would protect himself against the worst that can happen by using the minimax criterion.

5 The Minimax Criterion

If we apply the minimax criterion to the illustration of Section 3, dealing with the coin that is either two-headed or balanced with heads on one side and tails on the other, we find from the table on the previous page with d_2 and d_4 deleted that for d_1 the maximum risk is $\frac{1}{2}$, for d_3 the maximum risk is 1, and, hence, the one that minimizes the maximum risk is d_1.

EXAMPLE 8

Use the minimax criterion to estimate the parameter θ of a binomial distribution on the basis of the random variable X, the observed number of successes in n trials, when the decision function is of the form

$$d(x) = \frac{x+a}{n+b}$$

where a and b are constants, and the loss function is given by

$$L\left(\frac{x+a}{n+b}, \theta\right) = c\left(\frac{x+a}{n+b} - \theta\right)^2$$

where c is a positive constant.

Solution

The problem is to find the values of a and b that will minimize the corresponding risk function after it has been maximized with respect to θ. After all, we have control over the choice of a and b, while Nature (our presumed opponent) has control over the choice of θ.

Since $E(X) = n\theta$ and $E(X^2) = n\theta(1-\theta+n\theta)$ it follows that

$$R(d,\theta) = E\left[c\left(\frac{X+a}{n+b} - \theta\right)^2\right]$$
$$= \frac{c}{(n+b)^2}[\theta^2(b^2-n) + \theta(n-2ab) + a^2]$$

and, using calculus, we could find the value of θ that maximizes this expression and then minimize $R(d,\theta)$ for this value of θ with respect to a and b. This is not particularly difficult, but it is left to the reader in Exercise 6 as it involves some tedious algebraic detail.

To simplify the work in a problem of this kind, we can often use the **equalizer principle**, according to which (under fairly general conditions) the risk function of a minimax decision rule is a constant; for instance, it tells us that in Example 8 the risk function should not depend on the value of θ.† To justify this principle, at least intuitively, observe that in Example 6 the minimax strategy of Player A leads to an expected loss of \$3.41 regardless of whether Player B chooses Strategy 1 or Strategy 2.

To make the risk function of Example 8 independent of θ, the coefficients of θ and θ^2 must both equal 0 in the expression for $R(d,\theta)$. This yields $b^2 - n = 0$ and $n - 2ab = 0$, and, hence, $a = \frac{1}{2}\sqrt{n}$ and $b = \sqrt{n}$. Thus, the minimax decision function is given by

$$d(x) = \frac{x + \frac{1}{2}\sqrt{n}}{n+\sqrt{n}}$$

†The exact conditions under which the equalizer principle holds are given in the book by T. S. Ferguson listed among the references at the end of this chapter.

and if we actually obtained 39 successes in 100 trials, we would estimate the parameter θ of this binomial distribution as

$$d(39) = \frac{39 + \frac{1}{2}\sqrt{100}}{100 + \sqrt{100}} = 0.40$$

6 The Bayes Criterion

To apply the Bayes criterion in the illustration of Section 3, the one dealing with the coin that is either two-headed or balanced with heads on one side and tails on the other, we will have to assign probabilities to the two strategies of Nature, θ_1 and θ_2. If we assign θ_1 and θ_2, respectively, the probabilities p and $1 - p$, it can be seen from the second table in Section 3 that for d_1 the Bayes risk is

$$0 \cdot p + \frac{1}{2} \cdot (1 - p) = \frac{1}{2} \cdot (1 - p)$$

and that for d_3 the Bayes risk is

$$1 \cdot p + 0 \cdot (1 - p) = p$$

It follows that the Bayes risk of d_1 is less than that of d_3 (and d_1 is to be preferred to d_3) when $p > \frac{1}{3}$ and that the Bayes risk of d_3 is less than that of d_1 (and d_3 is to be preferred to d_1) when $p < \frac{1}{3}$. When $p = \frac{1}{3}$, the two Bayes risks are equal, and we can use either d_1 or d_3.

EXAMPLE 9

With reference to Example 7, suppose that the parameter of the uniform density is looked upon as a random variable with the probability density

$$h(\theta) = \begin{cases} \theta \cdot e^{-\theta} & \text{for } \theta > 0 \\ 0 & \text{elsewhere} \end{cases}$$

If there is no restriction on the form of the decision function and the loss function is quadratic, that is, its values are given by

$$L[d(x), \theta] = c\{d(x) - \theta\}^2$$

find the decision function that minimizes the Bayes risk.

Solution

Since Θ is now a random variable, we look upon the original probability density as the conditional density

$$f(x|\theta) = \begin{cases} \frac{1}{\theta} & \text{for } 0 < x < \theta \\ 0 & \text{elsewhere} \end{cases}$$

and, letting $f(x, \theta) = f(x|\theta) \cdot h(\theta)$ we get

$$f(x,\theta) = \begin{cases} e^{-\theta} & \text{for } 0 < x < \theta \\ 0 & \text{elsewhere} \end{cases}$$

As the reader will be asked to verify in Exercise 8, this yields

$$g(x) = \begin{cases} e^{-x} & \text{for } x > 0 \\ 0 & \text{elsewhere} \end{cases}$$

for the marginal density of X and

$$\varphi(\theta|x) = \begin{cases} e^{x-\theta} & \theta > x \\ 0 & \text{elsewhere} \end{cases}$$

for the conditional density of Θ given $X = x$.

Now, the Bayes risk $E[R(d, \Theta)]$ that we shall want to minimize is given by the double integral

$$\int_0^\infty \left\{ \int_0^\theta c[d(x) - \theta]^2 f(x|\theta)\, dx \right\} h(\theta)\, d\theta$$

which can also be written as

$$\int_0^\infty \left\{ \int_0^\infty c[d(x) - \theta]^2 \varphi(\theta|x)\, d\theta \right\} g(x)\, dx$$

making use of the fact that $f(x|\theta) \cdot h(\theta) = \varphi(\theta|x) \cdot g(x)$ and changing the order of integration. To minimize this double integral, we must choose $d(x)$ for each x so that the integral

$$\int_x^\infty c[d(x) - \theta]^2 \varphi(\theta|x)\, d\theta = \int_x^\infty c[d(x) - \theta]^2 e^{x-\theta}\, d\theta$$

is as small as possible. Differentiating with respect to $d(x)$ and putting the derivative equal to 0, we get

$$2ce^x \cdot \int_x^\infty [d(x) - \theta]e^{-\theta}\, d\theta = 0$$

This yields

$$d(x) \cdot \int_x^\infty e^{-\theta}\, d\theta - \int_x^\infty \theta e^{-\theta}\, d\theta = 0$$

and, finally,

$$d(x) = \frac{\int_x^\infty \theta e^{-\theta}\, d\theta}{\int_x^\infty e^{-\theta}\, d\theta} = \frac{(x+1)e^{-x}}{e^{-x}} = x + 1$$

Thus, if the observation we get is $x = 5$, this decision function gives the Bayes estimate $5 + 1 = 6$ for the parameter of the original uniform density.

Exercises

3. With reference to the illustration in Section 3, show that even if the coin is flipped n times, there are only two admissible decision functions. Also, construct a table showing the values of the risk function corresponding to these two decision functions and the two states of Nature.

4. With reference to Example 7, show that if the losses are proportional to the squared errors instead of their absolute values, the risk function becomes

$$R(d,\theta) = \frac{c\theta^2}{3}(k^2 - 3k + 3)$$

and its minimum is at $k = \frac{3}{2}$.

5. A statistician has to decide on the basis of a single observation whether the parameter θ of the density

$$f(x) = \begin{cases} \frac{2x}{\theta^2} & \text{for } 0 < x < \theta \\ 0 & \text{elsewhere} \end{cases}$$

equals θ_1 or θ_2, where $\theta_1 < \theta_2$. If he decides on θ_1 when the observed value is less than the constant k, on θ_2 when the observed value is greater than or equal to the constant k, and he is fined C dollars for making the wrong decision, which value of k will minimize the maximum risk?

6. Find the value of θ that maximizes the risk function of Example 8, and then find the values of a and b that minimize the risk function for that value of θ. Compare the results with those given in Section 6.

7. If we assume in Example 8 that Θ is a random variable having a uniform density with $\alpha = 0$ and $\beta = 1$, show that the Bayes risk is given by

$$\frac{c}{(n+b)^2}\left[\frac{1}{3}(b^2 - n) + \frac{1}{2}(n - 2ab) + a^2\right]$$

Also show that this Bayes risk is a minimum when $a = 1$ and $b = 2$, so that the optimum Bayes decision rule is given by $d(x) = \frac{x+1}{n+2}$.

8. Verify the results given on the previous page for the marginal density of X and the conditional density of Θ given $X = x$.

9. Suppose that we want to estimate the parameter θ of the geometric distribution on the basis of a single observation. If the loss function is given by

$$L[d(x),\theta] = c\{d(x) - \theta\}^2$$

and Θ is looked upon as a random variable having the uniform density $h(\theta) = 1$ for $0 < \theta < 1$ and $h(\theta) = 0$ elsewhere, duplicate the steps in Example 9 to show that

(a) the conditional density of Θ given $X = x$ is

$$\varphi(\theta|x) = \begin{cases} x(x+1)\theta(1-\theta)^{x-1} & \text{for } 0 < \theta < 1 \\ 0 & \text{elsewhere} \end{cases}$$

(b) the Bayes risk is minimized by the decision function

$$d(x) = \frac{2}{x+2}$$

(*Hint*: Make use of the fact that the integral of any beta density is equal to 1.)

7 The Theory in Practice

When Prof. A. Wald (1902–1950) first developed the ideas of decision theory, it was intended to deal with the assumption of normality and the arbitrariness of the choice of levels of significance in statistical testing of hypotheses. However, statistical decision theory requires the choice of a loss function as well as a decision criterion, and sometimes the mathematics can be cumbersome. Perhaps it is for these reasons that decision theory is not often employed in applications. However, this theory is a remarkable contribution to statistical thinking and, in the opinion of the authors, it should be used more often.

In this section we offer an example of how some of the ideas of decision theory can be used in **acceptance sampling**. Acceptance sampling is a process whereby a random sample is taken from a lot of manufactured product, and the units in the sample are inspected to make a decision whether to accept or reject the lot. If the number of defective units in the sample exceeds a certain limit (the "acceptance number"), the entire lot is rejected, otherwise it is accepted and sent to the warehouse, or to a distributor for eventual sale. If the lot is "rejected," it is rarely scrapped; instead it is "detailed," that is, it is inspected further and efforts are made to cull out the defective units. The following example shows how elements of decision theory can be applied to such a process.

EXAMPLE 10

Suppose a manufacturer incurs warranty costs of C_w for every defective unit shipped and it costs C_d to detail an entire lot. The sampling inspection procedure is to inspect n items chosen at random from a lot containing N units, and to make the decision to accept or reject on the basis of the number of defective units found in the sample. Two strategies are to be compared, as follows:

Number of Sample Defectives, x	*Strategy 1*	*Strategy 2*
0	Accept	Accept
1	Accept	Reject
2	Accept	Reject
3 or more	Reject	Reject

In other words, the acceptance number is 2 under the first strategy, and 0 under the second.

(a) Find the risk function for these two strategies.

(b) Under what conditions is either strategy preferable?

Solution

The decision function d_1 accepts the lot if x, the number of defective units found in the sampling inspection, does not exceed 2, and rejects the lot otherwise. The decision function d_2 accepts the lot if $x = 0$ and rejects it otherwise. Thus, the loss functions are

$$\begin{aligned} L(d_1, \theta) &= C_w \cdot x \cdot P(x = 0, 1, 2|\theta) + C_d \cdot P(x > 2|\theta) \\ &= C_w \cdot x \cdot B(2; n, \theta) + C_d \cdot [1 - B(2; n, \theta)] \\ L(d_2, \theta) &= C_w \cdot x \cdot P(x = 0|\theta) + C_d \cdot P(x > 0|\theta) \\ &= C_w \cdot x \cdot B(0; n, \theta) + C_d \cdot [1 - B(0; n, \theta)] \end{aligned}$$

where $B(x; n, \theta)$ represents the cumulative binomial distribution having the parameters n and θ. The corresponding risk functions are found by taking the expected values of the loss functions with respect to x, obtaining

$$\begin{aligned} R(d_1, \theta) &= C_w \cdot n\theta \cdot B(2; n, \theta) + C_d \cdot [1 - B(2; n, \theta)] \\ R(d_2, \theta) &= C_w \cdot n\theta \cdot B(0; n, \theta) + C_d \cdot [1 - B(0; n, \theta)] \end{aligned}$$

Either the minimax or the Bayes criterion could be used to choose between the two decision functions. However, if we use the minimax criterion, we need to maximize the risk functions with respect to θ and then minimize the results. This is a somewhat daunting task for this example, and we shall not attempt it here. On the other hand, use of the Bayes criterion requires that we assume a prior distribution for θ, thus introducing a new assumption that may not be warranted. It is not too difficult, however, to examine the difference between the two risk functions as a function of θ and to determine for which values of θ one is associated with less risk than the other. Experience with the proportions of defectives in prior lots can guide us in determining for which "reasonable" values of θ we should compare the two risks.

To illustrate, suppose the sample size is chosen to be $n = 10$, the warranty cost per defective unit shipped is $C_w = \$100$, and the cost of detailing a rejected lot is $C_d = \$2{,}000$. The risk functions become

$$R(d_1, \theta) = 1{,}000 \cdot \theta \cdot B(2; 10, \theta) + 2{,}000 \cdot [1 - B(2; 10, \theta)]$$

$$R(d_2, \theta) = 1{,}000 \cdot \theta \cdot B(0; 10, \theta) + 2{,}000 \cdot [1 - B(0; 10, \theta)]$$

Collecting coefficients of $B(2; 10, \theta)$ in the first equation and $B(2; 10, \theta)$ in the second, then subtracting, we obtain

$$\delta(\theta) = R(d_1, \theta) - R(d_2, \theta) = (1{,}000\theta - 2{,}000)[B(2; 10, \theta) - B(0; 10, \theta)]$$

Since $\theta \leq 1$, the quantity $(1{,}000\theta - 2{,}000) \leq 0$. Also, it is straight forward to show that $B(2; 10, \theta) \geq B(0; 10, \theta)$. Thus, $\delta(\theta)$ is never positive and, since the risk for Strategy 1 is less than or equal to that for Strategy 2 for all values of θ, we choose Strategy 1, for which the acceptance number is 2.

Applied Exercises SECS. 1–2

10. With reference to Example 1, what decision would minimize the manufacturer's expected loss if he felt that
(a) the odds for a recession are 3 to 2;
(b) the odds for a recession are 7 to 4?

11. With reference to Example 1, would the manufacturer's decision remain the same if
(a) the \$164,000 profit is replaced by a \$200,000 profit and the odds are 2 to 1 that there will be a recession;
(b) the \$40,000 loss is replaced by a \$60,000 loss and the odds are 3 to 2 that there will be a recession?

12. Ms. Cooper is planning to attend a convention in Honolulu, and she must send in her room reservation immediately. The convention is so large that the activities are held partly in Hotel X and partly in Hotel Y, and Ms. Cooper does not know whether the particular session she wants to attend will be held at Hotel X or Hotel Y. She is planning to stay only one night, which would cost her \$66.00 at Hotel X and \$62.40 at Hotel Y, and it will cost her an extra \$6.00 for cab fare if she stays at the wrong hotel.
(a) If Ms. Cooper feels that the odds are 3 to 1 that the session she wants to attend will be held at Hotel X, where should she make her reservation so as to minimize her expected cost?
(b) If Ms. Cooper feels that the odds are 5 to 1 that the session she wants to attend will be held at Hotel X, where should she make her reservation so as to minimize her expected cost?

13. A truck driver has to deliver a load of lumber to one of two construction sites, which are, respectively, 27 and 33 miles from the lumberyard, but he has misplaced the order telling him where the load of lumber should go. The two construction sites are 12 miles apart, and, to complicate matters, the telephone at the lumberyard is out of order. Where should he go first if he wants to minimize the distance he can expect to drive and he feels that
(a) the odds are 5 to 1 that the lumber should go to the construction site that is 33 miles from the lumberyard;
(b) the odds are 2 to 1 that the lumber should go to the construction site that is 33 miles from the lumberyard;
(c) the odds are 3 to 1 that the lumber should go to the construction site that is 33 miles from the lumberyard?

14. Basing their decisions on pessimism as in Example 2, where should
(a) Ms. Cooper of Exercise 12 make her reservation;
(b) the truck driver of Exercise 13 go first?

15. Basing their decisions on optimism (that is, maximizing maximum gains or minimizing minimum losses), what decisions should be reached by
(a) the manufacturer of Example 1;
(b) Ms. Cooper of Exercise 12;
(c) the truck driver of Exercise 13?

16. Suppose that the manufacturer of Example 1 is the kind of person who always worries about losing out on a good deal. For instance, he finds that if he delays expansion and economic conditions remain good, he will lose out by \$84,000 (the difference between the \$164,000 profit that he would have made if he had decided to

expand right away and the $80,000 profit that he will actually make). Referring to this quantity as an **opportunity loss**, or **regret**, find
(a) the opportunity losses that correspond to the other three possibilities;
(b) the decision that would minimize the manufacturer's maximum loss of opportunity.

17. With reference to the definition of Exercise 16, find the decisions that will minimize the maximum opportunity loss of
(a) Ms. Cooper of Exercise 12;
(b) the truck driver of Exercise 13.

18. With reference to Example 1, suppose that the manufacturer has the option of hiring an infallible forecaster for $15,000 to find out for certain whether there will be a recession. Based on the original 2 to 1 odds that there will be a recession, would it be worthwhile for the manufacturer to spend this $15,000?

19. Each of the following is the payoff matrix (the payments Player A makes to Player B) for a zero-sum two-person game. Eliminate all dominated strategies and determine the optimum strategy for each player as well as the value of the game:

(a)

3	−2
5	7

(b)

14	11
16	−2

(c)

−5	0	3
−6	−3	−3
−12	−1	1

(d)

7	10	8
8	8	11
7	5	9

20. Each of the following is the payoff matrix of a zero-sum two-person game. Find the saddle point (or saddle points) and the value of each game:

(a)

−1	5	−2
0	3	1
−2	−4	5

(b)

3	2	4	9
4	4	4	3
5	6	5	6
5	7	5	9

21. A small town has two service stations, which share the town's market for gasoline. The owner of Station A is debating whether to give away free glasses to her customers as part of a promotional scheme, and the owner of Station B is debating whether to give away free steak knives. They know (from similar situations elsewhere) that if Station A gives away free glasses and Station B does not give away free steak knives, Station A's share of the market will increase by 6 percent; if Station B gives away free steak knives and Station A does not give away free glasses, Station B's share of the market will increase by 8 percent; and if both stations give away the respective items, Station B's share of the market will increase by 3 percent.
(a) Present this information in the form of a payoff table in which the entries are Station A's losses in its share of the market.
(b) Find optimum strategies for the owners of the two stations.

22. Verify the two probabilities $\frac{4}{17}$ and $\frac{13}{17}$, which we gave in Section 2, for the randomized strategy of Player B.

23. The following is the payoff matrix of a 2×2 zero-sum two-person game:

3	−4
−3	1

(a) What randomized strategy should Player A use so as to minimize his maximum expected loss?
(b) What randomized strategy should Player B use so as to maximize her minimum expected gain?
(c) What is the value of the game?

24. With reference to Exercise 12, what randomized strategy will minimize Ms. Cooper's maximum expected cost?

25. A country has two airfields with installations worth $2,000,000 and $10,000,000, respectively, of which it can defend only one against an attack by its enemy. The enemy, on the other hand, can attack only one of these airfields and take it successfully only if it is left undefended. Considering the "payoff" to the country to be the total value of the installations it holds after the attack, find the optimum strategy of the country as well as that of its enemy and the value of the "game."

26. Two persons agree to play the following game: The first writes either 1 or 4 on a slip of paper, and at the same time the second writes either 0 or 3 on another slip of paper. If the sum of the two numbers is odd, the first wins this amount in dollars; otherwise, the second wins $2.
(a) Construct the payoff matrix in which the payoffs are the first person's losses.
(b) What randomized decision procedure should the first person use so as to minimize her maximum expected loss?
(c) What randomized decision procedure should the second person use so as to maximize his minimum expected gain?

27. There are two gas stations in a certain block, and the owner of the first station knows that if neither station lowers its prices, he can expect a net profit of $100 on any given day. If he lowers his prices while the other station does not, he can expect a net profit of $140; if he does not lower his prices but the other station does, he can expect a net profit of $70; and if both stations participate in this "price war," he can expect a net profit of $80. The owners of the two gas stations decide independently what prices to charge on any given day, and it is assumed that they cannot change their prices after they discover those charged by the other.

(a) Should the owner of the first gas station charge his regular prices or should he lower them if he wants to maximize his minimum net profit?

(b) Assuming that the profit figures for the first gas station apply also to the second gas station, how might the owners of the gas stations collude so that each could expect a net profit of $105?

Note that this "game" is not zero-sum, so that the possibility of collusion opens entirely new possibilities.

SECS. 3–6

28. A statistician has to decide on the basis of one observation whether the parameter θ of a Bernoulli distribution is 0, $\frac{1}{2}$, or 1; her loss in dollars (a penalty that is deducted from her fee) is 100 times the absolute value of her error.

(a) Construct a table showing the nine possible values of the loss function.

(b) List the nine possible decision functions and construct a table showing all the values of the corresponding risk function.

(c) Show that five of the decision functions are not admissible and that, according to the minimax criterion, the remaining decision functions are all equally good.

(d) Which decision function is best, according to the Bayes criterion, if the three possible values of the parameter θ are regarded as equally likely?

29. A statistician has to decide on the basis of two observations whether the parameter θ of a binomial distribution is $\frac{1}{4}$ or $\frac{1}{2}$; his loss (a penalty that is deducted from his fee) is $160 if he is wrong.

(a) Construct a table showing the four possible values of the loss function.

(b) List the eight possible decision functions and construct a table showing all the values of the corresponding risk function.

(c) Show that three of the decision functions are not admissible.

(d) Find the decision function that is best according to the minimax criterion.

(e) Find the decision function that is best according to the Bayes criterion if the probabilities assigned to $\theta = \frac{1}{4}$ and $\theta = \frac{1}{2}$ are, respectively, $\frac{2}{3}$ and $\frac{1}{3}$.

SEC. 7

30. A manufacturer produces an item consisting of two components, which must both work for the item to function properly. The cost of returning one of the items to the manufacturer for repairs is α dollars, the cost of inspecting one of the components is β dollars, and the cost of repairing a faulty component is φ dollars. She can ship each item without inspection with the guarantee that it will be put into perfect working condition at her factory in case it does not work; she can inspect both components and repair them if necessary; or she can randomly select one of the components and ship the item with the original guarantee if it works, or repair it and also check the other component.

(a) Construct a table showing the manufacturer's expected losses corresponding to her three "strategies" and the three "states" of Nature that 0, 1, or 2 of the components do not work.

(b) What should the manufacturer do if $\alpha = \$25.00$, $\varphi = \$10.00$, and she wants to minimize her maximum expected losses?

(c) What should the manufacturer do to minimize her Bayes risk if $\alpha = \$10.00$, $\beta = \$12.00$, $\varphi = \$30.00$, and she feels that the probabilities for 0, 1, and 2 defective components are, respectively, 0.70, 0.20, and 0.10?

31. Rework Example 10, changing the first strategy to an acceptance number of 1, instead of 2.

32. With reference to Example 10, for what values of C_w and C_d will Strategy 2 be preferred?

References

Some fairly elementary material on the theory of games and decision theory can be found in

CHERNOFF, H., and MOSES, L. E., *Elementary Decision Theory*. Mineola, N.Y.: Dover Publications, Inc. (Republication of 1959 edition),

DRESHER, M., *Games of Strategy: Theory and Applications*. Upper Saddle River, N.J.: Prentice Hall, 1961,

HAMBURG, M., *Statistical Analysis for Decision Making*, 4th ed. Orlando, Fla.: Harcourt Brace Jovanovich, 1988,

MCKINSEY, J. C. C., *Introduction to the Theory of Games*. New York: McGraw-Hill Book Company, 1952,

OWEN, G., *Game Theory*. Philadelphia: W. B. Saunders Company, 1968,

WILLIAMS, J. D., *The Compleat Strategyst*. New York: McGraw-Hill Book Company, 1954, and more advanced treatments in,

BICKEL, P. J., and DOKSUM, K. A., *Mathematical Statistics: Basic Ideas and Selected Topics*. Upper Saddle River, N.J.: Prentice Hall, 1977,

FERGUSON, T. S., *Mathematical Statistics: A Decision Theoretic Approach*. New York: Academic Press, Inc., 1967,

WALD, A., *Statistical Decision Functions*. New York: John Wiley & Sons, Inc., 1950.

Answers to Odd-Numbered Exercises

1 n.

3

	d_1	d_2
θ_1	0	1
θ_2	$\frac{1}{2n}$	0

5 $\dfrac{\theta_1\theta_2}{\sqrt{\theta_1^2+\theta_2^2}}$.

11 (a) The decision would be reversed. **(b)** The decision would be the same.

13 (a) He should go to the construction site that is 33 miles from the lumberyard. **(b)** He should go to the construction site that is 27 miles from the lumberyard. **(c)** It does not matter.

15 (a) He should expand his plant capacity now. **(b)** She should choose Hotel Y. **(c)** He should go to the construction site that is 27 miles from the lumberyard.

17 (a) She should choose Hotel Y. **(b)** He should go to the construction site that is 27 miles from the lumberyard.

19 (a) The optimum strategies are I and 2 and the value of the game is 5. **(b)** The optimum strategies are II and 1 and the value is 11. **(c)** The optimum strategies are I and 1 and the value is −5. **(d)** The optimum strategies are I and 2 and the value is 8.

21 (a) The payoffs are 0 and −6 for the first row of the table and 8 and 3 for the second row of the table. **(b)** The optimum strategies are for Station A to give away the glasses and for Station B to give away the knives.

23 (a) $\frac{5}{11}$ and $\frac{6}{11}$; **(b)** $\frac{4}{11}$ and $\frac{7}{11}$; **(c)** $-\frac{9}{11}$.

25 The defending country should randomize its strategies with probabilities $\frac{1}{6}$ and $\frac{5}{6}$, and the enemy should randomize its strategies with probabilities $\frac{5}{6}$ and $\frac{1}{6}$; the value is $10,333,333.

27 (a) He should lower the prices. **(b)** They could accomplish this by lowering their prices on alternate days.

29 (a) The values of the first row are 0 and 160; those of the second row are 160 and 0.

(b) $d_1(0)=\frac{1}{4}, d_1(1)=\frac{1}{4}, d_1(2)=\frac{1}{4}, d_2(0)=\frac{1}{4},$
$d_2(1)=\frac{1}{4}, d_2(2)=\frac{1}{2}, d_3(0)=\frac{1}{4}, d_3(1)=\frac{1}{2}, d_3(2)=\frac{1}{4},$
$d_4(0)=\frac{1}{4}, d_4(1)=\frac{1}{2}, d_4(2)=\frac{1}{2}, d_5(0)=\frac{1}{2}, d_5(1)=\frac{1}{4},$
$d_5(2)=\frac{1}{2}, d_6(0)=\frac{1}{2}, d_6(1)=\frac{1}{4}, d_6(2)=\frac{1}{2}, d_7(0)=\frac{1}{2},$
$d_7(1)=\frac{1}{2}, d_7(2)=\frac{1}{2}, d_8(0)=\frac{1}{2}, d_8(1)=\frac{1}{2}, d_8(2)=\frac{1}{2};$
(d) d_4; **(e)** d_2.

Point Estimation

1 Introduction

Traditionally, problems of statistical inference are divided into **problems of estimation** and **tests of hypotheses**, though actually they are all decision problems and, hence, could be handled by the unified approach. The main difference between the two kinds of problems is that in problems of estimation we must determine the value of a parameter (or the values of several parameters) from a possible continuum of alternatives, whereas in tests of hypotheses we must decide whether to accept or reject a specific value or a set of specific values of a parameter (or those of several parameters).

> **DEFINITION 1. POINT ESTIMATION.** *Using the value of a sample statistic to estimate the value of a population parameter is called* ***point estimation****. We refer to the value of the statistic as a* ***point estimate****.*

For example, if we use a value of $\overline{X}$ to estimate the mean of a population, an observed sample proportion to estimate the parameter θ of a binomial population, or a value of S^2 to estimate a population variance, we are in each case using a point estimate of the parameter in question. These estimates are called point estimates because in each case a single number, or a single point on the real axis, is used to estimate the parameter.

Correspondingly, we refer to the statistics themselves as **point estimators**. For instance, $\overline{X}$ may be used as a point estimator of μ, in which case $\overline{x}$ is a point estimate of this parameter. Similarly, S^2 may be used as a point estimator of σ^2, in which case s^2 is a point estimate of this parameter. Here we used the word "point" to distinguish between these estimators and estimates and the **interval estimators** and **interval estimates**.

Since estimators are random variables, one of the key problems of point estimation is to study their sampling distributions. For instance, when we estimate the variance of a population on the basis of a random sample, we can hardly expect that the value of S^2 we get will actually equal σ^2, but it would be reassuring, at least, to know whether we can expect it to be close. Also, if we must decide whether to use a sample mean or a sample median to estimate the mean of a population, it would be important to know, among other things, whether $\overline{X}$ or $\widetilde{X}$ is more likely to yield a value that is actually close.

Various statistical properties of estimators can thus be used to decide which estimator is most appropriate in a given situation, which will expose us to the smallest risk, which will give us the most information at the lowest cost, and so forth. The particular properties of estimators that we shall discuss in Sections 2 through 6 are **unbiasedness**, **minimum variance**, **efficiency**, **consistency**, **sufficiency**, and **robustness**.

2 Unbiased Estimators

Perfect decision functions do not exist, and in connection with problems of estimation this means that there are no perfect estimators that always give the right answer. Thus, it would seem reasonable that an estimator should do so at least on the average; that is, its expected value should equal the parameter that it is supposed to estimate. If this is the case, the estimator is said to be **unbiased**; otherwise, it is said to be **biased**. Formally, this concept is expressed by means of the following definition.

DEFINITION 2. UNBIASED ESTIMATOR. *A statistic $\hat{\Theta}$ is an* ***unbiased estimator*** *of the parameter θ of a given distribution if and only if $E(\hat{\Theta}) = \theta$ for all possible values of θ.*

The following are some examples of unbiased and biased estimators.

EXAMPLE 1

Definition 2 requires that $E(\Theta) = \theta$ for all possible values of θ. To illustrate why this statement is necessary, show that unless $\theta = \frac{1}{2}$, the minimax estimator of the binomial parameter θ is biased.

Solution

Since $E(X) = n\theta$, it follows that

$$E\left(\frac{X + \frac{1}{2}\sqrt{n}}{n + \sqrt{n}}\right) = \frac{E\left(X + \frac{1}{2}\sqrt{n}\right)}{n + \sqrt{n}} = \frac{n\theta + \frac{1}{2}\sqrt{n}}{n + \sqrt{n}}$$

and it can easily be seen that this quantity does not equal θ unless $\theta = \frac{1}{2}$.

EXAMPLE 2

If X has the binomial distribution with the parameters n and θ, show that the sample proportion, $\frac{X}{n}$, is an unbiased estimator of θ.

Solution

Since $E(X) = n\theta$, it follows that

$$E\left(\frac{X}{n}\right) = \frac{1}{n} \cdot E(X) = \frac{1}{n} \cdot n\theta = \theta$$

and hence that $\frac{X}{n}$ is an unbiased estimator of θ.

EXAMPLE 3

If $X_1, X_2, \ldots, X_n$ constitute a random sample from the population given by

$$f(x) = \begin{cases} e^{-(x-\delta)} & \text{for } x > \delta \\ 0 & \text{elsewhere} \end{cases}$$

show that $\overline{X}$ is a biased estimator of δ.

Solution
Since the mean of the population is

$$\mu = \int_{\delta}^{\infty} x \cdot e^{-(x-\delta)} dx = 1 + \delta$$

it follows from the theorem "If $\overline{X}$ is the mean of a random sample of size n taken without replacement from a finite population of size N with the mean μ and the variance σ^2, then $E(\overline{X}) = \mu$ and $\text{var}(\overline{X}) = \frac{\sigma^2}{n} \cdot \frac{N-n}{N-1}$" that $E(\overline{X}) = 1 + \delta \neq \delta$ and hence that $\overline{X}$ is a biased estimator of δ.

When $\hat{\Theta}$, based on a sample of size n from a given population, is a biased estimator of θ, it may be of interest to know the extent of the **bias**, given by

$$b_n(\theta) = E(\hat{\Theta}) - \theta$$

Thus, for Example 1 the bias is

$$\frac{n\theta + \frac{1}{2}\sqrt{n}}{n + \sqrt{n}} - \theta = \frac{\frac{1}{2} - \theta}{\sqrt{n} + 1}$$

and it can be seen that it tends to be small when θ is close to $\frac{1}{2}$ and also when n is large.

DEFINITION 3. ASYMPTOTICALLY UNBIASED ESTIMATOR. *Letting* $b_n(\theta) = E(\hat{\Theta}) - \theta$ *express the* ***bias*** *of an estimator* $\hat{\Theta}$ *based on a random sample of size* n *from a given distribution, we say that* $\hat{\Theta}$ *is an* ***asymptotically unbiased estimator*** *of* θ *if and only if*

$$\lim_{n\to\infty} b_n(\theta) = 0$$

As far as Example 3 is concerned, the bias is $(1+\delta) - \delta = 1$, but here there is something we can do about it. Since $E(\overline{X}) = 1 + \delta$, it follows that $E(\overline{X} - 1) = \delta$ and hence that $\overline{X} - 1$ is an unbiased estimator of δ. The following is another example where a minor modification of an estimator leads to an estimator that is unbiased.

EXAMPLE 4

If $X_1, X_2, \ldots, X_n$ constitute a random sample from a uniform population with $\alpha = 0$, show that the largest sample value (that is, the nth order statistic, Y_n) is a biased estimator of the parameter β. Also, modify this estimator of β to make it unbiased.

Solution

Substituting into the formula for $g_n(y_n) = \begin{cases} \frac{n}{\theta} \cdot e^{-y_n/\theta}[1 - e^{-y_n/\theta}]^{n-1} & \text{for } y_n > 0 \\ 0 & \text{elsewhere} \end{cases}$,

we find that the sampling distribution of Y_n is given by

$$g_n(y_n) = n \cdot \frac{1}{\beta} \cdot \left(\int_0^{y_n} \frac{1}{\beta} dx \right)^{n-1} = \frac{n}{\beta^n} \cdot y_n^{n-1}$$

for $0 < y_n < \beta$ and $g_n(y_n) = 0$ elsewhere, and hence that

$$E(Y_n) = \frac{n}{\beta^n} \cdot \int_0^{\beta} y_n^n \, dy_n = \frac{n}{n+1} \cdot \beta$$

Thus, $E(Y_n) \neq \beta$ and the nth order statistic is a biased estimator of the parameter β. However, since

$$E\left(\frac{n+1}{n} \cdot Y_n \right) = \frac{n+1}{n} \cdot \frac{n}{n+1} \cdot \beta = \beta$$

it follows that $\frac{n+1}{n}$ times the largest sample value is an unbiased estimator of the parameter β.

As unbiasedness is a desirable property of an estimator, we can explain why we divided by $n - 1$ and not by n when we defined the sample variance: It makes S^2 an unbiased estimator of σ^2 for random samples from infinite populations.

THEOREM 1. If S^2 is the variance of a random sample from an infinite population with the finite variance σ^2, then $E(S^2) = \sigma^2$.

Proof By definition of sample mean and sample variance,

$$E(S^2) = E\left[\frac{1}{n-1} \cdot \sum_{i=1}^{n} (X_i - \overline{X})^2 \right] = \frac{1}{n-1} \cdot E\left[\sum_{i=1}^{n} \{(X_i - \mu) - (\overline{X} - \mu)\}^2 \right] = \frac{1}{n-1} \cdot \left[\sum_{i=1}^{n} E\{(X_i - \mu)^2\} - n \cdot E\{(\overline{X} - \mu)^2\} \right]$$

Then, since $E\{(X_i - \mu)^2\} = \sigma^2$ and $E\{(\overline{X} - \mu)^2\} = \frac{\sigma^2}{n}$, it follows that

$$E(S^2) = \frac{1}{n-1} \cdot \left[\sum_{i=1}^{n} \sigma^2 - n \cdot \frac{\sigma^2}{n} \right] = \sigma^2$$

Although S^2 is an unbiased estimator of the variance of an infinite population, it is not an unbiased estimator of the variance of a finite population, and in neither case is S an unbiased estimator of σ. The bias of S as an estimator of σ is discussed, among others, in the book by E. S. Keeping listed among the references at the end of this chapter.

The discussion of the preceding paragraph illustrates one of the difficulties associated with the concept of unbiasedness. It may not be retained under functional transformations; that is, if $\hat{\Theta}$ is an unbiased estimator of θ, it does not necessarily follow that $\omega(\hat{\Theta})$ is an unbiased estimator of $\omega(\theta)$. Another difficulty associated with the concept of unbiasedness is that unbiased estimators are not necessarily unique. For instance, in Example 6 we shall see that $\frac{n+1}{n} \cdot Y_n$ is not the only unbiased estimator of the parameter β of Example 4, and in Exercise 8 we shall see that $\overline{X} - 1$ is not the only unbiased estimator of the parameter δ of Example 3.

3 Efficiency

If we have to choose one of several unbiased estimators of a given parameter, we usually take the one whose sampling distribution has the smallest variance. The estimator with the smaller variance is "more reliable."

DEFINITION 4. MINIMUM VARIANCE UNBIASED ESTIMATOR. *The estimator for the parameter θ of a given distribution that has the smallest variance of all unbiased estimators for θ is called the **minimum variance unbiased estimator**, or the **best unbiased estimator** for θ.*

If $\hat{\Theta}$ is an unbiased estimator of θ, it can be shown under very general conditions (referred to in the references at the end of the chapter) that the variance of $\hat{\Theta}$ must satisfy the inequality

$$\text{var}(\hat{\Theta}) \geqq \frac{1}{n \cdot E\left[\left(\frac{\partial \ln f(X)}{\partial \theta}\right)^2\right]}$$

where $f(x)$ is the value of the population density at x and n is the size of the random sample. This inequality, the **Cramér–Rao inequality**, leads to the following result.

THEOREM 2. If $\hat{\Theta}$ is an unbiased estimator of θ and

$$\text{var}(\hat{\Theta}) = \frac{1}{n \cdot E\left[\left(\frac{\partial \ln f(X)}{\partial \theta}\right)^2\right]}$$

then $\hat{\Theta}$ is a minimum variance unbiased estimator of θ.

Here, the quantity in the denominator is referred to as the **information** about θ that is supplied by the sample (see also Exercise 19). Thus, the smaller the variance is, the greater the information.

EXAMPLE 5

Show that $\overline{X}$ is a minimum variance unbiased estimator of the mean μ of a normal population.

Solution

Since

$$f(x) = \frac{1}{\sigma\sqrt{2\pi}} \cdot e^{-\frac{1}{2}\left(\frac{x-\mu}{\sigma}\right)^2} \quad \text{for } -\infty < x < \infty$$

it follows that

$$\ln f(x) = -\ln \sigma\sqrt{2\pi} - \frac{1}{2}\left(\frac{x-\mu}{\sigma}\right)^2$$

so that

$$\frac{\partial \ln f(x)}{\partial \mu} = \frac{1}{\sigma}\left(\frac{x-\mu}{\sigma}\right)$$

and hence

$$E\left[\left(\frac{\partial \ln f(X)}{\partial \mu}\right)^2\right] = \frac{1}{\sigma^2} \cdot E\left[\left(\frac{X-\mu}{\sigma}\right)^2\right] = \frac{1}{\sigma^2} \cdot 1 = \frac{1}{\sigma^2}$$

Thus,

$$\frac{1}{n \cdot E\left[\left(\frac{\partial \ln f(X)}{\partial \mu}\right)^2\right]} = \frac{1}{n \cdot \frac{1}{\sigma^2}} = \frac{\sigma^2}{n}$$

and since $\overline{X}$ is unbiased and $\text{var}(\overline{X}) = \frac{\sigma^2}{n}$, it follows that $\overline{X}$ is a minimum variance unbiased estimator of μ.

It would be erroneous to conclude from this example that $\overline{X}$ is a minimum variance unbiased estimator of the mean of any population. Indeed, in Exercise 3 the reader will be asked to verify that this is not so for random samples of size $n = 3$ from the continuous uniform population with $\alpha = \theta - \frac{1}{2}$ and $\beta = \theta + \frac{1}{2}$.

As we have indicated, unbiased estimators of one and the same parameter are usually compared in terms of the size of their variances. If $\hat{\Theta}_1$ and $\hat{\Theta}_2$ are two unbiased estimators of the parameter θ of a given population and the variance of $\hat{\Theta}_1$ is less than the variance of $\hat{\Theta}_2$, we say that $\hat{\Theta}_1$ is **relatively more efficient** than $\hat{\Theta}_2$. Also, we use the ratio

$$\frac{\text{var}(\hat{\Theta}_1)}{\text{var}(\hat{\Theta}_2)}$$

as a measure of the efficiency of $\hat{\Theta}_2$ relative to $\hat{\Theta}_1$.

EXAMPLE 6

In Example 4 we showed that if $X_1, X_2, \ldots, X_n$ constitute a random sample from a uniform population with $\alpha = 0$, then $\frac{n+1}{n} \cdot Y_n$ is an unbiased estimator of β.

(a) Show that $2\overline{X}$ is also an unbiased estimator of β.

(b) Compare the efficiency of these two estimators of β.

Solution

(a) Since the mean of the population is $\mu = \frac{\beta}{2}$ according to the theorem "The mean and the variance of the uniform distribution are given by $\mu = \frac{\alpha+\beta}{2}$ and $\sigma^2 = \frac{1}{12}(\beta - \alpha)^2$" it follows from the theorem "If $X_1, X_2, \ldots, X_n$ constitute a random sample from an infinite population with the mean μ and the variance σ^2, then $E(\overline{X}) = \mu$ and $\text{var}(\overline{X}) = \frac{\sigma^2}{n}$" that $E(\overline{X}) = \frac{\beta}{2}$ and hence that $E(2\overline{X}) = \beta$. Thus, $2\overline{X}$ is an unbiased estimator of β.

(b) First we must find the variances of the two estimators. Using the sampling distribution of Y_n and the expression for $E(Y_n)$ given in Example 4, we get

$$E(Y_n^2) = \frac{n}{\beta^n} \cdot \int_0^\beta y_n^{n+1} \, dy_n = \frac{n}{n+2} \cdot \beta^2$$

and

$$\text{var}(Y_n) = \frac{n}{n+2} \cdot \beta^2 - \left(\frac{n}{n+1} \cdot \beta \right)^2$$

If we leave the details to the reader in Exercise 27, it can be shown that

$$\text{var}\left(\frac{n+1}{n} \cdot Y_n \right) = \frac{\beta^2}{n(n+2)}$$

Since the variance of the population is $\sigma^2 = \frac{\beta^2}{12}$ according to the first stated theorem in the example, it follows from the above (second) theorem that $\text{var}(\overline{X}) = \frac{\beta^2}{12n}$ and hence that

$$\text{var}(2\overline{X}) = 4 \cdot \text{var}(\overline{X}) = \frac{\beta^2}{3n}$$

Therefore, the efficiency of $2\overline{X}$ relative to $\frac{n+1}{n} \cdot Y_n$ is given by

$$\frac{\text{var}\left(\frac{n+1}{n} \cdot Y_n \right)}{\text{var}(2\overline{X})} = \frac{\frac{\beta^2}{n(n+2)}}{\frac{\beta^2}{3n}} = \frac{3}{n+2}$$

and it can be seen that for $n > 1$ the estimator based on the nth order statistic is much more efficient than the other one. For $n = 10$, for example, the relative efficiency is only 25 percent, and for $n = 25$ it is only 11 percent.

EXAMPLE 7

When the mean of a normal population is estimated on the basis of a random sample of size $2n+1$, what is the efficiency of the median relative to the mean?

Solution

From the theorem on the previous page we know that $\overline{X}$ is unbiased and that

$$\text{var}(\overline{X}) = \frac{\sigma^2}{2n+1}$$

As far as $\widetilde{X}$ is concerned, it is unbiased by virtue of the symmetry of the normal distribution about its mean, and for large samples

$$\text{var}(\widetilde{X}) = \frac{\pi\sigma^2}{4n}$$

Thus, for large samples, the efficiency of the median relative to the mean is approximately

$$\frac{\text{var}(\overline{X})}{\text{var}(\widetilde{X})} = \frac{\dfrac{\sigma^2}{2n+1}}{\dfrac{\pi\sigma^2}{4n}} = \frac{4n}{\pi(2n+1)}$$

and the **asymptotic efficiency** of the median with respect to the mean is

$$\lim_{n\to\infty} \frac{4n}{\pi(2n+1)} = \frac{2}{\pi}$$

or about 64 percent.

The result of the preceding example may be interpreted as follows: For large samples, the mean requires only 64 percent as many observations as the median to estimate μ with the same reliability.

It is important to note that we have limited our discussion of relative efficiency to unbiased estimators. If we included biased estimators, we could always assure ourselves of an estimator with zero variance by letting its values equal the same constant regardless of the data that we may obtain. Therefore, if $\hat{\Theta}$ is not an unbiased estimator of a given parameter θ, we judge its merits and make efficiency comparisons on the basis of the **mean square error** $E[(\hat{\Theta}-\theta)^2]$ instead of the variance of $\hat{\Theta}$.

Exercises

1. If $X_1, X_2, \ldots, X_n$ constitute a random sample from a population with the mean μ, what condition must be imposed on the constants $a_1, a_2, \ldots, a_n$ so that

$$a_1X_1 + a_2X_2 + \cdots + a_nX_n$$

is an unbiased estimator of μ?

2. If $\hat{\Theta}_1$ and $\hat{\Theta}_2$ are unbiased estimators of the same parameter θ, what condition must be imposed on the constants k_1 and k_2 so that

$$k_1\hat{\Theta}_1 + k_2\hat{\Theta}_2$$

is also an unbiased estimator of θ?

3. This question has been intentionally omitted for this edition.

4. This question has been intentionally omitted for this edition.

5. Given a random sample of size n from a population that has the known mean μ and the finite variance σ^2, show that

$$\frac{1}{n}\cdot\sum_{i=1}^{n}(X_i-\mu)^2$$

is an unbiased estimator of σ^2.

6. This question has been intentionally omitted for this edition.

7. Show that $\frac{X+1}{n+2}$ is a biased estimator of the binomial parameter θ. Is this estimator asymptotically unbiased?

8. With reference to Example 3, find an unbiased estimator of δ based on the smallest sample value (that is, on the first order statistic, Y_1).

9. With reference to Example 4, find an unbiased estimator of β based on the smallest sample value (that is, on the first order statistic, Y_1).

10. If $X_1, X_2, \ldots, X_n$ constitute a random sample from a normal population with $\mu = 0$, show that

$$\sum_{i=1}^{n} \frac{X_i^2}{n}$$

is an unbiased estimator of σ^2.

11. If X is a random variable having the binomial distribution with the parameters n and θ, show that $n \cdot \frac{X}{n} \cdot \left(1 - \frac{X}{n}\right)$ is a biased estimator of the variance of X.

12. If a random sample of size n is taken without replacement from the finite population that consists of the positive integers $1, 2, \ldots, k$, show that

(a) the sampling distribution of the nth order statistic, Y_n, is given by

$$f(y_n) = \frac{\binom{y_n - 1}{n-1}}{\binom{k}{n}}$$

for $y_n = n, \ldots, k$;

(b) $\frac{n+1}{n} \cdot Y_n - 1$ is an unbiased estimator of k.

See also Exercise 80.

13. Show that if $\hat{\Theta}$ is an unbiased estimator of θ and $\text{var}(\hat{\Theta}) \neq 0$, then $\hat{\Theta}^2$ is not an unbiased estimator of θ^2.

14. Show that the sample proportion $\frac{X}{n}$ is a minimum variance unbiased estimator of the binomial parameter θ. (*Hint*: Treat $\frac{X}{n}$ as the mean of a random sample of size n from a Bernoulli population with the parameter θ.)

15. Show that the mean of a random sample of size n is a minimum variance unbiased estimator of the parameter λ of a Poisson population.

16. If $\hat{\Theta}_1$ and $\hat{\Theta}_2$ are independent unbiased estimators of a given parameter θ and $\text{var}(\hat{\Theta}_1) = 3 \cdot \text{var}(\hat{\Theta}_2)$, find the constants a_1 and a_2 such that $a_1\hat{\Theta}_1 + a_2\hat{\Theta}_2$ is an unbiased estimator with minimum variance for such a linear combination.

17. Show that the mean of a random sample of size n from an exponential population is a minimum variance unbiased estimator of the parameter θ.

18. Show that for the unbiased estimator of Example 4, $\frac{n+1}{n} \cdot Y_n$, the Cramér–Rao inequality is not satisfied.

19. The information about θ in a random sample of size n is also given by

$$-n \cdot E\left[\frac{\partial^2 \ln f(X)}{\partial \theta^2}\right]$$

where $f(x)$ is the value of the population density at x, provided that the extremes of the region for which $f(x) \neq 0$ do not depend on θ. The derivation of this formula takes the following steps:

(a) Differentiating the expressions on both sides of

$$\int f(x)\, dx = 1$$

with respect to θ, show that

$$\int \frac{\partial \ln f(x)}{\partial \theta} \cdot f(x)\, dx = 0$$

by interchanging the order of integration and differentiation.

(b) Differentiating again with respect to θ, show that

$$E\left[\left(\frac{\partial \ln f(X)}{\partial \theta}\right)^2\right] = -E\left[\frac{\partial^2 \ln f(X)}{\partial \theta^2}\right]$$

20. Rework Example 5 using the alternative formula for the information given in Exercise 19.

21. If $\overline{X}_1$ is the mean of a random sample of size n from a normal population with the mean μ and the variance σ_1^2, $\overline{X}_2$ is the mean of a random sample of size n from a normal population with the mean μ and the variance σ_2^2, and the two samples are independent, show that

(a) $\omega \cdot \overline{X}_1 + (1-\omega) \cdot \overline{X}_2$, where $0 \leqq \omega \leqq 1$, is an unbiased estimator of μ;

(b) the variance of this estimator is a minimum when

$$\omega = \frac{\sigma_2^2}{\sigma_1^2 + \sigma_2^2}$$

22. With reference to Exercise 21, find the efficiency of the estimator of part (a) with $\omega = \frac{1}{2}$ relative to this estimator with

$$\omega = \frac{\sigma_2^2}{\sigma_1^2 + \sigma_2^2}$$

23. If $\overline{X}_1$ and $\overline{X}_2$ are the means of independent random samples of sizes n_1 and n_2 from a normal population with the mean μ and the variance σ^2, show that the variance of the unbiased estimator

$$\omega \cdot \overline{X}_1 + (1-\omega) \cdot \overline{X}_2$$

is a minimum when $\omega = \dfrac{n_1}{n_1+n_2}$.

24. With reference to Exercise 23, find the efficiency of the estimator with $\omega = \frac{1}{2}$ relative to the estimator with $\omega = \dfrac{n_1}{n_1+n_2}$.

25. If X_1, X_2, and X_3 constitute a random sample of size $n = 3$ from a normal population with the mean μ and the variance σ^2, find the efficiency of $\dfrac{X_1+2X_2+X_3}{4}$ relative to $\dfrac{X_1+X_2+X_3}{3}$ as estimates of μ.

26. If X_1 and X_2 constitute a random sample of size $n = 2$ from an exponential population, find the efficiency of $2Y_1$ relative to $\overline{X}$, where Y_1 is the first order statistic and $2Y_1$ and $\overline{X}$ are both unbiased estimators of the parameter θ.

27. Verify the result given for $\text{var}\left(\dfrac{n+1}{n} \cdot Y_n\right)$ in Example 6.

28. With reference to Example 3, we showed that $\overline{X}-1$ is an unbiased estimator of δ, and in Exercise 8 the reader was asked to find another unbiased estimator of δ based on the smallest sample value. Find the efficiency of the first of these two estimators relative to the second.

29. With reference to Exercise 12, show that $2\overline{X}-1$ is also an unbiased estimator of k, and find the efficiency of this estimator relative to the one of part (b) of Exercise 12 for

(a) $n = 2$; **(b)** $n = 3$.

30. Since the variances of the mean and the midrange are not affected if the same constant is added to each observation, we can determine these variances for random samples of size 3 from the uniform population

$$f(x) = \begin{cases} 1 & \text{for } \theta - \frac{1}{2} < x < \theta + \frac{1}{2} \\ 0 & \text{elsewhere} \end{cases}$$

by referring instead to the uniform population

$$f(x) = \begin{cases} 1 & \text{for } 0 < x < 1 \\ 0 & \text{elsewhere} \end{cases}$$

Show that $E(X) = \frac{1}{2}, E(X^2) = \frac{1}{3}$, and $\text{var}(X) = \frac{1}{12}$ for this population so that for a random sample of size $n = 3$, $\text{var}(\widetilde{X}) = \frac{1}{36}$.

31. Show that if $\hat{\Theta}$ is a biased estimator of θ, then

$$E[(\hat{\Theta}-\theta)^2] = \text{var}(\hat{\Theta}) + [b(\theta)]^2$$

32. If $\hat{\Theta}_1 = \dfrac{X}{n}$, $\hat{\Theta}_2 = \dfrac{X+1}{n+2}$, and $\hat{\Theta}_3 = \dfrac{1}{3}$ are estimators of the parameter θ of a binomial population and $\theta = \frac{1}{2}$, for what values of n is

(a) the mean square error of $\hat{\Theta}_2$ less than the variance of $\hat{\Theta}_1$;

(b) the mean square error of $\hat{\Theta}_3$ less than the variance of $\hat{\Theta}_1$?

4 Consistency

In the preceding section we assumed that the variance of an estimator, or its mean square error, is a good indication of its chance fluctuations. The fact that these measures may not provide good criteria for this purpose is illustrated by the following example: Suppose that we want to estimate on the basis of one observation the parameter θ of the population given by

$$f(x) = \omega \cdot \frac{1}{\sigma\sqrt{2\pi}} \cdot e^{-\frac{1}{2}\left(\frac{x-\theta}{\sigma}\right)^2} + (1-\omega) \cdot \frac{1}{\pi} \cdot \frac{1}{1+(x-\theta)^2}$$

for $-\infty < x < \infty$ and $0 < \omega < 1$. Evidently, this population is a combination of a normal population with the mean θ and the variance σ^2 and a Cauchy population with $\alpha = \theta$ and $\beta = 1$. Now, if ω is very close to 1, say, $\omega = 1 - 10^{-100}$, and σ is very small, say, $\sigma = 10^{-100}$, the probability that a random variable having this distribution will take on a value that is very close to θ, and hence is a very good estimate of θ, is practically 1. Yet, since the variance of the Cauchy distribution does not exist, neither will the variance of this estimator.

The example of the preceding paragraph is a bit farfetched, but it suggests that we pay more attention to the probabilities with which estimators will take on values that are close to the parameters that they are supposed to estimate. Basing our argument on Chebyshev's theorem, when $n \to \infty$ the probability approaches 1 that the sample proportion $\frac{X}{n}$ will take on a value that differs from the binomial parameter θ by less than any arbitrary constant $c > 0$. Also using Chebyshev's theorem, we see that when $n \to \infty$ the probability approaches 1 that $\overline{X}$ will take on a value that differs from the mean of the population sampled by less than any arbitrary constant $c > 0$.

In both of these examples we are practically assured that, for large n, the estimators will take on values that are very close to the respective parameters. Formally, this concept of "closeness" is expressed by means of the following definition of **consistency**.

DEFINITION 5. CONSISTENT ESTIMATOR. *The statistic $\hat{\Theta}$ is a **consistent estimator** of the parameter θ of a given distribution if and only if for each $c > 0$*

$$\lim_{n \to \infty} P(|\hat{\Theta} - \theta| < c) = 1$$

Note that consistency is an **asymptotic property**, that is, a limiting property of an estimator. Informally, Definition 5 says that when n is sufficiently large, we can be practically certain that the error made with a consistent estimator will be less than any small preassigned positive constant. The kind of convergence expressed by the limit in Definition 5 is generally called **convergence in probability**.

Based on Chebyshev's theorem, $\frac{X}{n}$ is a consistent estimator of the binomial parameter θ and $\overline{X}$ is a consistent estimator of the mean of a population with a finite variance. In practice, we can often judge whether an estimator is consistent by using the following sufficient condition, which, in fact, is an immediate consequence of Chebyshev's theorem.

THEOREM 3. If $\hat{\Theta}$ is an unbiased estimator of the parameter θ and $\text{var}(\hat{\Theta}) \to 0$ as $n \to \infty$, then $\hat{\Theta}$ is a consistent estimator of θ.

EXAMPLE 8

Show that for a random sample from a normal population, the sample variance S^2 is a consistent estimator of σ^2.

Solution

Since S^2 is an unbiased estimator of σ^2 in accordance with Theorem 3, it remains to be shown that $\text{var}(S^2) \to 0$ as $n \to \infty$. Referring to the theorem "the random variable $\frac{(n-1)S^2}{\sigma^2}$ has a chi-square distribution with $n - 1$ degrees of freedom", we find that for a random sample from a normal population

$$\text{var}(S^2) = \frac{2\sigma^4}{n-1}$$

It follows that $\text{var}(S^2) \to 0$ as $n \to \infty$, and we have thus shown that S^2 is a consistent estimator of the variance of a normal population.

It is of interest to note that Theorem 3 also holds if we substitute "asymptotically unbiased" for "unbiased." This is illustrated by the following example.

EXAMPLE 9

With reference to Example 3, show that the smallest sample value (that is, the first order statistic Y_1) is a consistent estimator of the parameter δ.

Solution

Substituting into the formula for $g_1(y_1)$, we find that the sampling distribution of Y_1 is given by

$$\begin{aligned} g_1(y_1) &= n \cdot e^{-(y_1-\delta)} \cdot \left[\int_{y_1}^{\infty} e^{-(x-\delta)}\, dx \right]^{n-1} \\ &= n \cdot e^{-n(y_1-\delta)} \end{aligned}$$

for $y_1 > \delta$ and $g_1(y_1) = 0$ elsewhere. Based on this result, it can easily be shown that $E(Y_1) = \delta + \frac{1}{n}$ and hence that Y_1 is an asymptotically unbiased estimator of δ. Furthermore,

$$\begin{aligned} P(|Y_1 - \delta| < c) &= P(\delta < Y_1 < \delta + c) \\ &= \int_{\delta}^{\delta+c} n \cdot e^{-n(y_1-\delta)}\, dy_1 \\ &= 1 - e^{-nc} \end{aligned}$$

Since $\lim_{n \to \infty} (1 - e^{-nc}) = 1$, it follows from Definition 5 that Y_1 is a consistent estimator of δ.

Theorem 3 provides a sufficient condition for the consistency of an estimator. It is not a necessary condition because consistent estimators need not be unbiased, or even asymptotically unbiased. This is illustrated by Exercise 41.

5 Sufficiency

An estimator $\hat{\Theta}$ is said to be **sufficient** if it utilizes all the information in a sample relevant to the estimation of θ, that is, if all the knowledge about θ that can be gained from the individual sample values and their order can just as well be gained from the value of $\hat{\Theta}$ alone.

Formally, we can describe this property of an estimator by referring to the conditional probability distribution or density of the sample values given $\hat{\Theta} = \hat{\theta}$, which is given by

$$f(x_1, x_2, \ldots, x_n | \hat{\theta}) = \frac{f(x_1, x_2, \ldots, x_n, \hat{\theta})}{g(\hat{\theta})} = \frac{f(x_1, x_2, \ldots, x_n)}{g(\hat{\theta})}$$

If it depends on θ, then particular values of $X_1, X_2, \ldots, X_n$ yielding $\hat{\Theta} = \hat{\theta}$ will be more probable for some values of θ than for others, and the knowledge of these sample values will help in the estimation of θ. On the other hand, if it does not depend on θ, then particular values of $X_1, X_2, \ldots, X_n$ yielding $\hat{\Theta} = \hat{\theta}$ will be just as likely for any value of θ, and the knowledge of these sample values will be of no help in the estimation of θ.

DEFINITION 6. SUFFICIENT ESTIMATOR. *The statistic $\hat{\Theta}$ is a **sufficient estimator** of the parameter θ of a given distribution if and only if for each value of $\hat{\Theta}$ the conditional probability distribution or density of the random sample $X_1, X_2, \ldots, X_n$, given $\hat{\Theta} = \theta$, is independent of θ.*

EXAMPLE 10

If $X_1, X_2, \ldots, X_n$ constitute a random sample of size n from a Bernoulli population, show that

$$\hat{\Theta} = \frac{X_1 + X_2 + \cdots + X_n}{n}$$

is a sufficient estimator of the parameter θ.

Solution

By the definition "**BERNOULLI DISTRIBUTION.** *A random variable X has a **Bernoulli distribution** and it is referred to as a Bernoulli random variable if and only if its probability distribution is given by* $f(x;\theta) = \theta^x(1-\theta)^{1-x}$ for $x = 0, 1$",

$$f(x_i;\theta) = \theta^{x_i}(1-\theta)^{1-x_i} \qquad \text{for } x_i = 0, 1$$

so that

$$\begin{aligned} f(x_1, x_2, \ldots, x_n) &= \prod_{i=1}^{n} \theta^{x_i}(1-\theta)^{1-x_i} \\ &= \theta^{\sum_{i=1}^{n} x_i}(1-\theta)^{n-\sum_{i=1}^{n} x_i} \\ &= \theta^x(1-\theta)^{n-x} \\ &= \theta^{n\hat{\theta}}(1-\theta)^{n-n\hat{\theta}} \end{aligned}$$

for $x_i = 0$ or 1 and $i = 1, 2, \ldots, n$. Also, since

$$X = X_1 + X_2 + \cdots + X_n$$

is a binomial random variable with the parameters θ and n, its distribution is given by

$$b(x; n, \theta) = \binom{n}{x}\theta^x(1-\theta)^{n-x}$$

and the transformation-of-variable technique yields

$$g(\hat{\theta}) = \binom{n}{n\hat{\theta}}\theta^{n\hat{\theta}}(1-\theta)^{n-n\hat{\theta}} \qquad \text{for } \hat{\theta} = 0, \frac{1}{n}, \ldots, 1$$

Now, substituting into the formula for $f(x_1, x_2, \ldots, x_n|\hat{\theta})$ on the previous page, we get

$$\frac{f(x_1, x_2, \ldots, x_n, \hat{\theta})}{g(\hat{\theta})} = \frac{f(x_1, x_2, \ldots, x_n)}{g(\hat{\theta})}$$

$$= \frac{\theta^{n\hat{\theta}}(1-\theta)^{n-n\hat{\theta}}}{\binom{n}{n\hat{\theta}}\theta^{n\hat{\theta}}(1-\theta)^{n-n\hat{\theta}}}$$

$$= \frac{1}{\binom{n}{n\hat{\theta}}}$$

$$= \frac{1}{\binom{n}{x}}$$

$$= \frac{1}{\binom{n}{x_1 + x_2 + \cdots + x_n}}$$

for $x_i = 0$ or 1 and $i = 1, 2, \ldots, n$. Evidently, this does not depend on θ and we have shown, therefore, that $\hat{\Theta} = \dfrac{X}{n}$ is a sufficient estimator of θ.

EXAMPLE 11

Show that $Y = \frac{1}{6}(X_1 + 2X_2 + 3X_3)$ is not a sufficient estimator of the Bernoulli parameter θ.

Solution

Since we must show that

$$f(x_1, x_2, x_3|y) = \frac{f(x_1, x_2, x_3, y)}{g(y)}$$

is not independent of θ for some values of X_1, X_2, and X_3, let us consider the case where $x_1 = 1, x_2 = 1$, and $x_3 = 0$. Thus, $y = \frac{1}{6}(1 + 2 \cdot 1 + 3 \cdot 0) = \frac{1}{2}$ and

$$f\left(1, 1, 0|Y = \frac{1}{2}\right) = \frac{P\left(X_1 = 1, X_2 = 1, X_3 = 0, Y = \frac{1}{2}\right)}{P\left(Y = \frac{1}{2}\right)}$$

$$= \frac{f(1, 1, 0)}{f(1, 1, 0) + f(0, 0, 1)}$$

where

$$f(x_1, x_2, x_3) = \theta^{x_1 + x_2 + x_3}(1-\theta)^{3-(x_1+x_2+x_3)}$$

for $x_1 = 0$ or 1 and $i = 1, 2, 3$. Since $f(1,1,0) = \theta^2(1-\theta)$ and $f(0,0,1) = \theta(1-\theta)^2$, it follows that

$$f\left(1,1,0|Y = \frac{1}{2}\right) = \frac{\theta^2(1-\theta)}{\theta^2(1-\theta)+\theta(1-\theta)^2} = \theta$$

and it can be seen that this conditional probability depends on θ. We have thus shown that $Y = \frac{1}{6}(X_1+2X_2+3X_3)$ is not a sufficient estimator of the parameter θ of a Bernoulli population.

Because it can be very tedious to check whether a statistic is a sufficient estimator of a given parameter based directly on Definition 6, it is usually easier to base it instead on the following **factorization theorem**.

THEOREM 4. The statistic $\hat{\Theta}$ is a sufficient estimator of the parameter θ if and only if the joint probability distribution or density of the random sample can be factored so that

$$f(x_1, x_2, \ldots, x_n; \theta) = g(\hat{\theta}, \theta) \cdot h(x_1, x_2, \ldots, x_n)$$

where $g(\hat{\theta}, \theta)$ depends only on $\hat{\theta}$ and θ, and $h(x_1, x_2, \ldots, x_n)$ does not depend on θ.

A proof of this theorem may be found in more advanced texts; see, for instance, the book by Hogg and Tanis listed among the references at the end of this chapter. Here, let us illustrate the use of Theorem 4 by means of the following example.

EXAMPLE 12

Show that $\overline{X}$ is a sufficient estimator of the mean μ of a normal population with the known variance σ^2.

Solution

Making use of the fact that

$$f(x_1, x_2, \ldots, x_n; \mu) = \left(\frac{1}{\sigma\sqrt{2\pi}}\right)^n \cdot e^{-\frac{1}{2}\cdot\sum_{i=1}^{n}\left(\frac{x_i-\mu}{\sigma}\right)^2}$$

and that

$$\begin{aligned}\sum_{i=1}^{n}(x_i-\mu)^2 &= \sum_{i=1}^{n}[(x_i-\overline{x})-(\mu-\overline{x})]^2 \\ &= \sum_{i=1}^{n}(x_i-\overline{x})^2 + \sum_{i=1}^{n}(\overline{x}-\mu)^2 \\ &= \sum_{i=1}^{n}(x_i-\overline{x})^2 + n(\overline{x}-\mu)^2\end{aligned}$$

we get

$$f(x_1, x_2, \ldots, x_n; \mu) = \left\{ \frac{\sqrt{n}}{\sigma\sqrt{2\pi}} \cdot e^{-\frac{1}{2}\left(\frac{\bar{x}-\mu}{\sigma/\sqrt{n}}\right)^2} \right\}$$

$$\times \left\{ \frac{1}{\sqrt{n}} \left(\frac{1}{\sigma\sqrt{2\pi}} \right)^{n-1} \cdot e^{-\frac{1}{2} \cdot \sum_{i=1}^{n} \left(\frac{x_i - \bar{x}}{\sigma}\right)^2} \right\}$$

where the first factor on the right-hand side depends only on the estimate $\bar{x}$ and the population mean μ, and the second factor does not involve μ. According to Theorem 4, it follows that $\overline{X}$ is a sufficient estimator of the mean μ of a normal population with the known variance σ^2.

Based on Definition 6 and Theorem 4, respectively, we have presented two ways of checking whether a statistic $\hat{\Theta}$ is a sufficient estimator of a given parameter θ. As we already said, the factorization theorem usually leads to easier solutions; but if we want to show that a statistic $\hat{\Theta}$ is not a sufficient estimator of a given parameter θ, it is nearly always easier to proceed with Definition 6. This was illustrated by Example 11.

Let us also mention the following important property of sufficient estimators. If $\hat{\Theta}$ is a sufficient estimator of θ, then any single-valued function $Y = u(\hat{\Theta})$, not involving θ, is also a sufficient estimator of θ, and therefore of $u(\theta)$, provided $y = u(\hat{\theta})$ can be solved to give the single-valued inverse $\hat{\theta} = w(y)$. This follows directly from Theorem 4, since we can write

$$f(x_1, x_2, \ldots, x_n; \theta) = g[w(y), \theta] \cdot h(x_1, x_2, \ldots, x_n)$$

where $g[w(y), \theta]$ depends only on y and θ. If we apply this result to Example 10, where we showed that $\hat{\Theta} = \dfrac{X}{n}$ is a sufficient estimator of the Bernoulli parameter θ, it follows that $X = X_1 + X_2 + \cdots + X_n$ is also a sufficient estimator of the mean $\mu = n\theta$ of a binomial population.

6 Robustness

In recent years, special attention has been paid to a statistical property called **robustness.** It is indicative of the extent to which estimation procedures (and, as we shall see later, other methods of inference) are adversely affected by violations of underlying assumptions. In other words, an estimator is said to be **robust** if its sampling distribution is not seriously affected by violations of assumptions. Such violations are often due to outliers caused by outright errors made, say, in reading instruments or recording the data or by mistakes in experimental procedures. They may also pertain to the nature of the populations sampled or their parameters. For instance, when estimating the average useful life of a certain electronic component, we may think that we are sampling an exponential population, whereas actually we are sampling a Weibull population, or when estimating the average income of a certain age group, we may use a method based on the assumption that we are sampling a normal population, whereas actually the population (income distribution) is highly skewed. Also, when estimating the difference between the average weights of two kinds of frogs, the difference between the mean I.Q.'s of two ethnic groups, and in general the difference $\mu_1 - \mu_2$ between the means of two populations, we may be assuming that the two populations have the same variance σ^2, whereas in reality $\sigma_1^2 \neq \sigma_2^2$.

As should be apparent, most questions of robustness are difficult to answer; indeed, much of the language used in the preceding paragraph is relatively imprecise. After all, what do we mean by "not seriously affected"? Furthermore, when we speak of violations of underlying assumptions, it should be clear that some violations are more serious than others. When it comes to questions of robustness, we are thus faced with all sorts of difficulties, mathematically and otherwise, and for the most part they can be resolved only by computer simulations.

Exercises

33. Use Definition 5 to show that Y_1, the first order statistic, is a consistent estimator of the parameter α of a uniform population with $\beta = \alpha + 1$.

34. With reference to Exercise 33, use Theorem 3 to show that $Y_1 - \frac{1}{n+1}$ is a consistent estimator of the parameter α.

35. With reference to the uniform population of Example 4, use the definition of consistency to show that Y_n, the nth order statistic, is a consistent estimator of the parameter β.

36. If $X_1, X_2, \ldots, X_n$ constitute a random sample of size n from an exponential population, show that $\overline{X}$ is a consistent estimator of the parameter θ.

37. With reference to Exercise 36, is X_n a consistent estimator of the parameter θ?

38. Show that the estimator of Exercise 21 is consistent.

39. Substituting "asymptotically unbiased" for "unbiased" in Theorem 3, show that $\frac{X+1}{n+2}$ is a consistent estimator of the binomial parameter θ.

40. Substituting "asymptotically unbiased" for "unbiased" in Theorem 3, use this theorem to rework Exercise 35.

41. To show that an estimator can be consistent without being unbiased or even asymptotically unbiased, consider the following estimation procedure: To estimate the mean of a population with the finite variance σ^2, we first take a random sample of size n. Then we randomly draw one of n slips of paper numbered from 1 through n, and if the number we draw is $2, 3, \ldots$, or n, we use as our estimator the mean of the random sample; otherwise, we use the estimate n^2. Show that this estimation procedure is

(a) consistent;

(b) neither unbiased nor asymptotically unbiased.

42. If $X_1, X_2, \ldots, X_n$ constitute a random sample of size n from an exponential population, show that $\overline{X}$ is a sufficient estimator of the parameter θ.

43. If X_1 and X_2 are independent random variables having binomial distributions with the parameters θ and n_1 and θ and n_2, show that $\frac{X_1+X_2}{n_1+n_2}$ is a sufficient estimator of θ.

44. In reference to Exercise 43, is $\frac{X_1+2X_2}{n_1+2n_2}$ a sufficient estimator of θ?

45. After referring to Example 4, is the nth order statistic, Y_n, a sufficient estimator of the parameter β?

46. If X_1 and X_2 constitute a random sample of size $n = 2$ from a Poisson population, show that the mean of the sample is a sufficient estimator of the parameter λ.

47. If X_1, X_2, and X_3 constitute a random sample of size $n = 3$ from a Bernoulli population, show that $Y = X_1 + 2X_2 + X_3$ is not a sufficient estimator of θ. (*Hint*: Consider special values of X_1, X_2, and X_3.)

48. If $X_1, X_2, \ldots, X_n$ constitute a random sample of size n from a geometric population, show that $Y = X_1 + X_2 + \cdots + X_n$ is a sufficient estimator of the parameter θ.

49. Show that the estimator of Exercise 5 is a sufficient estimator of the variance of a normal population with the known mean μ.

7 The Method of Moments

As we have seen in this chapter, there can be many different estimators of one and the same parameter of a population. Therefore, it would seem desirable to have some general method, or methods, that yield estimators with as many desirable properties as possible. In this section and in Section 8 we shall present two such methods, the **method of moments**, which is historically one of the oldest methods, and the **method of maximum likelihood**. Furthermore, **Bayesian estimation** will be treated briefly in Section 9.

The method of moments consists of equating the first few moments of a population to the corresponding moments of a sample, thus getting as many equations as are needed to solve for the unknown parameters of the population.

DEFINITION 7. SAMPLE MOMENTS. *The **kth sample moment** of a set of observations* $x_1, x_2, \ldots, x_n$ *is the mean of their* k*th powers and it is denoted by* m'_k*; symbolically,*

$$m'_k = \frac{\sum_{i=1}^{n} x_i^k}{n}$$

Thus, if a population has r parameters, the method of moments consists of solving the system of equations

$$m'_k = \mu'_k \quad k = 1, 2, \ldots, r$$

for the r parameters.

EXAMPLE 13

Given a random sample of size n from a uniform population with $\beta = 1$, use the method of moments to obtain a formula for estimating the parameter α.

Solution

The equation that we shall have to solve is $m'_1 = \mu'_1$, where $m'_1 = \overline{x}$ and $\mu'_1 = \frac{\alpha+\beta}{2} = \frac{\alpha+1}{2}$. Thus,

$$\overline{x} = \frac{\alpha+1}{2}$$

and we can write the estimate of α as

$$\hat{\alpha} = 2\overline{x} - 1$$

EXAMPLE 14

Given a random sample of size n from a gamma population, use the method of moments to obtain formulas for estimating the parameters α and β.

Solution

The system of equations that we shall have to solve is

$$m'_1 = \mu'_1 \quad \text{and} \quad m'_2 = \mu'_2$$

where $\mu'_1 = \alpha\beta$ and $\mu'_2 = \alpha(\alpha+1)\beta^2$. Thus,

$$m'_1 = \alpha\beta \quad \text{and} \quad m'_2 = \alpha(\alpha+1)\beta^2$$

and, solving for α and β, we get the following formulas for estimating the two parameters of the gamma distribution:

$$\hat{\alpha} = \frac{(m'_1)^2}{m'_2 - (m'_1)^2} \quad \text{and} \quad \hat{\beta} = \frac{m'_2 - (m'_1)^2}{m'_1}$$

Since $m'_1 = \frac{\sum_{i=1}^{n} x_i}{n} = \overline{x}$ and $m'_2 = \frac{\sum_{i=1}^{n} x_i^2}{n}$, we can write

$$\hat{\alpha} = \frac{n\overline{x}^2}{\sum_{i=1}^{n}(x_i - \overline{x})^2} \quad \text{and} \quad \hat{\beta} = \frac{\sum_{i=1}^{n}(x_i - \overline{x})^2}{n\overline{x}}$$

in terms of the original observations.

In these examples we were concerned with the parameters of a specific population. It is important to note, however, that when the parameters to be estimated are the moments of the population, then the method of moments can be used without any knowledge about the nature, or functional form, of the population.

8 The Method of Maximum Likelihood

In two papers published early in the last century, R. A. Fisher proposed a general method of estimation called the **method of maximum likelihood**. He also demonstrated the advantages of this method by showing that it yields sufficient estimators whenever they exist and that maximum likelihood estimators are asymptotically minimum variance unbiased estimators.

To help to understand the principle on which the method of maximum likelihood is based, suppose that four letters arrive in somebody's morning mail, but unfortunately one of them is misplaced before the recipient has a chance to open it. If, among the remaining three letters, two contain credit-card billings and the other one does not, what might be a good estimate of k, the total number of credit-card billings among the four letters received? Clearly, k must be two or three, and if we assume that each letter had the same chance of being misplaced, we find that the probability of the observed data (two of the three remaining letters contain credit-card billings) is

$$\frac{\binom{2}{2}\binom{2}{1}}{\binom{4}{3}} = \frac{1}{2}$$

for $k = 2$ and

$$\frac{\binom{3}{2}\binom{1}{1}}{\binom{4}{3}} = \frac{3}{4}$$

for $k = 3$. Therefore, if we choose as our estimate of k the value that maximizes the probability of getting the observed data, we obtain $k = 3$. We call this estimate a **maximum likelihood estimate**, and the method by which it was obtained is called the method of maximum likelihood.

Thus, the essential feature of the method of maximum likelihood is that we look at the sample values and then choose as our estimates of the unknown parameters the values for which the probability or probability density of getting the sample values is a maximum. In what follows, we shall limit ourselves to the one-parameter case; but, as we shall see in Example 18, the general idea applies also when there are several unknown parameters. In the discrete case, if the observed sample values are $x_1, x_2, \ldots, x_n$, the probability of getting them is

$$P(X_1 = x_1, X_2 = x_2, \ldots, X_n = x_n) = f(x_1, x_2, \ldots, x_n; \theta)$$

which is just the value of the joint probability distribution of the random variables $X_1, X_2, \ldots, X_n$ at $X_1 = x_1, X_2 = x_2, \ldots, X_n = x_n$. Since the sample values have been observed and are therefore fixed numbers, we regard $f(x_1, x_2, \ldots, x_n; \theta)$ as a value of a function of θ, and we refer to this function as the **likelihood function**. An analogous definition applies when the random sample comes from a continuous population, but in that case $f(x_1, x_2, \ldots, x_n; \theta)$ is the value of the joint probability density of the random variables $X_1, X_2, \ldots, X_n$ at $X_1 = x_1, X_2 = x_2, \ldots, X_n = x_n$.

DEFINITION 8. MAXIMUM LIKELIHOOD ESTIMATOR. *If $x_1, x_2, \ldots, x_n$ are the values of a random sample from a population with the parameter θ, the* ***likelihood function*** *of the sample is given by*

$$L(\theta) = f(x_1, x_2, \ldots, x_n; \theta)$$

for values of θ within a given domain. Here, $f(x_1, x_2, \ldots, x_n; \theta)$ is the value of the joint probability distribution or the joint probability density of the random variables $X_1, X_2, \ldots, X_n$ at $X_1 = x_1, X_2 = x_2, \ldots, X_n = x_n$. We refer to the value of θ that maximizes $L(\theta)$ as the ***maximum likelihood estimator*** *of θ.*

EXAMPLE 15

Given x "successes" in n trials, find the maximum likelihood estimate of the parameter θ of the corresponding binomial distribution.

Solution

To find the value of θ that maximizes

$$L(\theta) = \binom{n}{x} \theta^x (1-\theta)^{n-x}$$

it will be convenient to make use of the fact that the value of θ that maximizes $L(\theta)$ will also maximize

$$\ln L(\theta) = \ln \binom{n}{x} + x \cdot \ln \theta + (n-x) \cdot \ln(1-\theta)$$

Thus, we get

$$\frac{d[\ln L(\theta)]}{d\theta} = \frac{x}{\theta} - \frac{n-x}{1-\theta}$$

and, equating this derivative to 0 and solving for θ, we find that the likelihood function has a maximum at $\theta = \frac{x}{n}$. This is the maximum likelihood estimate of the binomial parameter θ, and we refer to $\hat{\Theta} = \frac{X}{n}$ as the corresponding **maximum likelihood estimator.**

EXAMPLE 16

If $x_1, x_2, \ldots, x_n$ are the values of a random sample from an exponential population, find the maximum likelihood estimator of its parameter θ.

Solution

Since the likelihood function is given by

$$\begin{aligned} L(\theta) &= f(x_1, x_2, \ldots, x_n; \theta) \\ &= \prod_{i=1}^{n} f(x_i; \theta) \\ &= \left(\frac{1}{\theta}\right)^n \cdot e^{-\frac{1}{\theta}\left(\sum_{i=1}^{n} x_i\right)} \end{aligned}$$

differentiation of $\ln L(\theta)$ with respect to θ yields

$$\frac{d[\ln L(\theta)]}{d\theta} = -\frac{n}{\theta} + \frac{1}{\theta^2} \cdot \sum_{i=1}^{n} x_i$$

Equating this derivative to zero and solving for θ, we get the maximum likelihood estimate

$$\hat{\theta} = \frac{1}{n} \cdot \sum_{i=1}^{n} x_i = \overline{x}$$

Hence, the maximum likelihood estimator is $\hat{\Theta} = \overline{X}$.

Now let us consider an example in which straightforward differentiation cannot be used to find the maximum value of the likelihood function.

EXAMPLE 17

If $x_1, x_2, \ldots, x_n$ are the values of a random sample of size n from a uniform population with $\alpha = 0$ (as in Example 4), find the maximum likelihood estimator of β.

Solution

The likelihood function is given by

$$L(\beta) = \prod_{i=1}^{n} f(x_i; \beta) = \left(\frac{1}{\beta}\right)^n$$

for β greater than or equal to the largest of the x's and 0 otherwise. Since the value of this likelihood function increases as β decreases, we must make β as small as possible, and it follows that the maximum likelihood estimator of β is Y_n, the nth order statistic.

Comparing the result of this example with that of Example 4, we find that maximum likelihood estimators need not be unbiased. However, the ones of Examples 15 and 16 were unbiased.

The method of maximum likelihood can also be used for the simultaneous estimation of several parameters of a given population. In that case we must find the values of the parameters that jointly maximize the likelihood function.

EXAMPLE 18

If $X_1, X_2, \ldots, X_n$ constitute a random sample of size n from a normal population with the mean μ and the variance σ^2, find joint maximum likelihood estimates of these two parameters.

Solution

Since the likelihood function is given by

$$L(\mu, \sigma^2) = \prod_{i=1}^{n} n(x_i; \mu, \sigma)$$

$$= \left(\frac{1}{\sigma\sqrt{2\pi}}\right)^n \cdot e^{-\frac{1}{2\sigma^2} \cdot \sum_{i=1}^{n} (x_i - \mu)^2}$$

partial differentiation of $\ln L(\mu, \sigma^2)$ with respect to μ and σ^2 yields

$$\frac{\partial[\ln L(\mu, \sigma^2)]}{\partial\mu} = \frac{1}{\sigma^2} \cdot \sum_{i=1}^{n} (x_i - \mu)$$

and

$$\frac{\partial[\ln L(\mu, \sigma^2)]}{\partial\sigma^2} = -\frac{n}{2\sigma^2} + \frac{1}{2\sigma^4} \cdot \sum_{i=1}^{n} (x_i - \mu)^2$$

Equating the first of these two partial derivatives to zero and solving for μ, we get

$$\hat{\mu} = \frac{1}{n} \cdot \sum_{i=1}^{n} x_i = \overline{x}$$

and equating the second of these partial derivatives to zero and solving for σ^2 after substituting $\mu = \overline{x}$, we get

$$\hat{\sigma}^2 = \frac{1}{n} \cdot \sum_{i=1}^{n} (x_i - \overline{x})^2$$

It should be observed that we did not show that $\hat{\sigma}$ is a maximum likelihood estimate of σ, only that $\hat{\sigma}^2$ is a maximum likelihood estimate of σ^2. However, it can be shown (see reference at the end of this chapter) that maximum likelihood estimators have the **invariance property** that if $\hat{\Theta}$ is a maximum likelihood estimator of θ and the function given by $g(\theta)$ is continuous, then $g(\hat{\Theta})$ is also a maximum likelihood estimator of $g(\theta)$. It follows that

$$\hat{\sigma} = \sqrt{\frac{1}{n} \cdot \sum_{i=1}^{n} (x_i - \bar{x})^2}$$

which differs from s in that we divide by n instead of $n-1$, is a maximum likelihood estimate of σ.

In Examples 15, 16, and 18 we maximized the logarithm of the likelihood function instead of the likelihood function itself, but this is by no means necessary. It just so happened that it was convenient in each case.

Exercises

50. If $X_1, X_2, \ldots, X_n$ constitute a random sample from a population with the mean μ and the variance σ^2, use the method of moments to find estimators for μ and σ^2.

51. Given a random sample of size n from an exponential population, use the method of moments to find an estimator of the parameter θ.

52. Given a random sample of size n from a uniform population with $\alpha = 0$, find an estimator for β by the method of moments.

53. Given a random sample of size n from a Poisson population, use the method of moments to obtain an estimator for the parameter λ.

54. Given a random sample of size n from a beta population with $\beta = 1$, use the method of moments to find a formula for estimating the parameter α.

55. If $X_1, X_2, \ldots, X_n$ constitute a random sample of size n from a population given by

$$f(x;\theta) = \begin{cases} \dfrac{2(\theta - x)}{\theta^2} & \text{for } 0 < x < \theta \\ 0 & \text{elsewhere} \end{cases}$$

find an estimator for θ by the method of moments.

56. If $X_1, X_2, \ldots, X_n$ constitute a random sample of size n from a population given by

$$g(x;\theta,\delta) = \begin{cases} \dfrac{1}{\theta} \cdot e^{-\frac{x-\delta}{\theta}} & \text{for } x > \delta \\ 0 & \text{elsewhere} \end{cases}$$

find estimators for δ and θ by the method of moments. This distribution is sometimes referred to as the **two-parameter exponential distribution**, and for $\theta = 1$ it is the distribution of Example 3.

57. Given a random sample of size n from a continuous uniform population, use the method of moments to find formulas for estimating the parameters α and β.

58. Consider N independent random variables having identical binomial distributions with the parameters θ and $n = 3$. If n_0 of them take on the value 0, n_1 take on the value 1, n_2 take on the value 2, and n_3 take on the value 3, use the method of moments to find a formula for estimating θ.

59. Use the method of maximum likelihood to rework Exercise 53.

60. Use the method of maximum likelihood to rework Exercise 54.

61. If $X_1, X_2, \ldots, X_n$ constitute a random sample of size n from a gamma population with $\alpha = 2$, use the method of maximum likelihood to find a formula for estimating β.

62. Given a random sample of size n from a normal population with the known mean μ, find the maximum likelihood estimator for σ.

63. If $X_1, X_2, \ldots, X_n$ constitute a random sample of size n from a geometric population, find formulas for estimating its parameter θ by using

(a) the method of moments;

(b) the method of maximum likelihood.

64. Given a random sample of size n from a Rayleigh population, find an estimator for its parameter α by the method of maximum likelihood.

65. Given a random sample of size n from a Pareto population, use the method of maximum likelihood to find a formula for estimating its parameter α.

66. Use the method of maximum likelihood to rework Exercise 56.

67. Use the method of maximum likelihood to rework Exercise 57.

68. Use the method of maximum likelihood to rework Exercise 58.

69. Given a random sample of size n from a gamma population with the known parameter α, find the maximum likelihood estimator for
(a) β; **(b)** $\tau = (2\beta - 1)^2$.

70. If $V_1, V_2, \ldots, V_n$ and $W_1, W_2, \ldots, W_n$ are independent random samples of size n from normal populations with the means $\mu_1 = \alpha + \beta$ and $\mu_2 = \alpha - \beta$ and the common variance $\sigma^2 = 1$, find maximum likelihood estimators for α and β.

71. If $V_1, V_2, \ldots, V_{n_1}$ and $W_1, W_2, \ldots, W_{n_2}$ are independent random samples of sizes n_1 and n_2 from normal populations with the means μ_1 and μ_2 and the common variance σ^2, find maximum likelihood estimators for μ_1, μ_2, and σ^2.

72. Let $X_1, X_2, \ldots, X_n$ be a random sample of size n from the uniform population given by

$$f(x;\theta) = \begin{cases} 1 & \text{for } \theta - \frac{1}{2} < x < \theta + \frac{1}{2} \\ 0 & \text{elsewhere} \end{cases}$$

Show that if Y_1 and Y_n are the first and nth order statistic, any estimator $\hat{\Theta}$ such that

$$Y_n - \frac{1}{2} \leqq \hat{\Theta} \leqq Y_1 + \frac{1}{2}$$

can serve as a maximum likelihood estimator of θ. This shows that maximum likelihood estimators need not be unique.

73. With reference to Exercise 72, check whether the following estimators are maximum likelihood estimators of θ:
(a) $\frac{1}{2}(Y_1 + Y_n)$; **(b)** $\frac{1}{3}(Y_1 + 2Y_2)$.

9 Bayesian Estimation[†]

So far we have assumed in this chapter that the parameters that we want to estimate are unknown constants; in Bayesian estimation the parameters are looked upon as random variables having **prior distributions**, usually reflecting the strength of one's belief about the possible values that they can assume.

The main problem of Bayesian estimation is that of combining prior feelings about a parameter with direct sample evidence, and this is accomplished by determining $\varphi(\theta|x)$, the conditional density of Θ given $X = x$. In contrast to the prior distribution of Θ, this conditional distribution (which also reflects the direct sample evidence) is called the **posterior distribution** of Θ. In general, if $h(\theta)$ is the value of the prior distribution of Θ at θ and we want to combine the information that it conveys with direct sample evidence about Θ, for instance, the value of a statistic $W = u(X_1, X_2, \ldots, X_n)$, we determine the posterior distribution of Θ by means of the formula

$$\varphi(\theta|w) = \frac{f(\theta, w)}{g(w)} = \frac{h(\theta) \cdot f(w|\theta)}{g(w)}$$

Here $f(w|\theta)$ is the value of the sampling distribution of W given $\Theta = \theta$ at w, $f(\theta, w)$ is the value of the joint distribution of Θ and W at θ and w, and $g(w)$ is the value of the marginal distribution of W at w. Note that the preceding formula for $\varphi(\theta|w)$ is, in fact, an extension of Bayes' theorem to the continuous case. Hence, the term "Bayesian estimation."

[†] This section may be omitted with no loss of continuity.

Once the posterior distribution of a parameter has been obtained, it can be used to make estimates, or it can be used to make probability statements about the parameter, as will be illustrated in Example 20. Although the method we have described has extensive applications, we shall limit our discussion here to inferences about the parameter Θ of a binomial population and the mean of a normal population; inferences about the parameter of a Poisson population are treated in Exercise 77.

THEOREM 5. If X is a binomial random variable and the prior distribution of Θ is a beta distribution with the parameters α and β, then the posterior distribution of Θ given $X = x$ is a beta distribution with the parameters $x+\alpha$ and $n-x+\beta$.

Proof For $\Theta = \theta$ we have

$$f(x|\theta) = \binom{n}{x}\theta^x(1-\theta)^{n-x} \quad \text{for } x = 0, 1, 2, \ldots, n$$

$$h(\theta) = \begin{cases} \dfrac{\Gamma(\alpha+\beta)}{\Gamma(\alpha)\cdot\Gamma(\beta)}\cdot\theta^{\alpha-1}(1-\theta)^{\beta-1} & \text{for } 0 < \theta < 1 \\ 0 & \text{elsewhere} \end{cases}$$

and hence

$$f(\theta, x) = \frac{\Gamma(\alpha+\beta)}{\Gamma(\alpha)\cdot\Gamma(\beta)}\cdot\theta^{\alpha-1}(1-\theta)^{\beta-1}\times\binom{n}{x}\theta^x(1-\theta)^{n-x}$$

$$= \binom{n}{x}\frac{\Gamma(\alpha+\beta)}{\Gamma(\alpha)\cdot\Gamma(\beta)}\cdot\theta^{x+\alpha-1}(1-\theta)^{n-x+\beta-1}$$

for $0 < \theta < 1$ and $x = 0, 1, 2, \ldots, n$, and $f(\theta, x) = 0$ elsewhere. To obtain the marginal density of X, let us make use of the fact that the integral of the beta density from 0 to 1 equals 1; that is,

$$\int_0^1 x^{\alpha-1}(1-x)^{\beta-1}dx = \frac{\Gamma(\alpha)\cdot\Gamma(\beta)}{\Gamma(\alpha+\beta)}$$

Thus, we get

$$g(x) = \binom{n}{x}\cdot\frac{\Gamma(\alpha+\beta)}{\Gamma(\alpha)\cdot\Gamma(\beta)}\cdot\frac{\Gamma(\alpha+x)\cdot\Gamma(n-x+\beta)}{\Gamma(n+\alpha+\beta)}$$

for $x = 0, 1, \ldots, n$, and hence

$$\varphi(\theta|x) = \frac{\Gamma(n+\alpha+\beta)}{\Gamma(\alpha+x)\cdot\Gamma(n-x+\beta)}\cdot\theta^{x+\alpha-1}(1-\theta)^{n-x+\beta-1}$$

for $0 < \theta < 1$, and $\varphi(\theta|x) = 0$ elsewhere. As can be seen by inspection, this is a beta density with the parameters $x+\alpha$ and $n-x+\beta$.

To make use of this theorem, let us refer to the result that (under very general conditions) the mean of the posterior distribution minimizes the Bayes risk when the loss function is quadratic, that is, when the loss function is given by

$$L[d(x), \theta] = c[d(x) - \theta]^2$$

where c is a positive constant. Since the posterior distribution of Θ is a beta distribution with parameters $x + \alpha$ and $n - x + \beta$, it follows from the theorem "The mean and the variance of the beta distribution are given by $\mu = \frac{\alpha}{\alpha+\beta}$ and $\sigma^2 = \frac{\alpha\beta}{(\alpha+\beta)^2(\alpha+\beta+1)}$" that

$$E(\Theta|x) = \frac{x + \alpha}{\alpha + \beta + n}$$

is a value of an estimator of θ that minimizes the Bayes risk when the loss function is quadratic and the prior distribution of Θ is of the given form.

EXAMPLE 19

Find the mean of the posterior distribution as an estimate of the "true" probability of a success if 42 successes are obtained in 120 binomial trials and the prior distribution of Θ is a beta distribution with $\alpha = \beta = 40$.

Solution

Substituting $x = 42, n = 120, \alpha = 40$, and $\beta = 40$ into the formula for $E(\Theta|x)$, we get

$$E(\Theta|42) = \frac{42 + 40}{40 + 40 + 120} = 0.41$$

Note that without knowledge of the prior distribution of Θ, the minimum variance unbiased estimate of θ (see Exercise 14) would be the sample proportion

$$\hat{\theta} = \frac{x}{n} = \frac{42}{120} = 0.35$$

THEOREM 6. If $\overline{X}$ is the mean of a random sample of size n from a normal population with the known variance σ^2 and the prior distribution of M (capital Greek *mu*) is a normal distribution with the mean μ_0 and the variance σ_0^2, then the posterior distribution of M given $\overline{X} = \overline{x}$ is a normal distribution with the mean μ_1 and the variance σ_1^2, where

$$\mu_1 = \frac{n\overline{x}\sigma_0^2 + \mu_0\sigma^2}{n\sigma_0^2 + \sigma^2} \quad \text{and} \quad \frac{1}{\sigma_1^2} = \frac{n}{\sigma^2} + \frac{1}{\sigma_0^2}$$

Proof For $\text{M} = \mu$ we have

$$f(\overline{x}|\mu) = \frac{\sqrt{n}}{\sigma\sqrt{2\pi}} \cdot e^{-\frac{1}{2}\left(\frac{\overline{x}-\mu}{\sigma/\sqrt{n}}\right)^2} \quad \text{for } -\infty < \overline{x} < \infty$$

and

$$h(\mu) = \frac{1}{\sigma_0\sqrt{2\pi}} \cdot e^{-\frac{1}{2}\left(\frac{\mu-\mu_0}{\sigma_0}\right)^2} \qquad \text{for } -\infty < \mu < \infty$$

so that

$$\varphi(\mu|\overline{x}) = \frac{h(\mu)\cdot f(\overline{x}|\mu)}{g(\overline{x})}$$

$$= \frac{\sqrt{n}}{2\pi\sigma\sigma_0 g(\overline{x})} \cdot e^{-\frac{1}{2}\left(\frac{\overline{x}-\mu}{\sigma/\sqrt{n}}\right)^2 - \frac{1}{2}\left(\frac{\mu-\mu_0}{\sigma_0}\right)^2} \qquad \text{for } -\infty < \mu < \infty$$

Now, if we collect powers of μ in the exponent of e, we get

$$-\frac{1}{2}\left(\frac{n}{\sigma^2}+\frac{1}{\sigma_0^2}\right)\mu^2 + \left(\frac{n\overline{x}}{\sigma^2}+\frac{\mu_0}{\sigma_0^2}\right)\mu - \frac{1}{2}\left(\frac{n\overline{x}^2}{\sigma^2}+\frac{\mu_0^2}{\sigma_0^2}\right)$$

and if we let

$$\frac{1}{\sigma_1^2} = \frac{n}{\sigma^2}+\frac{1}{\sigma_0^2} \quad \text{and} \quad \mu_1 = \frac{n\overline{x}\sigma_0^2+\mu_0\sigma^2}{n\sigma_0^2+\sigma^2}$$

factor out $-\frac{1}{2\sigma_1^2}$, and complete the square, the exponent of e in the expression for $\varphi(\mu|\overline{x})$ becomes

$$-\frac{1}{2\sigma_1^2}(\mu-\mu_1)^2 + R$$

where R involves $n, \overline{x}, \mu_0, \sigma$, and σ_0, but not μ. Thus, the posterior distribution of M becomes

$$\varphi(\mu|\overline{x}) = \frac{\sqrt{n}\cdot e^R}{2\pi\sigma\sigma_0 g(\overline{x})} \cdot e^{-\frac{1}{2\sigma_1^2}(\mu-\mu_1)^2} \qquad \text{for } -\infty < \mu < \infty$$

which is easily identified as a normal distribution with the mean μ_1 and the variance σ_1^2. Hence, it can be written as

$$\varphi(\mu|\overline{x}) = \frac{1}{\sigma_1\sqrt{2\pi}} \cdot e^{-\frac{1}{2}\left(\frac{\mu-\mu_1}{\sigma_1}\right)^2} \qquad \text{for } -\infty < \mu < \infty$$

where μ_1 and σ_1 are defined above. Note that we did not have to determine $g(\overline{x})$ as it was absorbed in the constant in the final result.

EXAMPLE 20

A distributor of soft-drink vending machines feels that in a supermarket one of his machines will sell on the average $\mu_0 = 738$ drinks per week. Of course, the mean will vary somewhat from market to market, and the distributor feels that this variation is measured by the standard deviation $\sigma_0 = 13.4$. As far as a machine placed in a particular market is concerned, the number of drinks sold will vary from week to

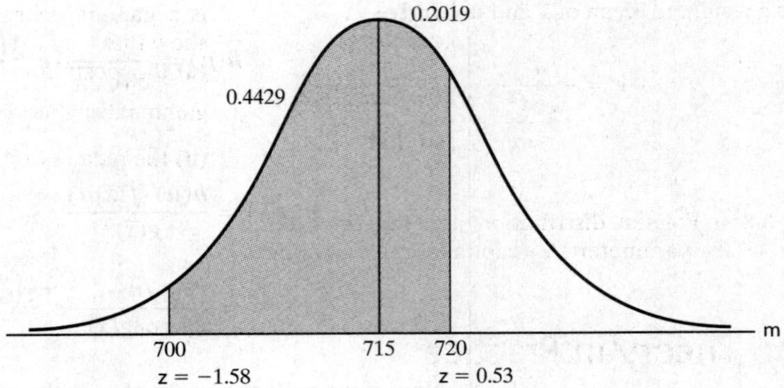

Figure 1. Diagram for Example 20.

week, and this variation is measured by the standard deviation $\sigma = 42.5$. If one of the distributor's machines put into a new supermarket averaged $\bar{x} = 692$ during the first 10 weeks, what is the probability (the distributor's personal probability) that for this market the value of M is actually between 700 and 720?

Solution

Assuming that the population sampled is approximately normal and that it is reasonable to treat the prior distribution of M as a normal distribution with the mean μ_0 and the standard deviation $\sigma_0 = 13.4$, we find that substitution into the two formulas of Theorem 6 yields

$$\mu_1 = \frac{10.692(13.4)^2 + 738(42.5)^2}{10(13.4)^2 + (42.5)^2} = 715$$

and

$$\frac{1}{\sigma_1^2} = \frac{10}{(42.5)^2} + \frac{1}{(13.4)^2} = 0.0111$$

so that $\sigma_1^2 = 90.0$ and $\sigma_1 = 9.5$. Now, the answer to our question is given by the area of the shaded region of Figure 1, that is, the area under the standard normal curve between

$$z = \frac{700 - 715}{9.5} = -1.58 \quad \text{and} \quad z = \frac{720 - 715}{9.5} = 0.53$$

Thus, the probability that the value of M is between 700 and 720 is $0.4429 + 0.2019 = 0.6448$, or approximately 0.645.

Exercises

74. This question has been intentionally omitted for this edition.

75. This question has been intentionally omitted for this edition.

76. Show that the mean of the posterior distribution of M given in Theorem 6 can be written as

$$\mu_1 = w \cdot \bar{x} + (1 - w) \cdot \mu_0$$

that is, as a weighted mean of $\overline{x}$ and μ_0, where

$$w = \frac{n}{n + \frac{\sigma^2}{\sigma_0^2}}$$

77. If X has a Poisson distribution and the prior distribution of its parameter Λ (capital Greek *lambda*) is a gamma distribution with the parameters α and β, show that

(a) the posterior distribution of Λ given $X = x$ is a gamma distribution with the parameters $\alpha + x$ and $\frac{\beta}{\beta+1}$;

(b) the mean of the posterior distribution of Λ is

$$\mu_1 = \frac{\beta(\alpha + x)}{\beta + 1}$$

10 The Theory in Practice

The sample mean, $\overline{x}$, is most frequently used to estimate the mean of a distribution from a random sample taken from that distribution. It has been shown to be the minimum variance unbiased estimator as well as a sufficient estimator for the mean of a normal distribution. It is at least asymptotically unbiased as an estimator for the mean of most frequently encountered distributions.

In spite of these desirable properties of the sample mean as an estimator for a population mean, we know that the sample mean will never equal the population mean. Let us examine the error we make when using $\overline{x}$ to estimate μ, $E = |\overline{x} - \mu|$. If the sample size, n, is large, the quantity

$$\frac{\overline{x} - \mu}{\sigma/\sqrt{n}}$$

is a value of a random variable having approximately the standard normal distribution. Thus, we can state with probability $1 - \alpha$ that

$$\frac{|\overline{x} - \mu|}{\sigma/\sqrt{n}} \le z_{\alpha/2}$$

or

$$E \le z_{\alpha/2} \frac{\sigma}{\sqrt{n}}$$

EXAMPLE 21

A pollster wishes to estimate the percentage of voters who favor a certain candidate. She wishes to be sure with a probability of 0.95 that the error in the resulting estimate will not exceed 3 percent. How many registered voters should she interview?

Solution

We shall use the normal approximation to the binomial distribution, assuming that n will turn out to be large. As per the theorem "If X has a binomial distribution with the parameters n and θ and $Y = \frac{X}{n}$, then $E(Y) = \theta$ and $\sigma_Y^2 = \frac{\theta(1-\theta)}{n}$" we know that $\sigma_{X/n}^2 = \frac{\theta(1-\theta)}{n}$, where θ is the parameter of the binomial distribution. Since this quantity is maximized when $\theta = \frac{1}{2}$, the maximum value of σ is $\frac{1}{2\sqrt{n}}$. Since the maximum error is to be 0.03, the inequality for E can be written as

$$E \le z_{\alpha/2} \cdot \frac{1}{2\sqrt{n}}$$

Noting that $z_{\alpha/2} = z_{.025} = 1.96$, and solving this inequality for n, we obtain for the sample size that assures, with probability 0.95, that the error in the resulting estimate will not exceed 3 percent

$$n \leq \frac{z^2\alpha/2}{4E^2} = \frac{(1.96)^2}{4(.03)^2} = 1{,}068$$

(Note that, in performing this calculation, we always round *up* to the nearest integer.)

It should not be surprising in view of this result that most such polls use sample sizes of about 1,000.

Another consideration related to the accuracy of a sample estimate deals with the concept of **sampling bias**. Sampling bias occurs when a sample is chosen that does not accurately represent the population from which it is taken. For example, a national poll based on automobile registrations in each of the 50 states probably is biased, because it omits people who do not own cars. Such people may well have different opinions than those who do. A sample of product stored on shelves in a warehouse is likely to be biased if all units of product in the sample were selected from the bottom shelf. Ambient conditions, such as temperature and humidity, may well have affected the top-shelf units differently than those on the bottom shelf.

The mean square error defined above can be viewed as the expected squared error loss encountered when we estimate the parameter θ with the point estimator $\hat{\Theta}$. We can write

$$\begin{aligned}
\text{MSE}(\hat{\Theta}) &= E(\hat{\Theta} - \theta)^2 \\
&= E[\hat{\Theta} - E(\hat{\Theta}) + E(\hat{\Theta}) - \theta]^2 \\
&= E[\hat{\Theta} - E(\hat{\Theta})]^2 + [E(\hat{\Theta}) - \theta]^2 + 2\{E[\hat{\Theta} - E(\hat{\Theta})][E(\hat{\Theta}) - \theta]\}
\end{aligned}$$

The first term of the cross product, $E[\hat{\Theta} - E(\hat{\Theta})] = E(\hat{\Theta}) - E(\hat{\Theta}) = 0$, and we are left with

$$\text{MSE}(\hat{\Theta}) = E[\hat{\Theta} - E(\hat{\Theta})]^2 + [E(\hat{\Theta}) - \theta]^2$$

The first term is readily seen to be the variance of $\hat{\Theta}$ and the second term is the square of the bias, the difference between the expected value of the estimate of the parameter θ and its true value. Thus, we can write

$$\text{MSE}(\hat{\Theta}) = \sigma^2_{\hat{\Theta}} + [\text{Bias}]^2$$

While it is possible to estimate the variance of $\hat{\Theta}$ in most applications, the sampling bias usually is unknown. Great care should be taken to avoid, or at least minimize sampling bias, for it can be much greater than the sampling variance $\sigma^2_{\hat{\Theta}}$. This can be done by carefully calibrating all instruments to be used in measuring the sample units, by eliminating human subjectivity as much as possible, and by assuring that the method of sampling is appropriately randomized over the entire population for which sampling estimates are to be made. These and other related issues are more thoroughly discussed in the book by Hogg and Tanis, referenced at the end of this chapter.

Applied Exercises

SECS. 1–3

78. Independent random samples of sizes n_1 and n_2 are taken from a normal population with the mean μ and the variance σ^2. If $n_1 = 25, n_2 = 50, \bar{x}_1 = 27.6$, and $\bar{x}_2 = 38.1$, estimate μ using the estimator of Exercise 23.

79. Random samples of size n are taken from normal populations with the mean μ and the variances $\sigma_1^2 = 4$ and $\sigma_2^2 = 9$. If $\bar{x}_1 = 26.0$ and $\bar{x}_2 = 32.5$, estimate μ using the estimator of part (b) of Exercise 21.

80. A country's military intelligence knows that an enemy built certain new tanks numbered serially from 1 to k. If three of these tanks are captured and their serial numbers are 210, 38, and 155, use the estimator of part (b) of Exercise 12 to estimate k.

SECS. 4–8

81. On 12 days selected at random, a city's consumption of electricity was 6.4, 4.5, 10.8, 7.2, 6.8, 4.9, 3.5, 16.3, 4.8, 7.0, 8.8, and 5.4 million kilowatt-hours. Assuming that these data may be looked upon as a random sample from a gamma population, use the estimators obtained in Example 14 to estimate the parameters α and β.

82. Certain radial tires had useful lives of 35,200, 41,000, 44,700, 38,600, and 41,500 miles. Assuming that these data can be looked upon as a random sample from an exponential population, use the estimator obtained in Exercise 51 to estimate the parameter θ.

83. The size of an animal population is sometimes estimated by the **capture–recapture method.** In this method, n_1 of the animals are captured in the area under consideration, tagged, and released. Later, n_2 of the animals are captured, X of them are found to be tagged, and this information is used to estimate N, the total number of animals of the given kind in the area under consideration. If $n_1 = 3$ rare owls are captured in a section of a forest, tagged, and released, and later $n_2 = 4$ such owls are captured and only one of them is found to be tagged, estimate N by the method of maximum likelihood. (*Hint*: Try $N = 9, 10, 11, 12, 13$, and 14.)

84. Among six measurements of the boiling point of a silicon compound, the size of the error was 0.07, 0.03, 0.14, 0.04, 0.08, and 0.03°C. Assuming that these data can be looked upon as a random sample from the population of Exercise 55, use the estimator obtained there by the method of moments to estimate the parameter θ.

85. Not counting the ones that failed immediately, certain light bulbs had useful lives of 415, 433, 489, 531, 466, 410, 479, 403, 562, 422, 475, and 439 hours. Assuming that these data can be looked upon as a random sample from a two-parameter exponential population, use the estimators obtained in Exercise 56 to estimate the parameters δ and θ.

86. Rework Exercise 85 using the estimators obtained in Exercise 66 by the method of maximum likelihood.

87. Data collected over a number of years show that when a broker called a random sample of eight of her clients, she got a busy signal 6.5, 10.6, 8.1, 4.1, 9.3, 11.5, 7.3, and 5.7 percent of the time. Assuming that these figures can be looked upon as a random sample from a continuous uniform population, use the estimators obtained in Exercise 57 to estimate the parameters α and β.

88. Rework Exercise 87 using the estimators obtained in Exercise 67.

89. In a random sample of the teachers in a large school district, their annual salaries were \$23,900, \$21,500, \$26,400, \$24,800, \$33,600, \$24,500, \$29,200, \$36,200, \$22,400, \$21,500, \$28,300, \$26,800, \$31,400, \$22,700, and \$23,100. Assuming that these data can be looked upon as a random sample from a Pareto population, use the estimator obtained in Exercise 65 to estimate the parameter α.

90. Every time Mr. Jones goes to the race track he bets on three races. In a random sample of 20 of his visits to the race track, he lost all his bets 11 times, won once 7 times, and won twice on 2 occasions. If θ is the probability that he will win any one of his bets, estimate it by using the maximum likelihood estimator obtained in Exercise 68.

91. On 20 very cold days, a farmer got her tractor started on the first, third, fifth, first, second, first, third, seventh, second, fourth, fourth, eighth, first, third, sixth, fifth, second, first, sixth, and second try. Assuming that these data can be looked upon as a random sample from a geometric population, estimate its parameter θ by either of the methods of Exercise 63.

92. The I.Q.'s of 10 teenagers belonging to one ethnic group are 98, 114, 105, 101, 123, 117, 106, 92, 110, and 108, whereas those of 6 teenagers belonging to another ethnic group are 122, 105, 99, 126, 114, and 108. Assuming that these data can be looked upon as independent random samples from normal populations with the means μ_1 and μ_2 and the common variance σ^2, estimate these parameters by means of the maximum likelihood estimators obtained in Exercise 71.

SEC. 9

93. The output of a certain integrated-circuit production line is checked daily by inspecting a sample of 100 units. Over a long period of time, the process has maintained a yield of 80 percent, that is, a proportion defective of 20 percent, and the variation of the proportion defective from day to day is measured by a standard deviation of 0.04. If on a certain day the sample contains

38 defectives, find the mean of the posterior distribution of Θ as an estimate of that day's proportion defective. Assume that the prior distribution of Θ is a beta distribution.

94. Records of a university (collected over many years) show that on the average 74 percent of all incoming freshmen have I.Q.'s of at least 115. Of course, the percentage varies somewhat from year to year, and this variation is measured by a standard deviation of 3 percent. If a sample check of 30 freshmen entering the university in 2003 showed that only 18 of them have I.Q.'s of at least 115, estimate the true proportion of students with I.Q.'s of at least 115 in that freshman class using
(a) only the prior information;
(b) only the direct information;
(c) the result of Exercise 74 to combine the prior information with the direct information.

95. With reference to Example 20, find $P(712 < M < 725|\bar{x} = 692)$.

96. A history professor is making up a final examination that is to be given to a very large group of students. His feelings about the average grade that they should get is expressed subjectively by a normal distribution with the mean $\mu_0 = 65.2$ and the standard deviation $\sigma_0 = 1.5$.
(a) What prior probability does the professor assign to the actual average grade being somewhere on the interval from 63.0 to 68.0?
(b) What posterior probability would he assign to this event if the examination is tried on a random sample of 40 students whose grades have a mean of 72.9 and a standard deviation of 7.4? Use $s = 7.4$ as an estimate of σ.

97. An office manager feels that for a certain kind of business the daily number of incoming telephone calls is a random variable having a Poisson distribution, whose parameter has a prior gamma distribution with $\alpha = 50$ and $\beta = 2$. Being told that one such business had 112 incoming calls on a given day, what would be her estimate of that particular business's average daily number of incoming calls if she considers
(a) only the prior information;
(b) only the direct information;
(c) both kinds of information and the theory of Exercise 77?

SEC. 10

98. How large a random sample is required from a population whose standard deviation is 4.2 so that the sample estimate of the mean will have an error of at most 0.5 with a probability of 0.99?

99. A random sample of 36 resistors is taken from a production line manufacturing resistors to a specification of 40 ohms. Assuming a standard deviation of 1 ohm, is this sample adequate to ensure, with 95 percent probability, that the sample mean will be within 1.5 ohms of the mean of the population of resistors being produced?

100. Sections of sheet metal of various lengths are lined up on a conveyor belt that moves at a constant speed. A sample of these sections is taken for inspection by taking whatever section is passing in front of the inspection station at each five-minute interval. If the purpose is to estimate the number of defects per section in the population of all such manufactured sections, explain how this sampling procedure could be biased.

101. Comment on the sampling bias (if any) of a poll taken by asking how people will vote in an election if the sample is confined to the person claiming to be the head of household.

References

Various properties of sufficient estimators are discussed in

LEHMANN, E. L., *Theory of Point Estimation*. New York: John Wiley & Sons, Inc., 1983,

WILKS, S. S., *Mathematical Statistics*. New York: John Wiley & Sons, Inc., 1962,

and a proof of Theorem 4 may be found in

HOGG, R. V., and TANIS, E. A., *Probability and Statistical Inference*, 6th ed. Upper Saddle River, N.J.: Prentice Hall, 1995.

Important properties of maximum likelihood estimators are discussed in

KEEPING, E. S., *Introduction to Statistical Inference*. Princeton, N.J.: D. Van Nostrand Co., Inc., 1962,

and a derivation of the Cramér–Rao inequality, as well as the most general conditions under which it applies, may be found in

RAO, C. R., *Advanced Statistical Methods in Biometric Research*. New York: John Wiley & Sons, Inc., 1952.

Answers to Odd-Numbered Exercises

1 $\sum_{i=1}^{n} a_i = 1$.

9 $(n+1)Y_1$.

25 $\frac{8}{9}$.

29 **(a)** $\frac{3}{4}$; **(b)** $\frac{3}{5}$.

37 Yes.

45 Yes.

51 $\hat{\theta} = m'_1$.

53 $\hat{\lambda} = m'_1$.

55 $\hat{\theta} = 3m'_1$.

57 $\hat{\beta} = m'_1 + \sqrt{3[m'_2 - (m'_1)^2]}$.

59 $\hat{\lambda} = \bar{x}$.

61 $\hat{\beta} = \frac{\bar{x}}{2}$.

63 **(a)** $\hat{\theta} = \frac{1}{\bar{x}}$; **(b)** $\hat{\theta} = \frac{1}{\bar{x}}$.

65 $\hat{\alpha} = \dfrac{n}{\sum_{i=1}^{n} \ln x_i}$.

67 $\hat{\alpha} = y_1, \hat{\beta} = y_n$.

69 **(a)** $\hat{\beta} = \dfrac{\bar{x}}{\alpha}$; $\hat{\tau} = \left(\dfrac{2\bar{x}}{\alpha} - 1\right)^2$.

71 $\mu'_1 = \bar{v}$; $\mu'_2 = \bar{v}, \hat{\sigma}^2 = \dfrac{\sum(v-\bar{v})^2 + \sum(w-\bar{w})^2}{n_1 + n_2}$.

73 **(a)** Yes; **(b)** No.

75 $\mu = \frac{1}{2}$; $\sigma^2 = \frac{1}{18}$; symmetrical about $x = \frac{1}{2}$.

79 $\hat{\mu} = 28$.

81 $\hat{\alpha} = 4.627$ and $\hat{\beta} = 1.556$.

83 $N = 11$ or 12.

85 $\hat{\theta} = 47.69$ and $\hat{\delta} = 412.64$.

87 $\hat{\alpha} = 3.83$ and $\hat{\beta} = 11.95$.

91 $\hat{\theta} = 0.30$.

93 $E(\Theta|38) = 0.29$.

95 0.4786.

97 **(a)** $\hat{\mu} = 100$; **(b)** $\hat{\mu} = 112$; **(c)** $\hat{\mu} = 108$.

99 Yes.

INTERVAL ESTIMATION

1 Introduction

Although point estimation is a common way in which estimates are expressed, it leaves room for many questions. For instance, it does not tell us on how much information the estimate is based, nor does it tell us anything about the possible size of the error. Thus, we might have to supplement a point estimate $\hat{\theta}$ of θ with the size of the sample and the value of $\text{var}(\hat{\Theta})$ or with some other information about the sampling distribution of $\hat{\Theta}$. As we shall see, this will enable us to appraise the possible size of the error.

Alternatively, we might use **interval estimation**. An interval estimate of θ is an interval of the form $\hat{\theta}_1 < \theta < \hat{\theta}_2$, where $\hat{\theta}_1$ and $\hat{\theta}_2$ are values of appropriate random variables $\hat{\Theta}_1$ and $\hat{\Theta}_2$.

DEFINITION 1. CONFIDENCE INTERVAL. *If $\hat{\theta}_1$ and $\hat{\theta}_2$ are values of the random variables $\hat{\Theta}_1$ and $\hat{\Theta}_2$ such that*

$$P(\hat{\Theta}_1 < \theta < \hat{\Theta}_2) = 1 - \alpha$$

for some specified probability $1 - \alpha$, we refer to the interval

$$\hat{\theta}_1 < \theta < \hat{\theta}_2$$

*as a $(1 - \alpha)100\%$ **confidence interval** for θ. The probability $1 - \alpha$ is called the **degree of confidence**, and the endpoints of the interval are called the lower and upper **confidence limits**.*

For instance, when $\alpha = 0.05$, the degree of confidence is 0.95 and we get a 95% confidence interval.

It should be understood that, like point estimates, interval estimates of a given parameter are not unique. This is illustrated by Exercises 2 and 3 and also in Section 2, where we show that, based on a single random sample, there are various confidence intervals for μ, all having the same degree of confidence $1 - \alpha$. As was the case in point estimation, methods of interval estimation are judged by their various statistical properties. For instance, one desirable property is to have the length of a $(1 - \alpha)100\%$ confidence interval as short as possible; another desirable property is to have the expected length, $E(\hat{\Theta}_2 - \hat{\Theta}_1)$ as small as possible.

From Chapter 11 of *John E. Freund's Mathematical Statistics with Applications*, Eighth Edition. Irwin Miller, Marylees Miller.

2 The Estimation of Means

To illustrate how the possible size of errors can be appraised in point estimation, suppose that the mean of a random sample is to be used to estimate the mean of a normal population with the known variance σ^2. By the theorem, "If χ is the mean of a random sample of size n from a normal population with the mean μ and the variance σ^2, its sampling distribution is a normal distribution with the mean μ and the variance σ^2/n", the sampling distribution of $\overline{X}$ for random samples of size n from a normal population with the mean μ and the variance σ^2 is a normal distribution with

$$\mu_{\overline{x}} = \mu \quad \text{and} \quad \sigma_{\overline{x}}^2 = \frac{\sigma^2}{n}$$

Thus, we can write

$$P(|Z| < z_{\alpha/2}) = 1 - \alpha$$

where

$$Z = \frac{\overline{X} - \mu}{\sigma/\sqrt{n}}$$

and $z_{\alpha/2}$ is such that the integral of the standard normal density from $z_{\alpha/2}$ to ∞ equals $\alpha/2$. It follows that

$$P\left(\left|\overline{X} - \mu\right| < z_{\alpha/2} \cdot \frac{\sigma}{\sqrt{n}}\right) = 1 - \alpha$$

or, in words, we have the following theorem.

THEOREM 1. If $\overline{X}$, the mean of a random sample of size n from a normal population with the known variance σ^2, is to be used as an estimator of the mean of the population, the probability is $1 - \alpha$ that the error will be less than $z_{\alpha/2} \cdot \dfrac{\sigma}{\sqrt{n}}$.

EXAMPLE 1

A team of efficiency experts intends to use the mean of a random sample of size $n = 150$ to estimate the average mechanical aptitude of assembly-line workers in a large industry (as measured by a certain standardized test). If, based on experience, the efficiency experts can assume that $\sigma = 6.2$ for such data, what can they assert with probability 0.99 about the maximum error of their estimate?

Solution

Substituting $n = 150$, $\sigma = 6.2$, and $z_{0.005} = 2.575$ into the expression for the maximum error, we get

$$2.575 \cdot \frac{6.2}{\sqrt{150}} = 1.30$$

Thus, the efficiency experts can assert with probability 0.99 that their error will be less than 1.30.

Suppose now that these efficiency experts actually collect the necessary data and get $\bar{x} = 69.5$. Can they still assert with probability 0.99 that the error of their estimate, $\bar{x} = 69.5$, is less than 1.30? After all, $\bar{x} = 69.5$ differs from the true (population) mean by less than 1.30 or it does not, and they have no way of knowing whether it is one or the other. Actually, they can, but it must be understood that the 0.99 probability applies to the method that they used to get their estimate and calculate the maximum error (collecting the sample data, determining the value of $\bar{x}$, and using the formula of Theorem 1) and not directly to the parameter that they are trying to estimate.

To clarify this distinction, it has become the custom to use the word "confidence" here instead of "probability." *In general, we make probability statements about future values of random variables (say, the potential error of an estimate) and confidence statements once the data have been obtained.* Accordingly, we should have said in our example that the efficiency experts can be 99% confident that the error of their estimate, $\bar{x} = 69.5$, is less than 1.30.

To construct a confidence-interval formula for estimating the mean of a normal population with the known variance σ^2, we return to the probability

$$P\left(|\overline{X} - \mu| < z_{\alpha/2} \cdot \frac{\sigma}{\sqrt{n}}\right) = 1 - \alpha$$

the previous page, which we now write as

$$P\left(\overline{X} - z_{\alpha/2} \cdot \frac{\sigma}{\sqrt{n}} < \mu < \overline{X} + z_{\alpha/2} \cdot \frac{\sigma}{\sqrt{n}}\right) = 1 - \alpha$$

From this result, we have the following theorem.

THEOREM 2. If $\bar{x}$ is the value of the mean of a random sample of size n from a normal population with the known variance σ^2, then

$$\bar{x} - z_{\alpha/2} \cdot \frac{\sigma}{\sqrt{n}} < \mu < \bar{x} + z_{\alpha/2} \cdot \frac{\sigma}{\sqrt{n}}$$

is a $(1-\alpha)100\%$ confidence interval for the mean of the population.

EXAMPLE 2

If a random sample of size $n = 20$ from a normal population with the variance $\sigma^2 = 225$ has the mean $\bar{x} = 64.3$, construct a 95% confidence interval for the population mean μ.

Solution

Substituting $n = 20$, $\bar{x} = 64.3$, $\sigma = 15$, and $z_{0.025} = 1.96$ into the confidence-interval formula of Theorem 2, we get

$$64.3 - 1.96 \cdot \frac{15}{\sqrt{20}} < \mu < 64.3 + 1.96 \cdot \frac{15}{\sqrt{20}}$$

which reduces to

$$57.7 < \mu < 70.9$$

As we pointed out earlier, confidence-interval formulas are not unique. This may be seen by changing the confidence-interval formula of Theorem 2 to

$$\overline{x} - z_{2\alpha/3} \cdot \frac{\sigma}{\sqrt{n}} < \mu < \overline{x} + z_{\alpha/3} \cdot \frac{\sigma}{\sqrt{n}}$$

or to the **one-sided** $(1-\alpha)100\%$ **confidence-interval** formula

$$\mu < \overline{x} + z_{\alpha} \cdot \frac{\sigma}{\sqrt{n}}$$

Alternatively, we could base a confidence interval for μ on the sample median or, say, the midrange.

Strictly speaking, Theorems 1 and 2 require that we are dealing with a random sample from a normal population with the known variance σ^2. However, by virtue of the central limit theorem, these results can also be used for random samples from nonnormal populations provided that n is sufficiently large; that is, $n \geqq 30$. In that case, we may also substitute for σ the value of the sample standard deviation.

EXAMPLE 3

An industrial designer wants to determine the average amount of time it takes an adult to assemble an "easy-to-assemble" toy. Use the following data (in minutes), a random sample, to construct a 95% confidence interval for the mean of the population sampled:

17	13	18	19	17	21	29	22	16	28	21	15
26	23	24	20	8	17	17	21	32	18	25	22
16	10	20	22	19	14	30	22	12	24	28	11

Solution

Substituting $n = 36, \overline{x} = 19.92, z_{0.025} = 1.96$, and $s = 5.73$ for σ into the confidence-interval formula of Theorem 2, we get

$$19.92 - 1.96 \cdot \frac{5.73}{\sqrt{36}} < \mu < 19.92 + 1.96 \cdot \frac{5.73}{\sqrt{36}}$$

Thus, the 95% confidence limits are 18.05 and 21.79 minutes.

When we are dealing with a random sample from a normal population, $n < 30$, and σ is unknown, Theorems 1 and 2 cannot be used. Instead, we make use of the fact that

$$T = \frac{\overline{X} - \mu}{S/\sqrt{n}}$$

is a random variable having the t distribution with $n-1$ degrees of freedom. Substituting $\dfrac{\overline{X} - \mu}{S/\sqrt{n}}$ for T in

$$P(-t_{\alpha/2,n-1} < T < t_{\alpha/2,n-1}) = 1 - \alpha$$

we get the following confidence interval for μ.

THEOREM 3. If $\overline{x}$ and s are the values of the mean and the standard deviation of a random sample of size n from a normal population, then

$$\overline{x} - t_{\alpha/2,n-1} \cdot \frac{s}{\sqrt{n}} < \mu < \overline{x} + t_{\alpha/2,n-1} \cdot \frac{s}{\sqrt{n}}$$

is a $(1-\alpha)100\%$ confidence interval for the mean of the population.

Since this confidence-interval formula is used mainly when n is small, less than 30, we refer to it as a small-sample confidence interval for μ.

EXAMPLE 4

A paint manufacturer wants to determine the average drying time of a new interior wall paint. If for 12 test areas of equal size he obtained a mean drying time of 66.3 minutes and a standard deviation of 8.4 minutes, construct a 95% confidence interval for the true mean μ.

Solution

Substituting $\overline{x} = 66.3, s = 8.4$, and $t_{0.025,11} = 2.201$ (from Table IV of "Statistical Tables"), the 95% confidence interval for μ becomes

$$66.3 - 2.201 \cdot \frac{8.4}{\sqrt{12}} < \mu < 66.3 + 2.201 \cdot \frac{8.4}{\sqrt{12}}$$

or simply

$$61.0 < \mu < 71.6$$

This means that we can assert with 95% confidence that the interval from 61.0 minutes to 71.6 minutes contains the true average drying time of the paint.

The method by which we constructed confidence intervals in this section consisted essentially of finding a suitable random variable whose values are determined by the sample data as well as the population parameters, yet whose distribution does not involve the parameter we are trying to estimate. This was the case, for example, when we used the random variable

$$Z = \frac{\overline{X} - \mu}{\sigma/\sqrt{n}}$$

whose values cannot be calculated without knowledge of μ, but whose distribution for random samples from normal populations, the standard normal distribution, does not involve μ. This method of confidence-interval construction is called the **pivotal method** and it is widely used, but there exist more general methods, such as the one discussed in the book by Mood, Graybill, and Boes referred to at the end of this chapter.

3 The Estimation of Differences Between Means

For independent random samples from normal populations

$$Z = \frac{(\overline{X}_1 - \overline{X}_2) - (\mu_1 - \mu_2)}{\sqrt{\dfrac{\sigma_1^2}{n_1} + \dfrac{\sigma_2^2}{n_2}}}$$

has the standard normal distribution. If we substitute this expression for Z into

$$P(-z_{\alpha/2} < Z < z_{\alpha/2}) = 1 - \alpha$$

the pivotal method yields the following confidence-interval formula for $\mu_1 - \mu_2$.

THEOREM 4. If $\overline{x}_1$ and $\overline{x}_2$ are the values of the means of independent random samples of sizes n_1 and n_2 from normal populations with the known variances σ_1^2 and σ_2^2, then

$$(\overline{x}_1 - x_2) - z_{\alpha/2} \cdot \sqrt{\frac{\sigma_1^2}{n_1} + \frac{\sigma_2^2}{n_2}} < \mu_1 - \mu_2 < (\overline{x}_1 - \overline{x}_2) + z_{\alpha/2} \cdot \sqrt{\frac{\sigma_1^2}{n_1} + \frac{\sigma_2^2}{n_2}}$$

is a $(1 - \alpha)100\%$ confidence interval for the difference between the two population means.

By virtue of the central limit theorem, this confidence-interval formula can also be used for independent random samples from nonnormal populations with known variances when n_1 and n_2 are large, that is, when $n_1 \geqq 30$ and $n_2 \geqq 30$.

EXAMPLE 5

Construct a 94% confidence interval for the difference between the mean lifetimes of two kinds of light bulbs, given that a random sample of 40 light bulbs of the first kind lasted on the average 418 hours of continuous use and 50 light bulbs of the second kind lasted on the average 402 hours of continuous use. The population standard deviations are known to be $\sigma_1 = 26$ and $\sigma_2 = 22$.

Solution

For $\alpha = 0.06$, we find from Table III of "Statistical Tables" that $z_{0.03} = 1.88$. Therefore, the 94% confidence interval for $\mu_1 - \mu_2$ is

$$(418 - 402) - 1.88 \cdot \sqrt{\frac{26^2}{40} + \frac{22^2}{50}} < \mu_1 - \mu_2 < (418 - 402) + 1.88 \cdot \sqrt{\frac{26^2}{40} + \frac{22^2}{50}}$$

which reduces to

$$6.3 < \mu_1 - \mu_2 < 25.7$$

Hence, we are 94% confident that the interval from 6.3 to 25.7 hours contains the actual difference between the mean lifetimes of the two kinds of light bulbs. The fact that both confidence limits are positive suggests that on the average the first kind of light bulb is superior to the second kind.

To construct a $(1 - \alpha)100\%$ confidence interval for the difference between two means when $n_1 \geqq 30$, $n_2 \geqq 30$, but σ_1 and σ_2 are unknown, we simply substitute

s_1 and s_2 for σ_1 and σ_2 and proceed as before. When σ_1 and σ_2 are unknown and either or both of the samples are small, the procedure for estimating the difference between the means of two normal populations is not straightforward unless it can be assumed that $\sigma_1 = \sigma_2$. If $\sigma_1 = \sigma_2 = \sigma$, then

$$Z = \frac{(\overline{X}_1 - \overline{X}_2) - (\mu_1 - \mu_2)}{\sigma\sqrt{\frac{1}{n_1} + \frac{1}{n_2}}}$$

is a random variable having the standard normal distribution, and σ^2 can be estimated by **pooling** the squared deviations from the means of the two samples. In Exercise 9 the reader will be asked to verify that the resulting **pooled estimator**

$$S_p^2 = \frac{(n_1 - 1)S_1^2 + (n_2 - 1)S_2^2}{n_1 + n_2 - 2}$$

is, indeed, an unbiased estimator of σ^2. Now, by the two theorems, "If $\overline{X}$ and S^2 are the mean and the variance of a random sample of size n from a normal population with the mean μ and the standard deviation σ, then **1.** $\overline{X}$ and S^2 are independent; **2.** the random variable $\frac{(n-1)S^2}{\sigma^2}$ has a chi-square distribution with $n-1$ degrees of freedom. If $X_1, X_2, \ldots, X_n$ are independent random variables having chi-square distributions with $\nu_1, \nu_2, \ldots, \nu_n$ degrees of freedom, then $Y = \sum_{i=1}^{n} X_i$ has the chi-square distribution with $\nu_1 + \nu_2 + \cdots + \nu_n$ degrees of freedom" the independent random variables

$$\frac{(n_1 - 1)S_1^2}{\sigma^2} \quad \text{and} \quad \frac{(n_2 - 1)S_2^2}{\sigma^2}$$

have chi-square distributions with $n_1 - 1$ and $n_2 - 1$ degrees of freedom, and their sum

$$Y = \frac{(n_1 - 1)S_1^2}{\sigma^2} + \frac{(n_2 - 1)S_2^2}{\sigma^2} = \frac{(n_1 + n_2 - 2)S_p^2}{\sigma^2}$$

has a chi-square distribution with $n_1 + n_2 - 2$ degrees of freedom. Since it can be shown that the above random variables Z and Y are independent (see references at the end of this chapter)

$$T = \frac{Z}{\sqrt{\frac{Y}{n_1 + n_2 - 2}}}$$

$$= \frac{(\overline{X}_1 - \overline{X}_2) - (\mu_1 - \mu_2)}{S_p\sqrt{\frac{1}{n_1} + \frac{1}{n_2}}}$$

has a t distribution with $n_1 + n_2 - 2$ degrees of freedom. Substituting this expression for T into

$$P(-t_{\alpha/2,n-1} < T < t_{\alpha/2,n-1}) = 1 - \alpha$$

we arrive at the following $(1 - \alpha)100\%$ confidence interval for $\mu_1 - \mu_2$.

THEOREM 5. If $\bar{x}_1, \bar{x}_2, s_1$, and s_2 are the values of the means and the standard deviations of independent random samples of sizes n_1 and n_2 from normal populations with equal variances, then

$$(\bar{x}_1 - \bar{x}_2) - t_{\alpha/2, n_1+n_2-2} \cdot s_p\sqrt{\frac{1}{n_1} + \frac{1}{n_2}} < \mu_1 - \mu_2$$

$$< (\bar{x}_1 - \bar{x}_2) + t_{\alpha/2, n_1+n_2-2} \cdot s_p\sqrt{\frac{1}{n_1} + \frac{1}{n_2}}$$

is a $(1-\alpha)100\%$ confidence interval for the difference between the two population means.

Since this confidence-interval formula is used mainly when n_1 and/or n_2 are small, less than 30, we refer to it as a small-sample confidence interval for $\mu_1 - \mu_2$.

EXAMPLE 6

A study has been made to compare the nicotine contents of two brands of cigarettes. Ten cigarettes of Brand A had an average nicotine content of 3.1 milligrams with a standard deviation of 0.5 milligram, while eight cigarettes of Brand B had an average nicotine content of 2.7 milligrams with a standard deviation of 0.7 milligram. Assuming that the two sets of data are independent random samples from normal populations with equal variances, construct a 95% confidence interval for the difference between the mean nicotine contents of the two brands of cigarettes.

Solution

First we substitute $n_1 = 10, n_2 = 8, s_1 = 0.5$, and $s_2 = 0.7$ into the formula for s_p, and we get

$$s_p = \sqrt{\frac{9(0.25) + 7(0.49)}{16}} = 0.596$$

Then, substituting this value together with $n_1 = 10$, $n_2 = 8$, $\bar{x}_1 = 3.1$, $\bar{x}_2 = 2.7$, and $t_{0.025,16} = 2.120$ (from Table IV of "Statistical Tables") into the confidence-interval formula of Theorem 5, we find that the required 95% confidence interval is

$$(3.1 - 2.7) - 2.120(0.596)\sqrt{\frac{1}{10} + \frac{1}{8}} < \mu_1 - \mu_2$$

$$< (3.1 - 2.7) + 2.120(0.596)\sqrt{\frac{1}{10} + \frac{1}{8}}$$

which reduces to

$$-0.20 < \mu_1 - \mu_2 < 1.00$$

Thus, the 95% confidence limits are -0.20 and 1.00 milligrams; but observe that since this includes $\mu_1 - \mu_2 = 0$, we cannot conclude that there is a real difference between the average nicotine contents of the two brands of cigarettes.

Exercises

1. If x is a value of a random variable having an exponential distribution, find k so that the interval from 0 to kx is a $(1-\alpha)100\%$ confidence interval for the parameter θ.

2. If x_1 and x_2 are the values of a random sample of size 2 from a population having a uniform density with $\alpha = 0$ and $\beta = \theta$, find k so that

$$0 < \theta < k(x_1 + x_2)$$

is a $(1-\alpha)100\%$ confidence interval for θ when

(a) $\alpha \leqq \frac{1}{2}$; **(b)** $\alpha > \frac{1}{2}$.

3. This question has been intentionally omitted for this edition.

4. Show that the $(1-\alpha)100\%$ confidence interval

$$\bar{x} - z_{\alpha/2} \cdot \frac{\sigma}{\sqrt{n}} < \mu < \bar{x} + z_{\alpha/2} \cdot \frac{\sigma}{\sqrt{n}}$$

is shorter than the $(1-\alpha)100\%$ confidence interval

$$\bar{x} - z_{2\alpha/3} \cdot \frac{\sigma}{\sqrt{n}} < \mu < \bar{x} + z_{\alpha/3} \cdot \frac{\sigma}{\sqrt{n}}$$

5. Show that among all $(1-\alpha)100\%$ confidence intervals of the form

$$\bar{x} - z_{k\alpha} \cdot \frac{\sigma}{\sqrt{n}} < \mu < \bar{x} + z_{(1-k)\alpha} \cdot \frac{\sigma}{\sqrt{n}}$$

the one with $k = 0.5$ is the shortest.

6. Show that if $\bar{x}$ is used as a point estimate of μ and σ is known, the probability is $1-\alpha$ that $|\bar{x} - \mu|$, the absolute value of our error, will not exceed a specified amount e when

$$n = \left[z_{\alpha/2} \cdot \frac{\sigma}{e} \right]^2$$

(If it turns out that $n < 30$, this formula cannot be used unless it is reasonable to assume that we are sampling a normal population.)

7. Modify Theorem 1 so that it can be used to appraise the maximum error when σ^2 is unknown. (Note that this method can be used only after the data have been obtained.)

8. State a theorem analogous to Theorem 1, which enables us to appraise the maximum error in using $\bar{x}_1 - \bar{x}_2$ as an estimate of $\mu_1 - \mu_2$ under the conditions of Theorem 4.

9. Show that S_p^2 is an unbiased estimator of σ^2 and find its variance under the conditions of Theorem 5.

10. This question has been intentionally omitted for this edition.

4 The Estimation of Proportions

In many problems we must estimate proportions, probabilities, percentages, or rates, such as the proportion of defectives in a large shipment of transistors, the probability that a car stopped at a road block will have faulty lights, the percentage of schoolchildren with I.Q.'s over 115, or the mortality rate of a disease. In many of these it is reasonable to assume that we are sampling a binomial population and, hence, that our problem is to estimate the binomial parameter θ. Thus, we can make use of the fact that for large n the binomial distribution can be approximated with a normal distribution; that is,

$$Z = \frac{X - n\theta}{\sqrt{n\theta(1-\theta)}}$$

can be treated as a random variable having approximately the standard normal distribution. Substituting this expression for Z into

$$P(-z_{\alpha/2} < Z < z_{\alpha/2}) = 1 - \alpha$$

we get

$$P\left(-z_{\alpha/2} < \frac{X - n\theta}{\sqrt{n\theta(1-\theta)}} < z_{\alpha/2} \right) = 1 - \alpha$$

and the two inequalities

$$-z_{\alpha/2} < \frac{x - n\theta}{\sqrt{n\theta(1-\theta)}} \quad \text{and} \quad \frac{x - n\theta}{\sqrt{n\theta(1-\theta)}} < z_{\alpha/2}$$

whose solution will yield $(1-\alpha)100\%$ confidence limits for θ. Leaving the details of this to the reader in Exercise 11, let us give here instead a large-sample approximation by rewriting $P(-z_{\alpha/2} < Z < z_{\alpha/2}) = 1-\alpha$, with $\frac{X - n\theta}{\sqrt{n\theta(1-\theta)}}$ substituted for Z, as

$$P\left(\hat{\Theta} - z_{\alpha/2} \cdot \sqrt{\frac{\theta(1-\theta)}{n}} < \theta < \hat{\Theta} + z_{\alpha/2} \cdot \sqrt{\frac{\theta(1-\theta)}{n}}\right) = 1-\alpha$$

where $\hat{\Theta} = \frac{X}{n}$. Then, if we substitute $\hat{\theta}$ for θ inside the radicals, which is a further approximation, we obtain the following theorem.

THEOREM 6. If X is a binomial random variable with the parameters n and θ, n is large, and $\hat{\theta} = \frac{x}{n}$, then

$$\hat{\theta} - z_{\alpha/2} \cdot \sqrt{\frac{\hat{\theta}(1-\hat{\theta})}{n}} < \theta < \hat{\theta} + z_{\alpha/2} \cdot \sqrt{\frac{\hat{\theta}(1-\hat{\theta})}{n}}$$

is an approximate $(1-\alpha)100\%$ confidence interval for θ.

EXAMPLE 7

In a random sample, 136 of 400 persons given a flu vaccine experienced some discomfort. Construct a 95% confidence interval for the true proportion of persons who will experience some discomfort from the vaccine.

Solution

Substituting $n = 400, \hat{\theta} = \frac{136}{400} = 0.34$, and $z_{0.025} = 1.96$ into the confidence-interval formula of Theorem 6, we get

$$0.34 - 1.96\sqrt{\frac{(0.34)(0.66)}{400}} < \theta < 0.34 + 1.96\sqrt{\frac{(0.34)(0.66)}{400}}$$

$$0.294 < \theta < 0.386$$

or, rounding to two decimals, $0.29 < \theta < 0.39$.

Using the same approximations that led to Theorem 6, we can also obtain the following theorem.

THEOREM 7. If $\hat{\theta} = \frac{x}{n}$ is used as an estimate of θ, we can assert with $(1-\alpha)100\%$ confidence that the error is less than

$$z_{\alpha/2} \cdot \sqrt{\frac{\hat{\theta}(1-\hat{\theta})}{n}}$$

EXAMPLE 8

A study is made to determine the proportion of voters in a sizable community who favor the construction of a nuclear power plant. If 140 of 400 voters selected at random favor the project and we use $\hat{\theta} = \frac{140}{400} = 0.35$ as an estimate of the actual proportion of all voters in the community who favor the project, what can we say with 99% confidence about the maximum error?

Solution
Substituting $n = 400, \hat{\theta} = 0.35$, and $z_{0.005} = 2.575$ into the formula of Theorem 7, we get

$$2.575 \cdot \sqrt{\frac{(0.35)(0.65)}{400}} = 0.061$$

or 0.06 rounded to two decimals. Thus, if we use $\hat{\theta} = 0.35$ as an estimate of the actual proportion of voters in the community who favor the project, we can assert with 99% confidence that the error is less than 0.06.

---

5 The Estimation of Differences Between Proportions

In many problems we must estimate the difference between the binomial parameters θ_1 and θ_2 on the basis of independent random samples of sizes n_1 and n_2 from two binomial populations. This would be the case, for example, if we want to estimate the difference between the proportions of male and female voters who favor a certain candidate for governor of Illinois.

If the respective numbers of successes are X_1 and X_2 and the corresponding sample proportions are denoted by $\hat{\Theta}_1 = \frac{X_1}{n_1}$ and $\hat{\Theta}_2 = \frac{X_2}{n_2}$, let us investigate the sampling distribution of $\hat{\Theta}_1 - \hat{\Theta}_2$, which is an obvious estimator of $\theta_1 - \theta_2$. Let's take

$$E(\hat{\Theta}_1 - \hat{\Theta}_2) = \theta_1 - \theta_2$$

and

$$\text{var}(\hat{\Theta}_1 - \hat{\Theta}_2) = \frac{\theta_1(1-\theta_1)}{n_1} + \frac{\theta_2(1-\theta_2)}{n_2}$$

and since, for large samples, X_1 and X_2, and hence also their difference, can be approximated with normal distributions, it follows that

$$Z = \frac{(\hat{\Theta}_1 - \hat{\Theta}_2) - (\theta_1 - \theta_2)}{\sqrt{\dfrac{\theta_1(1-\theta_1)}{n_1} + \dfrac{\theta_2(1-\theta_2)}{n_2}}}$$

is a random variable having approximately the standard normal distribution. Substituting this expression for Z into $P(-z_{\alpha/2} < Z < z_{\alpha/2}) = 1-\alpha$, we arrive at the following result.

THEOREM 8. If X_1 is a binomial random variable with the parameters n_1 and θ_1, X_2 is a binomial random variable with the parameters n_2 and θ_2, n_1 and n_2 are large, and $\hat{\theta}_1 = \dfrac{x_1}{n_1}$ and $\hat{\theta}_2 = \dfrac{x_2}{n_2}$, then

$$(\hat{\theta}_1 - \hat{\theta}_2) - z_{\alpha/2} \cdot \sqrt{\frac{\hat{\theta}_1(1-\hat{\theta}_1)}{n_1} + \frac{\hat{\theta}_2(1-\hat{\theta}_2)}{n_2}} < \theta_1 - \theta_2$$

$$< (\hat{\theta}_1 - \hat{\theta}_2) + z_{\alpha/2} \cdot \sqrt{\frac{\hat{\theta}_1(1-\hat{\theta}_1)}{n_1} + \frac{\hat{\theta}_2(1-\hat{\theta}_2)}{n_2}}$$

is an approximate $(1-\alpha)100\%$ confidence interval for $\theta_1 - \theta_2$.

EXAMPLE 9

If 132 of 200 male voters and 90 of 150 female voters favor a certain candidate running for governor of Illinois, find a 99% confidence interval for the difference between the actual proportions of male and female voters who favor the candidate.

Solution

Substituting $\hat{\theta}_1 = \frac{132}{200} = 0.66$, $\hat{\theta}_2 = \frac{90}{150} = 0.60$, and $z_{0.005} = 2.575$ into the confidence-interval formula of Theorem 8, we get

$$(0.66 - 0.60) - 2.575\sqrt{\frac{(0.66)(0.34)}{200} + \frac{(0.60)(0.40)}{150}} < \theta_1 - \theta_2$$

$$< (0.66 - 0.60) + 2.575\sqrt{\frac{(0.66)(0.34)}{200} + \frac{(0.60)(0.40)}{150}}$$

which reduces to

$$-0.074 < \theta_1 - \theta_2 < 0.194$$

Thus, we are 99% confident that the interval from -0.074 to 0.194 contains the difference between the actual proportions of male and female voters who favor the candidate. Observe that this includes the possibility of a zero difference between the two proportions.

Exercises

11. By solving

$$-z_{\alpha/2} = \frac{x - n\theta}{\sqrt{n\theta(1-\theta)}} \quad \text{and} \quad \frac{x - n\theta}{\sqrt{n\theta(1-\theta)}} = z_{\alpha/2}$$

for θ, show that

$$\frac{x + \frac{1}{2} \cdot z_{\alpha/2}^2 \pm z_{\alpha/2}\sqrt{\frac{x(n-x)}{n} + \frac{1}{4} \cdot z_{\alpha/2}^2}}{n + z_{\alpha/2}^2}$$

are $(1-\alpha)100\%$ confidence limits for θ.

12. Use the formula of Theorem 7 to demonstrate that we can be at least $(1-\alpha)100\%$ confident that the error we make is less than e when we use a sample proportion $\hat{\theta} = \frac{x}{n}$ with

$$n = \frac{z_{\alpha/2}^2}{4e^2}$$

as an estimate of θ.

13. Find a formula for n analogous to that of Exercise 12 when it is known that θ must lie on the interval from θ' to θ''.

14. Fill in the details that led from the Z statistic on the previous page, substituted into $P(-z_{\alpha/2} < Z < z_{\alpha/2}) = 1 - \alpha$, to the confidence-interval formula of Theorem 8.

15. Find a formula for the maximum error analogous to that of Theorem 7 when we use $\hat{\theta}_1 - \hat{\theta}_2$ as an estimate of $\theta_1 - \theta_2$.

16. Use the result of Exercise 15 to show that when $n_1 = n_2 = n$, we can be at least $(1-\alpha)100\%$ confident that the error that we make when using $\hat{\theta}_1 - \hat{\theta}_2$ as an estimate of $\theta_1 - \theta_2$ is less than e when

$$n = \frac{z_{\alpha/2}^2}{2e^2}$$

6 The Estimation of Variances

Given a random sample of size n from a normal population, we can obtain a $(1-\alpha)100\%$ confidence interval for σ^2 by making use of the theorem referred under Section 3, according to which

$$\frac{(n-1)S^2}{\sigma^2}$$

is a random variable having a chi-square distribution with $n-1$ degrees of freedom. Thus,

$$P\left[\chi^2_{1-\alpha/2,\,n-1} < \frac{(n-1)S^2}{\sigma^2} < \chi^2_{\alpha/2,\,n-1}\right] = 1 - \alpha$$

$$P\left[\frac{(n-1)S^2}{\chi^2_{\alpha/2,\,n-1}} < \sigma^2 < \frac{(n-1)S^2}{\chi^2_{1-\alpha/2,\,n-1}}\right] = 1 - \alpha$$

Thus, we obtain the following theorem.

THEOREM 9. If s^2 is the value of the variance of a random sample of size n from a normal population, then

$$\frac{(n-1)s^2}{\chi^2_{\alpha/2,\,n-1}} < \sigma^2 < \frac{(n-1)s^2}{\chi^2_{1-\alpha/2,\,n-1}}$$

is a $(1-\alpha)100\%$ confidence interval for σ^2.

Corresponding $(1-\alpha)100\%$ confidence limits for σ can be obtained by taking the square roots of the confidence limits for σ^2.

EXAMPLE 10

In 16 test runs the gasoline consumption of an experimental engine had a standard deviation of 2.2 gallons. Construct a 99% confidence interval for σ^2, which measures the true variability of the gasoline consumption of the engine.

Solution

Assuming that the observed data can be looked upon as a random sample from a normal population, we substitute $n = 16$ and $s = 2.2$, along with $\chi^2_{0.005,15} = 32.801$ and $\chi^2_{0.995,15} = 4.601$, obtained from Table V of "Statistical Tables", into the confidence-interval formula of Theorem 9, and we get

$$\frac{15(2.2)^2}{32.801} < \sigma^2 < \frac{15(2.2)^2}{4.601}$$

or

$$2.21 < \sigma^2 < 15.78$$

To get a corresponding 99% confidence interval for σ, we take square roots and get $1.49 < \sigma < 3.97$.

7 The Estimation of the Ratio of Two Variances

If S_1^2 and S_2^2 are the variances of independent random samples of sizes n_1 and n_2 from normal populations, then, according to the theorem, "If S_1^2 and S_2^2 are the variances of independent random samples of sizes n_1 and n_2 from normal populations with the variances σ_1^2 and σ_2^2, then $F = \frac{S_1^2/\sigma_1^2}{S_2^2/\sigma_2^2} = \frac{\sigma_2^2 S_1^2}{\sigma_1^2 S_2^2}$ is a random variable having an F distribution with $n_1 - 1$ and $n_2 - 1$ degrees of freedom",

$$F = \frac{\sigma_2^2 S_1^2}{\sigma_1^2 S_2^2}$$

is a random variable having an F distribution with $n_1 - 1$ and $n_2 - 1$ degrees of freedom. Thus, we can write

$$P\left(f_{1-\alpha/2,n_1-1,n_2-1} < \frac{\sigma_2^2 S_1^2}{\sigma_1^2 S_2^2} < f_{\alpha/2,n_1-1,n_2-1}\right) = 1-\alpha$$

Since

$$f_{1-\alpha/2,n_1-1,n_2-1} = \frac{1}{f_{\alpha/2,n_2-1,n_1-1}}$$

we have the following result.

Theorem 10. If s_1^2 and s_2^2 are the values of the variances of independent random samples of sizes n_1 and n_2 from normal populations, then

$$\frac{s_1^2}{s_2^2} \cdot \frac{1}{f_{\alpha/2, n_1-1, n_2-1}} < \frac{\sigma_1^2}{\sigma_2^2} < \frac{s_1^2}{s_2^2} \cdot f_{\alpha/2, n_2-1, n_1-1}$$

is a $(1-\alpha)100\%$ confidence interval for $\frac{\sigma_1^2}{\sigma_2^2}$.

Corresponding $(1-\alpha)100\%$ confidence limits for $\frac{\sigma_1}{\sigma_2}$ can be obtained by taking the square roots of the confidence limits for $\frac{\sigma_1^2}{\sigma_2^2}$.

EXAMPLE 11

With reference to Example 6, find a 98% confidence interval for $\frac{\sigma_1^2}{\sigma_2^2}$.

Solution

Substituting $n_1 = 10, n_2 = 8, s_1 = 0.5, s_2 = 0.7$, and $f_{0.01,9,7} = 6.72$ and $f_{0.01,7,9} = 5.61$ from Table VI of "Statistical Tables", we get

$$\frac{0.25}{0.49} \cdot \frac{1}{6.72} < \frac{\sigma_1^2}{\sigma_2^2} < \frac{0.25}{0.49} \cdot 5.61$$

or

$$0.076 < \frac{\sigma_1^2}{\sigma_2^2} < 2.862$$

Since the interval obtained here includes the possibility that the ratio is 1, there is no real evidence against the assumption of equal population variances in Example 6.

Exercises

17. If it can be assumed that the binomial parameter θ assumes a value close to zero, upper confidence limits of the form $\theta < C$ are often useful. For a random sample of size n, the one-sided interval

$$\theta < \frac{1}{2n} \chi^2_{\alpha, 2(x+1)}$$

has a confidence level closely approximating $(1-\alpha)$ 100%. Use this formula to find a 99% upper confidence limit for the proportion of defectives produced by a process if a sample of 200 units contains three defectives.

18. Fill in the details that led from the probability in Section 6 to the confidence-interval formula of Theorem 10.

19. For large n, the sampling distribution of S is sometimes approximated with a normal distribution having the mean σ and the variance $\frac{\sigma^2}{2n}$. Show that this approximation leads to the following $(1-\alpha)100\%$ large-sample confidence interval for σ:

$$\frac{s}{1 + \frac{z_{\alpha/2}}{\sqrt{2n}}} < \sigma < \frac{s}{1 - \frac{z_{\alpha/2}}{\sqrt{2n}}}$$

8 The Theory in Practice

In the examples of this chapter we showed a number of details about substitutions into the various formulas and subsequent calculations. In practice, none of this is really necessary, because there is an abundance of software that requires only that we enter the original **raw** (untreated) **data** into our computer together with the appropriate commands. To illustrate, consider the following example.

EXAMPLE 12

To study the durability of a new paint for white center lines, a highway department painted test strips across heavily traveled roads in eight different locations, and electronic counters showed that they deteriorated after having been crossed by (to the nearest hundred) 142,600, 167,800, 136,500, 108,300, 126,400, 133,700, 162,000, and 149,400 cars. Construct a 95% confidence interval for the average amount of traffic (car crossings) that this paint can withstand before it deteriorates.

Solution

The computer printout of Figure 1 shows that the desired confidence interval is

$$124{,}758 < \mu < 156{,}917$$

car crossings. It also shows the sample size, the mean of the data, their standard deviation, and the estimated standard error of the mean, SE MEAN, which is given by $\frac{s}{\sqrt{n}}$.

```
DATA> 142600 167800 136500 108300 126400 133700 162000 149400
DATA> tint 95 c1

One-Sample T: C1

Variable        N      Mean     StDev   SE Mean          95.0% CI
C1              8    140838     19228      6798   (  124751,  156924)
MTB >
```

Figure 1. Computer printout for Example 12.

As used in this example, computers enable us to do more efficiently—faster, more cheaply, and almost automatically—what was done previously by means of desk calculators, hand-held calculators, or even by hand. However, dealing with a sample of size $n = 8$, the example cannot very well do justice to the power of computers to handle enormous sets of data and perform calculations not even deemed possible until recent years. Also, our example does not show how computers can summarize the output as well the input and the results as well as the original data in various kinds of graphs and charts, which allow for methods of analysis that were not available in the past.

All this is important, but it does not do justice to the phenomenal impact that computers have had on statistics. Among other things, computers can be used to tabulate or graph functions (say, the t, F, or χ^2 distributions) and thus give the investigator a clear understanding of underlying models and make it possible to study the effects of violations of assumptions. Also important is the use of computers in simulating values of random variables (that is, sampling all kinds of populations) when a formal mathematical approach is not feasible. This provides an important tool when we study the appropriateness of statistical models.

In the applied exercises that follow, the reader is encouraged to use a statistical computer program as much as possible.

Applied Exercises SECS. 1–3

20. A district official intends to use the mean of a random sample of 150 sixth graders from a very large school district to estimate the mean score that all the sixth graders in the district would get if they took a certain arithmetic achievement test. If, based on experience, the official knows that $\sigma = 9.4$ for such data, what can she assert with probability 0.95 about the maximum error?

21. With reference to Exercise 20, suppose that the district official takes her sample and gets $\bar{x} = 61.8$. Use all the given information to construct a 99% confidence interval for the mean score of all the sixth graders in the district.

22. A medical research worker intends to use the mean of a random sample of size $n = 120$ to estimate the mean blood pressure of women in their fifties. If, based on experience, he knows that $\sigma = 10.5$ mm of mercury, what can he assert with probability 0.99 about the maximum error?

23. With reference to Exercise 22, suppose that the research worker takes his sample and gets $\bar{x} = 141.8$ mm of mercury. Construct a 98% confidence interval for the mean blood pressure of women in their fifties.

24. A study of the annual growth of certain cacti showed that 64 of them, selected at random in a desert region, grew on the average 52.80 mm with a standard deviation of 4.5 mm. Construct a 99% confidence interval for the true average annual growth of the given kind of cactus.

25. To estimate the average time required for certain repairs, an automobile manufacturer had 40 mechanics, a random sample, timed in the performance of this task. If it took them on the average 24.05 minutes with a standard deviation of 2.68 minutes, what can the manufacturer assert with 95% confidence about the maximum error if he uses $\bar{x} = 24.05$ minutes as an estimate of the actual mean time required to perform the given repairs?

26. This question has been intentionally omitted for this edition.

27. Use the modification suggested in Exercise 26 to rework Exercise 21, given that there are 900 sixth graders in the school district.

28. An efficiency expert wants to determine the average amount of time it takes a pit crew to change a set of four tires on a race car. Use the formula for n in Exercise 6 to determine the sample size that is needed so that the efficiency expert can assert with probability 0.95 that the sample mean will differ from μ, the quantity to be estimated, by less than 2.5 seconds. It is known from previous studies that $\sigma = 12.2$ seconds.

29. In a study of television viewing habits, it is desired to estimate the average number of hours that teenagers spend watching per week. If it is reasonable to assume that $\sigma = 3.2$ hours, how large a sample is needed so that it will be possible to assert with 95% confidence that the sample mean is off by less than 20 minutes. (*Hint*: Refer to Exercise 6.)

30. The length of the skulls of 10 fossil skeletons of an extinct species of bird has a mean of 5.68 cm and a standard deviation of 0.29 cm. Assuming that such measurements are normally distributed, find a 95% confidence interval for the mean length of the skulls of this species of bird.

31. A major truck stop has kept extensive records on various transactions with its customers. If a random sample of 18 of these records shows average sales of 63.84 gallons of diesel fuel with a standard deviation of 2.75 gallons, construct a 99% confidence interval for the mean of the population sampled.

32. A food inspector, examining 12 jars of a certain brand of peanut butter, obtained the following percentages of impurities: 2.3, 1.9, 2.1, 2.8, 2.3, 3.6, 1.4, 1.8, 2.1, 3.2, 2.0, and 1.9. Based on the modification of Theorem 1 of Exercise 7, what can she assert with 95% confidence about the maximum error if she uses the mean of this sample as an estimate of the average percentage of impurities in this brand of peanut butter?

33. Independent random samples of sizes $n_1 = 16$ and $n_2 = 25$ from normal populations with $\sigma_1 = 4.8$ and $\sigma_2 = 3.5$ have the means $\bar{x}_1 = 18.2$ and $\bar{x}_2 = 23.4$. Find a 90% confidence interval for $\mu_1 - \mu_2$.

34. A study of two kinds of photocopying equipment shows that 61 failures of the first kind of equipment took on the average 80.7 minutes to repair with a standard deviation of 19.4 minutes, whereas 61 failures of the second kind of equipment took on the average 88.1 minutes to repair with a standard deviation of 18.8 minutes. Find a 99% confidence interval for the difference between the true average amounts of time it takes to repair failures of the two kinds of photocopying equipment.

35. Twelve randomly selected mature citrus trees of one variety have a mean height of 13.8 feet with a standard deviation of 1.2 feet, and 15 randomly selected mature citrus trees of another variety have a mean height of 12.9 feet with a standard deviation of 1.5 feet. Assuming that the random samples were selected from normal populations with equal variances, construct a 95% confidence interval for the difference between the true average heights of the two kinds of citrus trees.

36. The following are the heat-producing capacities of coal from two mines (in millions of calories per ton):

Mine A: 8,500, 8,330, 8,480, 7,960, 8,030
Mine B: 7,710, 7,890, 7,920, 8,270, 7,860

Assuming that the data constitute independent random samples from normal populations with equal variances, construct a 99% confidence interval for the difference between the true average heat-producing capacities of coal from the two mines.

37. To study the effect of alloying on the resistance of electric wires, an engineer plans to measure the resistance of $n_1 = 35$ standard wires and $n_2 = 45$ alloyed wires. If it can be assumed that $\sigma_1 = 0.004$ ohm and $\sigma_2 = 0.005$ ohm for such data, what can she assert with 98% confidence about the maximum error if she uses $\bar{x}_1 - \bar{x}_2$ as an estimate of $\mu_1 - \mu_2$? (*Hint*: Use the result of Exercise 8.)

SECS. 4–5

38. A sample survey at a supermarket showed that 204 of 300 shoppers regularly use coupons. Use the large-sample confidence-interval formula of Theorem 6 to construct a 95% confidence interval for the corresponding true proportion.

39. With reference to Exercise 38, what can we say with 99% confidence about the maximum error if we use the observed sample proportion as an estimate of the proportion of all shoppers in the population sampled who use coupons?

40. In a random sample of 250 television viewers in a large city, 190 had seen a certain controversial program. Construct a 99% confidence interval for the corresponding true proportion using
(a) the large-sample confidence-interval formula of Theorem 6;
(b) the confidence limits of Exercise 11.

41. With reference to Exercise 40, what can we say with 95% confidence about the maximum error if we use the observed sample proportion as an estimate of the corresponding true proportion?

42. Among 100 fish caught in a certain lake, 18 were inedible as a result of chemical pollution. Construct a 99% confidence interval for the corresponding true proportion.

43. In a random sample of 120 cheerleaders, 54 had suffered moderate to severe damage to their voices. With 90% confidence, what can we say about the maximum error if we use the sample proportion $\frac{54}{120} = 0.45$ as an estimate of the true proportion of cheerleaders who are afflicted in this way?

44. In a random sample of 300 persons eating lunch at a department store cafeteria, only 102 had dessert. If we use $\frac{102}{300} = 0.34$ as an estimate of the corresponding true proportion, with what confidence can we assert that our error is less than 0.05?

45. A private opinion poll is engaged by a politician to estimate what proportion of her constituents favor the decriminalization of certain minor narcotics violations. Use the formula of Exercise 12 to determine how large a sample the poll will have to take to be at least 95% confident that the sample proportion is off by less than 0.02.

46. Use the result of Exercise 13 to rework Exercise 45, given that the poll has reason to believe that the true proportion does not exceed 0.30.

47. Suppose that we want to estimate what proportions of all drivers exceed the legal speed limit on a certain stretch of road between Los Angeles and Bakersfield. Use the formula of Exercise 12 to determine how large a sample we will need to be at least 99% confident that the resulting estimate, the sample proportion, is off by less than 0.04.

48. Use the result of Exercise 13 to rework Exercise 47, given that we have good reason to believe that the proportion we are trying to estimate is at least 0.65.

49. In a random sample of visitors to a famous tourist attraction, 84 of 250 men and 156 of 250 women bought souvenirs. Construct a 95% confidence interval for the difference between the true proportions of men and women who buy souvenirs at this tourist attraction.

50. Among 500 marriage license applications chosen at random in a given year, there were 48 in which the woman was at least one year older than the man, and among 400 marriage license applications chosen at random six years later, there were 68 in which the woman was at least one year older than the man. Construct a 99% confidence interval for the difference between the corresponding true proportions of marriage license applications in which the woman was at least one year older than the man.

51. With reference to Exercise 50, what can we say with 98% confidence about the maximum error if we use the difference between the observed sample proportions as an estimate of the difference between the corresponding true proportions? (*Hint*: Use the result of Exercise 15.)

52. Suppose that we want to determine the difference between the proportions of the customers of a donut chain in North Carolina and Vermont who prefer the chain's donuts to those of all its competitors. Use the formula of Exercise 16 to determine the size of the samples that are needed to be at least 95% confident that the difference between the two sample proportions is off by less than 0.05.

SECS. 6–7

53. With reference to Exercise 30, construct a 95% confidence interval for the true variance of the skull length of the given species of bird.

54. With reference to Exercise 32, construct a 90% confidence interval for the standard deviation of the population sampled, that is, for the percentage of impurities in the given brand of peanut butter.

55. With reference to Exercise 24, use the large-sample confidence-interval formula of Exercise 19 to construct a 99% confidence interval for the standard deviation of the annual growth of the given kind of cactus.

56. With reference to Exercise 25, use the large-sample confidence-interval formula of Exercise 19 to construct a 98% confidence interval for the standard deviation of the time it takes a mechanic to perform the given task.

57. With reference to Exercise 34, construct a 98% confidence interval for the ratio of the variances of the two populations sampled.

58. With reference to Exercise 35, construct a 98% confidence interval for the ratio of the variances of the two populations sampled.

59. With reference to Exercise 36, construct a 90% confidence interval for the ratio of the variances of the two populations sampled.

SEC. 8

60. Twenty pilots were tested in a flight simulator, and the time for each to complete a certain corrective action was measured in seconds, with the following results:

5.2	5.6	7.6	6.8	4.8	5.7	9.0	6.0	4.9	7.4
6.5	7.9	6.8	4.3	8.5	3.6	6.1	5.8	6.4	4.0

Use a computer program to find a 95% confidence interval for the mean time to take corrective action.

61. The following are the compressive strengths (given to the nearest 10 psi) of 30 concrete samples.

4890	4830	5490	4820	5230	4960	5040	5060	4500	5260
4600	4630	5330	5160	4950	4480	5310	4730	4710	4390
4820	4550	4970	4740	4840	4910	4880	5200	5150	4890

Use a computer program to find a 90% confidence interval for the standard deviation of these compressive strengths.

References

A general method for obtaining confidence intervals is given in

MOOD, A. M., GRAYBILL, F. A., and BOES, D. C., *Introduction to the Theory of Statistics*, 3rd ed. New York: McGraw-Hill Book Company, 1974,

and further criteria for judging the relative merits of confidence intervals may be found in

LEHMANN, E. L., *Testing Statistical Hypotheses*. New York: John Wiley & Sons, Inc., 1959,

and in other advanced texts on mathematical statistics. Special tables for constructing 95% and 99% confidence intervals for proportions are given in the *Biometrika Tables*. For a proof of the independence of the random variables Z and Y in Section 3, see

BRUNK, H. D., *An Introduction to Mathematical Statistics*, 3rd ed. Lexington, Mass.: Xerox Publishing Co., 1975.

Answers to Odd-Numbered Exercises

1 $k = \dfrac{-1}{\ln(1-\alpha)}$.

3 $c = \dfrac{1 \pm \sqrt{1-\alpha}}{\alpha}$.

7 Substitute $t_{\alpha/2,n-1}\dfrac{s}{\sqrt{n}}$ for $z_{\alpha/2}\dfrac{\sigma}{\sqrt{n}}$.

9 $\dfrac{2\sigma^4}{(n_1 + n_2 - \lambda)}$.

13 $n = \theta^*(1-\theta^*)\dfrac{z^2_{\alpha/2}}{e^2}$, where θ^* is the value on the interval from θ' to θ^n closest to $\frac{1}{2}$.

15 $E < z_{\alpha/2}\sqrt{\dfrac{\hat{\theta}_1(1-\hat{\theta}_1)}{n_1} + \dfrac{\hat{\theta}_2(1-\hat{\theta}_2)}{n_2}}$.

17 0.050.

21 $59.82 < \mu < 63.78$.

23 $139.57 < \mu < 144.03$.

25 0.83 minute.

27 $59.99 < \mu < 63.61$.

29 355.

31 $61.96 < \mu < 65.72$ gallons.

33 $-7.485 < \mu_1 - \mu_2 < -2.915$.

35 $-1.198 < \mu_1 - \mu_2 < 1.998$ feet.

37 0.0023 ohm.

39 0.069.

41 0.053.

43 0.075.

45 $n = 2,401$.

47 $n = 1,037$.

49 $-0.372 < \theta_1 - \theta_2 < -0.204$.

51 0.053.

53 $0.04 < \sigma^2 < 0.28$.

55 $3.67 < \sigma < 5.83$.

57 $0.58 < \dfrac{\sigma_1^2}{\sigma_2^2} < 1.96$.

59 $0.233 < \dfrac{\sigma_1^2}{\sigma_2^2} < 9.506$.

61 $227.7 < \sigma < 352.3$.

Hypothesis Testing

1 Introduction

If an engineer has to decide on the basis of sample data whether the true average lifetime of a certain kind of tire is at least 42,000 miles, if an agronomist has to decide on the basis of experiments whether one kind of fertilizer produces a higher yield of soybeans than another, and if a manufacturer of pharmaceutical products has to decide on the basis of samples whether 90 percent of all patients given a new medication will recover from a certain disease, these problems can all be translated into the language of **statistical tests of hypotheses**. In the first case we might say that the engineer has to test the hypothesis that θ, the parameter of an exponential population, is at least 42,000; in the second case we might say that the agronomist has to decide whether $\mu_1 > \mu_2$, where μ_1 and μ_2 are the means of two normal populations; and in the third case we might say that the manufacturer has to decide whether θ, the parameter of a binomial population, equals 0.90. In each case it must be assumed, of course, that the chosen distribution correctly describes the experimental conditions; that is, the distribution provides the correct **statistical model**.

As in the preceding examples, most tests of statistical hypotheses concern the parameters of distributions, but sometimes they also concern the type, or nature, of the distributions themselves. For instance, in the first of our three examples the engineer may also have to decide whether he is actually dealing with a sample from an exponential population or whether his data are values of random variables having, say, the Weibull distribution.

DEFINITION 1. STATISTICAL HYPOTHESIS. *An assertion or conjecture about the distribution of one or more random variables is called a* ***statistical hypothesis****. If a statistical hypothesis completely specifies the distribution, it is called a* ***simple hypothesis****; if not, it is referred to as a* ***composite hypothesis****.*

A simple hypothesis must therefore specify not only the functional form of the underlying distribution, but also the values of all parameters. Thus, in the third of the above examples, the one dealing with the effectiveness of the new medication, the hypothesis $\theta = 0.90$ is simple, assuming, of course, that we specify the sample size and that the population is binomial. However, in the first of the preceding examples the hypothesis is composite since $\theta \geqq 42{,}000$ does not assign a specific value to the parameter θ.

To be able to construct suitable criteria for testing statistical hypotheses, it is necessary that we also formulate **alternative hypotheses**. For instance, in the example dealing with the lifetimes of the tires, we might formulate the alternative hypothesis that the parameter θ of the exponential population is less than 42,000; in the example dealing with the two kinds of fertilizer, we might formulate the alternative hypothesis $\mu_1 = \mu_2$; and in the example dealing with the new medication, we might formulate the alternative hypothesis that the parameter θ of the given binomial population is only 0.60, which is the disease's recovery rate without the new medication.

The concept of simple and composite hypotheses applies also to alternative hypotheses, and in the first example we can now say that we are testing the composite hypothesis $\theta \geqq 42{,}000$ against the **composite alternative** $\theta < 42{,}000$, where θ is the parameter of an exponential population. Similarly, in the second example we are testing the composite hypothesis $\mu_1 > \mu_2$ against the composite alternative $\mu_1 = \mu_2$, where μ_1 and μ_2 are the means of two normal populations, and in the third example we are testing the simple hypothesis $\theta = 0.90$ against the **simple alternative** $\theta = 0.60$, where θ is the parameter of a binomial population for which n is given.

Frequently, statisticians formulate as their hypotheses the exact opposite of what they may want to show. For instance, if we want to show that the students in one school have a higher average I.Q. than those in another school, we might formulate the hypothesis that there is no difference: the hypothesis $\mu_1 = \mu_2$. With this hypothesis we know what to expect, but this would not be the case if we formulated the hypothesis $\mu_1 > \mu_2$, at least not unless we specify the actual difference between μ_1 and μ_2.

Similarly, if we want to show that one kind of ore has a higher percentage content of uranium than another kind of ore, we might formulate the hypothesis that the two percentages are the same; and if we want to show that there is a greater variability in the quality of one product than there is in the quality of another, we might formulate the hypothesis that there is no difference; that is, $\sigma_1 = \sigma_2$. In view of the assumptions of "no difference," hypotheses such as these led to the term **null hypothesis**, but nowadays this term is applied to any hypothesis that we may want to test.

Symbolically, we shall use the symbol H_0 for the null hypothesis that we want to test and H_1 or H_A for the alternative hypothesis. Problems involving more than two hypotheses, that is, problems involving several alternative hypotheses, tend to be quite complicated.

2 Testing a Statistical Hypothesis

The testing of a statistical hypothesis is the application of an explicit set of rules for deciding on the basis of a random sample whether to accept the null hypothesis or to reject it in favor of the alternative hypothesis. Suppose, for example, that a statistician wants to test the null hypothesis $\theta = \theta_0$ against the alternative hypothesis $\theta = \theta_1$. In order to make a choice, he will generate sample data by conducting an experiment and then compute the value of a **test statistic**, which will tell him what action to take for each possible outcome of the sample space. The test procedure, therefore, partitions the possible values of the test statistic into two subsets: an **acceptance region** for H_0 and a **rejection region** for H_0.

The procedure just described can lead to two kinds of errors. For instance, if the true value of the parameter θ is θ_0 and the statistician incorrectly concludes that $\theta = \theta_1$, he is committing an error referred to as a **type I error**. On the other hand, if the true value of the parameter θ is θ_1 and the statistician incorrectly concludes that $\theta = \theta_0$, he is committing a second kind of error referred to as a **type II error**.

DEFINITION 2. TYPE I AND TYPE II ERRORS.

1. *Rejection of a null hypothesis when it is true is called a* ***type I error****. The probability of committing a type I error is denoted by* α.
2. *Acceptance of the null hypothesis when it is false is called a* ***type II error****. The probability of committing a type II error is denoted by* β.

Definition 2 is more readily visualized with the aid for the following table:

	H_0 is true	H_0 is false
Accept H_0	No error	Type II error probability $= \beta$
Reject H_0	Type I error probability $= \alpha$	No error

DEFINITION 3. CRITICAL REGION. *It is customary to refer to the rejection region for* H_0 *as the* ***critical region*** *of a test. The probability of obtaining a value of the test statistic inside the critical region when* H_0 *is true is called the* ***size*** *of the critical region. Thus, the size of the critical region is just the probability* α *of committing a type I error. This probability is also called the* ***level of significance*** *of the test* (see the last part of Section 5).

EXAMPLE 1

Suppose that the manufacturer of a new medication wants to test the null hypothesis $\theta = 0.90$ against the alternative hypothesis $\theta = 0.60$. His test statistic is X, the observed number of successes (recoveries) in 20 trials, and he will accept the null hypothesis if $x > 14$; otherwise, he will reject it. Find α and β.

Solution

The acceptance region for the null hypothesis is $x = 15, 16, 17, 18, 19$, and 20, and, correspondingly, the rejection region (or critical region) is $x = 0, 1, 2, \ldots, 14$. Therefore, from the Binomial Probabilities table of "Statistical Tables",

$$\alpha = P(X \leqq 14; \theta = 0.90) = 0.0114$$

and

$$\beta = P(X > 14; \theta = 0.60) = 0.1255$$

A good test procedure is one in which both α and β are small, thereby giving us a good chance of making the correct decision. The probability of a type II error in Example 1 is rather high, but this can be reduced by appropriately changing the critical region. For instance, if we use the acceptance region $x > 15$ in this example so that the critical region is $x \leqq 15$, it can easily be checked that this would make $\alpha = 0.0433$ and $\beta = 0.0509$. Thus, although the probability of a type II error is reduced, the probability of a type I error has become larger. The only way in which we can reduce the probabilities of both types of errors is to increase the size of the sample, but as long as n is held fixed, this inverse relationship between the probabilities of type I and type II errors is typical of statistical decision procedures. In other words, if the probability of one type of error is reduced, that of the other type of error is increased.

EXAMPLE 2

Suppose that we want to test the null hypothesis that the mean of a normal population with $\sigma^2 = 1$ is μ_0 against the alternative hypothesis that it is μ_1, where $\mu_1 > \mu_0$. Find the value of K such that $\overline{x} > K$ provides a critical region of size $\alpha = 0.05$ for a random sample of size n.

Solution

Referring to Figure 1 and the Standard Normal Distribution table of "Statistical Tables", we find that $z = 1.645$ corresponds to an entry of 0.4500 and hence that

$$1.645 = \frac{K - \mu_0}{1/\sqrt{n}}$$

It follows that

$$K = \mu_0 + \frac{1.645}{\sqrt{n}}$$

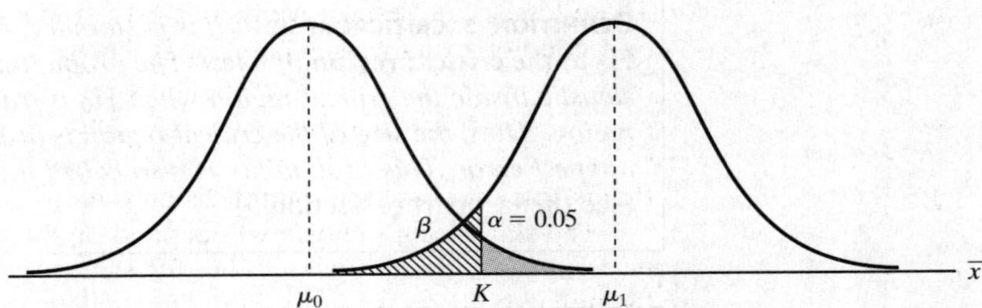

Figure 1. Diagram for Examples 2 and 3.

EXAMPLE 3

With reference to Example 2, determine the minimum sample size needed to test the null hypothesis $\mu_0 = 10$ against the alternative hypothesis $\mu_1 = 11$ with $\beta \leqq 0.06$.

Solution

Since β is given by the area of the ruled region of Figure 1, we get

$$\beta = P\left(\overline{X} < 10 + \frac{1.645}{\sqrt{n}};\ \mu = 11\right)$$

$$= P\left[Z < \frac{\left(10 + \frac{1.645}{\sqrt{n}}\right) - 11}{1/\sqrt{n}}\right]$$

$$= P(Z < -\sqrt{n} + 1.645)$$

and since $z = 1.555$ corresponds to an entry of $0.5000 - 0.06 = 0.4400$ in the Standard Normal Distribution table of "Statistical Tables", we set $-\sqrt{n} + 1.645$ equal to -1.555. It follows that $\sqrt{n} = 1.645 + 1.555 = 3.200$ and $n = 10.24$, or 11 rounded up to the nearest integer.

3 Losses and Risks†

The concepts of loss functions and risk functions also play an important part in the theory of hypothesis testing. In the decision theory approach to testing the null hypothesis that a population parameter θ equals θ_0 against the alternative that it equals θ_1, the statistician either takes the action a_0 and accepts the null hypothesis, or takes the action a_1 and accepts the alternative hypothesis. Depending on the true "state of Nature" and the action that she takes, her losses are shown in the following table:

		Statistician	
		a_0	a_1
Nature	θ_0	$L(a_0, \theta_0)$	$L(a_1, \theta_0)$
	θ_1	$L(a_0, \theta_1)$	$L(a_1, \theta_1)$

These losses can be positive or negative (reflecting penalties or rewards), and the only condition that we shall impose is that

$$L(a_0, \theta_0) < L(a_1, \theta_0) \quad \text{and} \quad L(a_1, \theta_1) < L(a_0, \theta_1)$$

that is, in either case the right decision is more profitable than the wrong one.

The statistician's choice will depend on the outcome of an experiment and the decision function d, which tells her for each possible outcome what action to take. If the null hypothesis is true and the statistician accepts the alternative hypothesis, that is, if the value of the parameter is θ_0 and the statistician takes action a_1, she commits a type I error; correspondingly, if the value of the parameter is θ_1 and the statistician takes action a_0, she commits a type II error. For the decision function d, we shall let $\alpha(d)$ denote the probability of committing a type I error and $\beta(d)$ the probability of committing a type II error. The values of the risk function are thus

$$\begin{aligned} R(d, \theta_0) &= [1 - \alpha(d)]L(a_0, \theta_0) + \alpha(d)L(a_1, \theta_0) \\ &= L(a_0, \theta_0) + \alpha(d)[L(a_1, \theta_0) - L(a_0, \theta_0)] \end{aligned}$$

and

$$\begin{aligned} R(d, \theta_1) &= \beta(d)L(a_0, \theta_1) + [1 - \beta(d)]L(a_1, \theta_1) \\ &= L(a_1, \theta_1) + \beta(d)[L(a_0, \theta_1) - L(a_1, \theta_1)] \end{aligned}$$

where, by assumption, the quantities in brackets are both positive. It is apparent from this (and should, perhaps, have been obvious from the beginning) that to minimize the risks the statistician must choose a decision function that, in some way, keeps the probabilities of both types of errors as small as possible.

If we could assign prior probabilities to θ_0 and θ_1 and if we knew the exact values of all the losses $L(a_j, \theta_i)$, we could calculate the Bayes risk and look for the decision function that minimizes this risk. Alternatively, if we looked upon Nature as a malevolent opponent, we could use the minimax criterion and choose the decision function that minimizes the maximum risk; but this is not a very realistic approach in most practical situations.

†This section may be omitted.

4 The Neyman–Pearson Lemma

In the theory of hypothesis testing that is nowadays referred to as "classical" or "traditional," the **Neyman–Pearson theory**, we circumvent the dependence between probabilities of type I and type II errors by limiting ourselves to test statistics for which the probability of a type I error is less than or equal to some constant α. In other words, we restrict ourselves to critical regions of size less than or equal to α. (We must allow for the critical region to be of size less than α to take care of discrete random variables, where it may be impossible to find a test statistic for which the size of the critical region is exactly equal to α.) For all practical purposes, then, we hold the probability of a type I error fixed and look for the test statistic that minimizes the probability of a type II error or, equivalently, that maximizes the quantity $1 - \beta$.

DEFINITION 4. THE POWER OF A TEST. *When testing the null hypothesis* H_0: $\theta = \theta_0$ *against the alternative hypothesis* H_1: $\theta = \theta_1$, *the quantity* $1 - \beta$ *is referred to as the* ***power*** *of the test at* $\theta = \theta_1$. *A critical region for testing a simple null hypothesis* H_0: $\theta = \theta_0$ *against a simple alternative hypothesis* H_1: $\theta = \theta_1$ *is said to be a* ***best critical region*** *or a* ***most powerful critical region*** *if the power of the test is a maximum at* $\theta = \theta_1$.

To construct a most powerful critical region in this kind of situation, we refer to the likelihoods of a random sample of size n from the population under consideration when $\theta = \theta_0$ and $\theta = \theta_1$. Denoting these likelihoods by L_0 and L_1, we thus have

$$L_0 = \prod_{i=1}^{n} f(x_i; \theta_0) \quad \text{and} \quad L_1 = \prod_{i=1}^{n} f(x_i; \theta_1)$$

Intuitively speaking, it stands to reason that $\frac{L_0}{L_1}$ should be small for sample points inside the critical region, which lead to type I errors when $\theta = \theta_0$ and to correct decisions when $\theta = \theta_1$; similarly, it stands to reason that $\frac{L_0}{L_1}$ should be large for sample points outside the critical region, which lead to correct decisions when $\theta = \theta_0$ and type II errors when $\theta = \theta_1$. The fact that this argument does, indeed, guarantee a most powerful critical region is proved by the following theorem.

THEOREM 1. (*Neyman–Pearson Lemma*) If C is a critical region of size α and k is a constant such that

$$\frac{L_0}{L_1} \leqq k \quad \text{inside } C$$

and

$$\frac{L_0}{L_1} \geqq k \quad \text{outside } C$$

then C is a most powerful critical region of size α for testing $\theta = \theta_0$ against $\theta = \theta_1$.

Proof Suppose that C is a critical region satisfying the conditions of the theorem and that D is some other critical region of size α. Thus,

$$\int_C \cdots \int L_0\, dx = \int_D \cdots \int L_0\, dx = \alpha$$

where dx stands for $dx_1, dx_2, \ldots, dx_n$, and the two multiple integrals are taken over the respective n-dimensional regions C and D. Now, making use of the fact that C is the union of the disjoint sets $C \cap D$ and $C \cap D'$, while D is the union of the disjoint sets $C \cap D$ and $C' \cap D$, we can write

$$\int_{C\cap D} \cdots \int L_0\, dx + \int_{C\cap D'} \cdots \int L_0\, dx = \int_{C\cap D} \cdots \int L_0\, dx + \int_{C'\cap D} \cdots \int L_0\, dx = \alpha$$

and hence

$$\int_{C\cap D'} \cdots \int L_0\, dx = \int_{C'\cap D} \cdots \int L_0\, dx$$

Then, since $L_1 \geqq L_0/k$ inside C and $L_1 \leqq L_0/k$ outside C, it follows that

$$\int_{C\cap D'} \cdots \int L_1\, dx \geqq \int_{C\cap D'} \cdots \int \frac{L_0}{k}\, dx = \int_{C'\cap D} \cdots \int \frac{L_0}{k}\, dx \geqq \int_{C'\cap D} \cdots \int L_1\, dx$$

and hence that

$$\int_{C\cap D'} \cdots \int L_1\, dx \geqq \int_{C'\cap D} \cdots \int L_1\, dx$$

Finally,

$$\begin{aligned}\int_C \cdots \int L_1\, dx &= \int_{C\cap D} \cdots \int L_1\, dx + \int_{C\cap D'} \cdots \int L_1\, dx \\ &\geqq \int_{C\cap D} \cdots \int L_1\, dx + \int_{C'\cap D} \cdots \int L_1\, dx = \int_D \cdots \int L_1\, dx\end{aligned}$$

so that

$$\int_C \cdots \int L_1\, dx \geqq \int_D \cdots \int L_1\, dx$$

and this completes the proof of Theorem 1.

The final inequality states that for the critical region C the probability of *not* committing a type II error is greater than or equal to the corresponding probability for any other critical region of size α. (For the discrete case the proof is the same, with summations taking the place of integrals.)

EXAMPLE 4

A random sample of size n from a normal population with $\sigma^2 = 1$ is to be used to test the null hypothesis $\mu = \mu_0$ against the alternative hypothesis $\mu = \mu_1$, where $\mu_1 > \mu_0$. Use the Neyman–Pearson lemma to find the most powerful critical region of size α.

Solution

The two likelihoods are

$$L_0 = \left(\frac{1}{\sqrt{2\pi}}\right)^n \cdot e^{-\frac{1}{2}\Sigma(x_i-\mu_0)^2} \quad \text{and} \quad L_1 = \left(\frac{1}{\sqrt{2\pi}}\right)^n \cdot e^{-\frac{1}{2}\Sigma(x_i-\mu_1)^2}$$

where the summations extend from $i = 1$ to $i = n$, and after some simplification their ratio becomes

$$\frac{L_0}{L_1} = e^{\frac{n}{2}(\mu_1^2-\mu_0^2)+(\mu_0-\mu_1)\cdot\Sigma x_i}$$

Thus, we must find a constant k and a region C of the sample space such that

$$e^{\frac{n}{2}(\mu_1^2-\mu_0^2)+(\mu_0-\mu_1)\cdot\Sigma x_i} \leqq k \quad \text{inside } C$$

$$e^{\frac{n}{2}(\mu_1^2-\mu_0^2)+(\mu_0-\mu_1)\cdot\Sigma x_i} \geqq k \quad \text{outside } C$$

and after taking logarithms, subtracting $\frac{n}{2}(\mu_1^2 - \mu_0^2)$, and dividing by the negative quantity $n(\mu_0 - \mu_1)$, these two inequalities become

$$\bar{x} \geqq K \quad \text{inside } C$$

$$\bar{x} \leqq K \quad \text{outside } C$$

where K is an expression in k, n, μ_0, and μ_1.

In actual practice, constants like K are determined by making use of the size of the critical region and appropriate statistical theory. In our case (see Example 2) we obtain $K = \mu_0 + z_\alpha \cdot \frac{1}{\sqrt{n}}$. Thus, the most powerful critical region of size α for testing the null hypothesis $\mu = \mu_0$ against the alternative $\mu = \mu_1$ (with $\mu_1 > \mu_0$) for the given normal population is

$$\bar{x} \geqq \mu_0 + z_\alpha \cdot \frac{1}{\sqrt{n}}$$

and it should be noted that it does not depend on μ_1. This is an important property, to which we shall refer again in Section 5.

Note that we derived the critical region here without first mentioning that the test statistic is to be $\overline{X}$. Since the specification of a critical region thus defines the corresponding test statistic, and vice versa, these two terms, "critical region" and "test statistic," are often used interchangeably in the language of statistics.

Exercises

1. Decide in each case whether the hypothesis is simple or composite:
(a) the hypothesis that a random variable has a gamma distribution with $\alpha = 3$ and $\beta = 2$;
(b) the hypothesis that a random variable has a gamma distribution with $\alpha = 3$ and $\beta \neq 2$;
(c) the hypothesis that a random variable has an exponential density;
(d) the hypothesis that a random variable has a beta distribution with the mean $\mu = 0.50$.

2. Decide in each case whether the hypothesis is simple or composite:
(a) the hypothesis that a random variable has a Poisson distribution with $\lambda = 1.25$;
(b) the hypothesis that a random variable has a Poisson distribution with $\lambda > 1.25$;
(c) the hypothesis that a random variable has a normal distribution with the mean $\mu = 100$;
(d) the hypothesis that a random variable has a negative binomial distribution with $k = 3$ and $\theta < 0.60$.

3. A single observation of a random variable having a hypergeometric distribution with $N = 7$ and $n = 2$ is used to test the null hypothesis $k = 2$ against the alternative hypothesis $k = 4$. If the null hypothesis is rejected if and only if the value of the random variable is 2, find the probabilities of type I and type II errors.

4. With reference to Example 1, what would have been the probabilities of type I and type II errors if the acceptance region had been $x > 16$ and the corresponding rejection region had been $x \leqq 16$?

5. A single observation of a random variable having a geometric distribution is used to test the null hypothesis $\theta = \theta_0$ against the alternative hypothesis $\theta = \theta_1 > \theta_0$. If the null hypothesis is rejected if and only if the observed value of the random variable is greater than or equal to the positive integer k, find expressions for the probabilities of type I and type II errors.

6. A single observation of a random variable having an exponential distribution is used to test the null hypothesis that the mean of the distribution is $\theta = 2$ against the alternative that it is $\theta = 5$. If the null hypothesis is accepted if and only if the observed value of the random variable is less than 3, find the probabilities of type I and type II errors.

7. Let X_1 and X_2 constitute a random sample from a normal population with $\sigma^2 = 1$. If the null hypothesis $\mu = \mu_0$ is to be rejected in favor of the alternative hypothesis $\mu = \mu_1 > \mu_0$ when $\bar{x} > \mu_0 + 1$, what is the size of the critical region?

8. A single observation of a random variable having a uniform density with $\alpha = 0$ is used to test the null hypothesis $\beta = \beta_0$ against the alternative hypothesis $\beta = \beta_0 + 2$. If the null hypothesis is rejected if and only if the random variable takes on a value greater than $\beta_0 + 1$, find the probabilities of type I and type II errors.

9. Let X_1 and X_2 constitute a random sample of size 2 from the population given by

$$f(x;\theta) = \begin{cases} \theta x^{\theta-1} & \text{for } 0 < x < 1 \\ 0 & \text{elsewhere} \end{cases}$$

If the critical region $x_1 x_2 \geqq \frac{3}{4}$ is used to test the null hypothesis $\theta = 1$ against the alternative hypothesis $\theta = 2$, what is the power of this test at $\theta = 2$?

10. Show that if $\mu_1 < \mu_0$ in Example 4, the Neyman–Pearson lemma yields the critical region

$$\bar{x} \leqq \mu_0 - z_\alpha \cdot \frac{1}{\sqrt{n}}$$

11. A random sample of size n from an exponential population is used to test the null hypothesis $\theta = \theta_0$ against the alternative hypothesis $\theta = \theta_1 > \theta_0$. Use the Neyman–Pearson lemma to find the most powerful critical region of size α.

12. Use the Neyman–Pearson lemma to indicate how to construct the most powerful critical region of size α to test the null hypothesis $\theta = \theta_0$, where θ is the parameter of a binomial distribution with a given value of n, against the alternative hypothesis $\theta = \theta_1 < \theta_0$.

13. With reference to Exercise 12, if $n = 100$, $\theta_0 = 0.40$, $\theta_1 = 0.30$, and α is as large as possible without exceeding 0.05, use the normal approximation to the binomial distribution to find the probability of committing a type II error.

14. A single observation of a random variable having a geometric distribution is to be used to test the null hypothesis that its parameter equals θ_0 against the alternative that it equals $\theta_1 > \theta_0$. Use the Neyman–Pearson lemma to find the best critical region of size α.

15. Given a random sample of size n from a normal population with $\mu = 0$, use the Neyman–Pearson lemma to construct the most powerful critical region of size α to test the null hypothesis $\sigma = \sigma_0$ against the alternative $\sigma = \sigma_1 > \sigma_0$.

16. Suppose that in Example 1 the manufacturer of the new medication feels that the odds are 4 to 1 that with this medication the recovery rate from the disease is 0.90

rather than 0.60. With these odds, what are the probabilities that he will make a wrong decision if he uses the decision function

(a) $d_1(x) = \begin{cases} a_0 & \text{for } x > 14 \\ a_1 & \text{for } x \leqq 14 \end{cases}$

(b) $d_2(x) = \begin{cases} a_0 & \text{for } x > 15 \\ a_1 & \text{for } x \leqq 15 \end{cases}$

(c) $d_3(x) = \begin{cases} a_0 & \text{for } x > 16 \\ a_1 & \text{for } x \leqq 16 \end{cases}$

5 The Power Function of a Test

In Example 1 we were able to give unique values for the probabilities of committing type I and type II errors because we were testing a simple hypothesis against a simple alternative. In actual practice, it is relatively rare, however, that simple hypotheses are tested against simple alternatives; usually one or the other, or both, are composite. For instance, in Example 1 it might well have been more realistic to test the null hypothesis that the recovery rate from the disease is $\theta \geqq 0.90$ against the alternative hypothesis $\theta < 0.90$, that is, the alternative hypothesis that the new medication is not as effective as claimed.

When we deal with composite hypotheses, the problem of evaluating the merits of a test criterion, or critical region, becomes more involved. In that case we have to consider the probabilities $\alpha(\theta)$ of committing a type I error for all values of θ within the domain specified under the null hypothesis H_0 and the probabilities $\beta(\theta)$ of committing a type II error for all values of θ within the domain specified under the alternative hypothesis H_1. It is customary to combine the two sets of probabilities in the following way.

DEFINITION 5. POWER FUNCTION. *The* ***power function*** *of a test of a statistical hypothesis* H_0 *against an alternative hypothesis* H_1 *is given by*

$$\pi(\theta) = \begin{cases} \alpha(\theta) & \text{for values of } \theta \text{ assumed under } H_0 \\ 1 - \beta(\theta) & \text{for values of } \theta \text{ assumed under } H_1 \end{cases}$$

Thus, the values of the power function are the probabilities of rejecting the null hypothesis H_0 for various values of the parameter θ. Observe also that for values of θ assumed under H_0, the power function gives the probability of committing a type I error, and for values of θ assumed under H_1, it gives the probability of *not* committing a type II error.

EXAMPLE 5

With reference to Example 1, suppose that we had wanted to test the null hypothesis $\theta \geqq 0.90$ against the alternative hypothesis $\theta < 0.90$. Investigate the power function corresponding to the same test criterion as in Exercises 3 and 4, where we accept the null hypothesis if $x > 14$ and reject it if $x \leqq 14$. As before, x is the observed number of successes (recoveries) in $n = 20$ trials.

Solution

Choosing values of θ for which the respective probabilities, $\alpha(\theta)$ or $\beta(\theta)$, are available from the Binomial Probabilities table of "Statistical Tables", we find the probabilities $\alpha(\theta)$ of getting at most 14 successes for $\theta = 0.90$ and 0.95 and the probabilities $\beta(\theta)$ of getting more than 14 successes for $\theta = 0.85, 0.80, \ldots, 0.50$. These are shown in the following table, together with the corresponding values of the power function, $\pi(\theta)$:

θ	*Probability of type I error* $\alpha(\theta)$	*Probability of type II error* $\beta(\theta)$	*Probability of rejecting* H_0 $\pi(\theta)$
0.95	0.0003		0.0003
0.90	0.0114		0.0114
0.85		0.9326	0.0674
0.80		0.8042	0.1958
0.75		0.6171	0.3829
0.70		0.4163	0.5837
0.65		0.2455	0.7545
0.60		0.1255	0.8745
0.55		0.0553	0.9447
0.50		0.0207	0.9793

The graph of this power function is shown in Figure 2. Of course, it applies only to the decision criterion of Example 1, the critical region $x \leqq 14$; but it is of interest to note how it compares with the power function of a corresponding ideal (infallible) test criterion, given by the dashed lines of Figure 2.

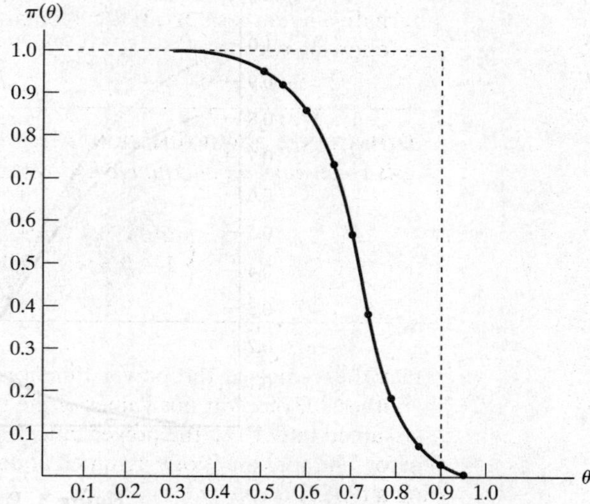

Figure 2. Diagram for Example 5.

Power functions play a very important role in the evaluation of statistical tests, particularly in the comparison of several critical regions that might all be used to test a given null hypothesis against a given alternative. Incidentally, if we had plotted in Figure 2 the probabilities of accepting H_0 (instead of those of rejecting H_0), we would have obtained the **operating characteristic curve**, **OC-curve**, of the given critical region. In other words, the values of the operating characteristic function, used mainly in industrial applications, are given by $1 - \pi(\theta)$.

In Section 4 we indicated that in the Neyman–Pearson theory of testing hypotheses we hold α, the probability of a type I error, fixed, and this requires that the null hypothesis H_0 be a simple hypothesis, say, $\theta = \theta_0$. As a result, the power function of any test of this null hypothesis will pass through the point (θ_0, α), the only point at which the value of a power function is the probability of making an error. This facilitates the comparison of the power functions of several critical regions, which are all designed to test the simple null hypothesis $\theta = \theta_0$ against a composite alternative, say, the alternative hypothesis $\theta \neq \theta_0$. To illustrate, consider Figure 3, giving the power functions of three different critical regions, or test criteria, designed for this purpose. Since for each value of θ, except θ_0, the values of power functions are probabilities of making correct decisions, it is desirable to have them as close to 1 as possible. Thus, it can be seen by inspection that the critical region whose power function is given by the dotted curve of Figure 3 is preferable to the critical region whose power function is given by the curve that is dashed. The probability of not committing a type II error with the first of these critical regions always exceeds that of the second, and we say that the first critical region is **uniformly more powerful** than the second; also, the second critical region is said to be **inadmissible.**

The same clear-cut distinction is not possible if we attempt to compare the critical regions whose power functions are given by the dotted and solid curves of Figure 3; in this case the first one is preferable for $\theta < \theta_0$, while the other is preferable for $\theta > \theta_0$. In situations like this we need further criteria for comparing power functions, for instance that of Exercise 27. Note that if the alternative hypothesis had been $\theta > \theta_0$, the critical region whose power function is given by the solid curve would have been uniformly more powerful than the critical region whose power function is given by the dotted curve.

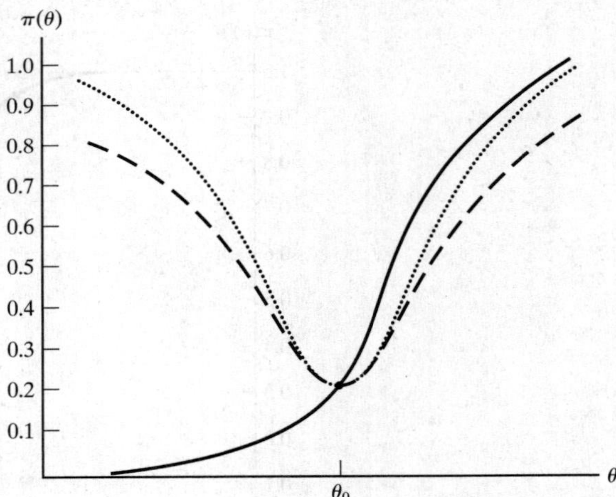

Figure 3. Power functions.

In general, when we test a simple hypothesis against a composite alternative, we specify α, the probability of a type I error, and refer to one critical region of size α as uniformly more powerful than another if the values of its power function are always greater than or equal to those of the other, with the strict inequality holding for at least one value of the parameter under consideration.

DEFINITION 6. UNIFORMLY MOST POWERFUL CRITICAL REGION (TEST). *If, for a given problem, a critical region of size α is uniformly more powerful than any other critical region of size α, it is said to be a* ***uniformly most powerful critical region****, or a* ***uniformly most powerful test****.*

Unfortunately, uniformly most powerful critical regions rarely exist when we test a simple hypothesis against a composite alternative. Of course, when we test a simple hypothesis against a simple alternative, a most powerful critical region of size α, as defined in Section 4, is, in fact, uniformly most powerful.

Until now we have always assumed that the acceptance of H_0 is equivalent to the rejection of H_1, and vice versa, but this is not the case, for example, in **multistage** or **sequential tests**, where the alternatives are to accept H_0, to accept H_1, or to defer the decision until more data have been obtained. It is also not the case in **tests of significance**, where the alternative to rejecting H_0 is reserving judgment instead of accepting H_0. For instance, if we want to test the null hypothesis that a coin is perfectly balanced against the alternative that this is not the case, and 100 tosses yield 57 heads and 43 tails, this will not enable us to reject the null hypothesis when $\alpha = 0.05$ (see Exercise 42). However, since we obtained quite a few more heads than the 50 that we can expect for a balanced coin, we may well be reluctant to accept the null hypothesis as true. To avoid this, we can say that the difference between 50 and 57, the number of heads that we expected and the number of heads that we obtained, may reasonably be attributed to chance, or we can say that this difference is not large enough to reject the null hypothesis. In either case, we do not really commit ourselves one way or the other, and as long as we do not actually accept the null hypothesis, we cannot commit a type II error. It is mainly in connection with tests of this kind that we refer to the probability of a type I error as the **level of significance**.

6 Likelihood Ratio Tests

The Neyman–Pearson lemma provides a means of constructing most powerful critical regions for testing a simple null hypothesis against a simple alternative hypothesis, but it does not always apply to composite hypotheses. We shall now present a general method for constructing critical regions for tests of composite hypotheses that in most cases have very satisfactory properties. The resulting tests, called **likelihood ratio tests**, are based on a generalization of the method of Section 4, but they are not necessarily uniformly most powerful. We shall discuss this method here with reference to tests concerning one parameter θ and continuous populations, but all our arguments can easily be extended to the multiparameter case and to discrete populations.

To illustrate the likelihood ratio technique, let us suppose that $X_1, X_2, \ldots, X_n$ constitute a random sample of size n from a population whose density at x is $f(x; \theta)$ and that Ω is the set of values that can be taken on by the parameter θ. We often refer to Ω as the **parameter space** for θ. The null hypothesis we shall want to test is

$$H_0\colon \quad \theta \in \omega$$

and the alternative hypothesis is

$$H_1\colon \quad \theta \in \omega'$$

where ω is a subset of Ω and ω' is the complement of ω with respect to Ω. Thus, the parameter space for θ is partitioned into the disjoint sets ω and ω'; according to the null hypothesis, θ is an element of the first set, and according to the alternative hypothesis, it is an element of the second set. In most problems Ω is either the set of all real numbers, the set of all positive real numbers, some interval of real numbers, or a discrete set of real numbers.

When H_0 and H_1 are both simple hypotheses, ω and ω' each have only one element, and in Section 4 we constructed tests by comparing the likelihoods L_0 and L_1. In the general case, where at least one of the two hypotheses is composite, we compare instead the two quantities max L_0 and max L, where max L_0 is the maximum value of the likelihood function for all values of θ in ω, and max L is the maximum value of the likelihood function for all values of θ in Ω. In other words, if we have a random sample of size n from a population whose density at x is $f(x; \theta)$, $\hat{\theta}$ is the maximum likelihood estimate of θ subject to the restriction that θ must be an element of ω, and $\hat{\hat{\theta}}$ is the maximum likelihood estimate of θ for all values of θ in Ω, then

$$\max L_0 = \prod_{i=1}^{n} f(x_i; \hat{\theta})$$

and

$$\max L = \prod_{i=1}^{n} f(x_i; \hat{\hat{\theta}})$$

These quantities are both values of random variables, since they depend on the observed values $x_1, x_2, \ldots, x_n$, and their ratio

$$\lambda = \frac{\max L_0}{\max L}$$

is referred to as a value of the **likelihood ratio statistic** Λ (capital Greek *lambda*).

Since max L_0 and max L are both values of a likelihood function and therefore are never negative, it follows that $\lambda \geqq 0$; also, since ω is a subset of the parameter space Ω, it follows that $\lambda \leqq 1$. When the null hypothesis is false, we would expect max L_0 to be small compared to max L, in which case λ would be close to zero. On the other hand, when the null hypothesis is true and $\theta \in \omega$, we would expect max L_0 to be close to max L, in which case λ would be close to 1. A likelihood ratio test states, therefore, that the null hypothesis H_0 is rejected if and only if λ falls in a critical region of the form $\lambda \leqq k$, where $0 < k < 1$. To summarize, we have the following definition.

DEFINITION 7. LIKELIHOOD RATIO TEST. *If ω and ω' are complementary subsets of the parameter space Ω and if the **likelihood ratio statistic***

$$\lambda = \frac{\max L_0}{\max L}$$

where max L_0 *and* max L *are the maximum values of the **likelihood function** for all values of θ in ω and Ω, respectively, then the critical region*

$$\lambda \leq k$$

*where $0 < k < 1$, defines a **likelihood ratio test** of the null hypothesis $\theta \in \omega$ against the alternative hypothesis $\theta \in \omega'$.*

If H_0 is a simple hypothesis, k is chosen so that the size of the critical region equals α; if H_0 is composite, k is chosen so that the probability of a type I error is less than or equal to α for all θ in ω, and equal to α, if possible, for at least one value of θ in ω. Thus, if H_0 is a simple hypothesis and $g(\lambda)$ is the density of Λ at λ when H_0 is true, then k must be such that

$$P(\Lambda \leqq k) = \int_0^k g(\lambda)d\lambda = \alpha$$

In the discrete case, the integral is replaced by a summation, and k is taken to be the largest value for which the sum is less than or equal to α.

EXAMPLE 6

Find the critical region of the likelihood ratio test for testing the null hypothesis

$$H_0: \quad \mu = \mu_0$$

against the composite alternative

$$H_1: \quad \mu \neq \mu_0$$

on the basis of a random sample of size n from a normal population with the known variance σ^2.

Solution

Since ω contains only μ_0, it follows that $\hat{\mu} = \mu_0$, and since Ω is the set of all real numbers, it follows that $\hat{\hat{\mu}} = \bar{x}$. Thus,

$$\max L_0 = \left(\frac{1}{\sigma\sqrt{2\pi}}\right)^n \cdot e^{-\frac{1}{2\sigma^2}\cdot\Sigma(x_i-\mu_0)^2}$$

and

$$\max L = \left(\frac{1}{\sigma\sqrt{2\pi}}\right)^n \cdot e^{-\frac{1}{2\sigma^2}\cdot\Sigma(x_i-\bar{x})^2}$$

where the summations extend from $i = 1$ to $i = n$, and the value of the likelihood ratio statistic becomes

$$\lambda = \frac{e^{-\frac{1}{2\sigma^2}\cdot\Sigma(x_i-\mu_0)^2}}{e^{-\frac{1}{2\sigma^2}\cdot\Sigma(x_i-\bar{x})^2}} = e^{-\frac{n}{2\sigma^2}(\bar{x}-\mu_0)^2}$$

after suitable simplifications, which the reader will be asked to verify in Exercise 19. Hence, the critical region of the likelihood ratio test is

$$e^{-\frac{n}{2\sigma^2}(\bar{x}-\mu_0)^2} \leqq k$$

and, after taking logarithms and dividing by $-\frac{n}{2\sigma_2}$, it becomes

$$(\bar{x}-\mu_0)^2 \geqq -\frac{2\sigma^2}{n} \cdot \ln k$$

or

$$|\bar{x} - \mu_0| \geqq K$$

where K will have to be determined so that the size of the critical region is α. Note that $\ln k$ is negative in view of the fact that $0 < k < 1$.

Since $\overline{X}$ has a normal distribution with the mean μ_0 and the variance $\frac{\sigma^2}{n}$, we find that the critical region of this likelihood ratio test is

$$|\bar{x} - \mu_0| \geqq z_{\alpha/2} \cdot \frac{\sigma}{\sqrt{n}}$$

or, equivalently,

$$|z| \geqq z_{\alpha/2}$$

where

$$z = \frac{\bar{x} - \mu_0}{\sigma/\sqrt{n}}$$

In other words, the null hypothesis must be rejected when Z takes on a value greater than or equal to $z_{\alpha/2}$ or a value less than or equal to $-z_{\alpha/2}$.

In the preceding example it was easy to find the constant that made the size of the critical region equal to α, because we were able to refer to the known distribution of $\overline{X}$ and did not have to derive the distribution of the likelihood ratio statistic Λ itself. Since the distribution of Λ is usually quite complicated, which makes it difficult to evaluate k, it is often preferable to use the following approximation, whose proof is referred to at the end of this chapter.

> **THEOREM 2.**† For large n, the distribution of $-2 \cdot \ln \Lambda$ approaches, under very general conditions, the chi-square distribution with 1 degree of freedom.

We should add that this theorem applies only to the one-parameter case; if the population has more than one unknown parameter upon which the null hypothesis imposes r restrictions, the number of degrees of freedom in the chi-square approximation to the distribution of $-2 \cdot \ln \Lambda$ is equal to r. For instance, if we want to test the null hypothesis that the unknown mean and variance of a normal population are μ_0 and σ_0^2 against the alternative hypothesis that $\mu \neq \mu_0$ and $\sigma^2 \neq \sigma_0^2$, the number of degrees of freedom in the chi-square approximation to the distribution of $-2 \cdot \ln \Lambda$ would be 2; the two restrictions are $\mu = \mu_0$ and $\sigma^2 = \sigma_0^2$.

Since small values of λ correspond to large values of $-2 \cdot \ln \lambda$, we can use Theorem 2 to write the critical region of this approximate likelihood ratio test as

$$-2 \cdot \ln \lambda \geqq \chi^2_{\alpha,1}$$

In connection with Example 6 we find that

$$-2 \cdot \ln \lambda = \frac{n}{\sigma^2}(\bar{x} - \mu_0)^2 = \left(\frac{\bar{x} - \mu_0}{\sigma/\sqrt{n}}\right)^2$$

---

†For a statement of the conditions under which Theorem 2 is true and for a proof of this theorem, see the references at the end of this chapter.

which actually *is* a value of a random variable having the chi-square distribution with 1 degree of freedom.

As we indicated in Section 6, the likelihood ratio technique will generally produce satisfactory results. That this is not always the case is illustrated by the following example, which is somewhat out of the ordinary.

EXAMPLE 7

On the basis of a single observation, we want to test the simple null hypothesis that the probability distribution of X is

x	1	2	3	4	5	6	7
$f(x)$	$\frac{1}{12}$	$\frac{1}{12}$	$\frac{1}{12}$	$\frac{1}{4}$	$\frac{1}{6}$	$\frac{1}{6}$	$\frac{1}{6}$

against the composite alternative that the probability distribution is

x	1	2	3	4	5	6	7
$g(x)$	$\frac{a}{3}$	$\frac{b}{3}$	$\frac{c}{3}$	$\frac{2}{3}$	0	0	0

where $a+b+c=1$. Show that the critical region obtained by means of the likelihood ratio technique is inadmissible.

Solution

The composite alternative hypothesis includes all the probability distributions that we get by assigning different values from 0 to 1 to a, b, and c, subject only to the restriction that $a+b+c=1$. To determine λ for each value of x, we first let $x=1$. For this value we get $\max L_0 = \frac{1}{12}$, $\max L = \frac{1}{3}$ (corresponding to $a=1$), and hence $\lambda = \frac{1}{4}$. Determining λ for the other values of x in the same way, we get the results shown in the following table:

x	1	2	3	4	5	6	7
λ	$\frac{1}{4}$	$\frac{1}{4}$	$\frac{1}{4}$	$\frac{3}{8}$	1	1	1

If the size of the critical region is to be $\alpha = 0.25$, we find that the likelihood ratio technique yields the critical region for which the null hypothesis is rejected when $\lambda = \frac{1}{4}$, that is, when $x=1, x=2$, or $x=3$; clearly, $f(1)+f(2)+f(3) = \frac{1}{12}+\frac{1}{12}+\frac{1}{12} = 0.25$. The corresponding probability of a type II error is given by $g(4)+g(5)+g(6)+g(7)$, and hence it equals $\frac{2}{3}$.

Now let us consider the critical region for which the null hypothesis is rejected only when $x=4$. Its size is also $\alpha = 0.25$ since $f(4) = \frac{1}{4}$, but the corresponding probability of a type II error is

$$g(1)+g(2)+g(3)+g(5)+g(6)+g(7) = \frac{a}{3}+\frac{b}{3}+\frac{c}{3}+0+0+0$$
$$= \frac{1}{3}$$

Since this is less than $\frac{2}{3}$, the critical region obtained by means of the likelihood ratio technique is inadmissible.

Exercises

17. With reference to Exercise 3, suppose that we had wanted to test the null hypothesis $k \leqq 2$ against the alternative hypothesis $k > 2$. Find the probabilities of
(a) type I errors for $k = 0, 1$, and 2;
(b) type II errors for $k = 4, 5, 6$, and 7.
Also plot the graph of the corresponding power function.

18. With reference to Example 5, suppose that we reject the null hypothesis if $x \leqq 15$ and accept it if $x > 15$. Calculate $\pi(\theta)$ for the same values of θ as in the table in Section 5 and plot the graph of the power function of this test criterion.

19. In the solution of Example 6, verify the step that led to

$$\lambda = e^{-\frac{n}{2\sigma^2}(\bar{x}-\mu_0)^2}$$

20. The number of successes in n trials is to be used to test the null hypothesis that the parameter θ of a binomial population equals $\frac{1}{2}$ against the alternative that it does not equal $\frac{1}{2}$.
(a) Find an expression for the likelihood ratio statistic.
(b) Use the result of part (a) to show that the critical region of the likelihood ratio test can be written as

$$x \cdot \ln x + (n-x) \cdot \ln(n-x) \geqq K$$

where x is the observed number of successes.
(c) Study the graph of $f(x) = x \cdot \ln x + (n-x) \cdot \ln(n-x)$, in particular its minimum and its symmetry, to show that the critical region of this likelihood ratio test can also be written as

$$\left| x - \frac{n}{2} \right| \geqq K$$

where K is a constant that depends on the size of the critical region.

21. A random sample of size n is to be used to test the null hypothesis that the parameter θ of an exponential population equals θ_0 against the alternative that it does not equal θ_0.
(a) Find an expression for the likelihood ratio statistic.
(b) Use the result of part (a) to show that the critical region of the likelihood ratio test can be written as

$$\bar{x} \cdot e^{-\bar{x}/\theta_0} \leqq K$$

22. This question has been intentionally omitted for this edition.

23. For the likelihood ratio statistic of Exercise 22, show that $-2 \cdot \ln \lambda$ approaches t^2 as $n \to \infty$. [*Hint*: Use the infinite series for $\ln(1+x)$.]

24. Given a random sample of size n from a normal population with unknown mean and variance, find an expression for the likelihood ratio statistic for testing the null hypothesis $\sigma = \sigma_0$ against the alternative hypothesis $\sigma \neq \sigma_0$.

25. Independent random samples of sizes $n_1, n_2, \ldots$, and n_k from k normal populations with unknown means and variances are to be used to test the null hypothesis $\sigma_1^2 = \sigma_2^2 = \cdots = \sigma_k^2$ against the alternative that these variances are not all equal.
(a) Show that under the null hypothesis the maximum likelihood estimates of the means μ_1 and the variances σ_i^2 are

$$\hat{\mu}_i = \bar{x}_i \quad \text{and} \quad \hat{\sigma}_i^2 = \sum_{i=1}^{k} \frac{(n_i - 1)s_i^2}{n}$$

where $n = \sum_{i=1}^{k} n_i$, while without restrictions the maximum likelihood estimates of the means μ_i and the variances σ_i^2 are

$$\hat{\hat{\mu}}_i = \bar{x}_i \quad \text{and} \quad \hat{\hat{\sigma}}_i^2 = \frac{(n_i - 1)s_i^2}{n_i}$$

(b) Using the results of part (a), show that the likelihood ratio statistic can be written as

$$\lambda = \frac{\prod_{i=1}^{k} \left[\frac{(n_i - 1)s_i^2}{n_i} \right]^{n_i/2}}{\left[\sum_{i=1}^{k} \frac{(n_i - 1)s_i^2}{n} \right]^{n/2}}$$

26. Show that for $k = 2$ the likelihood ratio statistic of Exercise 25 can be expressed in terms of the ratio of the two sample variances and that the likelihood ratio test can, therefore, be based on the F distribution.

27. When we test a simple null hypothesis against a composite alternative, a critical region is said to be **unbiased** if the corresponding power function takes on its minimum value at the value of the parameter assumed under the null hypothesis. In other words, a critical region is unbiased if the probability of rejecting the null hypothesis is least when the null hypothesis is true. Given a single observation of the random variable X having the density

$$f(x) = \begin{cases} 1 + \theta^2 \left(\frac{1}{2} - x \right) & \text{for } 0 < x < 1 \\ 0 & \text{elsewhere} \end{cases}$$

where $-1 \leqq \theta \leqq 1$, show that the critical region $x \leqq \alpha$ provides an unbiased critical region of size α for testing the null hypothesis $\theta = 0$ against the alternative hypothesis $\theta \neq 0$.

7 The Theory in Practice

The applied exercises that follow are intended to give the reader some practical experience with the theory of this chapter.

Applied Exercises SECS. 1–4

28. An airline wants to test the null hypothesis that 60 percent of its passengers object to smoking inside the plane. Explain under what conditions they would be committing a type I error and under what conditions they would be committing a type II error.

29. A doctor is asked to give an executive a thorough physical checkup to test the null hypothesis that he will be able to take on additional responsibilities. Explain under what conditions the doctor would be committing a type I error and under what conditions he would be committing a type II error.

30. The average drying time of a manufacturer's paint is 20 minutes. Investigating the effectiveness of a modification in the chemical composition of her paint, the manufacturer wants to test the null hypothesis $\mu = 20$ minutes against a suitable alternative, where μ is the average drying time of the modified paint.

(a) What alternative hypothesis should the manufacturer use if she does not want to make the modification in the chemical composition of the paint unless it decreases the drying time?

(b) What alternative hypothesis should the manufacturer use if the new process is actually cheaper and she wants to make the modification unless it increases the drying time of the paint?

31. A city police department is considering replacing the tires on its cars with a new brand tires. If μ_1 is the average number of miles that the old tires last and μ_2 is the average number of miles that the new tires will last, the null hypothesis to be tested is $\mu_1 = \mu_2$.

(a) What alternative hypothesis should the department use if it does not want to use the new tires unless they are definitely proved to give better mileage? In other words, the burden of proof is put on the new tires, and the old tires are to be kept unless the null hypothesis can be rejected.

(b) What alternative hypothesis should the department use if it is anxious to get the new tires unless they actually give poorer mileage than the old tires? Note that now the burden of proof is on the old tires, which will be kept only if the null hypothesis can be rejected.

(c) What alternative hypothesis should the department use so that rejection of the null hypothesis can lead either to keeping the old tires or to buying the new ones?

32. A botanist wishes to test the null hypothesis that the average diameter of the flowers of a particular plant is 9.6 cm. He decides to take a random sample of size $n = 80$ and accept the null hypothesis if the mean of the sample falls between 9.3 cm and 9.9 cm; if the mean of this sample falls outside this interval, he will reject the null hypothesis. What decision will he make and will it be in error if

(a) he gets a sample mean of 10.2 cm and $\mu = 9.6$ cm;

(b) he gets a sample mean of 10.2 cm and $\mu = 9.8$ cm;

(c) he gets a sample mean of 9.2 cm and $\mu = 9.6$ cm;

(d) he gets a sample mean of 9.2 cm and $\mu = 9.8$ cm?

33. An education specialist is considering the use of instructional material on compact discs for a special class of third-grade students with reading disabilities. Students in this class are given a standardized test in May of the school year, and μ_1 is the average score obtained on these tests after many years of experience. Let μ_2 be the average score for students using the discs, and assume that high scores are desirable.

(a) What null hypothesis should the education specialist use?

(b) What alternative hypothesis should be used if the specialist does not want to adopt the new discs unless they improve the standardized test scores?

(c) What alternative hypothesis should be used if the specialist wants to adopt the new discs unless they worsen the standardized test scores?

34. Suppose that we want to test the null hypothesis that an antipollution device for cars is effective.

(a) Explain under what conditions we would commit a type I error and under what conditions we would commit a type II error.

(b) Whether an error is a type I error or a type II error depends on how we formulate the null hypothesis. Rephrase the null hypothesis so that the type I error becomes a type II error, and vice versa.

35. A biologist wants to test the null hypothesis that the mean wingspan of a certain kind of insect is 12.3 mm against the alternative that it is not 12.3 mm. If she takes a random sample and decides to accept the null hypothesis if and only if the mean of the sample falls between 12.0 mm and 12.6 mm, what decision will she make if she gets $\bar{x} = 12.9$ mm and will it be in error if

(a) $\mu = 12.5$ mm; **(b)** $\mu = 12.3$ mm?

36. An employee of a bank wants to test the null hypothesis that on the average the bank cashes 10 bad checks per day against the alternative that this figure is too small. If he takes a random sample and decides to reject the null hypothesis if and only if the mean of the sample exceeds 12.5, what decision will he make if he gets $\bar{x} = 11.2$, and will it be in error if
(a) $\lambda = 11.5$; **(b)** $\lambda = 10.0$?
Here λ is the mean of the Poisson population being sampled.

37. Rework Example 3 with
(a) $\beta = 0.03$; **(b)** $\beta = 0.01$.

38. Suppose that we want to test the null hypothesis that a certain kind of tire will last, on the average, 35,000 miles against the alternative hypothesis that it will last, on the average, 45,000 miles. Assuming that we are dealing with a random variable having an exponential distribution, we specify the sample size and the probability of a type I error and use the Neyman–Pearson lemma to construct a critical region. Would we get the same critical region if we change the alternative hypothesis to
(a) $\theta_1 = 50{,}000$ miles; **(b)** $\theta_1 > 35{,}000$ miles?

SECS. 5–6

39. A single observation is to be used to test the null hypothesis that the mean waiting time between tremors recorded at a seismological station (the mean of an exponential population) is $\theta = 10$ hours against the alternative that $\theta \neq 10$ hours. If the null hypothesis is to be rejected if and only if the observed value is less than 8 or greater than 12, find
(a) the probability of a type I error;
(b) the probabilities of type II errors when $\theta = 2, 4, 6, 8, 12, 16$, and 20.
Also plot the power function of this test criterion.

40. A random sample of size 64 is to be used to test the null hypothesis that for a certain age group the mean score on an achievement test (the mean of a normal population with $\sigma^2 = 256$) is less than or equal to 40.0 against the alternative that it is greater than 40.0. If the null hypothesis is to be rejected if and only if the mean of the random sample exceeds 43.5, find
(a) the probabilities of type I errors when $\mu = 37.0, 38.0, 39.0$, and 40.0;
(b) the probabilities of type II errors when $\mu = 41.0, 42.0, 43.0, 44.0, 45.0, 46.0, 47.0$, and 48.0.
Also plot the power function of this test criterion.

41. The sum of the values obtained in a random sample of size $n = 5$ is to be used to test the null hypothesis that on the average there are more than two accidents per week at a certain intersection (that $\lambda > 2$ for this Poisson population) against the alternative hypothesis that on the average the number of accidents is two or less. If the null hypothesis is to be rejected if and only if the sum of the observations is five or less, find
(a) the probabilities of type I errors when $\lambda = 2.2, 2.4, 2.6, 2.8$, and 3.0;
(b) the probabilities of type II errors when $\lambda = 2.0, 1.5, 1.0$, and 0.5.
Also, plot the graph of the power function of this test criterion.

42. Verify the statement in Section 5 that 57 heads and 43 tails in 100 flips of a coin do not enable us to reject the null hypothesis that the coin is perfectly balanced (against the alternative that it is not perfectly balanced) at the 0.05 level of significance. (*Hint*: Use the normal approximation to the binomial distribution.)

43. To compare the variations in weight of four breeds of dogs, researchers took independent random samples of sizes $n_1 = 8, n_2 = 10, n_3 = 6$, and $n_4 = 8$, and got $s_1^2 = 16, s_2^2 = 25, s_3^2 = 12$, and $s_4^2 = 24$. Assuming that the populations sampled are normal, use the formula of part (b) of Exercise 25 to calculate $-2 \cdot \ln \lambda$ and test the null hypothesis $\sigma_1^2 = \sigma_2^2 = \sigma_3^2 = \sigma_4^2$ at the 0.05 level of significance. Explain why the number of degrees of freedom for this approximate chi-square test is 3.

44. The times to failure of certain electronic components in accelerate environment tests are 15, 28, 3, 12, 42, 19, 20, 2, 25, 30, 62, 12, 18, 16, 44, 65, 33, 51, 4, and 28 minutes. Looking upon these data as a random sample from an exponential population, use the results of Exercise 21 and Theorem 2 to test the null hypothesis $\theta = 15$ minutes against the alternative hypothesis $\theta \neq 15$ minutes at the 0.05 level of significance. (Use $\ln 1.763 = 0.570$.)

References

Discussions of various properties of likelihood ratio tests, particularly their large-sample properties, and a proof of Theorem 2 may be found in most advanced textbooks on the theory of statistics, for example, in

LEHMANN, E. L., *Testing Statistical Hypotheses*, 2nd ed. New York: John Wiley & Sons, Inc., 1986,

WILKS, S. S., *Mathematical Statistics*. New York: John Wiley & Sons, Inc., 1962.

Much of the original research done in this area is reproduced in

Selected Papers in Statistics and Probability by Abraham Wald. Stanford, Calif.: Stanford University Press, 1957.

Answers to Odd-Numbered Exercises

1 (a) Simple; **(b)** composite; **(c)** composite; **(d)** composite.

3 $\alpha = \frac{1}{21}$ and $\beta = \frac{5}{7}$.

5 $\alpha = (1-\theta_0)^{k-1}$ and $\beta = 1-(1-\theta_1)^{k-1}$.

7 $\alpha = 0.08$.

9 $1-\beta = 0.114$.

11 $\sum_{i=1}^{n} x_i \geq K$, where K can be determined by making use of the fact that $\sum_{i=1}^{n} X_i$ has the gamma distribution with $\alpha = n$ and $\beta = \theta_0$.

13 $\beta = 0.37$.

15 $\sum_{i=1}^{n} x_i^2 \geq K$, where K can be determined by making use of the formula for the sum of n terms of a geometric distribution.

17 (a) $0, 0, \frac{1}{21}$; **(b)** $\frac{5}{7}, \frac{11}{21}, \frac{2}{7}$, 0.

21 (a) $\lambda = \left(\frac{\bar{x}}{\theta_0}\right)^n e^{-(n\bar{x}/\theta_0+n)}$

31 (a) The alternative hypothesis is $\mu_2 > \mu_1$; **(b)** The alternative hypothesis is $\mu_1 > \mu_2$. **(c)** The alternative hypothesis is $\mu_1 \neq \mu_2$.

33 (a) The null hypothesis is $\mu_1 = \mu_2$. **(b)** The alternative hypothesis is $\mu_2 > \mu_1$. **(c)** The alternative hypothesis is $\mu_2 < \mu_1$.

35 (a) Correctly reject the null hypothesis. **(b)** Erroneously reject the null hypothesis.

39 (a) 0.852; **(b)** 0.016, 0.086, 0.129, 0.145, 0.144, 0.134, and 0.122.

41 (a) 0.0375, 0.0203, 0.0107, 0.0055, and 0.0027; **(b)** 0.9329, 0.7585, 0.3840, and 0.0420.

43 $-2 \cdot \ln \lambda = 1.424$; the null hypothesis cannot be rejected.

Tests of Hypothesis Involving Means, Variances, and Proportions

1 Introduction

In this chapter we shall present some of the standard tests that are most widely used in applications. Most of these tests, at least those based on known population distributions, can be obtained by the likelihood ratio technique.

> **DEFINITION 1. TEST OF SIGNIFICANCE.** *A statistical test which specifies a simple null hypothesis, the size of the critical region, α, and a composite alternative hypothesis is called a* ***test of significance****. In such a test, α is referred to as the* ***level of significance****.*

To explain the terminology we shall use, let us first consider a situation in which we want to test the null hypothesis $H_0\colon \theta = \theta_0$ against the **two-sided alternative** hypothesis $H_1\colon \theta \neq \theta_0$. Since it appears reasonable to accept the null hypothesis when our point estimate $\hat{\theta}$ of θ is close to θ_0 and to reject it when $\hat{\theta}$ is much larger or much smaller than θ_0, it would be logical to let the critical region consist of both tails of the sampling distribution of our test statistic $\hat{\Theta}$. Such a test is referred to as a **two-tailed test**.

On the other hand, if we are testing the null hypothesis $H_0\colon \theta = \theta_0$ against the **one-sided alternative** $H_1\colon \theta < \theta_0$, it would seem reasonable to reject H_0 only when $\hat{\theta}$ is much smaller than θ_0. Therefore, in this case it would be logical to let the critical region consist only of the left-hand tail of the sampling distribution of $\hat{\Theta}$. Likewise, in testing $H_0\colon \theta = \theta_0$ against the one-sided alternative $H_1\colon \theta > \theta_0$, we reject H_0 only for large values of $\hat{\theta}$, and the critical region consists only of the right tail of the sampling distribution of $\hat{\Theta}$. Any test where the critical region consists only of one tail of the sampling distribution of the test statistic is called a **one-tailed test**.

For instance, for the two-sided alternative $\mu \neq \mu_0$, the likelihood ratio technique leads to a two-tailed test with the critical region

$$|\bar{x} - \mu_0| \geqq z_{\alpha/2} \cdot \frac{\sigma}{\sqrt{n}}$$

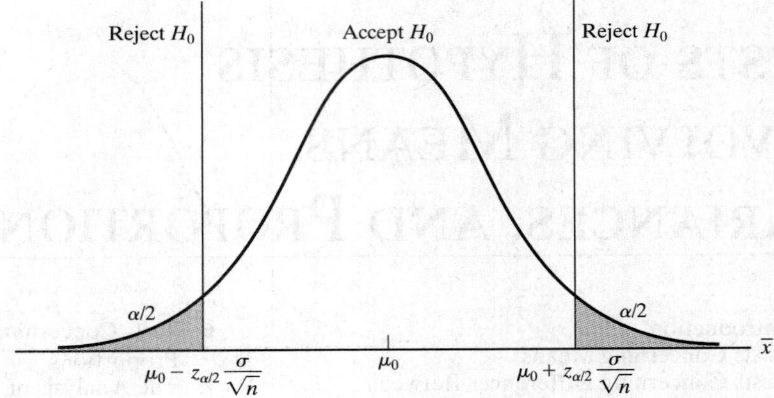

Figure 1. Critical region for two-tailed test.

or

$$\bar{x} \leqq \mu_0 - z_{\alpha/2}\cdot\frac{\sigma}{\sqrt{n}} \quad \text{and} \quad \bar{x} \geqq \mu_0 + z_{\alpha/2}\cdot\frac{\sigma}{\sqrt{n}}$$

As pictured in Figure 1, the null hypothesis $\mu = \mu_0$ is rejected if $\overline{X}$ takes on a value falling in either tail of its sampling distribution. Symbolically, this critical region can be written as $z \leqq -z_{\alpha/2}$ or $z \geqq z_{\alpha/2}$, where

$$z = \frac{\bar{x} - \mu_0}{\sigma/\sqrt{n}}$$

Had we used the one-sided alternative $\mu > \mu_0$, the likelihood ratio technique would have led to the one-tailed test whose critical region is pictured in Figure 2, and if we had used the one-sided alternative $\mu < \mu_0$, the likelihood ratio technique would have led to the one-tailed test whose critical region is pictured in Figure 3. It stands to reason that in the first case we would reject the null hypothesis only for values of $\overline{X}$ falling into the right-hand tail of its sampling distribution, and in the second case we would reject the null hypothesis only for values of $\overline{X}$ falling into the left-hand

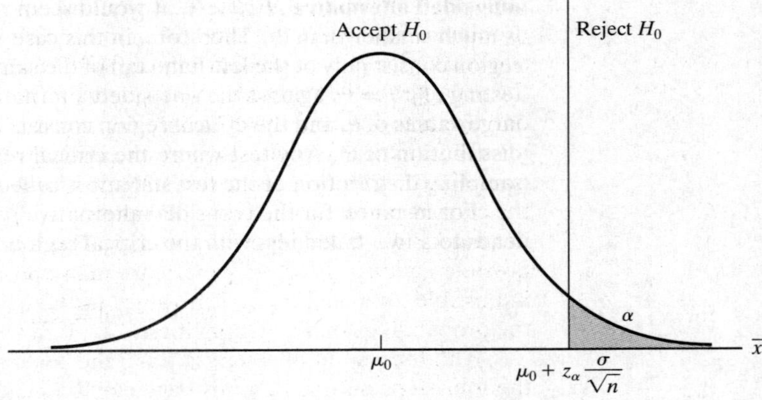

Figure 2. Critical region for one-tailed test ($H_1\colon \mu > \mu_0$).

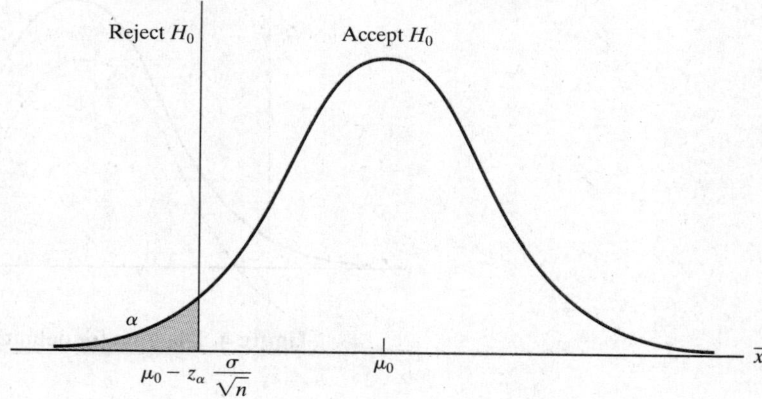

Figure 3. Critical region for one-tailed test (H_1: $\mu < \mu_0$).

tail of its sampling distribution. Symbolically, the corresponding critical regions can be written as $z \geqq z_\alpha$ and as $z \leqq -z_\alpha$, where z is as defined before. Although there are exceptions to this rule (see Exercise 1), two-sided alternatives usually lead to two-tailed tests and one-sided alternatives usually lead to one-tailed tests.

Traditionally, it has been the custom to outline tests of hypotheses by means of the following steps:

1. **Formulate H_0 and H_1, and specify α.**
2. **Using the sampling distribution of an appropriate test statistic, determine a critical region of size α.**
3. **Determine the value of the test statistic from the sample data.**
4. **Check whether the value of the test statistic falls into the critical region and, accordingly, reject the null hypothesis, or reserve judgment. (Note that we do not accept the null hypothesis because β, the probability of false acceptance, is not specified in a test of significance.)**

In Figures 1, 2, and 3, the dividing lines of the test criteria (that is, the **boundaries** of the critical regions, or the **critical values**) require knowledge of z_α or $z_{\alpha/2}$. These values are readily available from Table III of "Statistical Tables" (or more detailed tables of the standard normal distribution) for any level of significance α, but the problem is not always this simple. For instance, if the sampling distribution of the test statistic happens to be a t distribution, a chi-square distribution, or an F distribution, the usual tables will provide the necessary values of $t_\alpha, t_{\alpha/2}, \chi^2_\alpha, \chi^2_{\alpha/2}, F_\alpha$, or $F_{\alpha/2}$, but only for a few values of α. Mainly for this reason, it has been the custom to base tests of statistical hypotheses almost exclusively on the level of significance $\alpha = 0.05$ or $\alpha = 0.01$. This may seem very arbitrary, and of course it is, and this accounts for the current preference for using ***P*-values** (see Definition 2). Alternatively, we could use a decision-theory approach and thus take into account the consequences of all possible actions. However, "there are many problems in which it is difficult, if not impossible, to assign numerical values to the consequences of one's actions and to the probabilities of all eventualities."

With the advent of computers and the general availability of statistical software, the four steps outlined on this page may be modified to allow for more freedom in the choice of the level of significance α. With reference to the test for which the critical region is shown in Figure 2, we compare the shaded region of Figure 4 with

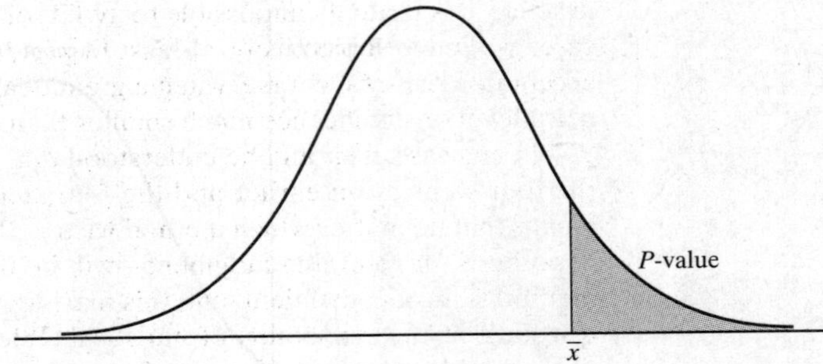

Figure 4. Diagram for definition of *P*-values.

α instead of comparing the observed value of $\overline{X}$ with the boundary of the critical region or the value of

$$Z = \frac{\overline{X} - \mu_0}{\sigma/\sqrt{n}}$$

with $z_{\alpha/2}$. In other words, we reject the null hypothesis if the shaded region of Figure 4 is less than or equal to α. This shaded region is referred to as the ***P*-value**, the **prob-value**, the **tail probability**, or the **observed level of significance** corresponding to $\bar{x}$, the observed value of $\overline{X}$. In fact, it is the probability $P(\overline{X} \geqq \bar{x})$ when the null hypothesis is true.

Correspondingly, when the alternative hypothesis is $\mu < \mu_0$ and the critical region is the one of Figure 3, the *P*-value is the probability $P(\overline{X} \leqq \bar{x})$ when the null hypothesis is true; and when the alternative hypothesis is $\mu \neq \mu_0$ and the critical region is the one of Figure 1, the *P*-value is $2P(\overline{X} \geqq \bar{x})$ or $2P(\overline{X} \leqq \bar{x})$, depending on whether $\bar{x}$ falls into the right-hand tail or the left-hand tail of the sampling distribution of $\overline{X}$. Here again we act as if the null hypothesis is true, or we withhold judgment.

More generally, we define *P*-values as follows.

> **DEFINITION 2. P-VALUE.** *Corresponding to an observed value of a test statistic, the* ***P-value*** *is the lowest level of significance at which the null hypothesis could have been rejected.*

With regard to this alternative approach to testing hypotheses, the first of the four steps on the previous page remains unchanged, the second step becomes

2′. Specify the test statistic.

the third step becomes

3′. Determine the value of the test statistic and the corresponding *P*-value from the sample data.

and the fourth step becomes

4′. Check whether the *P*-value is less than or equal to α and, accordingly, reject the null hypothesis, or reserve judgment.

As we pointed out earlier, this allows for more freedom in the choice of the level of significance, but it is difficult to conceive of situations in which we could justify using, say, $\alpha = 0.04$ rather than $\alpha = 0.05$ or $\alpha = 0.015$ rather than $\alpha = 0.01$. In practice, it is virtually impossible to avoid some element of arbitrariness, and in most cases we judge subjectively, at least in part, whether $\alpha = 0.05$ or $\alpha = 0.01$ reflects acceptable risks. Of course, when a great deal is at stake and it is practical, we might use a level of significance much smaller than $\alpha = 0.01$.

In any case, it should be understood that the two methods of testing hypotheses, the four steps given earlier and the four steps described here, are equivalent. This means that no matter which method we use, the ultimate decision—rejecting the null hypothesis, or reserving judgment—will be the same. In practice, we use whichever method is most convenient, and this may depend on the sampling distribution of the test statistic, the availability of statistical tables or computer software, and the nature of the problem (see, for instance, Example 8 and Exercise 57).

There are statisticians who prefer to avoid all problems relating to the choice of the level of significance. Limiting their role to data analysis, they do not specify α and omit step 4′. Of course, it is always desirable to have input from others (research workers or management) in formulating hypotheses and specifying α, but it would hardly seem reasonable to dump *P*-values into the laps of persons without adequate training in statistics and let them take it from there. To compound the difficulties, consider the temptation one might be exposed to when choosing α after having seen the *P*-value with which it is to be compared. Suppose, for instance, that an experiment yields a *P*-value of 0.036. If we are anxious to reject the null hypothesis and thus prove our point, it would be tempting to choose $\alpha = 0.05$; if we are anxious to accept the null hypothesis and thus prove our point, it would be tempting to choose $\alpha = 0.01$.

Nevertheless, in **exploratory data analysis**, where we are not really concerned with making inferences, *P*-values can be used as measures of the strength of evidence. Suppose, for instance, that in cancer research with two drugs, scientists get *P*-values of 0.0735 and 0.0021 for the effectiveness of these drugs in reducing the size of tumors. This suggests that there is more supporting evidence for the effectiveness of the second drug, or that the second drug "looks much more promising."

2 Tests Concerning Means

In this section we shall discuss the most widely used tests concerning the mean of a population, and in Section 3 we shall discuss the corresponding tests concerning the means of two populations. All the tests in this section are based on normal distribution theory, assuming either that the samples come from normal populations or that they are large enough to justify normal approximations; there are also some **nonparametric** alternatives to these tests, which do not require knowledge about the population or populations from which the samples are obtained.

Suppose that we want to test the null hypothesis $\mu = \mu_0$ against one of the alternatives $\mu \neq \mu_0$, $\mu > \mu_0$, or $\mu < \mu_0$ on the basis of a random sample of size n from a normal population with the known variance σ^2. The critical regions for the respective alternatives are $|z| \geqq z_{\alpha/2}$, $z \geqq z_\alpha$, and $z \leqq -z_\alpha$, where

$$z = \frac{\bar{x} - \mu_0}{\sigma/\sqrt{n}}$$

As we indicated in Section 1, the most commonly used levels of significance are 0.05 and 0.01, and as the reader can verify from Table III of "Statistical Tables", the corresponding values of z_α and $z_{\alpha/2}$ are $z_{0.05} = 1.645$, $z_{0.01} = 2.33$, $z_{0.025} = 1.96$, and $z_{0.005} = 2.575$.

EXAMPLE 1

Suppose that it is known from experience that the standard deviation of the weight of 8-ounce packages of cookies made by a certain bakery is 0.16 ounce. To check whether its production is under control on a given day, that is, to check whether the true average weight of the packages is 8 ounces, employees select a random sample of 25 packages and find that their mean weight is $\bar{x} = 8.091$ ounces. Since the bakery stands to lose money when $\mu > 8$ and the customer loses out when $\mu < 8$, test the null hypothesis $\mu = 8$ against the alternative hypothesis $\mu \neq 8$ at the 0.01 level of significance.

Solution

1. H_0: $\mu = 8$
   H_1: $\mu \neq 8$
   $\alpha = 0.01$
2. Reject the null hypothesis if $z \leqq -2.575$ or $z \geqq 2.575$, where

$$z = \frac{\bar{x} - \mu_0}{\sigma/\sqrt{n}}$$

3. Substituting $\bar{x} = 8.091$, $\mu_0 = 8$, $\sigma = 0.16$, and $n = 25$, we get

$$z = \frac{8.091 - 8}{0.16/\sqrt{25}} = 2.84$$

4. Since $z = 2.84$ exceeds 2.575, the null hypothesis must be rejected and suitable adjustments should be made in the production process.

Had we used the alternative approach described in Section 1, we would have obtained a P-value of 0.0046 (see Exercise 21), and since 0.0046 is less than 0.01, the conclusion would have been the same.

It should be noted that the critical region $z \geqq z_\alpha$ can also be used to test the null hypothesis $\mu = \mu_0$ against the simple alternative $\mu = \mu_1 > \mu_0$ or the composite null hypothesis $\mu \leqq \mu_0$ against the composite alternative $\mu > \mu_0$. In the first case we would be testing a simple hypothesis against a simple alternative, and in the second case α would be the maximum probability of committing a type I error for any value of μ assumed under the null hypothesis. Of course, similar arguments apply to the critical region $z \leqq -z_\alpha$.

When we are dealing with a large sample of size $n \geqq 30$ from a population that need not be normal but has a finite variance, we can use the central limit theorem to justify using the test for normal populations, and even when σ^2 is unknown we can approximate its value with s^2 in the computation of the test statistic. To illustrate the use of such an approximate **large-sample test**, consider the following example.

EXAMPLE 2

Suppose that 100 high-performance tires made by a certain manufacturer lasted on the average 21,819 miles with a standard deviation of 1,295 miles. Test the null hypothesis $\mu = 22{,}000$ miles against the alternative hypothesis $\mu < 22{,}000$ miles at the 0.05 level of significance.

Solution

1. H_0: $\mu = 22{,}000$
   H_1: $\mu < 22{,}000$
   $\alpha = 0.05$
2. Reject the null hypothesis if $z \leqq -1.645$, where

$$z = \frac{\bar{x} - \mu_0}{\sigma/\sqrt{n}}$$

3. Substituting $\bar{x} = 21{,}819$, $\mu_0 = 22{,}000$, $s = 1{,}295$ for σ, and $n = 100$, we get

$$z = \frac{21{,}819 - 22{,}000}{1{,}295/\sqrt{100}} = -1.40$$

4. Since $z = -1.40$ is greater than -1.645, the null hypothesis cannot be rejected; there is no convincing evidence that the tires are not as good as assumed under the null hypothesis.

Had we used the alternative approach described in Section 1, we would have btained a P-value of 0.0808 (see Exercise 22), which exceeds 0.05. As should have been expected, the conclusion is the same: The null hypothesis cannot be rejected.

When $n < 30$ and σ^2 is unknown, the test we have been discussing in this section cannot be used. However, for random samples from normal populations, the likelihood ratio technique yields a corresponding test based on

$$t = \frac{\bar{x} - \mu_0}{s/\sqrt{n}}$$

which is a value of a random variable having the t distribution with $n - 1$ degrees of freedom. Thus, critical regions of size α for testing the null hypothesis $\mu = \mu_0$ against the alternatives $\mu \neq \mu_0$, $\mu > \mu_0$, or $\mu < \mu_0$ are, respectively, $|t| \geqq t_{\alpha/2,n-1}, t \geqq t_{\alpha,n-1}$, and $t \leqq -t_{\alpha,n-1}$. Note that the comments made on the previous page in connection with the alternative hypothesis $\mu_1 > \mu_0$ and the test of the null hypothesis $\mu \leqq \mu_0$ against the alternative $\mu > \mu_0$ apply also in this case.

To illustrate this **one-sample t test**, as it is usually called, consider the following example.

EXAMPLE 3

The specifications for a certain kind of ribbon call for a mean breaking strength of 185 pounds. If five pieces randomly selected from different rolls have breaking strengths of 171.6, 191.8, 178.3, 184.9, and 189.1 pounds, test the null hypothesis $\mu = 185$ pounds against the alternative hypothesis $\mu < 185$ pounds at the 0.05 level of significance.

Solution

1. H_0: $\mu = 185$
   H_1: $\mu < 185$
   $\alpha = 0.05$

2. Reject the null hypothesis if $t \leqq -2.132$, where t is determined by means of the formula given above and 2.132 is the value of $t_{0.05,4}$.
3. First we calculate the mean and the standard deviation, getting $\bar{x} = 183.1$ and $s = 8.2$. Then, substituting these values together with $\mu_0 = 185$ and $n = 5$ into the formula for t, we get

$$t = \frac{183.1 - 185}{8.2/\sqrt{5}} = -0.51$$

4. Since $t = -0.49$ is greater than -2.132, the null hypothesis cannot be rejected. If we went beyond this and concluded that the rolls of ribbon from which the sample was selected meet specifications, we would, of course, be exposed to the unknown risk of committing a type II error.

3 Tests Concerning Differences Between Means

In many problems in applied research, we are interested in hypotheses concerning differences between the means of two populations. For instance, we may want to decide upon the basis of suitable samples whether men can perform a certain task as fast as women, or we may want to decide on the basis of an appropriate sample survey whether the average weekly food expenditures of families in one city exceed those of families in another city by at least $10.00.

Let us suppose that we are dealing with independent random samples of sizes n_1 and n_2 from two normal populations having the means μ_1 and μ_2 and the known variances σ_1^2 and σ_2^2 and that we want to test the null hypothesis $\mu_1 - \mu_2 = \delta$, where δ is a given constant, against one of the alternatives $\mu_1 - \mu_2 \neq \delta, \mu_1 - \mu_2 > \delta$, or $\mu_1 - \mu_2 < \delta$. Applying the likelihood ratio technique, we will arrive at a test based on $\bar{x}_1 - \bar{x}_2$, and the respective critical regions can be written as $|z| \geqq z_{\alpha/2}, z \geqq z_\alpha$, and $z \leqq -z_\alpha$, where

$$z = \frac{\bar{x}_1 - \bar{x}_2 - \delta}{\sqrt{\dfrac{\sigma_1^2}{n_1} + \dfrac{\sigma_2^2}{n_2}}}$$

When we deal with independent random samples from populations with unknown variances that may not even be normal, we can still use the test that we have just described with s_1 substituted for σ_1 and s_2 substituted for σ_2 as long as both samples are large enough to invoke the central limit theorem.

EXAMPLE 4

An experiment is performed to determine whether the average nicotine content of one kind of cigarette exceeds that of another kind by 0.20 milligram. If $n_1 = 50$ cigarettes of the first kind had an average nicotine content of $\bar{x}_1 = 2.61$ milligrams with a standard deviation of $s_1 = 0.12$ milligram, whereas $n_2 = 40$ cigarettes of the other kind had an average nicotine content of $\bar{x}_2 = 2.38$ milligrams with a standard deviation of $s_2 = 0.14$ milligram, test the null hypothesis $\mu_1 - \mu_2 = 0.20$ against the alternative hypothesis $\mu_1 - \mu_2 \neq 0.20$ at the 0.05 level of significance. Base the decision on the P-value corresponding to the value of the appropriate test statistic.

Solution

1. H_0: $\mu_1 - \mu_2 = 0.20$
   H_1: $\mu_1 - \mu_2 \neq 0.20$
   $\alpha = 0.05$

2′. Use the test statistic Z, where

$$z = \frac{\bar{x}_1 - \bar{x}_2 - \delta}{\sqrt{\dfrac{\sigma_1^2}{n_1} + \dfrac{\sigma_2^2}{n_2}}}$$

3′. Substituting $\bar{x}_1 = 2.61, \bar{x}_2 = 2.38, \delta = 0.20, s_1 = 0.12$ for $\sigma_1, s_2 = 0.14$ for $\sigma_2, n_1 = 50$, and $n_2 = 40$ into this formula, we get

$$z = \frac{2.61 - 2.38 - 0.20}{\sqrt{\dfrac{(0.12)^2}{50} + \dfrac{(0.14)^2}{40}}} = 1.08$$

The corresponding P-value is $2(0.5000 - 0.3599) = 0.2802$, where 0.3599 is the entry in Table III of "Statistical Tables" for $z = 1.08$.

4′. Since 0.2802 exceeds 0.05, the null hypothesis cannot be rejected; we say that the difference between $2.61 - 2.38 = 0.23$ and 0.20 is not significant. This means that the difference may well be attributed to chance.

When n_1 and n_2 are small and σ_1 and σ_2 are unknown, the test we have been discussing cannot be used. However, for independent random samples from two normal populations having the *same* unknown variance σ^2, the likelihood ratio technique yields a test based on

$$t = \frac{\bar{x}_1 - \bar{x}_2 - \delta}{s_p\sqrt{\dfrac{1}{n_1} + \dfrac{1}{n_2}}}$$

where

$$s_p^2 = \frac{(n_1 - 1)s_1^2 + (n_2 - 1)s_2^2}{n_1 + n_2 - 2}$$

Under the given assumptions and the null hypothesis $\mu_1 - \mu_2 = \delta$, this expression for t is a value of a random variable having the t distribution with $n_1 + n_2 - 2$ degrees of freedom. Thus, the appropriate critical regions of size α for testing the null hypothesis $\mu_1 - \mu_2 = \delta$ against the alternatives $\mu_1 - \mu_2 \neq \delta$, $\mu_1 - \mu_2 > \delta$, or $\mu_1 - \mu_2 < \delta$ under the given assumptions are, respectively, $|t| \geqq t_{\alpha/2, n_1+n_2-2}$, $t \geqq t_{\alpha, n_1+n_2-2}$, and $t \leqq -t_{\alpha, n_1+n_2-2}$. To illustrate this **two-sample *t* test**, consider the following problem.

EXAMPLE 5

In the comparison of two kinds of paint, a consumer testing service finds that four 1-gallon cans of one brand cover on the average 546 square feet with a standard deviation of 31 square feet, whereas four 1-gallon cans of another brand cover on the average 492 square feet with a standard deviation of 26 square feet. Assuming that the two populations sampled are normal and have equal variances, test the null hypothesis $\mu_1 - \mu_2 = 0$ against the alternative hypothesis $\mu_1 - \mu_2 > 0$ at the 0.05 level of significance.

Solution

1. H_0: $\mu_1 - \mu_2 = 0$
   H_1: $\mu_1 - \mu_2 > 0$
   $\alpha = 0.05$
2. Reject the null hypothesis if $t \geqq 1.943$, where t is calculated according to the formula given on the previous page and 1.943 is the value of $t_{0.05,6}$.
3. First calculating s_p, we get

$$s_p = \sqrt{\frac{3(31)^2 + 3(26)^2}{4+4-2}} = 28.609$$

and then substituting its value together with $\bar{x}_1 = 546, \bar{x}_2 = 492, \delta = 0$, and $n_1 = n_2 = 4$ into the formula for t, we obtain

$$t = \frac{546 - 492}{28.609\sqrt{\frac{1}{4} + \frac{1}{4}}} = 2.67$$

4. Since $t = 2.67$ exceeds 1.943, the null hypothesis must be rejected; we conclude that on the average the first kind of paint covers a greater area than the second.

Note that $n_1 = n_2$ in this example, so the formula for s_p^2 becomes

$$s_p^2 = \frac{1}{2}(s_1^2 + s_2^2)$$

Use of this formula would have simplified the calculations in this special case.

In Exercise 41 the reader will be asked to use suitable computer software to show that the P-value would have been 0.0185 in this example, and the conclusion would, of course, have been the same.

If the assumption of equal variances is untenable in a problem of this kind, there are several possibilities. A relatively simple one consists of randomly pairing the values obtained in the two samples and then looking upon their differences as a random sample of size n_1 or n_2, whichever is smaller, from a normal population that, under the null hypothesis, has the mean $\mu = \delta$. Then we test this null hypothesis against the appropriate alternative by means of the methods of Section 2. This is a good reason for having $n_1 = n_2$, but there exist alternative techniques for handling the case where $n_1 \neq n_2$ (one of these, the *Smith–Satterthwaite* test, is mentioned among the references at the end of the chapter).

So far we have limited our discussion to random samples that are independent, and the methods we have introduced in this section cannot be used, for example, to decide on the basis of weights "before and after" whether a certain diet is really effective or whether an observed difference between the average I.Q.'s of husbands and their wives is really significant. In both of these examples the samples are not independent because the data are actually *paired*. A common way of handling this kind of problem is to proceed as in the preceding paragraph, that is, to work with the differences between the *paired* measurements or observations. If n is large, we can then use the test described in Section 2 to test the null hypothesis $\mu_1 - \mu_2 = \delta$ against the appropriate alternative, and if n is small, we can use the t test described also in Section 2 provided the differences can be looked upon as a random sample from a normal population.

Exercises

1. Given a random sample of size n from a normal population with the known variance σ^2, show that the null hypothesis $\mu = \mu_0$ can be tested against the alternative hypothesis $\mu \neq \mu_0$ with the use of a one-tailed criterion based on the chi-square distribution.

2. Suppose that a random sample from a normal population with the known variance σ^2 is to be used to test the null hypothesis $\mu = \mu_0$ against the alternative hypothesis $\mu = \mu_1$, where $\mu_1 > \mu_0$, and that the probabilities of type I and type II errors are to have the preassigned values α and β. Show that the required size of the sample is given by

$$n = \frac{\sigma^2(z_\alpha + z_\beta)^2}{(\mu_1 - \mu_0)^2}$$

3. With reference to the preceding exercise, find the required size of the sample when $\sigma = 9$, $\mu_0 = 15$, $\mu_1 = 20$, $\alpha = 0.05$, and $\beta = 0.01$.

4. Suppose that independent random samples of size n from two normal populations with the known variances σ_1^2 and σ_2^2 are to be used to test the null hypothesis $\mu_1 - \mu_2 = \delta$ against the alternative hypothesis $\mu_1 - \mu_2 = \delta'$ and that the probabilities of type I and type II errors are to have the preassigned values α and β. Show that the required size of the sample is given by

$$n = \frac{(\sigma_1^2 + \sigma_2^2)(z_\alpha + z_\beta)^2}{(\delta - \delta')^2}$$

5. With reference to Exercise 4, find the required size of the samples when $\sigma_1 = 9, \sigma_2 = 13, \delta = 80, \delta' = 86, \alpha = 0.01$, and $\beta = 0.01$.

4 Tests Concerning Variances

There are several reasons why it is important to test hypotheses concerning the variances of populations. As far as direct applications are concerned, a manufacturer who has to meet rigid specifications will have to perform tests about the variability of his product, a teacher may want to know whether certain statements are true about the variability that he or she can expect in the performance of a student, and a pharmacist may have to check whether the variation in the potency of a medicine is within permissible limits. As far as indirect applications are concerned, tests about variances are often prerequisites for tests concerning other parameters. For instance, the two-sample t test described in Section 3 requires that the two population variances be equal, and in practice this means that we may have to check on the reasonableness of this assumption before we perform the test concerning the means.

The tests that we shall study in this section include a test of the null hypothesis that the variance of a normal population equals a given constant and the likelihood ratio test of the equality of the variances of two normal populations.

Given a random sample of size n from a normal population, we shall want to test the null hypothesis $\sigma^2 = \sigma_0^2$ against one of the alternatives $\sigma^2 \neq \sigma_0^2, \sigma^2 > \sigma_0^2$, or $\sigma^2 < \sigma_0^2$, and the likelihood ratio technique leads to a test based on s^2, the value of the sample variance. Based on theorem "If X_1 and X_2 are independent random variables, X_1 has a chi-square distribution with ν_1 degrees of freedom, and $X_1 + X_2$ has a chi-square distribution with $\nu > \nu_1$ degrees of freedom, then X_2 has a chi-square distribution with $\nu - \nu_1$ degrees of freedom", we can thus write the critical regions for testing the null hypothesis against the two one-sided alternatives as $\chi^2 \geqq \chi^2_{\alpha,n-1}$ and $\chi^2 \leqq \chi^2_{1-\alpha,n-1}$, where

$$\chi^2 = \frac{(n-1)s^2}{\sigma_0^2}$$

As far as the two-sided alternative is concerned, we reject the null hypothesis if $\chi^2 \geqq \chi^2_{\alpha/2,n-1}$ or $\chi^2 \leqq \chi^2_{1-\alpha/2,n-1}$, and the size of all these critical regions is, of course, equal to α.

EXAMPLE 6

Suppose that the uniformity of the thickness of a part used in a semiconductor is critical and that measurements of the thickness of a random sample of 18 such parts have the variance $s^2 = 0.68$, where the measurements are in thousandths of an inch. The process is considered to be under control if the variation of the thicknesses is given by a variance not greater than 0.36. Assuming that the measurements constitute a random sample from a normal population, test the null hypothesis $\sigma^2 = 0.36$ against the alternative hypothesis $\sigma^2 > 0.36$ at the 0.05 level of significance.

Solution

1. H_0: $\sigma^2 = 0.36$
   H_1: $\sigma^2 > 0.36$
   $\alpha = 0.05$
2. Reject the null hypothesis if $\chi^2 \geqq 27.587$, where

$$\chi^2 = \frac{(n-1)s^2}{\sigma_0^2}$$

   and 27.587 is the value of $\chi^2_{0.05,17}$.
3. Substituting $s^2 = 0.68$, $\sigma_0^2 = 0.36$, and $n = 18$, we get

$$\chi^2 = \frac{17(0.68)}{0.36} = 32.11$$

4. Since $\chi^2 = 32.11$ exceeds 27.587, the null hypothesis must be rejected and the process used in the manufacture of the parts must be adjusted.

Note that if α had been 0.01 in the preceding example, the null hypothesis could not have been rejected, since $\chi^2 = 32.11$ does not exceed $\chi^2_{0.01,17} = 33.409$. This serves to indicate again that the choice of the level of significance is something that must always be specified in advance, so we will be spared the temptation of choosing a value that happens to suit our purpose (see also Section 1).

The likelihood ratio statistic for testing the equality of the variances of two normal populations can be expressed in terms of the ratio of the two sample variances. Given independent random samples of sizes n_1 and n_2 from two normal populations with the variances σ_1^2 and σ_2^2, we thus find from the theorem "If S_1^2 and S_2^2 are the variances of independent random samples of sizes n_1 and n_2 from normal populations with the variances σ_1^2 and σ_2^2, then $F = \dfrac{S_1^2/\sigma_1^2}{S_2^2/\sigma_2^2} = \dfrac{\sigma_2^2 S_1^2}{\sigma_1^2 S_2^2}$ is a random variable having an F distribution with $n_1 - 1$ and $n_2 - 1$ degrees of freedom" that corresponding critical regions of size α for testing the null hypothesis $\sigma_1^2 = \sigma_2^2$ against the one-sided alternatives $\sigma_1^2 > \sigma_2^2$ or $\sigma_1^2 < \sigma_2^2$ are, respectively,

$$\frac{s_1^2}{s_2^2} \geqq f_{\alpha, n_1-1, n_2-1} \quad \text{and} \quad \frac{s_2^2}{s_1^2} \geqq f_{\alpha, n_2-1, n_1-1}$$

The appropriate critical region for testing the null hypothesis against the two-sided alternative $\sigma_1^2 \neq \sigma_2^2$ is

$$\frac{s_1^2}{s_2^2} \geqq f_{\alpha/2, n_1-1, n_2-1} \quad \text{if } s_1^2 \geqq s_2^2$$

and

$$\frac{s_2^2}{s_1^2} \geqq f_{\alpha/2, n_2-1, n_1-1} \quad \text{if } s_1^2 < s_2^2$$

Note that this test is based entirely on the right-hand tail of the F distribution, which is made possible by the fact that if the random variable X has an F distribution with ν_1 and ν_2 degrees of freedom, then $\dfrac{1}{X}$ has an F distribution with ν_2 and ν_1 degrees of freedom.

EXAMPLE 7

In comparing the variability of the tensile strength of two kinds of structural steel, an experiment yielded the following results: $n_1 = 13$, $s_1^2 = 19.2$, $n_2 = 16$, and $s_2^2 = 3.5$, where the units of measurement are 1,000 pounds per square inch. Assuming that the measurements constitute independent random samples from two normal populations, test the null hypothesis $\sigma_1^2 = \sigma_2^2$ against the alternative $\sigma_1^2 \neq \sigma_2^2$ at the 0.02 level of significance.

Solution

1. H_0: $\sigma_1^2 = \sigma_2^2$
   H_1: $\sigma_1^2 \neq \sigma_2^2$
   $\alpha = 0.02$
2. Since $s_1^2 \geqq s_2^2$, reject the null hypothesis if $\dfrac{s_1^2}{s_2^2} \geqq 3.67$, where 3.67 is the value of $f_{0.01,12,15}$.
3. Substituting $s_1^2 = 19.2$ and $s_2^2 = 3.5$, we get

$$\frac{s_1^2}{s_2^2} = \frac{19.2}{3.5} = 5.49$$

4. Since $f = 5.49$ exceeds 3.67, the null hypothesis must be rejected; we conclude that the variability of the tensile strength of the two kinds of steel is not the same.

Exercises

6. Making use of the fact that the chi-square distribution can be approximated with a normal distribution when ν, the number of degrees of freedom, is large, show that for large samples from normal populations

$$s^2 \geqq \sigma_0^2 \left[1 + z_\alpha \sqrt{\frac{2}{n-1}}\right]$$

is an approximate critical region of size α for testing the null hypothesis $\sigma^2 = \sigma_0^2$ against the alternative $\sigma^2 > \sigma_0^2$. Also construct corresponding critical regions for testing this null hypothesis against the alternatives $\sigma^2 < \sigma_0^2$ and $\sigma^2 \neq \sigma_0^2$.

7. This question has been intentionally omitted for this edition.

5 Tests Concerning Proportions

If an outcome of an experiment is the number of votes that a candidate receives in a poll, the number of imperfections found in a piece of cloth, the number of children who are absent from school on a given day, . . ., we refer to such data as **count data**. Appropriate models for the analysis of count data are the binomial distribution, the Poisson distribution, the multinomial distribution, and some of the other discrete distributions. In this section we shall present one of the most common tests based on count data, a test concerning the parameter θ of the binomial distribution. Thus, we might test on the basis of a sample whether the true proportion of cures from a certain disease is 0.90 or whether the true proportion of defectives coming off an assembly line is 0.02.

Let's take that the most powerful critical region for testing the null hypothesis $\theta = \theta_0$ against the alternative hypothesis $\theta = \theta_1 < \theta_0$, where θ is the parameter of a binomial population, is based on the value of X, the number of "successes" obtained in n trials. When it comes to composite alternatives, the likelihood ratio technique also yields tests based on the observed number of successes. In fact, if we want to test the null hypothesis $\theta = \theta_0$ against the one-sided alternative $\theta > \theta_0$, the critical region of size α of the likelihood ratio criterion is

$$x \geqq k_\alpha$$

where k_α is the smallest integer for which

$$\sum_{y=k_\alpha}^{n} b(y; n, \theta_0) \leqq \alpha$$

and $b(y; n, \theta_0)$ is the probability of getting y successes in n binomial trials when $\theta = \theta_0$. The size of this critical region, as well as the ones that follow, is thus as close as possible to α without exceeding it.

The corresponding critical region for testing the null hypothesis $\theta = \theta_0$ against the one-sided alternative $\theta < \theta_0$ is

$$x \leqq k'_\alpha$$

where k'_α is the largest integer for which

$$\sum_{y=0}^{k'_\alpha} b(y; n, \theta_0) \leqq \alpha$$

and, finally, the critical region for testing the null hypothesis $\theta = \theta_0$ against the two-sided alternative $\theta \neq \theta_0$ is

$$x \geqq k_{\alpha/2} \quad \text{or} \quad x \leqq k'_{\alpha/2}$$

We shall not illustrate this method of determining critical regions for tests concerning the binomial parameter θ because, in actual practice, it is much less tedious to base the decisions on P-values.

EXAMPLE 8

If $x = 4$ of $n = 20$ patients suffered serious side effects from a new medication, test the null hypothesis $\theta = 0.50$ against the alternative hypothesis $\theta \neq 0.50$ at the 0.05 level of significance. Here θ is the true proportion of patients suffering serious side effects from the new medication.

Solution

1. H_0: $\theta = 0.50$
H_1: $\theta \neq 0.50$
$\alpha = 0.05$

2′. Use the test statistic X, the observed number of successes.

3′. $x = 4$, and since $P(X \leqq 4) = 0.0059$, the P-value is $2(0.0059) = 0.0118$.

4′. Since the P-value, 0.0118, is less than 0.05, the null hypothesis must be rejected; we conclude that $\theta \neq 0.50$.

The tests we have described require the use of a table of binomial probabilities, regardless of whether we use the four steps discussed in Section 1. For $n \leqq 20$ we can use Table I of "Statistical Tables", and for values of n up to 100 we can use the tables in *Tables of the Binomial Probability Distribution* and *Binomial Tables*. Alternatively, for large values of n we can use the normal approximation to the binomial distribution and treat

$$z = \frac{x - n\theta}{\sqrt{n\theta(1-\theta)}}$$

as a value of a random variable having the standard normal distribution. For large n, we can thus test the null hypothesis $\theta = \theta_0$ against the alternatives $\theta \neq \theta_0, \theta > \theta_0$, or $\theta < \theta_0$ using, respectively, the critical regions $|z| \geqq z_{\alpha/2}, z \geqq z_\alpha$, and $z \leqq -z_\alpha$, where

$$z = \frac{x - n\theta_0}{\sqrt{n\theta_0(1-\theta_0)}}$$

or

$$z = \frac{\left(x \pm \frac{1}{2}\right) - n\theta_0}{\sqrt{n\theta_0(1-\theta_0)}}$$

if we use the continuity correction. We use the minus sign when x exceeds $n\theta_0$ and the plus sign when x is less than $n\theta_0$.

EXAMPLE 9

An oil company claims that less than 20 percent of all car owners have not tried its gasoline. Test this claim at the 0.01 level of significance if a random check reveals that 22 of 200 car owners have not tried the oil company's gasoline.

Solution

1. H_0: $\theta = 0.20$
   H_1: $\theta < 0.20$
   $\alpha = 0.01$
2. Reject the null hypothesis of $z \leqq -2.33$, where (without the continuity correction)
$$z = \frac{x - n\theta_0}{\sqrt{n\theta_0(1-\theta_0)}}$$
3. Substituting $x = 22, n = 200$, and $\theta_0 = 0.20$, we get
$$z = \frac{22 - 200(0.20)}{\sqrt{200(0.20)(0.80)}} = -3.18$$
4. Since $z = -3.18$ is less than -2.33, the null hypothesis must be rejected; we conclude that, as claimed, less than 20 percent of all car owners have not tried the oil company's gasoline.

Note that if we had used the continuity correction in the preceding example, we would have obtained $z = -3.09$ and the conclusion would have been the same.

6 Tests Concerning Differences Among *k* Proportions

In many problems in applied research, we must decide whether observed differences among sample proportions, or percentages, are significant or whether they can be attributed to chance. For instance, if 6 percent of the frozen chickens in a sample from one supplier fail to meet certain standards and only 4 percent in a sample from another supplier fail to meet the standards, we may want to investigate whether the difference between these two percentages is significant. Similarly, we may want to judge on the basis of sample data whether equal proportions of voters in four different cities favor a certain candidate for governor.

To indicate a general method for handling problems of this kind, suppose that $x_1, x_2, \ldots, x_k$ are observed values of k independent random variables $X_1, X_2, \ldots, X_k$ having binomial distributions with the parameters n_1 and θ_1, n_2 and $\theta_2, \ldots, n_k$ and θ_k. If the n's are sufficiently large, we can approximate the distributions of the independent random variables

$$Z_i = \frac{X_i - n_i\theta_i}{\sqrt{n_i\theta_i(1-\theta_i)}} \quad \text{for } i = 1, 2, \ldots, k$$

with standard normal distributions, and, according to the theorem "If $X_1, X_2, \ldots, X_n$ are independent random variables having standard normal distributions, then $Y = \sum_{i=1}^{n} X_i^2$ has the chi-square distribution with $\nu = n$ degrees of freedom", we can then look upon

$$\chi^2 = \sum_{i=1}^{k} \frac{(x_i - n_i\theta_i)^2}{n_i\theta_i(1-\theta_i)}$$

as a value of a random variable having the chi-square distribution with k degrees of freedom. To test the null hypothesis, $\theta_1 = \theta_2 = \cdots = \theta_k = \theta_0$ (against the alternative that at least one of the θ's does not equal θ_0), we can thus use the critical region $\chi^2 \geqq \chi^2_{\alpha,k}$, where

$$\chi^2 = \sum_{i=1}^{k} \frac{(x_i - n_i\theta_0)^2}{n_i\theta_0(1-\theta_0)}$$

When θ_0 is not specified, that is, when we are interested only in the null hypothesis $\theta_1 = \theta_2 = \cdots = \theta_k$, we substitute for θ the pooled estimate

$$\hat{\theta} = \frac{x_1 + x_2 + \cdots + x_k}{n_1 + n_2 + \cdots + n_k}$$

and the critical region becomes $\chi^2 \geqq \chi^2_{\alpha,k-1}$, where

$$\chi^2 = \sum_{i=1}^{k} \frac{(x_i - n_i\hat{\theta})^2}{n_i\hat{\theta}(1-\hat{\theta})}$$

The loss of 1 degree of freedom, that is, the change in the critical region from $\chi^2_{\alpha,k}$ to $\chi^2_{\alpha,k-1}$, is due to the fact that an estimate is substituted for the unknown parameter θ.

Let us now present an alternative formula for the chi-square statistic immediately above, which, as we shall see in Section 7, lends itself more rapidly to other applications. If we arrange the data as in the following table, let us refer to its entries as the **observed cell frequencies** f_{ij}, where the first subscript indicates the row and the second subscript indicates the column of this $k \times 2$ table.

	Successes	*Failures*
Sample 1	x_1	$n_1 - x_1$
Sample 2	x_2	$n_2 - x_2$
	...	...
Sample k	x_k	$n_k - x_k$

Under the null hypothesis $\theta_1 = \theta_2 = \cdots = \theta_k = \theta_0$ the **expected cell frequencies** for the first column are $n_i\theta_0$ for $i = 1, 2, \ldots, k$, and those for the second column are $n_i(1-\theta_0)$. When θ_0 is not known, we substitute for it, as before, the pooled estimate $\hat{\theta}$, and estimate the expected cell frequencies as

$$e_{i1} = n_i\hat{\theta} \quad \text{and} \quad e_{i2} = n_i(1-\hat{\theta})$$

for $i = 1, 2, \ldots, k$. It will be left to the reader to show in Exercise 8 that the chi-square statistic

$$\chi^2 = \sum_{i=1}^{k} \frac{(x_i - n_i\hat{\theta})^2}{n_i\hat{\theta}(1-\hat{\theta})}$$

can also be written as

$$\chi^2 = \sum_{i=1}^{k}\sum_{j=1}^{2}\frac{(f_{ij}-e_{ij})^2}{e_{ij}}$$

EXAMPLE 10

Determine, on the basis of the sample data shown in the following table, whether the true proportion of shoppers favoring detergent A over detergent B is the same in all three cities:

	Number favoring detergent A	*Number favoring detergent B*	
Los Angeles	232	168	400
San Diego	260	240	500
Fresno	197	203	400

Use the 0.05 level of significance.

Solution

1. H_0: $\theta_1 = \theta_2 = \theta_3$
   H_1: $\theta_1, \theta_2,$ and θ_3 are not all equal.
   $\alpha = 0.05$
2. Reject the null hypothesis if $\chi^2 \geqq 5.991$, where

$$\chi^2 = \sum_{i=1}^{3}\sum_{j=1}^{2}\frac{(f_{ij}-e_{ij})^2}{e_{ij}}$$

   and 5.991 is the value of $\chi^2_{0.05,2}$.
3. Since the pooled estimate of θ is

$$\hat{\theta} = \frac{232+260+197}{400+500+400} = \frac{689}{1{,}300} = 0.53$$

   the expected cell frequencies are

$$e_{11} = 400(0.53) = 212 \quad \text{and} \quad e_{12} = 400(0.47) = 188$$
$$e_{21} = 500(0.53) = 265 \quad \text{and} \quad e_{22} = 500(0.47) = 235$$
$$e_{31} = 400(0.53) = 212 \quad \text{and} \quad e_{32} = 400(0.47) = 188$$

   and substitution into the formula for χ^2 given previously yields

$$\chi^2 = \frac{(232-212)^2}{212} + \frac{(260-265)^2}{265} + \frac{(197-212)^2}{212} + \frac{(168-188)^2}{188} + \frac{(240-235)^2}{235} + \frac{(203-188)^2}{188}$$
$$= 6.48$$

4. Since $\chi^2 = 6.48$ exceeds 5.991, the null hypothesis must be rejected; in other words, the true proportions of shoppers favoring detergent A over detergent B in the three cities are not the same.

Exercises

8. Show that the two formulas for χ^2 in Section 6 are equivalent.

9. Modify the critical regions in Section 5 so that they can be used to test the null hypothesis $\lambda = \lambda_0$ against the alternative hypotheses $\lambda > \lambda_0$, $\lambda < \lambda_0$, and $\lambda \neq \lambda_0$ on the basis of n observations. Here λ is the parameter of the Poisson distribution.

10. With reference to Exercise 9, use Table II of "Statistical Tables" to find values corresponding to $k_{0.025}$ and $k'_{0.025}$ to test the null hypothesis $\lambda = 3.6$ against the alternative hypothesis $\lambda \neq 3.6$ on the basis of five observations. Use the 0.05 level of significance.

11. For $k = 2$, show that the χ^2 formula can be written as

$$\chi^2 = \frac{(n_1+n_2)(n_2x_1 - n_1x_2)^2}{n_1n_2(x_1+x_2)[(n_1+n_2)-(x_1+x_2)]}$$

12. Given large random samples from two binomial populations, show that the null hypothesis $\theta_1 = \theta_2$ can be tested on the basis of the statistic

$$z = \frac{\dfrac{x_1}{n_1} - \dfrac{x_2}{n_2}}{\sqrt{\hat{\theta}(1-\hat{\theta})\left(\dfrac{1}{n_1}+\dfrac{1}{n_2}\right)}}$$

where $\hat{\theta} = \dfrac{x_1+x_2}{n_1+n_2}$.

13. Show that the square of the expression for z in Exercise 12 equals

$$\chi^2 = \sum_{i=1}^{2} \frac{(x_i - n_i\hat{\theta})^2}{n_i\hat{\theta}(1-\hat{\theta})}$$

so that the two tests are actually equivalent when the alternative hypothesis is $\theta_1 \neq \theta_2$. Note that the test described in Exercise 12, but not the one based on the χ^2 statistic, can be used when the alternative hypothesis is $\theta_1 < \theta_2$ or $\theta_1 > \theta_2$.

7 The Analysis of an $r \times c$ Table

The method we shall describe in this section applies to two kinds of problems, which differ conceptually but are analyzed in the same way. In the first kind of problem we deal with samples from r multinomial populations, with each trial permitting c possible outcomes. This would be the case, for instance, when persons interviewed in five different precincts are asked whether they are for a candidate, against her, or undecided. Here $r = 5$ and $c = 3$.

It would also have been the case in Example 10 if each shopper had been asked whether he or she favors detergent A, favors detergent B, or does not care one way or the other. We might thus have obtained the results shown in the following $3\times$ 3 table:

	Number favoring detergent A	*Number favoring detergent B*	*Number indifferent*	
Los Angeles	174	93	133	400
San Diego	196	124	180	500
Fresno	148	105	147	400

The null hypothesis we would want to test in a problem like this is that we are sampling r identical multinomial populations. Symbolically, if θ_{ij} is the probability of the jth outcome for the ith population, we would want to test the null hypothesis

$$\theta_{1j} = \theta_{2j} = \cdots = \theta_{rj}$$

for $j = 1, 2, \ldots, c$. The alternative hypothesis would be that $\theta_{1j}, \theta_{2j}, \ldots,$ and θ_{rj} are not all equal for at least one value of j.

In the preceding example we dealt with three samples, whose fixed sizes were given by the row totals, 400, 500, and 400; on the other hand, the column totals were left to chance. In the other kind of problem where the method of this section applies, we are dealing with one sample and the row totals as well as the column totals are left to chance.

To give an example, let us consider the following table obtained in a study of the relationship, if any, of the I.Q.'s of persons who have gone through a large company's job-training program and their subsequent performance on the job:

		Performance			
		Poor	*Fair*	*Good*	
I.Q.	*Below average*	67	64	25	156
	Average	42	76	56	174
	Above average	10	23	37	70
		119	163	118	400

Here there is one sample of size 400, and the row totals as well as the column totals are left to chance.

DEFINITION 3. CONTINGENCY TABLE. *A table having* r *rows and* c *columns where each row represents* c *values of a non-numerical variable and each column represents* r *values of a different nonnumerical variable is called a* ***contingency table****. In such a table, the entries are count data (positive integers) and both the row and the column totals are left to chance. Such a table is assembled for the purpose of testing whether the row variable and the column variable are independent.*

The null hypothesis we shall want to test by means of the preceding table is that the on-the-job performance of persons who have gone through the training program is independent of their I.Q. Symbolically, if θ_{ij} is the probability that an item will fall into the cell belonging to the ith row and the jth column, $\theta_{i\cdot}$ is the probability that an item will fall into the ith row, and $\theta_{\cdot j}$ is the probability that an item will fall into the jth column, the null hypothesis we want to test is

$$\theta_{ij} = \theta_{i\cdot} \cdot \theta_{\cdot j}$$

for $i = 1, 2, \ldots, r$ and $j = 1, 2, \ldots, c$. Correspondingly, the alternative hypothesis is $\theta_{ij} \neq \theta_{i\cdot} \cdot \theta_{\cdot j}$ for at least one pair of values of i and j.

Since the method by which we analyze an $r \times c$ table is the same regardless of whether we are dealing with r samples from multinomial populations with c different outcomes or one sample from a multinomial population with rc different outcomes, let us discuss it here with regard to the latter. In Exercise 15 the reader will be asked to parallel the work for the first kind of problem.

In what follows, we shall denote the observed frequency for the cell in the ith row and the jth column by f_{ij}, the row totals by $f_{i\cdot}$, the column totals by $f_{\cdot j}$, and the grand total, the sum of all the cell frequencies, by f. With this notation, we estimate the probabilities $\theta_{i\cdot}$ and $\theta_{\cdot j}$ as

$$\hat{\theta}_{i\cdot} = \frac{f_{i\cdot}}{f} \quad \text{and} \quad \hat{\theta}_{\cdot j} = \frac{f_{\cdot j}}{f}$$

and under the null hypothesis of independence we get

$$e_{ij} = \hat{\theta}_{i\cdot} \cdot \hat{\theta}_{\cdot j} \cdot f = \frac{f_{i\cdot}}{f} \cdot \frac{f_{\cdot j}}{f} \cdot f = \frac{f_{i\cdot} \cdot f_{\cdot j}}{f}$$

for the expected frequency for the cell in the ith row and the jth column. Note that e_{ij} is thus obtained by *multiplying the total of the row to which the cell belongs by the total of the column to which it belongs and then dividing by the grand total.*

Once we have calculated the e_{ij}, we base our decision on the value of

$$\chi^2 = \sum_{i=1}^{r}\sum_{j=1}^{c} \frac{(f_{ij} - e_{ij})^2}{e_{ij}}$$

and reject the null hypothesis if it exceeds $\chi^2_{\alpha,(r-1)(c-1)}$.

The number of degrees of freedom is $(r-1)(c-1)$, and in connection with this let us make the following observation: Whenever expected cell frequencies in chi-square formulas are estimated on the basis of sample count data, the number of degrees of freedom is $s-t-1$, where s is the number of terms in the summation and t is the number of independent parameters replaced by estimates. When testing for differences among k proportions with the chi-square statistic of Section 6, we had $s = 2k$ and $t = k$, since we had to estimate the k parameters $\theta_1, \theta_2, \ldots, \theta_k$, and the number of degrees of freedom was $2k - k - 1 = k - 1$. When testing for independence with an $r \times c$ contingency table, we have $s = rc$ and $t = r + c - 2$, since the r parameters $\theta_{i\cdot}$ and the c parameters $\theta_{\cdot j}$ are not all independent: Their respective sums must equal 1. Thus, we get $s - t - 1 = rc - (r + c - 2) - 1 = (r-1)(c-1)$.

Since the test statistic that we have described has only approximately a chi-square distribution with $(r-1)(c-1)$ degrees of freedom, it is customary to use this test only when none of the e_{ij} is less than 5; sometimes this requires that we combine some of the cells with a corresponding loss in the number of degrees of freedom.

EXAMPLE 11

Use the data shown in the following table to test at the 0.01 level of significance whether a person's ability in mathematics is independent of his or her interest in statistics.

		Ability in mathematics		
		Low	*Average*	*High*
	Low	63	42	15
Interest in statistics	*Average*	58	61	31
	High	14	47	29

Solution

1. H_0: Ability in mathematics and interest in statistics are independent.
   H_1: Ability in mathematics and interest in statistics are not independent.
   $\alpha = 0.01$
2. Reject the null hypothesis if $\chi^2 \geqq 13.277$, where

$$\chi^2 = \sum_{i=1}^{r}\sum_{j=1}^{c}\frac{(f_{ij} - e_{ij})^2}{e_{ij}}$$

   and 13.277 is the value of $\chi^2_{0.01,4}$.
3. The expected frequencies for the first row are $\frac{120 \cdot 135}{360} = 45.0$, $\frac{120 \cdot 150}{360} = 50.0$, and $120 - 45.0 - 50.0 = 25.0$, where we made use of the fact that for each row or column the sum of the expected cell frequencies equals the sum of the corresponding observed frequencies (see Exercise 14). Similarly, the expected frequencies for the second row are 56.25, 62.5, and 31.25, and those for the third row (all obtained by subtraction from the column totals) are 33.75, 37.5, and 18.75. Then, substituting into the formula for χ^2 yields

$$\chi^2 = \frac{(63 - 45.0)^2}{45.0} + \frac{(42 - 50.0)^2}{50.0} + \cdots + \frac{(29 - 18.75)^2}{18.75}$$
$$= 32.14$$

4. Since $\chi^2 = 32.14$ exceeds 13.277, the null hypothesis must be rejected; we conclude that there is a relationship between a person's ability in mathematics and his or her interest in statistics.

A shortcoming of the chi-square analysis of an $r \times c$ table is that it does not take into account a possible ordering of the rows and/or columns. For instance, in Example 11, ability in mathematics, as well as interest in statistics, is ordered from low to average to high, and the value we get for χ^2 would remain the same if the rows and/or columns were interchanged among themselves. Also, the columns of the table in Section 7 reflect a definite ordering from favoring B (not favoring A) to being indifferent to favoring A, but in this case there is no specific ordering of the rows.

8 Goodness of Fit

The goodness-of-fit test considered here applies to situations in which we want to determine whether a set of data may be looked upon as a random sample from a population having a given distribution. To illustrate, suppose that we want to decide on the basis of the data (observed frequencies) shown in the following table whether the number of errors a compositor makes in setting a galley of type is a random variable having a Poisson distribution:

Number of errors	Observed frequencies f_i	Poisson probabilities with $\lambda = 3$	Expected frequencies e_i
0	18	0.0498	21.9
1	53	0.1494	65.7
2	103	0.2240	98.6
3	107	0.2240	98.6
4	82	0.1680	73.9
5	46	0.1008	44.4
6	18	0.0504	22.2
7	10	0.0216	9.5
8	2 } 3	0.0081	3.6 } 5.3
9	1	0.0038	1.7

Note that we have combined the last two classes in this table to create a new class with an expected frequency greater than 5.

To determine a corresponding set of expected frequencies for a random sample from a Poisson population, we first use the mean of the observed distribution to estimate the Poisson parameter λ, getting $\hat{\lambda} = \frac{1{,}341}{440} = 3.05$ or, approximately, $\hat{\lambda} = 3$. Then, copying the Poisson probabilities for $\lambda = 3$ from Table II of "Statistical Tables" (with the probability of 9 *or more* used instead of the probability of 9) and multiplying by 440, the total frequency, we get the expected frequencies shown in the right-hand column of the table. To test the null hypothesis that the observed frequencies constitute a random sample from a Poisson population, we must judge how good a fit, or how close an agreement, we have between the two sets of frequencies. In general, to test the null hypothesis H_0 that a set of observed data comes from a population having a specified distribution against the alternative that the population has some other distribution, we compute

$$\chi^2 = \sum_{i=1}^{m} \frac{(f_i - e_i)^2}{e_i}$$

and reject H_0 at the level of significance α if $\chi^2 \geqq \chi^2_{\alpha, m-t-1}$, where m is the number of terms in the summation and t is the number of independent parameters estimated on the basis of the sample data (see the discussion in Section 7). In the above illustration, $t = 1$ since only one parameter is estimated on the basis of the data, and the number of degrees of freedom is $m - 2$.

EXAMPLE 12

For the data in the table on this page, test at the 0.05 level of significance whether the number of errors the compositor makes in setting a galley of type is a random variable having a Poisson distribution.

Solution

(Since the expected frequencies corresponding to eight and nine errors are less than 5, the two classes are combined.)

1. H_0: Number of errors is a Poisson random variable.
   H_1: Number of errors is not a Poisson random variable.
   $\alpha = 0.05$

2. Reject the null hypothesis if $\chi^2 \geqq 14.067$, where

$$\chi^2 = \sum_{i=1}^{m} \frac{(f_i - e_i)^2}{e_i}$$

and 14.067 is the value of $\chi^2_{0.05,7}$.

3. Substituting into the formula for χ^2, we get

$$\chi^2 = \frac{(18-21.9)^2}{21.9} + \frac{(53-65.7)^2}{65.7} + \cdots + \frac{(3-5.3)^2}{5.3}$$
$$= 6.83$$

4. Since $\chi^2 = 6.83$ is less than 14.067, the null hypothesis cannot be rejected; indeed, the close agreement between the observed and expected frequencies suggests that the Poisson distribution provides a "good fit."

Exercises

14. Verify that if the expected cell frequencies are calculated in accordance with the rule in Section 7, their sum for any row or column equals the sum of the corresponding observed frequencies.

15. Show that the rule in Section 7 for calculating the expected cell frequencies applies also when we test the null hypothesis that we are sampling r populations with identical multinomial distributions.

16. Show that the following computing formula for χ^2 is equivalent to the formula in Section 7:

$$\chi^2 = \sum_{i=1}^{r}\sum_{j=1}^{c} \frac{f_{ij}^2}{e_{ij}} - f$$

17. Use the formula of Exercise 16 to recalculate χ^2 for Example 10.

18. If the analysis of a contingency table shows that there is a relationship between the two variables under consideration, the strength of this relationship may be measured by means of the **contingency coefficient**

$$C = \sqrt{\frac{\chi^2}{\chi^2 + f}}$$

where χ^2 is the value obtained for the test statistic, and f is the grand total as defined in Section 7. Show that

(a) for a 2×2 contingency table the maximum value of C is $\frac{1}{2}\sqrt{2}$;

(b) for a 3×3 contingency table the maximum value of C is $\frac{1}{3}\sqrt{6}$.

9 The Theory in Practice

Computer software exists for all the tests that we have discussed. Again, we have only to enter the original raw (untreated) data into our computer together with the appropriate command. To illustrate, consider the following example.

EXAMPLE 13

The following random samples are measurements of the heat-producing capacity (in millions of calories per ton) of specimens of coal from two mines:

Mine 1:	8,400	8,230	8,380	7,860	7,930
Mine 2:	7,510	7,690	7,720	8,070	7,660

Use the 0.05 level of significance to test whether the difference between the means of these two samples is significant.

Solution

The MINITAB computer printout in Figure 5 shows that the value of the test statistic is $t = 2.95$, the number of degrees of freedom is 7, and the P-value is 0.021.

Since 0.021 is less than 0.05, we conclude that the difference between the means of the two samples is significant at the 0.05 level of significance.

Two-Sample T-Test and CI: C1, C2

```
Two-sample T for C1 vs C2

    N  Mean  StDev  SE Mean
C1  5  8160    252      113
C2  5  7730    207       92

Difference = mu (C1) - mu (C2)
Estimate for difference:  430.000
95% CI for difference:  (85.543, 774.457)
T-Test of difference = 0 (vs not =): T-Value = 2.95  P-Value = 0.021  DF = 7
```

Figure 5. Computer printout for Example 13.

The impact of computers on statistics goes for Example 13, but we wanted to make the point that there exists software for all the standard testing procedures that we have discussed. The use of appropriate statistical computer software is recommended for many of the applied exercises that follow.

Applied Exercises SECS. 1–3

19. Based on certain data, a null hypothesis is rejected at the 0.05 level of significance. Would it also be rejected at the
(a) 0.01 level of significance;
(b) 0.10 level of significance?

20. In the test of a certain hypothesis, the P-value corresponding to the test statistic is 0.0316. Can the null hypothesis be rejected at the
(a) 0.01 level of significance;
(b) 0.05 level of significance;
(c) 0.10 level of significance?

21. With reference to Example 1, verify that the P-value corresponding to the observed value of the test statistic is 0.0046.

22. With reference to Example 2, verify that the P-value corresponding to the observed value of the test statistic is 0.0808.

23. With reference to Example 3, use suitable statistical software to find the P-value that corresponds to $t = -0.49$, where t is a value of a random variable having the t distribution with 4 degrees of freedom. Use this P-value to rework the example.

24. Test at the 0.05 level of significance whether the mean of a random sample of size $n = 16$ is "significantly less than 10" if the distribution from which the sample was taken is normal, $\bar{x} = 8.4$, and $\sigma = 3.2$. What are the null and alternative hypotheses for this test?

25. According to the norms established for a reading comprehension test, eighth graders should average 84.3 with a standard deviation of 8.6. If 45 randomly selected eighth graders from a certain school district averaged 87.8, use the four steps in the initial part of Section 1 to test the null hypothesis $\mu = 84.3$ against the alternative $\mu > 84.3$ at the 0.01 level of significance.

26. Rework Exercise 25, basing the decision on the P-value corresponding to the observed value of the test statistic.

27. The security department of a factory wants to know whether the true average time required by the night guard to walk his round is 30 minutes. If, in a random sample of 32 rounds, the night guard averaged 30.8

minutes with a standard deviation of 1.5 minutes, determine whether this is sufficient evidence to reject the null hypothesis $\mu = 30$ minutes in favor of the alternative hypothesis $\mu \neq 30$ minutes. Use the four steps in the initial part of Section 1 and the 0.01 level of significance.

28. Rework Exercise 27, basing the decision on the P-value corresponding to the observed value of the test statistic.

29. In 12 test runs over a marked course, a newly designed motorboat averaged 33.6 seconds with a standard deviation of 2.3 seconds. Assuming that it is reasonable to treat the data as a random sample from a normal population, use the four steps in the initial part of Section 1 to test the null hypothesis $\mu = 35$ against the alternative $\mu < 35$ at the 0.05 level of significance.

30. Five measurements of the tar content of a certain kind of cigarette yielded 14.5, 14.2, 14.4, 14.3, and 14.6 mg/cigarette. Assuming that the data are a random sample from a normal population, use the four steps in the initial part of Section 1 to show that at the 0.05 level of significance the null hypothesis $\mu = 14.0$ must be rejected in favor of the alternative $\mu \neq 14.0$.

31. With reference to Exercise 30, show that if the first measurement is recorded incorrectly as 16.0 instead of 14.5, this will reverse the result. Explain the apparent paradox that even though the difference between the sample mean and μ_0 has increased, it is no longer significant.

32. With reference to Exercise 30, use suitable statistical software to find the P-value that corresponds to the observed value of the test statistic. Use this P-value to rework the exercise.

33. If the same hypothesis is tested often enough, it is likely to be rejected at least once, even if it is true. A professor of biology, attempting to demonstrate this fact, ran white mice through a maze to determine if white mice ran the maze faster than the norm established by many previous tests involving various colors of mice.

(a) If the professor conducts this experiment once with several mice (using the 0.05 level of significance), what is the probability that he will come up with a "significant" result even if the color of the mouse does not affect its speed in running the maze?

(b) If the professor repeats the experiment with a new set of white mice, what is the probability that at least one of the experiments will yield a "significant" result even if the color of a mouse does not affect its maze-running speed?

(c) If the professor has 30 of his students independently run the same experiment, each with a different group of white mice, what is the probability that at least one of these experiments will come up "significant" even if mouse color plays no role in their maze-running speed?

34. An epidemiologist is trying to discover the cause of a certain kind of cancer. He studies a group of 10,000 people for five years, measuring 48 different "factors" involving eating habits, drinking habits, smoking, exercise, and so on. His object is to determine if there are any differences in the means of these factors (variables) between those who developed the given cancer and those who did not. He assumes that these variables are independent, even though there may be evidence to the contrary. In an effort to be cautiously conservative, he uses the 0.01 level of significance in all his statistical tests.

(a) What is the probability that one of these factors will be "associated with" the cancer, even if none of them is a causative factor?

(b) What is the probability that more than one of these factors will be associated with the cancer, even if none of them is a causative factor?

35. With reference to Example 4, for what values of $\bar{x}_1 - \bar{x}_2$ would the null hypothesis have been rejected? Also find the probabilities of type II errors with the given criterion if

(a) $\mu_1 - \mu_2 = 0.12$; **(b)** $\mu_1 - \mu_2 = 0.16$;

(c) $\mu_1 - \mu_2 = 0.24$; **(d)** $\mu_1 - \mu_2 = 0.28$.

36. A study of the number of business lunches that executives in the insurance and banking industries claim as deductible expenses per month was based on random samples and yielded the following results:

$$n_1 = 40 \quad \bar{x}_1 = 9.1 \quad s_1 = 1.9$$

$$n_2 = 50 \quad \bar{x}_2 = 8.0 \quad s_2 = 2.1$$

Use the four steps in the initial part of Section 1 and the 0.05 level of significance to test the null hypothesis $\mu_1 - \mu_2 = 0$ against the alternative hypothesis $\mu_1 - \mu_2 \neq 0$.

37. Rework Exercise 36, basing the decision on the P-value corresponding to the observed value of the test statistic.

38. Sample surveys conducted in a large county in a certain year and again 20 years later showed that originally the average height of 400 ten-year-old boys was 53.8 inches with a standard deviation of 2.4 inches, whereas 20 years later the average height of 500 ten-year-old boys was 54.5 inches with a standard deviation of 2.5 inches. Use the four steps in the initial part of Section 1 and the 0.05 level of significance to test the null hypothesis $\mu_1 - \mu_2 = -0.5$ against the alternative hypothesis $\mu_1 - \mu_2 < -0.5$.

39. Rework Exercise 38, basing the decision on the P-value corresponding to the observed value of the test statistic.

40. To find out whether the inhabitants of two South Pacific islands may be regarded as having the same racial ancestry, an anthropologist determines the cephalic indices of six adult males from each island, getting $\bar{x}_1 = 77.4$ and $\bar{x}_2 = 72.2$ and the corresponding standard deviations $s_1 = 3.3$ and $s_2 = 2.1$. Use the four steps in the initial part of Section 1 and the 0.01 level of significance to see whether the difference between the two sample means can reasonably be attributed to chance. Assume that the populations sampled are normal and have equal variances.

41. With reference to Example 5, use suitable statistical software to show that the P-value corresponding to $t = 2.67$ is 0.0185.

42. To compare two kinds of front-end designs, six of each kind were installed on a certain make of compact car. Then each car was run into a concrete wall at 5 miles per hour, and the following are the costs of the repairs (in dollars):

Design 1:	127	168	143	165	122	139
Design 2:	154	135	132	171	153	149

Use the four steps in the initial part of Section 1 to test at the 0.01 level of significance whether the difference between the means of these two samples is significant.

43. With reference to Exercise 42, use suitable statistical software to find the P-value corresponding to the observed value of the test statistic. Use this P-value to rework the exercise.

44. In a study of the effectiveness of certain exercises in weight reduction, a group of 16 persons engaged in these exercises for one month and showed the following results:

Weight before	*Weight after*	*Weight before*	*Weight after*
211	198	172	166
180	173	155	154
171	172	185	181
214	209	167	164
182	179	203	201
194	192	181	175
160	161	245	233
182	182	146	142

Use the 0.05 level of significance to test the null hypothesis $\mu_1 - \mu_2 = 0$ against the alternative hypothesis $\mu_1 - \mu_2 > 0$, and thus judge whether the exercises are effective in weight reduction.

45. The following are the average weekly losses of work-hours due to accidents in 10 industrial plants before and after a certain safety program was put into operation:

45 and 36, 73 and 60, 46 and 44, 124 and 119, 33 and 35,

57 and 51, 83 and 77, 34 and 29, 26 and 24, and 17 and 11

Use the four steps in the initial part of Section 1 and the 0.05 level of significance to test whether the safety program is effective.

46. With reference to Exercise 45, use suitable statistical software to find the P-value that corresponds to the observed value of the test statistic. Use this P-value to rework the exercise.

SEC. 4

47. Nine determinations of the specific heat of iron had a standard deviation of 0.0086. Assuming that these determinations constitute a random sample from a normal population, test the null hypothesis $\sigma = 0.0100$ against the alternative hypothesis $\sigma < 0.0100$ at the 0.05 level of significance.

48. In a random sample, the weights of 24 Black Angus steers of a certain age have a standard deviation of 238 pounds. Assuming that the weights constitute a random sample from a normal population, test the null hypothesis $\sigma = 250$ pounds against the two-sided alternative $\sigma \neq 250$ pounds at the 0.01 level of significance.

49. In a random sample, $s = 2.53$ minutes for the amount of time that 30 women took to complete the written test for their driver's licenses. At the 0.05 level of significance, test the null hypothesis $\sigma = 2.85$ minutes against the alternative hypothesis $\sigma < 2.85$ minutes. (Use the method described in the text.)

50. Use the method of Exercise 7 to rework Exercise 49.

51. Past data indicate that the standard deviation of measurements made on sheet metal stampings by experienced inspectors is 0.41 square inch. If a new inspector measures 50 stampings with a standard deviation of 0.49 square inch, use the method of Exercise 7 to test the null hypothesis $\sigma = 0.41$ square inch against the alternative hypothesis $\sigma > 0.41$ square inch at the 0.05 level of significance.

52. With reference to Exercise 51, find the P-value corresponding to the observed value of the test statistic and use it to decide whether the null hypothesis could have been rejected at the 0.015 level of significance.

53. With reference to Example 5, test the null hypothesis $\sigma_1 - \sigma_2 = 0$ against the alternative hypothesis $\sigma_1 - \sigma_2 > 0$ at the 0.05 level of significance.

54. With reference to Exercise 40, test at the 0.10 level of significance whether it is reasonable to assume that the two populations sampled have equal variances.

55. With reference to Exercise 42, test at the 0.02 level of significance whether it is reasonable to assume that the two populations sampled have equal variances.

SECS. 5–6

56. With reference to Example 8, show that the critical region is $x \leqq 5$ or $x \geqq 15$ and that, corresponding to this critical region, the level of significance is actually 0.0414.

57. It has been claimed that more than 40 percent of all shoppers can identify a highly advertised trademark. If, in a random sample, 10 of 18 shoppers were able to identify the trademark, test at the 0.05 level of significance whether the null hypothesis $\theta = 0.40$ can be rejected against the alternative hypothesis $\theta > 0.40$.

58. With reference to Exercise 57, find the critical region and the actual level of significance corresponding to this critical region.

59. A doctor claims that less than 30 percent of all persons exposed to a certain amount of radiation will feel any ill effects. If, in a random sample, only 1 of 19 persons exposed to such radiation felt any ill effects, test the null hypothesis $\theta = 0.30$ against the alternative hypothesis $\theta < 0.30$ at the 0.05 level of significance.

60. With reference to Exercise 59, find the critical region and the actual level of significance corresponding to this critical region.

61. In a random sample, 12 of 14 industrial accidents were due to unsafe working conditions. Use the 0.01 level of significance to test the null hypothesis $\theta = 0.40$ against the alternative hypothesis $\theta \neq 0.40$.

62. With reference to Exercise 61, find the critical region and the actual level of significance corresponding to this critical region.

63. In a random survey of 1,000 households in the United States, it is found that 29 percent of the households contained at least one member with a college degree. Does this finding refute the statement that the proportion of all such U.S. households is at least 35 percent? (Use the 0.05 level of significance.)

64. In a random sample of 12 undergraduate business students, 6 said that they will take advanced work in accounting. Use the 0.01 level of significance to test the null hypothesis $\theta = 0.20$, that is, 20 percent of all undergraduate business students will take advanced work in accounting, against the alternative hypothesis $\theta > 0.20$.

65. A food processor wants to know whether the probability is really 0.60 that a customer will prefer a new kind of packaging to the old kind. If, in a random sample, 7 of 18 customers prefer the new kind of packaging to the old kind, test the null hypothesis $\theta = 0.60$ against the alternative hypothesis $\theta \neq 0.60$ at the 0.05 level of significance.

66. In a random sample of 600 cars making a right turn at a certain intersection, 157 pulled into the wrong lane. Use the 0.05 level of significance to test the null hypothesis that the actual proportion of drivers who make this mistake at the given intersection is $\theta = 0.30$ against the alternative hypothesis $\theta \neq 0.30$.

67. The manufacturer of a spot remover claims that his product removes 90 percent of all spots. If, in a random sample, only 174 of 200 spots were removed with the manufacturer's product, test the null hypothesis $\theta = 0.90$ against the alternative hypothesis $\theta < 0.90$ at the 0.05 level of significance.

68. In random samples, 74 of 250 persons who watched a certain television program on a small TV set and 92 of 250 persons who watched the same program on a large set remembered 2 hours later what products were advertised. Use the χ^2 statistic to test the null hypothesis $\theta_1 = \theta_2$ against the alternative hypothesis $\theta_1 \neq \theta_2$ at the 0.01 level of significance.

69. Use the statistic of Exercise 12 to rework Exercise 68.

70. In random samples, 46 of 400 tulip bulbs from one nursery failed to bloom and 18 of 200 tulip bulbs from another nursery failed to bloom. Use the χ^2 statistic to test the null hypothesis $\theta_1 = \theta_2$ against the alternative hypothesis $\theta_1 \neq \theta_2$ at the 0.05 level of significance.

71. Use the statistic of Exercise 12 to rework Exercise 70, and verify that the square of the value obtained for z equals the value obtained for χ^2.

72. In a random sample of 200 persons who skipped breakfast, 82 reported that they experienced midmorning fatigue, and in a random sample of 300 persons who ate breakfast, 87 reported that they experienced midmorning fatigue. Use the method of Exercise 12 and the 0.05 level of significance to test the null hypothesis that there is no difference between the corresponding population proportions against the alternative hypothesis that midmorning fatigue is more prevalent among persons who skip breakfast.

73. If 26 of 200 tires of brand A failed to last 30,000 miles, whereas the corresponding figures for 200 tires of brands B, C, and D were 23, 15, and 32, test the null hypothesis that there is no difference in the durability of the four kinds of tires at the 0.05 level of significance.

74. In random samples of 250 persons with low incomes, 200 persons with average incomes, and 150 persons with high incomes, there were, respectively, 155, 118, and 87 who favor a certain piece of legislation. Use the 0.05 level of significance to test the null hypothesis $\theta_1 = \theta_2 = \theta_3$ (that the proportion of persons favoring the legislation is the same for all three income groups) against the alternative hypothesis that the three θ's are not all equal.

SECS. 7–8

75. Samples of an experimental material are produced by three different prototype processes and tested for compliance to a strength standard. If the tests showed the following results, can it be said at the 0.01 level of significance that the three processes have the same probability of passing this strength standard?

	Process A	*Process B*	*Process C*
Number passing test	45	58	49
Number failing test	21	15	35

76. In a study of parents' feelings about a required course in sex education, 360 parents, a random sample, are classified according to whether they have one, two, or three or more children in the school system and also whether they feel that the course is poor, adequate, or good. Based on the results shown in the following table, test at the 0.05 level of significance whether there is a relationship between parents' reaction to the course and the number of children that they have in the school system:

	Number of children		
	1	2	3 *or more*
Poor	48	40	12
Adequate	55	53	29
Good	57	46	20

77. Tests of the fidelity and the selectivity of 190 radios produced the results shown in the following table:

		Fidelity		
		Low	*Average*	*High*
	Low	7	12	31
Selectivity	*Average*	35	59	18
	High	15	13	0

Use the 0.01 level of significance to test the null hypothesis that fidelity is independent of selectivity.

78. The following sample data pertain to the shipments received by a large firm from three different vendors:

	Number rejected	*Number imperfect but acceptable*	*Number perfect*
Vendor A	12	23	89
Vendor B	8	12	62
Vendor C	21	30	119

Test at the 0.01 level of significance whether the three vendors ship products of equal quality.

79. Analyze the 3×3 table in the initial part of Section 1, which pertains to the responses of shoppers in three different cities with regard to two detergents. Use the 0.05 level of significance.

80. Four coins were tossed 160 times and 0, 1, 2, 3, or 4 heads showed, respectively, 19, 54, 58, 23, and 6 times. Use the 0.05 level of significance to test whether it is reasonable to suppose that the coins are balanced and randomly tossed.

81. It is desired to test whether the number of gamma rays emitted per second by a certain radioactive substance is a random variable having the Poisson distribution with $\lambda = 2.4$. Use the following data obtained for 300 1-second intervals to test this null hypothesis at the 0.05 level of significance:

Number of gamma rays	*Frequency*
0	19
1	48
2	66
3	74
4	44
5	35
6	10
7 or more	4

82. Each day, Monday through Saturday, a baker bakes three large chocolate cakes, and those not sold on the same day are given away to a food bank. Use the data shown in the following table to test at the 0.05 level of significance whether they may be looked upon as values of a binomial random variable:

Number of cakes sold	*Number of days*
0	1
1	16
2	55
3	228

83. The following is the distribution of the readings obtained with a Geiger counter of the number of particles emitted by a radioactive substance in 100 successive 40-second intervals:

Number of particles	*Frequency*
5–9	1
10–14	10
15–19	37
20–24	36
25–29	13
30–34	2
35–39	1

(a) Verify that the mean and the standard deviation of this distribution are $\bar{x} = 20$ and $s = 5$.

(b) Find the probabilities that a random variable having a normal distribution with $\mu = 20$ and $\sigma = 5$ will take on a value less than 9.5, between 9.5 and 14.5, between 14.5 and 19.5, between 19.5 and 24.5, between 24.5 and 29.5, between 29.5 and 34.5, and greater than 34.5.

(c) Find the expected normal curve frequencies for the various classes by multiplying the probabilities obtained in part (b) by the total frequency, and then test at the 0.05 level of significance whether the data may be looked upon as a random sample from a normal population.

SEC. 9

84. The following are the hours of operation to failure of 38 light bulbs.

150	389	345	310	20	310	175	376	334	340
332	331	327	344	328	341	325	2	311	320
256	315	55	345	111	349	245	367	81	327
355	309	375	316	336	278	396	287		

Use a suitable statistical computer program to test whether the mean failure time of such light bulbs is significantly less than 300 hours. Use the 0.01 level of significance.

85. The following are the drying times (minutes) of 40 sheets coated with polyurethane under two different ambient conditions.

Condition 1:	55.6	56.1	61.8	55.9	51.4	59.9	54.3	62.8	58.5	55.8
	58.3	60.2	54.2	50.1	57.1	57.5	63.6	59.3	60.9	61.8
Condition 2:	55.1	43.5	51.2	46.2	56.7	52.5	53.5	60.5	52.1	47.0
	53.0	53.8	51.6	53.6	42.9	52.0	55.1	57.1	62.8	54.8

Use a suitable statistical computer program to test whether there is a significant difference between the mean drying times under the two ambient conditions. Use the 0.05 level of significance.

86. Samples of three materials under consideration for the housing of machinery on a seagoing vessel are tested by means of a salt-spray test. Any sample that leaks when subject to a power spray is considered to have failed. The following are the test results:

	Material A	*Material B*	*Material C*
Number leaked	36	22	18
Number not leaked	63	45	29

Use a suitable statistical computer program to test at the 0.05 level of significance if the three materials have the same probability of leaking in this test.

References

The problem of determining the appropriate number of degrees of freedom for various uses of the chi-square statistic is discussed in

Cramér, H., *Mathematical Methods of Statistics*. Princeton, N.J.: Princeton University Press, 1946.

The *Smith–Satterthwaite* test of the null hypothesis that two normal populations with unequal variances have the same mean is given in

Johnson, R. A., *Miller and Freund's Probability and Statistics for Engineers*, 5th ed. Upper Saddle River, N.J.: Prentice Hall, 1994.

Additional comments relative to using the continuity correction for testing hypotheses concerning binomial parameters can be found in

Brownlee, K. A., *Statistical Theory and Methodology in Science and Engineering*, 2nd ed. New York: John Wiley & Sons, Inc., 1965.

Details about the analysis of contingency tables may be found in

Everitt, B. S., *The Analysis of Contingency Tables*. New York: John Wiley & Sons, Inc., 1977.

In recent years, research has been done on the analysis of $r \times c$ tables, where the categories represented by the rows and/or columns are ordered. This work is beyond the level of this chapter, but some introductory material may be found in

Agresti, A., *Analysis of Ordinal Categorical Data*. New York: John Wiley & Sons, Inc., 1984,

Agresti, A., *Categorical Data Analysis*. New York: John Wiley & Sons, Inc., 1990,

Goodman, L. A., *The Analysis of Cross-Classified Data Having Ordered Categories*. Cambridge, Mass.: Harvard University Press, 1984.

Answers to Odd-Numbered Exercises

1 Use the critical region $\frac{n(\bar{x}-\mu_0)^2}{\sigma^2} \geq \chi^2_{\alpha,1}$

3 $n = 52$.

5 $n = 151$.

9 The alternative hypothesis is $\lambda > \lambda_0$; reject the null hypothesis if $\sum_{i=1}^{n} x_i \geq k_\alpha$ where k_α is the smallest integer for which $\sum_{y=k_\alpha}^{\infty} p(y; n, \lambda_0) \leq \alpha$.

19 **(a)** No; **(b)** yes.

23 P-value $= 0.3249$; the null hypothesis cannot be rejected.

25 $z = 2.73$; the null hypothesis must be rejected.

27 $z = 3.02$; the null hypothesis must be rejected.

29 $t = -2.11$; the null hypothesis must be rejected.

31 s has also increased to 0.742.

33 **(a)** $P(\text{reject } H_0 | H_0 \text{ is true}) = 0.05$; **(b)** $P(\text{reject } H_0$ on experiment 1 or experiment 2 or both $|H_0 \text{ is true}) = 0.0975$; **(c)** $P(\text{reject } H_0$ on one or more of 30 experiments $|H_0 \text{ is true}) = 0.79$.

35 **(a)** $\beta = 0.18$; **(b)** $\beta = 0.71$; **(c)** $\beta = 0.71$; **(d)** $\beta = 0.18$.

37 The P-value is 0.0094; the null hypothesis must be rejected.

39 The P-value is 0.1112; the null hypothesis cannot be rejected.

43 The P-value is 0.61; the null hypothesis cannot be rejected.

45 $t = 4.03$; the null hypothesis must be rejected.

47 $\chi^2 = 5.92$; the null hypothesis cannot be rejected.

49 $\chi^2 = 22.85$; the null hypothesis cannot be rejected.

51 $z = 1.93$; the null hypothesis must be rejected.

53 $f = 1.42$; the null hypothesis cannot be rejected.

55 $f = 1.80$; the null hypothesis cannot be rejected.

57 The P-value is 0.1348; the null hypothesis cannot be rejected.

59 The P-value is 0.0104; the null hypothesis must be rejected.

61 The P-value is 0.0012; the null hypothesis must be rejected.

63 $z = -3.98$; the null hypothesis must be rejected; thus the statement is refuted.

65 The P-value is 0.1154; the null hypothesis cannot be rejected.

69 $z = -1.71$; the null hypothesis cannot be rejected.

73 $\chi^2 = 7.10$; the null hypothesis cannot be rejected.

75 $\chi^2 = 8.03$; the null hypothesis cannot be rejected.

77 $\chi^2 = 52.7$; the null hypothesis must be rejected.

79 $\chi^2 = 3.71$; the null hypothesis cannot be rejected.

81 $\chi^2 = 28.9$; the null hypothesis must be rejected.

83 **(b)** The probabilities are 0.0179, 0.1178, 0.3245, 0.3557, 0.1554, 0.0268, and 0.0019. **(c)** The expected frequencies are 1.8, 11.8, 32.4, 35.6, 15.5, 2.7, and 0.2; $\chi^2 = 1.46$; the null hypothesis cannot be rejected.

85 $t = 3.61$; the P-value = 0.0009; thus, the difference is significant at the 0.005 level of significance.

Regression and Correlation

1 Introduction

A major objective of many statistical investigations is to establish relationships that make it possible to predict one or more variables in terms of others. Thus, studies are made to predict the potential sales of a new product in terms of its price, a patient's weight in terms of the number of weeks he or she has been on a diet, family expenditures on entertainment in terms of family income, the per capita consumption of certain foods in terms of their nutritional values and the amount of money spent advertising them on television, and so forth.

Although it is, of course, desirable to be able to predict one quantity exactly in terms of others, this is seldom possible, and in most instances we have to be satisfied with predicting averages or expected values. Thus, we may not be able to predict exactly how much money Mr. Brown will make 10 years after graduating from college, but, given suitable data, we can predict the average income of a college graduate in terms of the number of years he has been out of college. Similarly, we can at best predict the average yield of a given variety of wheat in terms of data on the rainfall in July, and we can at best predict the average performance of students starting college in terms of their I.Q.'s.

Formally, if we are given the joint distribution of two random variables X and Y, and X is known to take on the value x, the basic problem of **bivariate regression** is that of determining the conditional mean $\mu_{Y|x}$, that is, the "average" value of Y for the given value of X. The term "regression," as it is used here, dates back to Francis Galton, who employed it to indicate certain relationships in the theory of heredity. In problems involving more than two random variables, that is, in **multiple regression**, we are concerned with quantities such as $\mu_{Z|x,y}$, the mean of Z for given values of X and Y, $\mu_{X_4|x_1,\, x_2,\, x_3}$, the mean of X_4 for given values of X_1, X_2, and X_3, and so on.

Definition 1. Bivariate Regression; Regression Equation. *If* f(x, y) *is the value of the joint density of two random variables* X *and* Y, ***bivariate regression*** *consists of determining the conditional density of* Y, *given* X = x *and then evaluating the integral*

$$\mu_{Y|x} = E(Y|x) = \int_{-\infty}^{\infty} y \cdot w(y|x)dy$$

The resulting equation is called the ***regression equation of*** **Y on** X. *Alternately, the* ***regression equation of*** X **on** Y *is given by*

$$\mu_{X|y} = E(X|y) = \int_{-\infty}^{\infty} x \cdot f(x|y)dy$$

In the discrete case, when we are dealing with probability distributions instead of probability densities, the integrals in the two regression equations given in Definition 1 are simply replaced by sums. When we do not know the joint probability density or distribution of the two random variables, or at least not all its parameters, the determination of $\mu_{Y|x}$ or $\mu_{X|y}$ becomes a problem of estimation based on sample data; this is an entirely different problem, which we shall discuss in Sections 3 and 4.

EXAMPLE 1

Given the two random variables X and Y that have the joint density

$$f(x, y) = \begin{cases} x \cdot e^{-x(1+y)} & \text{for } x > 0 \text{ and } y > 0 \\ 0 & \text{elsewhere} \end{cases}$$

find the regression equation of Y on X and sketch the regression curve.

Solution

Integrating out y, we find that the marginal density of X is given by

$$g(x) = \begin{cases} e^{-x} & \text{for } x > 0 \\ 0 & \text{elsewhere} \end{cases}$$

and hence the conditional density of Y given $X = x$ is given by

$$w(y|x) = \frac{f(x, y)}{g(x)} = \frac{x \cdot e^{-x(1+y)}}{e^{-x}} = x \cdot e^{-xy}$$

for $y > 0$ and $w(y|x) = 0$ elsewhere, which we recognize as an exponential density with $\theta = \frac{1}{x}$. Hence, by evaluating

$$\mu_{Y|x} = \int_{0}^{\infty} y \cdot x \cdot e^{-xy}\, dy$$

or by referring to the corollary of a theorem given here "The mean and the variance of the exponential distribution are given by $\mu = \theta$ and $\sigma^2 = \theta^2$," we find that the regression equation of Y on X is given by

$$\mu_{Y|x} = \frac{1}{x}$$

The corresponding regression curve is shown in Figure 1.

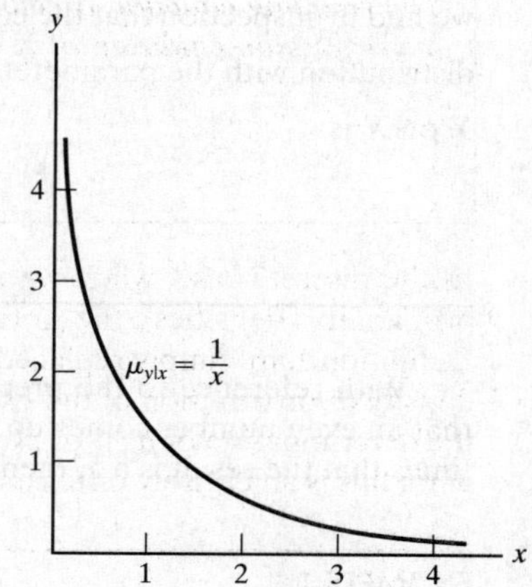

Figure 1. Regression curve of Example 1.

EXAMPLE 2

If X and Y have the multinomial distribution

$$f(x, y) = \binom{n}{x, y, n-x-y} \cdot \theta_1^x \theta_2^y (1-\theta_1-\theta_2)^{n-x-y}$$

for $x = 0, 1, 2, \ldots, n$, and $y = 0, 1, 2, \ldots, n$, with $x + y \leqq n$, find the regression equation of Y on X.

Solution

The marginal distribution of X is given by

$$g(x) = \sum_{y=0}^{n-x} \binom{n}{x, y, n-x-y} \cdot \theta_1^x \theta_2^y (1-\theta_1-\theta_2)^{n-x-y}$$
$$= \binom{n}{x} \theta_1^x (1-\theta_1)^{n-x}$$

for $x = 0, 1, 2, \ldots, n$, which we recognize as a binomial distribution with the parameters n and θ_1. Hence,

$$w(y|x) = \frac{f(x, y)}{g(x)} = \frac{\binom{n-x}{y} \theta_2^y (1-\theta_1-\theta_2)^{n-x-y}}{(1-\theta_1)^{n-x}}$$

for $y = 0, 1, 2, \ldots, n-x$, and, rewriting this formula as

$$w(y|x) = \binom{n-x}{y} \left(\frac{\theta_2}{1-\theta_1}\right)^y \left(\frac{1-\theta_1-\theta_2}{1-\theta_1}\right)^{n-x-y}$$

we find by inspection that the conditional distribution of Y given $X = x$ is a binomial distribution with the parameters $n - x$ and $\frac{\theta_2}{1-\theta_1}$, so that the regression equation of Y on X is

$$\mu_{Y|x} = \frac{(n-x)\theta_2}{1-\theta_1}$$

With reference to the preceding example, if we let X be the number of times that an even number comes up in 30 rolls of a balanced die and Y be the number of times that the result is a 5, then the regression equation becomes

$$\mu_{Y|x} = \frac{(30-x)\frac{1}{6}}{1-\frac{1}{2}} = \frac{1}{3}(30-x)$$

This stands to reason, because there are three equally likely possibilities, 1, 3, or 5, for each of the $30 - x$ outcomes that are not even.

EXAMPLE 3

If the joint density of X_1, X_2, and X_3 is given by

$$f(x_1, x_2, x_3) = \begin{cases} (x_1 + x_2)e^{-x_3} & \text{for } 0 < x_1 < 1, 0 < x_2 < 1, x_3 > 0 \\ 0 & \text{elsewhere} \end{cases}$$

find the regression equation of X_2 on X_1 and X_3.

Solution

The joint marginal density of X_1 and X_3 is given by

$$m(x_1, x_3) = \begin{cases} \left(x_1 + \frac{1}{2}\right) e^{-x_3} & \text{for } 0 < x_1 < 1, x_3 > 0 \\ 0 & \text{elsewhere} \end{cases}$$

Therefore,

$$\mu_{X_2|x_1,x_3} = \int_{-\infty}^{\infty} x_2 \cdot \frac{f(x_1, x_2, x_3)}{m(x_1, x_3)}\, dx_2 = \int_0^1 \frac{x_2(x_1 + x_2)}{\left(x_1 + \frac{1}{2}\right)}\, dx_2$$

$$= \frac{x_1 + \frac{2}{3}}{2x_1 + 1}$$

Note that the conditional expectation obtained in the preceding example depends on x_1 but not on x_3. This could have been expected, since there is a pairwise independence between X_2 and X_3.

2 Linear Regression

An important feature of Example 2 is that the regression equation is linear; that is, it is of the form

$$\mu_{Y|x} = \alpha + \beta x$$

where α and β are constants, called the **regression coefficients**. There are several reasons why linear regression equations are of special interest: First, they lend themselves readily to further mathematical treatment; then, they often provide good approximations to otherwise complicated regression equations; and, finally, in the case of the bivariate normal distribution, the regression equations are, in fact, linear.

To simplify the study of linear regression equations, let us express the regression coefficients α and β in terms of some of the lower moments of the joint distribution of X and Y, that is, in terms of $E(X) = \mu_1$, $E(Y) = \mu_2$, $\text{var}(X) = \sigma_1^2$, $\text{var}(Y) = \sigma_2^2$, and $\text{cov}(X, Y) = \sigma_{12}$. Then, also using the correlation coefficient

$$\rho = \frac{\sigma_{12}}{\sigma_1 \sigma_2}$$

we can prove the following results.

THEOREM 1. If the regression of Y on X is linear, then

$$\mu_{Y|x} = \mu_2 + \rho \frac{\sigma_2}{\sigma_1}(x - \mu_1)$$

and if the regression of X on Y is linear, then

$$\mu_{X|y} = \mu_1 + \rho \frac{\sigma_1}{\sigma_2}(y - \mu_2)$$

Proof Since $\mu_{Y|x} = \alpha + \beta x$, it follows that

$$\int y \cdot w(y|x)\, dy = \alpha + \beta x$$

and if we multiply the expression on both sides of this equation by $g(x)$, the corresponding value of the marginal density of X, and integrate on x, we obtain

$$\iint y \cdot w(y|x) g(x)\, dy\, dx = \alpha \int g(x)\, dx + \beta \int x \cdot g(x)\, dx$$

or

$$\mu_2 = \alpha + \beta \mu_1$$

since $w(y|x)g(x) = f(x, y)$. If we had multiplied the equation for $\mu_{Y|x}$ on both sides by $x \cdot g(x)$ before integrating on x, we would have obtained

$$\iint xy \cdot f(x,y)\,dy\,dx = \alpha \int x \cdot g(x)\,dx + \beta \int x^2 \cdot g(x)\,dx$$

or

$$E(XY) = \alpha\mu_1 + \beta E(X^2)$$

Solving $\mu_2 = \alpha + \beta\mu_1$ and $E(XY) = \alpha\mu_1 + \beta E(X^2)$ for α and β and making use of the fact that $E(XY) = \sigma_{12} + \mu_1\mu_2$ and $E(X^2) = \sigma_1^2 + \mu_1^2$, we find that

$$\alpha = \mu_2 - \frac{\sigma_{12}}{\sigma_1^2} \cdot \mu_1 = \mu_2 - \rho\frac{\sigma_2}{\sigma_1} \cdot \mu_1$$

and

$$\beta = \frac{\sigma_{12}}{\sigma_1^2} = \rho\frac{\sigma_2}{\sigma_1}$$

This enables us to write the linear regression equation of Y on X as

$$\mu_{Y|x} = \mu_2 + \rho\frac{\sigma_2}{\sigma_1}(x - \mu_1)$$

When the regression of X on Y is linear, similar steps lead to the equation

$$\mu_{X|y} = \mu_1 + \rho\frac{\sigma_1}{\sigma_2}(y - \mu_2)$$

It follows from Theorem 1 that if the regression equation is linear and $\rho = 0$, then $\mu_{Y|x}$ does not depend on x (or $\mu_{X|y}$ does not depend on y). When $\rho = 0$ and hence $\sigma_{12} = 0$, the two random variables X and Y are **uncorrelated**, and we can say that if two random variables are independent, they are also uncorrelated, but if two random variables are uncorrelated, they are not necessarily independent; the latter is again illustrated in Exercise 9.

The correlation coefficient and its estimates are of importance in many statistical investigations, and they will be discussed in some detail in Section 5. At this time, let us again point out that $-1 \leqq \rho \leqq +1$, as the reader will be asked to prove in Exercise 11, and the sign of ρ tells us directly whether the slope of a regression line is upward or downward.

3 The Method of Least Squares

In the preceding sections we have discussed the problem of regression only in connection with random variables having known joint distributions. In actual practice, there are many problems where a set of **paired data** gives the indication that the regression is linear, where we do not know the joint distribution of the random variables under consideration but, nevertheless, want to estimate the regression coefficients α and β. Problems of this kind are usually handled by the **method of least squares**, a method of curve fitting suggested early in the nineteenth century by the French mathematician Adrien Legendre.

To illustrate this technique, let us consider the following data on the number of hours that 10 persons studied for a French test and their scores on the test:

Hours studied x	*Test score* y
4	31
9	58
10	65
14	73
4	37
7	44
12	60
22	91
1	21
17	84

Plotting these data as in Figure 2, we get the impression that a straight line provides a reasonably good fit. Although the points do not all fall exactly on a straight line, the overall pattern suggests that the average test score for a given number of hours studied may well be related to the number of hours studied by means of an equation of the form $\mu_{Y|x} = \alpha + \beta x$.

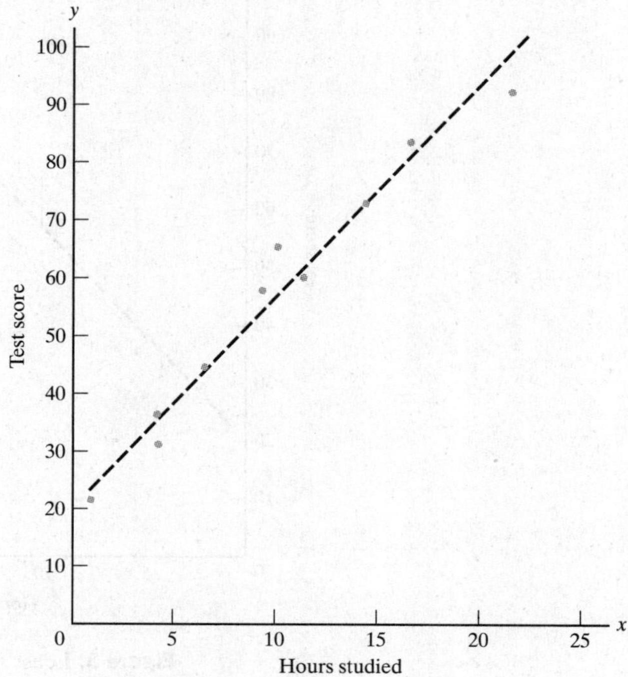

Figure 2. Data on hours studied and test scores.

Once we have decided in a given problem that the regression is approximately linear and the joint density of x and y is unknown, we face the problem of estimating the coefficients α and β from the sample data. In other words, we face the problem of obtaining estimates $\hat{\alpha}$ and $\hat{\beta}$ such that the estimated regression line $\hat{y} = \hat{\alpha} + \hat{\beta}x$ in some sense provides the best possible fit to the given data.

Denoting the vertical deviation from a point to the estimated regression line by e_i, as indicated in Figure 3, the least squares criterion on which we shall base this "goodness of fit" is defined as follows:

DEFINITION 2. LEAST SQUARES ESTIMATE. *If we are given a set of paired data*

$$\{(x_i, y_i);\ i = 1, 2, \ldots, n\}$$

The ***least squares estimates*** *of the regression coefficients in bivariate linear regression are those that make the quantity*

$$q = \sum_{i=1}^{n} e_i^2 = \sum_{i=1}^{n} [y_i - (\hat{\alpha} + \hat{\beta}x_i)]^2$$

a minimum with respect to $\hat{\alpha}$ and $\hat{\beta}$.

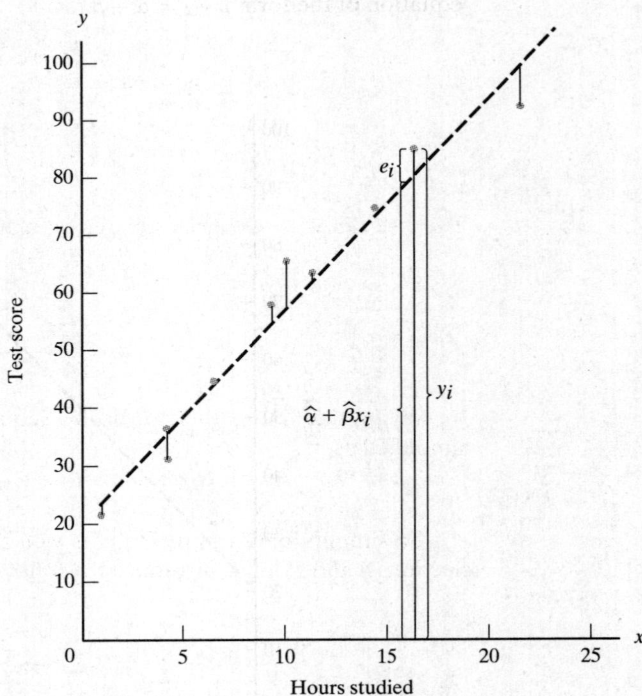

Figure 3. Least squares criterion.

Finding the minimum by differentiating partially with respect to $\hat{\alpha}$ and $\hat{\beta}$ and equating these partial derivatives to zero, we obtain

$$\frac{\partial q}{\partial \hat{\alpha}} = \sum_{i=1}^{n}(-2)[y_i - (\hat{\alpha} + \hat{\beta}x_i)] = 0$$

and

$$\frac{\partial q}{\partial \hat{\beta}} = \sum_{i=1}^{n}(-2)x_i[y_i - (\hat{\alpha} + \hat{\beta}x_i)] = 0$$

which yield the system of **normal equations**

$$\sum_{i=1}^{n} y_i = \hat{\alpha}n + \hat{\beta}\cdot\sum_{i=1}^{n} x_i$$

$$\sum_{i=1}^{n} x_i y_i = \hat{\alpha}\cdot\sum_{i=1}^{n} x_i + \hat{\beta}\cdot\sum_{i=1}^{n} x_i^2$$

Solving this system of equations by using determinants or the method of elimination, we find that the least squares estimate of β is

$$\hat{\beta} = \frac{n\left(\sum_{i=1}^{n} x_i y_i\right) - \left(\sum_{i=1}^{n} x_i\right)\left(\sum_{i=1}^{n} y_i\right)}{n\left(\sum_{i=1}^{n} x_i^2\right) - \left(\sum_{i=1}^{n} x_i\right)^2}$$

Then we can write the least squares estimate of α as

$$\hat{\alpha} = \frac{\sum_{i=1}^{n} y_i - \hat{\beta}\cdot\sum_{i=1}^{n} x_i}{n}$$

by solving the first of the two normal equations for $\hat{\alpha}$. This formula for $\hat{\alpha}$ can be simplified as

$$\hat{\alpha} = \overline{y} - \hat{\beta}\cdot\overline{x}$$

To simplify the formula for $\hat{\beta}$ as well as some of the formulas we shall meet in Sections 4 and 5, let us introduce the following notation:

$$S_{xx} = \sum_{i=1}^{n}(x_i - \overline{x})^2 = \sum_{i=1}^{n} x_i^2 - \frac{1}{n}\left(\sum_{i=1}^{n} x_i\right)^2$$

$$S_{yy} = \sum_{i=1}^{n}(y_i - \overline{y})^2 = \sum_{i=1}^{n} y_i^2 - \frac{1}{n}\left(\sum_{i=1}^{n} y_i\right)^2$$

and

$$S_{xy} = \sum_{i=1}^{n}(x_i - \overline{x})(y_i - \overline{y}) = \sum_{i=1}^{n} x_i y_i - \frac{1}{n}\left(\sum_{i=1}^{n} x_i\right)\left(\sum_{i=1}^{n} y_i\right)$$

We can thus write the following theorem.

THEOREM 2. Given the sample data $\{(x_i, y_i);\ i = 1, 2, \ldots, n\}$, the coefficients of the least squares line $\hat{y} = \hat{\alpha} + \hat{\beta}x$ are

$$\hat{\beta} = \frac{S_{xy}}{S_{xx}}$$

and

$$\hat{\alpha} = \overline{y} - \hat{\beta} \cdot \overline{x}$$

EXAMPLE 4

With reference to the data in the table in Section 3,

(a) find the equation of the least squares line that approximates the regression of the test scores on the number of hours studied;

(b) predict the average test score of a person who studied 14 hours for the test.

Solution

(a) Omitting the limits of summation for simplicity, we get $n = 10$, $\Sigma x = 100$, $\Sigma x^2 = 1{,}376$, $\Sigma y = 564$, and $\Sigma xy = 6{,}945$ from the data. Thus

$$S_{xx} = 1{,}376 - \frac{1}{10}(100)^2 = 376$$

and

$$S_{xy} = 6{,}945 - \frac{1}{10}(100)(564) = 1{,}305$$

Thus, $\hat{\beta} = \frac{1{,}305}{376} = 3.471$ and $\hat{\alpha} = \frac{564}{10} - 3.471 \cdot \frac{100}{10} = 21.69$, and the equation of the least squares line is

$$\hat{y} = 21.69 + 3.471x$$

(b) Substituting $x = 14$ into the equation obtained in part (a), we get

$$\hat{y} = 21.69 + 3.471(14) = 70.284$$

or $\hat{y} = 70$, rounded to the nearest unit.

Since we did not make any assumptions about the joint distribution of the random variables with which we were concerned in the preceding example, we cannot judge the "goodness" of the prediction obtained in part (b); also, we cannot judge the "goodness" of the estimates $\hat{\alpha} = 21.69$ and $\hat{\beta} = 3.471$ obtained in part (a). Problems like this will be discussed in Section 4.

The least squares criterion, or, in other words, the method of least squares, is used in many problems of curve fitting that are more general than the one treated in this section. Above all, it will be used in Sections 6 and 7 to estimate the coefficients of **multiple regression equations** of the form

$$\mu_{Y|x_1,\ldots,x_k} = \beta_0 + \beta_1 x_1 + \cdots + \beta_k x_k$$

Exercises

1. With reference to Example 1, show that the regression equation of X on Y is

$$\mu_{X|y} = \frac{2}{1+y}$$

Also sketch the regression curve.

2. Given the joint density

$$f(x,y) = \begin{cases} \frac{2}{5}(2x+3y) & \text{for } 0 < x < 1 \text{ and } 0 < y < 1 \\ 0 & \text{elsewhere} \end{cases}$$

find $\mu_{Y|x}$ and $\mu_{X|y}$.

3. Given the joint density

$$f(x,y) = \begin{cases} 6x & \text{for } 0 < x < y < 1 \\ 0 & \text{elsewhere} \end{cases}$$

find $\mu_{Y|x}$ and $\mu_{X|y}$.

4. Given the joint density

$$f(x,y) = \begin{cases} \dfrac{2x}{(1+x+xy)^3} & \text{for } x > 0 \text{ and } y > 0 \\ 0 & \text{elsewhere} \end{cases}$$

show that $\mu_{Y|x} = 1 + \frac{1}{x}$ and that $\text{var}(Y|x)$ does not exist.

5. This question has been intentionally omitted for this edition.

6. This question has been intentionally omitted for this edition.

7. Given the joint density

$$f(x,y) = \begin{cases} 2 & \text{for } 0 < y < x < 1 \\ 0 & \text{elsewhere} \end{cases}$$

show that

(a) $\mu_{Y|x} = \frac{x}{2}$ and $\mu_{X|y} = \frac{1+y}{2}$;

(b) $E(X^m Y^n) = \dfrac{2}{(n+1)(m+n+2)}$.

Also,

(c) verify the results of part (a) by substituting the values of μ_1, μ_2, σ_1, σ_2, and ρ, obtained with the formula of part (b), into the formulas of Theorem 1.

8. Given the joint density

$$f(x,y) = \begin{cases} 24xy & \text{for } x > 0, y > 0, \text{ and } x+y < 1 \\ 0 & \text{elsewhere} \end{cases}$$

show that $\mu_{Y|x} = \frac{2}{3}(1-x)$ and verify this result by determining the values of μ_1, μ_2, σ_1, σ_2, and ρ and by substituting them into the first formula of Theorem 1.

9. Given the joint density

$$f(x,y) = \begin{cases} 1 & \text{for } -y < x < y \text{ and } 0 < y < 1 \\ 0 & \text{elsewhere} \end{cases}$$

show that the random variables X and Y are uncorrelated but not independent.

10. Show that if $\mu_{Y|x}$ is linear in x and $\text{var}(Y|x)$ is constant, then $\text{var}(Y|x) = \sigma_2^2(1-\rho^2)$.

11. This question has been intentionally omitted for this edition.

12. Given the random variables X_1, X_2, and X_3 having the joint density $f(x_1, x_2, x_3)$, show that if the regression of X_3 on X_1 and X_2 is linear and written as

$$\mu_{X_3|x_1,x_2} = \alpha + \beta_1(x_1 - \mu_1) + \beta_2(x_2 - \mu_2)$$

then

$$\alpha = \mu_3$$

$$\beta_1 = \frac{\sigma_{13}\sigma_2^2 - \sigma_{12}\sigma_{23}}{\sigma_1^2\sigma_2^2 - \sigma_{12}^2}$$

$$\beta_2 = \frac{\sigma_{23}\sigma_1^2 - \sigma_{12}\sigma_{13}}{\sigma_1^2\sigma_2^2 - \sigma_{12}^2}$$

where $\mu_i = E(X_i)$, $\sigma_i^2 = \text{var}(X_i)$, and $\sigma_{ij} = \text{cov}(X_i, X_j)$. [*Hint*: Proceed as in Section 2, multiplying by $(x_1 - \mu_1)$ and $(x_2 - \mu_2)$, respectively, to obtain the second and third equations.]

13. Find the least squares estimate of the parameter β in the regression equation $\mu_{Y|x} = \beta x$.

14. Solve the normal equations in Section 3 simultaneously to show that

$$\hat{\alpha} = \frac{\left(\sum_{i=1}^{n} x_i^2\right)\left(\sum_{i=1}^{n} y_i\right) - \left(\sum_{i=1}^{n} x_i\right)\left(\sum_{i=1}^{n} x_i y_i\right)}{n\left(\sum_{i=1}^{n} x_i^2\right) - \left(\sum_{i=1}^{n} x_i\right)^2}$$

15. When the x's are equally spaced, the calculation of $\hat{\alpha}$ and $\hat{\beta}$ can be simplified by coding the x's by assigning them the values $\ldots, -3, -2, -1, 0, 1, 2, 3, \ldots$ when n is odd, or the values $\ldots, -5, -3, -1, 1, 3, 5, \ldots$ when n is even. Show that with this coding the formulas for $\hat{\alpha}$ and $\hat{\beta}$ become

$$\hat{\alpha} = \frac{\sum_{i=1}^{n} y_i}{n} \quad \text{and} \quad \hat{\beta} = \frac{\sum_{i=1}^{n} x_i y_i}{\sum_{i=1}^{n} x_i^2}$$

16. The method of least squares can be used to fit curves to data. Using the method of least squares, find the normal equations that provide least squares estimates of α, β, and γ when fitting a quadratic curve of the form $y = \alpha + \beta x + \gamma x^2$ to paired data.

4 Normal Regression Analysis

When we analyze a set of paired data $\{(x_i, y_i);\ 1, 2, \ldots, n\}$ by **regression analysis**, we look upon the x_i as constants and the y_i as values of corresponding independent random variables Y_i. This clearly differs from **correlation analysis**, which we shall take up in Section 5, where we look upon the x_i and the y_i as values of corresponding random variables X_i and Y_i. For example, if we want to analyze data on the ages and prices of used cars, treating the ages as known constants and the prices as values of random variables, this is a problem of regression analysis. On the other hand, if we want to analyze data on the height and weight of certain animals, and height and weight are both looked upon as random variables, this is a problem of correlation analysis.

This section will be devoted to some of the basic problems of **normal regression analysis**, where it is assumed that for each fixed x_i the conditional density of the corresponding random variable Y_i is the normal density

$$w(y_i|x_i) = \frac{1}{\sigma\sqrt{2\pi}} \cdot e^{-\frac{1}{2}\left[\frac{y_i - (\alpha + \beta x_i)}{\sigma}\right]^2} \qquad -\infty < y_i < \infty$$

where α, β, and σ are the same for each i. Given a random sample of such paired data, normal regression analysis concerns itself mainly with the estimation of σ and the regression coefficients α and β, with tests of hypotheses concerning these three parameters, and with predictions based on the estimated regression equation $\hat{y} = \hat{\alpha} + \hat{\beta}x$, where $\hat{\alpha}$ and $\hat{\beta}$ are estimates of α and β.

To obtain maximum likelihood estimates of the parameters α, β, and σ, we partially differentiate the likelihood function (or its logarithm, which is easier) with respect to α, β, and σ, equate the expressions to zero, and then solve the resulting system of equations. Thus, differentiating

$$\ln L = -n \cdot \ln \sigma - \frac{n}{2} \cdot \ln 2\pi - \frac{1}{2\sigma^2} \cdot \sum_{i=1}^{n} [y_i - (\alpha + \beta x_i)]^2$$

partially with respect to α, β, and σ and equating the expressions that we obtain to zero, we get

$$\frac{\partial \ln L}{\partial \alpha} = \frac{1}{\sigma^2} \cdot \sum_{i=1}^{n} [y_i - (\alpha + \beta x_i)] = 0$$

$$\frac{\partial \ln L}{\partial \beta} = \frac{1}{\sigma^2} \cdot \sum_{i=1}^{n} x_i [y_i - (\alpha + \beta x_i)] = 0$$

$$\frac{\partial \ln L}{\partial \sigma} = -\frac{n}{\sigma} + \frac{1}{\sigma^3} \cdot \sum_{i=1}^{n} [y_i - (\alpha + \beta x_i)]^2 = 0$$

Since the first two equations are equivalent to the two normal equations in an earlier page, the maximum likelihood estimates of α and β are identical with the least squares estimate of Theorem 2. Also, if we substitute these estimates of α and β into the equation obtained by equating $\frac{\partial \ln L}{\partial \sigma}$ to zero, it follows immediately that the maximum likelihood estimate of σ is given by

$$\hat{\sigma} = \sqrt{\frac{1}{n} \cdot \sum_{i=1}^{n} [y_i - (\hat{\alpha} + \hat{\beta} x_i)]^2}$$

This can also be written as

$$\hat{\sigma} = \sqrt{\frac{1}{n}(S_{yy} - \hat{\beta} \cdot S_{xy})}$$

as the reader will be asked to verify in Exercise 17.

Having obtained maximum likelihood estimators of the regression coefficients, let us now investigate their use in testing hypotheses concerning α and β and in constructing confidence intervals for these two parameters. Since problems concerning β are usually of more immediate interest than problems concerning α (β is the slope of the regression line, whereas α is merely the y-intercept; also, the null hypothesis $\beta = 0$ is equivalent to the null hypothesis $\rho = 0$), we shall discuss here some of the sampling theory relating to $\hat{\mathrm{B}}$, where B is the capital Greek letter *beta*. Corresponding theory relating to $\hat{\mathrm{A}}$, where A is the capital Greek letter *alpha*, will be treated in Exercises 20 and 22.

To study the sampling distribution of $\hat{\mathrm{B}}$, let us write

$$\hat{\mathrm{B}} = \frac{S_{xY}}{S_{xx}} = \frac{\sum_{i=1}^{n}(x_i - \overline{x})(Y_i - \overline{Y})}{S_{xx}}$$

$$= \sum_{i=1}^{n} \left(\frac{x_i - \overline{x}}{S_{xx}} \right) Y_i$$

which is seen to be a linear combination of the n independent normal random variables Y_i. $\hat{\mathrm{B}}$ itself has a normal distribution with the mean

$$E(\hat{\mathrm{B}}) = \sum_{i=1}^{n} \left[\frac{x_i - \overline{x}}{S_{xx}} \right] \cdot E(Y_i | x_i)$$

$$= \sum_{i=1}^{n} \left[\frac{x_i - \overline{x}}{S_{xx}} \right] (\alpha + \beta x_i) = \beta$$

and the variance

$$\text{var}(\hat{B}) = \sum_{i=1}^{n} \left[\frac{x_i - \bar{x}}{S_{xx}} \right]^2 \cdot \text{var}(Y_i | x_i)$$

$$= \sum_{i=1}^{n} \left[\frac{x_i - \bar{x}}{S_{xx}} \right]^2 \cdot \sigma^2 = \frac{\sigma^2}{S_{xx}}$$

In order to apply this theory to test hypotheses about β or construct confidence intervals for β, we shall have to use the following theorem.

THEOREM 3. Under the assumptions of normal regression analysis, $\frac{n\hat{\sigma}^2}{\sigma^2}$ is a value of a random variable having the chi-square distribution with $n-2$ degrees of freedom. Furthermore, this random variable and $\hat{B}$ are independent.

A proof of this theorem is referred to at the end of this chapter.

Making use of this theorem as well as the result proved earlier that $\hat{B}$ has a normal distribution with the mean β and the variance $\frac{\sigma^2}{S_{xx}}$, we find that the definition of the t distribution leads to the following theorem.

THEOREM 4. Under the assumptions of normal regression analysis,

$$t = \frac{\dfrac{\hat{\beta} - \beta}{\sigma / \sqrt{S_{xx}}}}{\sqrt{\dfrac{n\hat{\sigma}^2}{\sigma^2} / (n-2)}} = \frac{\hat{\beta} - \beta}{\hat{\sigma}} \sqrt{\frac{(n-2)S_{xx}}{n}}$$

is a value of a random variable having the t distribution with $n-2$ degrees of freedom.

Based on this statistic, let us now test a hypothesis about the regression coefficient β.

EXAMPLE 5

With reference to the data in the table in Section 3 pertaining to the amount of time that 10 persons studied for a certain test and the scores that they obtained, test the null hypothesis $\beta = 3$ against the alternative hypothesis $\beta > 3$ at the 0.01 level of significance.

Solution

1. H_0: $\beta = 3$
   H_1: $\beta > 3$
   $\alpha = 0.01$
2. Reject the null hypothesis if $t \geqq 2.896$, where t is determined in accordance with Theorem 4 and 2.896 is the value of $t_{0.01,8}$ obtained from the Table IV of "Statistical Tables."

3. Calculating $\sum y^2 = 36{,}562$ from the original data and copying the other quantities from Section 3, we get

$$S_{yy} = 36{,}562 - \frac{1}{10}(564)^2 = 4{,}752.4$$

and

$$\hat{\sigma} = \sqrt{\frac{1}{10}[4{,}752.4 - (3.471)(1{,}305)]} = 4.720$$

so that

$$t = \frac{3.471 - 3}{4.720}\sqrt{\frac{8 \cdot 376}{10}} = 1.73$$

4. Since $t = 1.73$ is less than 2.896, the null hypothesis cannot be rejected; we cannot conclude that on the average an extra hour of study will increase the score by more than 3 points.

Letting $\hat{\Sigma}$ be the random variable whose values are $\hat{\sigma}$, we have

$$P\left(-t_{\alpha/2,\, n-2} < \frac{\hat{\mathrm{B}} - \beta}{\hat{\Sigma}}\sqrt{\frac{(n-2)S_{xx}}{n}} < t_{\alpha/2,\, n-2}\right) = 1 - \alpha$$

according to Theorem 4. Writing this as

$$P\left[\hat{\mathrm{B}} - t_{\alpha/2,\, n-2} \cdot \hat{\Sigma}\sqrt{\frac{n}{(n-2)S_{xx}}} < \beta < \hat{\mathrm{B}} + t_{\alpha/2,\, n-2} \cdot \hat{\Sigma}\sqrt{\frac{n}{(n-2)S_{xx}}}\right] = 1 - \alpha$$

we arrive at the following confidence interval formula.

THEOREM 5. Under the assumptions of normal regression analysis,

$$\hat{\beta} - t_{\alpha/2,\, n-2} \cdot \hat{\sigma}\sqrt{\frac{n}{(n-2)S_{xx}}} < \beta < \hat{\beta} + t_{\alpha/2,\, n-2} \cdot \hat{\sigma}\sqrt{\frac{n}{(n-2)S_{xx}}}$$

is a $(1-\alpha)100\%$ confidence interval for the parameter β.

EXAMPLE 6

With reference to the same data as in Example 5, construct a 95% confidence interval for β.

Solution

Copying the various quantities from Example 4 and Section 4 and substituting them together with $t_{0.025,8} = 2.306$ into the confidence interval formula of Theorem 5, we get

$$3.471 - (2.306)(4.720)\sqrt{\frac{10}{8(376)}} < \beta < 3.471 + (2.306)(4.720)\sqrt{\frac{10}{8(376)}}$$

or

$$2.84 < \beta < 4.10$$

Since most realistically complex regression problems require fairly extensive calculations, they are virtually always done nowadays by using appropriate computer software. A printout obtained for our illustration using MINITAB software is shown in Figure 4; as can be seen, it provides not only the values of $\hat{\alpha}$ and $\hat{\beta}$ in the column headed COEFFICIENT, but also estimates of the standard deviations of the sampling distributions of $\hat{A}$ and $\hat{B}$ in the column headed ST. DEV. OF COEF. Had we used this printout in Example 5, we could have written the value of the t statistic directly as

$$t = \frac{3.471 - 3}{0.2723} = 1.73$$

and in Example 6 we could have written the confidence limits directly as $3.471 \pm (2.306)(0.2723)$.

```
MTB   > NAME C1 = 'X'
MTB   > NAME C2 = 'Y'
MTB   > SET C1
DATA  > 4   9   1Ø   14   4   7   12   22   1   17
MTB   > SET C2
DATA  > 31   58   65   73   37   44   6Ø   91   21   84
MTB   > REGR C2 1 C1

    THE REGRESSION EQUATION IS
    Y = 21.7 + 3.47 X

                                          ST. DEV.      T-RATIO =
    COLUMN          COEFFICIENT           OF COEF.      COEF/S.D.
                    21.693                3.194         6.79
    X                3.47Ø7               Ø.2723        12.74
```

Figure 4. Computer printout for Examples 4, 5, and 6.

Exercises

17. Making use of the fact that $\hat{\alpha} = \bar{y} - \hat{\beta}\bar{x}$ and $\hat{\beta} = \frac{S_{xy}}{S_{xx}}$, show that

$$\sum_{i=1}^{n}[y_i - (\hat{\alpha} + \hat{\beta}x_i)]^2 = S_{yy} - \hat{\beta}S_{xy}$$

18. Show that
(a) $\hat{\Sigma}^2$, the random variable corresponding to $\hat{\sigma}^2$, is not an unbiased estimator of σ^2;
(b) $S_e^2 = \frac{n \cdot \hat{\Sigma}^2}{n-2}$ is an unbiased estimator of σ^2.
The quantity s_e is often referred to as the **standard error of estimate**.

19. Using s_e (see Exercise 18) instead of $\hat{\sigma}$, rewrite
(a) the expression for t in Theorem 4;
(b) the confidence interval formula of Theorem 5.

20. Under the assumptions of normal regression analysis, show that
(a) the least squares estimate of α in Theorem 2 can be written in the form

$$\hat{\alpha} = \sum_{i=1}^{n}\left[\frac{S_{xx} + n\bar{x}^2 - n\bar{x}x_i}{nS_{xx}}\right]y_i$$

(b) $\hat{A}$ has a normal distribution with

$$E(\hat{A}) = \alpha \quad \text{and} \quad \text{var}(\hat{A}) = \frac{(S_{xx} + n\bar{x}^2)\sigma^2}{nS_{xx}}$$

21. This question has been intentionally omitted for this edition.

22. Use the result of part (b) of Exercise 20 to show that

$$z = \frac{(\hat{\alpha} - \alpha)\sqrt{nS_{xx}}}{\sigma\sqrt{S_{xx} + n\bar{x}^2}}$$

is a value of a random variable having the standard normal distribution. Also, use the first part of Theorem 3 and the fact that $\hat{A}$ and $\frac{n\hat{\Sigma}^2}{\sigma^2}$ are independent to show that

$$t = \frac{(\hat{\alpha} - \alpha)\sqrt{(n-2)S_{xx}}}{\hat{\sigma}\sqrt{S_{xx} + n\bar{x}^2}}$$

is a value of a random variable having the t distribution with $n - 2$ degrees of freedom.

23. Use the results of Exercises 20 and 21 and the fact that $E(\hat{B}) = \beta$ and $\text{var}(\hat{B}) = \frac{\sigma^2}{S_{xx}}$ to show that $\hat{Y}_0 = \hat{A} + \hat{B}x_0$ is a random variable having a normal distribution with the mean

$$\alpha + \beta x_0 = \mu_{Y|x_0}$$

and the variance

$$\sigma^2 \left[\frac{1}{n} + \frac{(x_0 - \bar{x})^2}{S_{xx}} \right]$$

Also, use the first part of Theorem 3 as well as the fact that $\hat{Y}_0$ and $\frac{n\hat{\Sigma}^2}{\sigma^2}$ are independent to show that

$$t = \frac{(\hat{y}_0 - \mu_{Y|x_0})\sqrt{n-2}}{\hat{\sigma}\sqrt{1 + \frac{n(x_0 - \bar{x})^2}{S_{xx}}}}$$

is a value of a random variable having the t distribution with $n - 2$ degrees of freedom.

24. Derive a $(1 - \alpha)100\%$ confidence interval for $\mu_{Y|x_0}$, the mean of Y at $x = x_0$, by solving the double inequality $-t_{\alpha/2,n-2} < t < t_{\alpha/2,n-2}$ with t given by the formula of Exercise 23.

25. Use the results of Exercises 20 and 21 and the fact that $E(\hat{B}) = \beta$ and $\text{var}(\hat{B}) = \frac{\sigma^2}{S_{xx}}$ to show that $Y_0 - (\hat{A} + \hat{B}x_0)$ is a random variable having a normal distribution with zero mean and the variance

$$\sigma^2 \left[1 + \frac{1}{n} + \frac{(x_0 - \bar{x})^2}{S_{xx}} \right]$$

Here Y_0 has a normal distribution with the mean $\alpha + \beta x_0$ and the variance σ^2; that is, Y_0 is a future observation of Y corresponding to $x = x_0$. Also, use the first part of Theorem 3 as well as the fact that $Y_0 - (\hat{A} + \hat{B}x_0)$ and $\frac{n\hat{\Sigma}^2}{\sigma^2}$ are independent to show that

$$t = \frac{[y_0 - (\hat{\alpha} + \hat{\beta}x_0)]\sqrt{n-2}}{\hat{\sigma}\sqrt{1 + n + \frac{n(x_0 - \bar{x})^2}{S_{xx}}}}$$

is a value of a random variable having the t distribution with $n - 2$ degrees of freedom.

26. Solve the double inequality $-t_{\alpha/2,n-2} < t < t_{\alpha/2,n-2}$ with t given by the formula of Exercise 25 so that the middle term is y_0 and the two limits can be calculated without knowledge of y_0. Note that although the resulting double inequality may be interpreted like a confidence interval, it is not designed to estimate a parameter; instead, it provides **limits of prediction** for a future observation of Y that corresponds to the (given or observed) value x_0.

5 Normal Correlation Analysis

In normal correlation analysis we drop the assumption that the x_i are fixed constants, analyzing the set of paired data $\{(x_i, y_i); i = 1, 2, \ldots, n\}$, where the x_i's and y_i's are values of a random sample from a bivariate normal population with the parameters μ_1, μ_2, σ_1, σ_2, and ρ. To estimate these parameters by the method of maximum likelihood, we shall have to maximize the likelihood

$$L = \prod_{i=1}^{n} f(x_i, y_i)$$

and to this end we shall have to differentiate L, or $\ln L$, partially with respect to μ_1, μ_2, σ_1, σ_2, and ρ, equate the resulting expressions to zero, and then solve the resulting system of equations for the five parameters. Leaving the details to the reader, let us merely state that when $\frac{\partial \ln L}{\partial \mu_1}$ and $\frac{\partial \ln L}{\partial \mu_2}$ are equated to zero, we get

$$-\frac{\sum_{i=1}^{n}(x_i-\mu_1)}{\sigma_1^2}+\frac{\rho\sum_{i=1}^{n}(y_i-\mu_2)}{\sigma_1\sigma_2}=0$$

and

$$-\frac{\rho\sum_{i=1}^{n}(x_i-\mu_1)}{\sigma_1\sigma_2}+\frac{\sum_{i=1}^{n}(y_i-\mu_2)}{\sigma_2^2}=0$$

Solving these two equations for μ_1 and μ_2, we find that the maximum likelihood estimates of these two parameters are

$$\hat{\mu}_1=\overline{x} \quad\text{and}\quad \hat{\mu}_2=\overline{y}$$

that is, the respective sample means. Subsequently, equating $\frac{\partial \ln L}{\partial \sigma_1}$, $\frac{\partial \ln L}{\partial \sigma_2}$, and $\frac{\partial \ln L}{\partial \rho}$ to zero and substituting $\overline{x}$ and $\overline{y}$ for μ_1 and μ_2, we obtain a system of equations whose solution is

$$\hat{\sigma}_1=\sqrt{\frac{\sum_{i=1}^{n}(x_i-\overline{x})^2}{n}}, \qquad \hat{\sigma}_2=\sqrt{\frac{\sum_{i=1}^{n}(y_i-\overline{y})^2}{n}}$$

$$\hat{\rho}=\frac{\sum_{i=1}^{n}(x_i-\overline{x})(y_i-\overline{y})}{\sqrt{\sum_{i=1}^{n}(x_i-\overline{x})^2}\sqrt{\sum_{i=1}^{n}(y_i-\overline{y})^2}}$$

(A detailed derivation of these maximum likelihood estimates is referred to at the end of this chapter.) It is of interest to note that the maximum likelihood estimates of σ_1 and σ_2 are identical with the one obtained for the standard deviation of the univariate normal distribution; they differ from the respective sample standard deviations s_1 and s_2 only by the factor $\sqrt{\frac{n-1}{n}}$.

The estimate $\hat{\rho}$, called the **sample correlation coefficient**, is usually denoted by the letter r, and its calculation is facilitated by using the following alternative, but equivalent, computing formula.

> **THEOREM 6.** If $\{(x_i, y_i);\ i = 1, 2, \ldots, n\}$ are the values of a random sample from a bivariate population, then
>
> $$r=\frac{S_{xy}}{\sqrt{S_{xx}\cdot S_{yy}}}$$

Since ρ measures the strength of the linear relationship between X and Y, there are many problems in which the estimation of ρ and tests concerning ρ are of special interest. When $\rho = 0$, the two random variables are uncorrelated, and, as we have already seen, in the case of the bivariate normal distribution this means that they are also independent. When ρ equals +1 or −1, it follows from the relationship

$$\sigma_{Y|x}^2 = \sigma^2 = \sigma_2^2(1 - \rho^2)$$

where $\sigma = 0$, and this means that there is a perfect linear relationship between X and Y. Using the invariance property of maximum likelihood estimators, we can write

$$\hat{\sigma}^2 = \hat{\sigma}_2^2(1 - r^2)$$

which not only provides an alternative computing formula for finding $\hat{\sigma}^2$, but also serves to tie together the concepts of regression and correlation. From this formula for $\hat{\sigma}^2$ it is clear that when $\hat{\sigma}^2 = 0$, that is, when the set of data points $\{(x_i, y_i);\ i = 1, 2, \ldots, n\}$ fall on a straight line, then r will equal +1 or −1. We take $r = +1$ when the line has a positive slope and $r = -1$ when it has a negative slope. In order to interpret values of r between 0 and +1 or 0 and −1, we solve the preceding equation for r^2 and multiply by 100, getting

$$100r^2 = \frac{\hat{\sigma}_2^2 - \hat{\sigma}^2}{\hat{\sigma}_2^2} \cdot 100$$

where $\hat{\sigma}_2^2$ measures the total variation of the y's, $\hat{\sigma}^2$ measures the conditional variation of the y's for fixed values of x, and hence $\hat{\sigma}_2^2 - \hat{\sigma}^2$ measures that part of the total variation of the y's that is accounted for by the relationship with x. *Thus, $100r^2$ is the percentage of the total variation of the* y*'s that is accounted for by the relationship with* x. For instance, when $r = 0.5$, then 25 percent of the variation of the y's is accounted for by the relationship with x; when $r = 0.7$, then 49 percent of the variation of the y's is accounted for by the relationship with x; and we might thus say that a correlation of $r = 0.7$ is almost "twice as strong" as a correlation of $r = 0.5$. Similarly, we might say that a correlation of $r = 0.6$ is "nine times as strong" as a correlation of $r = 0.2$.

EXAMPLE 7

Suppose that we want to determine on the basis of the following data whether there is a relationship between the time, in minutes, it takes a secretary to complete a certain form in the morning and in the late afternoon:

Morning x	*Afternoon* y
8.2	8.7
9.6	9.6
7.0	6.9
9.4	8.5
10.9	11.3
7.1	7.6
9.0	9.2
6.6	6.3
8.4	8.4
10.5	12.3

Compute and interpret the sample correlation coefficient.

Solution
From the data we get $n = 10$, $\Sigma x = 86.7$, $\Sigma x^2 = 771.35$, $\Sigma y = 88.8$, $\Sigma y^2 = 819.34$, and $\Sigma xy = 792.92$, so

$$S_{xx} = 771.35 - \frac{1}{10}(86.7)^2 = 19.661$$

$$S_{yy} = 819.34 - \frac{1}{10}(88.8)^2 = 30.796$$

$$S_{xy} = 792.92 - \frac{1}{10}(86.7)(88.8) = 23.024$$

and

$$r = \frac{23.024}{\sqrt{(19.661)(30.796)}} = 0.936$$

This is indicative of a positive association between the time it takes a secretary to perform the given task in the morning and in the late afternoon, and this is also apparent from the **scattergram** of Figure 5. Since $100r^2 = 100(0.936)^2 = 87.6$, we can say that almost 88 percent of the variation of the y's is accounted for by the implicit linear relationship with x.

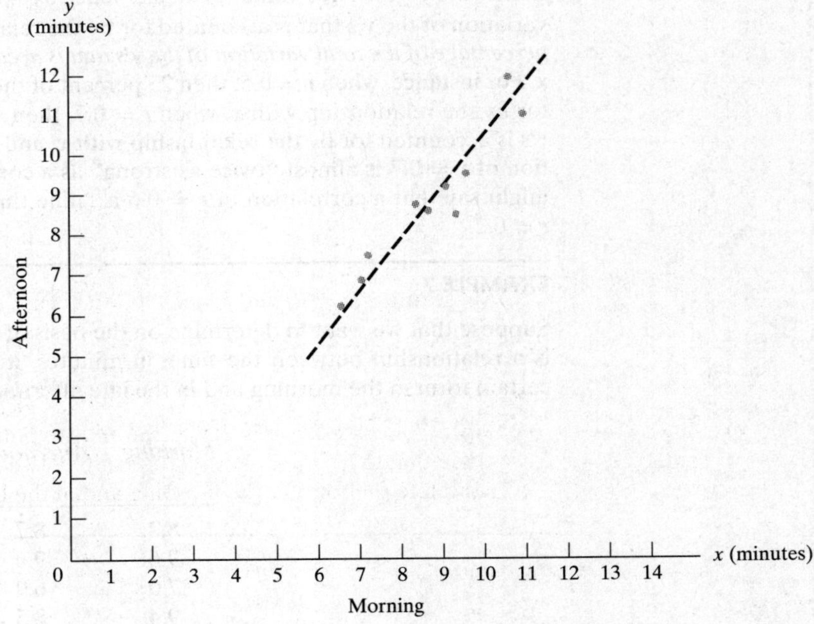

Figure 5. Scattergram of data of Example 7.

Since the sampling distribution of R for random samples from bivariate normal populations is rather complicated, it is common practice to base confidence intervals for ρ and tests concerning ρ on the statistic

$$\frac{1}{2} \cdot \ln \frac{1+R}{1-R}$$

whose distribution can be shown to be approximately normal with the mean $\frac{1}{2} \cdot \ln \frac{1+\rho}{1-\rho}$ and the variance $\frac{1}{n-3}$. Thus,

$$z = \frac{\frac{1}{2} \cdot \ln \frac{1+r}{1-r} - \frac{1}{2} \cdot \ln \frac{1+\rho}{1-\rho}}{\frac{1}{\sqrt{n-3}}}$$

$$= \frac{\sqrt{n-3}}{2} \cdot \ln \frac{(1+r)(1-\rho)}{(1-r)(1+\rho)}$$

can be looked upon as a value of a random variable having approximately the standard normal distribution. Using this approximation, we can test the null hypothesis $\rho = \rho_0$ against an appropriate alternative, as illustrated in Example 8, or we can calculate confidence intervals for ρ by the method suggested in Exercise 31.

EXAMPLE 8

With reference to Example 7, test the null hypothesis $\rho = 0$ against the alternative hypothesis $\rho \neq 0$ at the 0.01 level of significance.

Solution

1. H_0: $\rho = 0$
   H_1: $\rho \neq 0$
   $\alpha = 0.01$
2. Reject the null hypothesis if $z \leqq -2.575$ or $z \geqq 2.575$, where

$$z = \frac{\sqrt{n-3}}{2} \cdot \ln \frac{1+r}{1-r}$$

3. Substituting $n = 10$ and $r = 0.936$, we get

$$z = \frac{\sqrt{7}}{2} \cdot \ln \frac{1.936}{0.064} = 4.5$$

4. Since $z = 4.5$ exceeds 2.575, the null hypothesis must be rejected; we conclude that there is a linear relationship between the time it takes a secretary to complete the form in the morning and in the late afternoon.

Exercises

27. Verify the maximum likelihood estimates of $\mu_1, \mu_2, \sigma_1, \sigma_2$, and ρ given in Section 5.

28. Verify that the formula for t of Theorem 4 can be written as

$$t = \left(1 - \frac{\beta}{\hat{\beta}}\right) \frac{r\sqrt{n-2}}{\sqrt{1-r^2}}$$

29. Use the formula for t of Exercise 28 to derive the following $(1-\alpha)100\%$ confidence limits for β:

$$\hat{\beta}\left[1 \pm t_{\alpha/2, n-2} \cdot \frac{\sqrt{1-r^2}}{r\sqrt{n-2}}\right]$$

30. Use the formula for t of Exercise 28 to show that if the assumptions underlying normal regression analysis are met and $\beta = 0$, then R^2 has a beta distribution with the mean $\frac{1}{n-1}$.

31. By solving the double inequality $-z_{\alpha/2} \leqq z \leqq z_{\alpha/2}$ (with z given by the formula in the previous page) for ρ, derive a $(1-\alpha)100\%$ confidence interval formula for ρ.

32. In a random sample of n pairs of values of X and Y, (x_i, y_j) occurs f_{ij} times for $i = 1, 2, \ldots, r$ and $j = 1, 2, \ldots, c$. Letting $f_{i\cdot}$ denote the number of pairs where X takes on the value x_i and $f_{\cdot j}$ the number of pairs where Y takes on the value y_j, write a formula for the coefficient of correlation.

6 Multiple Linear Regression

Although there are many problems in which one variable can be predicted quite accurately in terms of another, it stands to reason that predictions should improve if one considers additional relevant information. For instance, we should be able to make better predictions of the performance of newly hired teachers if we consider not only their education, but also their years of experience and their personality. Also, we should be able to make better predictions of a new textbook's success if we consider not only the quality of the work, but also the potential demand and the competition.

Although many different formulas can be used to express regression relationships among more than two variables (see, for instance, Example 3), the most widely used are linear equations of the form

$$\mu_{Y|x_1, x_2, \ldots, x_k} = \beta_0 + \beta_1 x_1 + \beta_2 x_2 + \cdots + \beta_k x_k$$

This is partly a matter of mathematical convenience and partly due to the fact that many relationships are actually of this form or can be approximated closely by linear equations.

In the preceding equation, Y is the random variable whose values we want to predict in terms of given values of **the independent variables** $x_1, x_2, \ldots,$ and x_k, and the **multiple regression coefficients** $\beta_0, \beta_1, \beta_2, \ldots,$ and β_k are numerical constants that must be determined from observed data.

To illustrate, consider the following equation, which was obtained in a study of the demand for different meats:

$$\hat{y} = 3.489 - 0.090x_1 + 0.064x_2 + 0.019x_3$$

Here $\hat{y}$ denotes the estimated family consumption of federally inspected beef and pork in millions of pounds, x_1 denotes a composite retail price of beef in cents per pound, x_2 denotes a composite retail price of pork in cents per pound, and x_3 denotes family income as measured by a certain payroll index.

As in Section 3, where there was only one independent variable x, multiple regression coefficients are usually estimated by the method of least squares. For n data points

$$\{(x_{i1}, x_{i2}, \ldots, x_{ik}, y_i);\ i = 1, 2, \ldots, n\}$$

the least squares estimates of the β's are the values $\hat{\beta}_0, \hat{\beta}_1, \hat{\beta}_2, \ldots,$ and $\hat{\beta}_k$ for which the quantity

$$q = \sum_{i=1}^{n}[y_i - (\hat{\beta}_0 + \hat{\beta}_1 x_{i1} + \hat{\beta}_2 x_{i2} + \cdots + \hat{\beta}_k x_{ik})]^2$$

is a minimum. In this notation, x_{i1} is the ith value of the variable x_1, x_{i2} is the ith value of the variable x_2, and so on.

So, we differentiate partially with respect to the $\hat{\beta}$'s, and equating these partial derivatives to zero, we get

$$\frac{\partial q}{\partial \hat{\beta}_0} = \sum_{i=1}^{n}(-2)[y_i - (\hat{\beta}_0 + \hat{\beta}_1 x_{i1} + \hat{\beta}_2 x_{i2} + \cdots + \hat{\beta}_k x_{ik})] = 0$$

$$\frac{\partial q}{\partial \hat{\beta}_1} = \sum_{i=1}^{n}(-2)x_{i1}[y_i - (\hat{\beta}_0 + \hat{\beta}_1 x_{i1} + \hat{\beta}_2 x_{i2} + \cdots + \hat{\beta}_k x_{ik})] = 0$$

$$\frac{\partial q}{\partial \hat{\beta}_2} = \sum_{i=1}^{n}(-2)x_{i2}[y_i - (\hat{\beta}_0 + \hat{\beta}_1 x_{i1} + \hat{\beta}_2 x_{i2} + \cdots + \hat{\beta}_k x_{ik})] = 0$$

$$\cdots$$

$$\frac{\partial q}{\partial \hat{\beta}_k} = \sum_{i=1}^{n}(-2)x_{ik}[y_i - (\hat{\beta}_0 + \hat{\beta}_1 x_{i1} + \hat{\beta}_2 x_{i2} + \cdots + \hat{\beta}_k x_{ik})] = 0$$

and finally the $k+1$ normal equations

$$\begin{aligned}
\Sigma y &= \hat{\beta}_0 \cdot n + \hat{\beta}_1 \cdot \Sigma x_1 + \hat{\beta}_2 \cdot \Sigma x_2 + \cdots + \hat{\beta}_k \cdot \Sigma x_k \\
\Sigma x_1 y &= \hat{\beta}_0 \cdot \Sigma x_1 + \hat{\beta}_1 \cdot \Sigma x_1^2 + \hat{\beta}_2 \cdot \Sigma x_1 x_2 + \cdots + \hat{\beta}_k \cdot \Sigma x_1 x_k \\
\Sigma x_2 y &= \hat{\beta}_0 \cdot \Sigma x_2 + \hat{\beta}_1 \cdot \Sigma x_2 x_1 + \hat{\beta}_2 \cdot \Sigma x_2^2 + \cdots + \hat{\beta}_k \cdot \Sigma x_2 x_k \\
&\cdots \\
\Sigma x_k y &= \hat{\beta}_0 \cdot \Sigma x_k + \hat{\beta}_1 \cdot \Sigma x_k x_1 + \hat{\beta}_2 \cdot \Sigma x_k x_2 + \cdots + \hat{\beta}_k \cdot \Sigma x_k^2
\end{aligned}$$

Here we abbreviated our notation by writing $\sum_{i=1}^{n} x_{i1}$ as $\sum x_1$, $\sum_{i=1}^{n} x_{i1}x_{i2}$ as $\sum x_1 x_2$, and so on.

EXAMPLE 9

The following data show the number of bedrooms, the number of baths, and the prices at which a random sample of eight one-family houses sold in a certain large housing development:

Number of bedrooms x_1	*Number of baths* x_2	*Price (dollars)* y
3	2	292,000
2	1	264,600
4	3	317,500
2	1	265,500
3	2	302,000
2	2	275,500
5	3	333,000
4	2	307,500

Use the method of least squares to fit a linear equation that will enable us to predict the average sales price of a one-family house in the given housing development in terms of the number of bedrooms and the number of baths.

Solution
The quantities we need to substitute into the three normal equations are:

$$\sum x_1 y = 7{,}558{,}200 \text{ and } \sum x_2 y = 4{,}835{,}600$$

and we get

$$n = 8, \sum x_1 = 25, \sum x_2 = 16, \sum y = 2{,}357{,}600, \sum x_1^2 = 87, \sum x_1 x_2 = 55, \sum x_2^2 = 36$$

$$2{,}357{,}600 = 8\hat{\beta}_0 + 25\hat{\beta}_1 + 16\hat{\beta}_2$$
$$7{,}558{,}200 = 25\hat{\beta}_0 + 87\hat{\beta}_1 + 55\hat{\beta}_2$$
$$4{,}835{,}600 = 16\hat{\beta}_0 + 55\hat{\beta}_1 + 36\hat{\beta}_2$$

We could solve these equations by the method of elimination or by using determinants, but in view of the rather tedious calculations, such work is usually left to computers. Let us refer to the printout of Figure 6, which shows in the column headed "Coef" that $\hat{\beta}_0 = 224{,}929$, $\hat{\beta}_1 = 15{,}314$, and $\hat{\beta}_2 = 10{,}957$. The least squares equation becomes

$$\hat{y} = 224{,}929 + 15{,}314x_1 + 10{,}957x_2$$

and this tells us that (in the given housing development and at the time of this study) each bedroom adds on the average \$15,314 and each bath adds \$10,957 to the sales price of a house.

```
Regression Analysis: C3 versus C1, C2

The regression equation is
C3 = 224929 + 15314 C1 + 10957 C2

Predictor     Coef  SE Coef      T      P
Constant    224929     5016  44.84  0.000
C1           15314     2743   5.58  0.003
C2           10957     4086   2.68  0.044

S = 4444.45    R-Sq = 97.7%    R-Sq(adj) = 96.8%
```

Figure 6. Computer printout for Example 9.

EXAMPLE 10

Based on the result obtained in Example 9, predict the sales price of a three-bedroom house with two baths in the subject housing development.

Solution
Substituting $x_1 = 3$ and $x_2 = 2$ into the least squares equation obtained in the preceding example, we get

$$\hat{y} = 224{,}929 + 15{,}314 \cdot 3 + 10{,}957 \cdot 2$$
$$= \$292{,}785$$

Printouts like those of Figure 6 also provide information that is needed to make inferences about the multiple regression coefficients and to judge the merits of estimates or predictions based on the least squares equations. This corresponds to the work of Section 4, but we shall defer it until Section 7, where we shall study the whole problem of multiple linear regression in a much more compact notation.

7 Multiple Linear Regression (Matrix Notation)†

The model we are using in multiple linear regression lends itself uniquely to a unified treatment in matrix notation. This notation makes it possible to state general results in compact form and to utilize many results of matrix theory to great advantage. As is customary, we shall denote matrices by capital letters in boldface type.

We could introduce the matrix approach by expressing the sum of squares q (which we minimized in the preceding section by differentiating partially with respect to the $\hat{\beta}$'s) in matrix notation and take it from there, but leaving this to the reader in Exercise 33, let us begin here with the normal equations given earlier.

To express the normal equations in matrix notation, let us define the following three matrices:

$$\mathbf{X} = \begin{pmatrix} 1 & x_{11} & x_{12} & \cdots & x_{1k} \\ 1 & x_{21} & x_{22} & \cdots & x_{2k} \\ \cdots & \cdots & \cdots & \cdots & \cdots \\ 1 & x_{n1} & x_{n2} & \cdots & x_{nk} \end{pmatrix}$$

$$\mathbf{Y} = \begin{pmatrix} y_1 \\ y_2 \\ \cdot \\ \cdot \\ y_n \end{pmatrix} \quad \text{and} \quad \mathbf{B} = \begin{pmatrix} \hat{\beta}_0 \\ \hat{\beta}_1 \\ \cdot \\ \cdot \\ \hat{\beta}_k \end{pmatrix}$$

The first one, $\mathbf{X}$, is an $n \times (k+1)$ matrix consisting essentially of the given values of the x's, with the column of 1's appended to accommodate the constant terms. $\mathbf{Y}$ is an $n \times 1$ matrix (or column vector) consisting of the observed values of Y, and $\mathbf{B}$ is a $(k+1) \times 1$ matrix (or column vector) consisting of the least squares estimates of the regression coefficients.

Using these matrices, we can now write the following symbolic solution of the normal equations.

THEOREM 7. The least squares estimates of the multiple regression coefficients are given by

$$\mathbf{B} = (\mathbf{X}'\mathbf{X})^{-1}\mathbf{X}'\mathbf{Y}$$

where $\mathbf{X}'$ is the transpose of $\mathbf{X}$ and $(\mathbf{X}'\mathbf{X})^{-1}$ is the inverse of $\mathbf{X}'\mathbf{X}$.

† It is assumed for this section that the reader is familiar with the material ordinarily covered in a first course on matrix algebra.

Proof First we determine $\mathbf{X'X}, \mathbf{X'XB}$, and $\mathbf{X'Y}$, getting

$$\mathbf{X'X} = \begin{pmatrix} n & \Sigma x_1 & \Sigma x_2 & \cdots & \Sigma x_k \\ \Sigma x_1 & \Sigma x_1^2 & \Sigma x_1 x_2 & \cdots & \Sigma x_1 x_k \\ \Sigma x_2 & \Sigma x_2 x_1 & \Sigma x_2^2 & \cdots & \Sigma x_2 x_k \\ & & \cdots & & \\ \Sigma x_k & \Sigma x_k x_1 & \Sigma x_k x_2 & \cdots & \Sigma x_k^2 \end{pmatrix}$$

$$\mathbf{X'XB} = \begin{pmatrix} \hat{\beta}_0 \cdot n & + \hat{\beta}_1 \cdot \Sigma x_1 & + \hat{\beta}_2 \cdot \Sigma x_2 & + \cdots + & \hat{\beta}_k \cdot \Sigma x_k \\ \hat{\beta}_0 \cdot \Sigma x_1 & + \hat{\beta}_1 \cdot \Sigma x_1^2 & + \hat{\beta}_2 \cdot \Sigma x_1 x_2 & + \cdots + & \hat{\beta}_k \cdot \Sigma x_1 x_k \\ \hat{\beta}_0 \cdot \Sigma x_2 & + \hat{\beta}_1 \cdot \Sigma x_2 x_1 & + \hat{\beta}_2 \cdot \Sigma x_2^2 & + \cdots + & \hat{\beta}_k \cdot \Sigma x_2 x_k \\ & & \cdots & & \\ \hat{\beta}_0 \cdot \Sigma x_k & + \hat{\beta}_1 \cdot \Sigma x_k x_1 & + \hat{\beta}_2 \cdot \Sigma x_k x_2 & + \cdots + & \hat{\beta}_k \cdot \Sigma x_k^2 \end{pmatrix}$$

$$\mathbf{X'Y} = \begin{pmatrix} \Sigma y \\ \Sigma x_1 y \\ \Sigma x_2 y \\ \cdots \\ \Sigma x_k y \end{pmatrix}$$

Identifying the elements of $\mathbf{X'XB}$ as the expressions on the right-hand side of the normal equations given in an earlier page and those of $\mathbf{X'Y}$ as the expressions on the left-hand side, we can write

$$\mathbf{X'XB} = \mathbf{X'Y}$$

Multiplying on the left by $(\mathbf{X'X})^{-1}$, we get

$$(\mathbf{X'X})^{-1}\mathbf{X'XB} = (\mathbf{X'X})^{-1}\mathbf{X'Y}$$

and finally

$$\mathbf{B} = (\mathbf{X'X})^{-1}\mathbf{X'Y}$$

since $(\mathbf{X'X})^{-1}\mathbf{X'X}$ equals the $(k+1)\times(k+1)$ identity matrix $\mathbf{I}$ and by definition $\mathbf{IB} = \mathbf{B}$. We have assumed here that $\mathbf{X'X}$ is nonsingular so that its inverse exists.

EXAMPLE 11

With reference to Example 9, use Theorem 7 to determine the least squares estimates of the multiple regression coefficients.

Solution

Substituting $\sum x_1 = 25$, $\sum x_2 = 16$, $\sum x_1^2 = 87$, $\sum x_1 x_2 = 55$, $\sum x_2^2 = 36$, and $n = 8$ from Example 9 into the preceding expression for $\mathbf{X'X}$, we get

$$\mathbf{X'X} = \begin{pmatrix} 8 & 25 & 16 \\ 25 & 87 & 55 \\ 16 & 55 & 36 \end{pmatrix}$$

Then, the inverse of the matrix can be obtained by any one of a number of different techniques; using the one based on cofactors, we find that

$$(\mathbf{X}'\mathbf{X})^{-1} = \frac{1}{84} \cdot \begin{pmatrix} 107 & -20 & -17 \\ -20 & 32 & -40 \\ -17 & -40 & 71 \end{pmatrix}$$

where 84 is the value of $|\mathbf{X}'\mathbf{X}|$, the determinant of $\mathbf{X}'\mathbf{X}$.

Substituting $\sum y = 2,357,600$, $\sum x_1 y = 7,558,200$, and $\sum x_2 y = 4,835,600$ from Example 9 into the expression for $\mathbf{X}'\mathbf{Y}$ (given above), we then get

$$\mathbf{X}'\mathbf{Y} = \begin{pmatrix} 2{,}357{,}600 \\ 7{,}558{,}200 \\ 4{,}835{,}600 \end{pmatrix}$$

and finally

$$\hat{\mathbf{B}} = (\mathbf{X}'\mathbf{X})^{-1}\mathbf{X}'\mathbf{Y} = \frac{1}{84} \cdot \begin{pmatrix} 07 & -20 & -17 \\ -20 & 32 & -40 \\ -17 & -40 & 71 \end{pmatrix} \begin{pmatrix} 2{,}357{,}600 \\ 7{,}558{,}200 \\ 4{,}835{,}600 \end{pmatrix}$$

$$= \frac{1}{84} \cdot \begin{pmatrix} 18{,}894{,}000 \\ 1{,}286{,}400 \\ 920{,}400 \end{pmatrix}$$

$$= \begin{pmatrix} 224{,}929 \\ 15{,}314 \\ 10{,}957 \end{pmatrix}$$

where the $\hat{\beta}$'s are rounded to the nearest integer. Note that the results obtained here are identical with those shown on the computer printout of Figure 6.

Next, to generalize the work of Section 4, we assume that for $i = 1, 2, \ldots,$ and n, the Y_i are independent random variables having normal distributions with the means $\beta_0 + \beta_1 x_{i1} + \beta_2 x_{i2} + \cdots + \beta_k x_{ik}$ and the common standard deviation σ. Based on n data points

$$(x_{i1}, x_{i2}, \ldots, x_{ik}, y_i)$$

we can then make all sorts of inferences about the parameters of our model, the β's and σ, and judge the merits of estimates and predictions based on the estimated multiple regression equation.

Finding maximum likelihood estimates of the β's and σ is straightforward, as in Section 4, and it will be left to the reader in Exercise 33. The results are as follows: The maximum likelihood estimates of the β's equal the corresponding least squares estimates, so they are given by the elements of the $(k+1)\times 1$ column matrix

$$\mathbf{B} = (\mathbf{X}'\mathbf{X})^{-1}\mathbf{X}'\mathbf{Y}$$

The maximum likelihood estimate of σ is given by

$$\hat{\sigma} = \sqrt{\frac{1}{n} \cdot \sum_{i=1}^{n} [y_i - (\hat{\beta}_0 + \hat{\beta}_1 x_{i1} + \hat{\beta}_2 x_{i2} + \cdots + \hat{\beta}_k x_{ik})]^2}$$

where the $\hat{\beta}$'s are the maximum likelihood estimates of the β's and, as the reader will be asked to verify in Exercise 35, this estimator can also be written as

$$\hat{\sigma} = \sqrt{\frac{\mathbf{Y}'\mathbf{Y} - \mathbf{B}'\mathbf{X}'\mathbf{Y}}{n}}$$

in matrix notation.

EXAMPLE 12

Use the results of Example 11 to determine the value of $\hat{\sigma}$ for the data of Example 9.

Solution

First let us calculate $\mathbf{Y}'\mathbf{Y}$, which is simply $\sum_{i=1}^{n} y_i^2$ obtaining

$$\mathbf{Y}'\mathbf{Y} = (292{,}000)^2 + (264{,}600)^2 + \ldots + (307{,}500)^2$$
$$= 699{,}123{,}160{,}0001$$

Then, copying $\mathbf{B}$ and $\mathbf{X}'\mathbf{Y}$, we get

$$\mathbf{B}'\mathbf{X}'\mathbf{Y} = \frac{1}{84} \cdot (18{,}894{,}000 \quad 286{,}400 \quad 920{,}400) \begin{pmatrix} 637{,}000 \\ 7{,}558{,}200 \\ 4{,}835{,}600 \end{pmatrix}$$
$$= 699{,}024{,}394{,}285$$

It follows that

$$\hat{\sigma} = \sqrt{\frac{699{,}123{,}160{,}000 - 699{,}024{,}394{,}285}{8}}$$
$$= 3{,}514$$

It is of interest to note that the estimate that we obtained here, 3,514, does not equal the one shown in the computer printout of Figure 6. The estimate shown there, $S = 4{,}444$, is such that S^2 is an unbiased estimate of σ^2, analogous to the standard error of estimate that we defined earlier. It differs from $\hat{\sigma}$ in that we divide by $n - k - 1$ instead of n. If we had done so in our example, we would have obtained

$$s_e = \sqrt{\frac{699{,}123{,}160{,}000 - 699{,}024{,}394{,}285}{8 - 2 - 1}}$$
$$= 4{,}444$$

rounded to the nearest integer.

Proceeding as in Section 4, we investigate next the sampling distribution of the $\hat{B}_i$ for $i = 0, 1, \ldots, k$, and $\hat{\Sigma}$. Leaving the details to the reader, let us merely point out that arguments similar to those in Section 4 lead to the results that the $\hat{B}_i$ are linear combinations of the n independent random variables Y_i so that the $\hat{B}_i$ themselves have normal distributions. Furthermore, they are unbiased estimators, that is,

$$E(\hat{B}_i) = \beta_i \quad \text{for } i = 0, 1, \ldots, k$$

and their variances are given by

$$\text{var}(\hat{B}_i) = c_{ii}\sigma^2 \qquad \text{for } i = 0, 1, \ldots, k$$

Here c_{ij} is the element in the ith row and the jth column of the matrix $(\mathbf{X}'\mathbf{X})^{-1}$, with i and j taking on the values $0, 1, \ldots, k$.

Let us also state the result that, analogous to Theorem 3, the sampling distribution of $\frac{n\hat{\Sigma}^2}{\sigma^2}$, the random variable corresponding to $\frac{n\hat{\sigma}^2}{\sigma^2}$, is the chi-square distribution with $n-k-1$ degrees of freedom and that $\frac{n\hat{\Sigma}^2}{\sigma^2}$ and $\hat{B}_i$ are independent for $i = 0, 1, \ldots, k$. Combining all these results, we find that the definition of the t distribution leads to the following theorem.

THEOREM 8. Under the assumptions of normal multiple regression analysis,

$$t = \frac{\hat{\beta}_i - \beta_i}{\hat{\sigma} \cdot \sqrt{\frac{n|c_{ii}|}{n-k-1}}} \qquad \text{for } i = 0, 1, \ldots, k$$

are values of random variables having the t distribution with $n-k-1$ degrees of freedom.

Based on this theorem, let us now test a hypothesis about one of the multiple regression coefficients.

EXAMPLE 13

With reference to Example 9, test the null hypothesis $\beta_1 = \$9,500$ against the alternative hypothesis $\beta_1 > \$9,500$ at the 0.05 level of significance.

Solution

1. H_0: $\beta_1 = 9{,}500$
   H_1: $\beta_1 > 9{,}500$
   $\alpha = 0.05$
2. Reject the null hypothesis if $t \geq 2.015$, where t is determined in accordance with Theorem 8, and 2.015 is the value of $t_{0.05,5}$ obtained from the table of Values of $t\alpha, \nu$ of "Statistical Tables."
3. Substituting $n = 8$, $\hat{\beta}_1 = 15{,}314$, and $c_{11} = \frac{32}{84}$ from Example 11 and $\hat{\sigma} = 3{,}546$ from Example 12 into the formula for t, we get

$$t = \frac{15{,}314 - 9{,}500}{3{,}514\sqrt{\frac{8\cdot\left|\frac{32}{84}\right|}{5}}}$$
$$= \frac{5{,}814}{2{,}743}$$
$$= 2.12$$

4. Since $t = 2.12$ exceeds 2.015, the null hypothesis must be rejected; we conclude that on the average each bedroom adds more than \$9,500 to the sales price of such a house. (Note that the value of the denominator of the t statistic, 2,743, equals the second value in the column headed "SE Coef" in the computer printout of Figure 6.)

Analogous to Theorem 5, we can also use the t statistic of Theorem 8 to construct confidence intervals for regression coefficients (see Exercise 38).

Exercises

33. If $\mathbf{b}$ is the column vector of the β's, verify in matrix notation that $\mathbf{q} = (\mathbf{Y} - \mathbf{Xb})'(\mathbf{Y} - \mathbf{Xb})$ is a minimum when $\mathbf{b} = \mathbf{B} = (\mathbf{X}'\mathbf{X})^{-1}(\mathbf{X}'\mathbf{Y})$.

34. Verify that under the assumptions of normal multiple regression analysis
(a) the maximum likelihood estimates of the β's equal the corresponding least squares estimates;
(b) the maximum likelihood estimate of σ is

$$\hat{\sigma} = \sqrt{\frac{(\mathbf{Y} - \mathbf{XB})'(\mathbf{Y} - \mathbf{XB})}{n}}$$

35. Verify that the estimate of part (b) of Exercise 34 can also be written as

$$\hat{\sigma} = \sqrt{\frac{\mathbf{Y}'\mathbf{Y} - \mathbf{B}'\mathbf{X}'\mathbf{Y}}{n}}$$

36. Show that under the assumptions of normal multiple regression analysis
(a) $E(\hat{\mathrm{B}}_i) = \beta_i$ for $i = 0, 1, \ldots, k$;
(b) $\mathrm{var}(\hat{\mathrm{B}}_i) = c_{ii}\sigma^2$ for $i = 0, 1, \ldots, k$;
(c) $\mathrm{cov}(\hat{\mathrm{B}}_i, \hat{\mathrm{B}}_j) = c_{ij}\sigma^2$ for $i \neq j = 0, 1, \ldots, k$.

37. Show that for $k = 1$ the formulas of Exercise 36 are equivalent to those given in Section 4 and in Exercises 20 and 21.

38. Use the t statistic of Theorem 8 to construct a $(1 - \alpha)100\%$ confidence interval formula for β_i for $i = 0, 1, \ldots, k$.

39. If $x_{01}, x_{02}, \ldots, x_{0k}$ are given values of $x_1, x_2, \ldots, x_k$ and $\mathbf{X}_0$ is the column vector

$$\mathbf{X}_0 = \begin{pmatrix} 1 \\ x_{01} \\ x_{02} \\ \ldots \\ x_{0k} \end{pmatrix}$$

it can be shown that

$$t = \frac{\mathbf{B}'\mathbf{X}_0 - \mu_{Y|x_{01}, x_{02}, \ldots, x_{0k}}}{\hat{\sigma} \cdot \sqrt{\frac{n[\mathbf{X}_0'(\mathbf{X}'\mathbf{X})^{-1}\mathbf{X}_0]}{n-k-1}}}$$

is a value of a random variable having the t distribution with $n - k - 1$ degrees of freedom.
(a) Show that for $k = 1$ this statistic is equivalent to the one of Exercise 23.
(b) Derive a $(1 - \alpha)100\%$ confidence interval formula for

$$\mu_{Y|x_{01}, x_{02}, \ldots, x_{0k}}$$

40. With $x_{01}, x_{02}, \ldots, x_{0k}$ and $\mathbf{X}_0$ as defined in Exercise 39 and Y_0 being a random variable that has a normal distribution with the mean $\beta_0 + \beta_1 x_{01} + \cdots + \beta_k x_{0k}$ and the variance σ^2, it can be shown that

$$t = \frac{y_0 - \mathbf{B}'\mathbf{X}_0}{\hat{\sigma} \cdot \sqrt{\frac{n[1 + \mathbf{X}_0'(\mathbf{X}'\mathbf{X})^{-1}\mathbf{X}_0]}{n-k-1}}}$$

is a value of a random variable having the t distribution with $n - k - 1$ degrees of freedom.
(a) Show that for $k = 1$ this statistic is equivalent to the one of Exercise 25.
(b) Derive a formula for $(1 - \alpha)100\%$ limits of prediction for a future observation of Y_0.

8 The Theory in Practice

Multiple linear regression is used (and misused) widely in applications. In this section, we shall discuss some of the pitfalls presented by indiscriminate use of multiple regression analysis, and ways to deal with them. Specifically, we shall examine the problem of *multicollinearity*. In addition, we shall introduce methods for examining the *residuals* in a multiple regression analysis to check on the assumption of normality and other characteristics of the data.

To begin, let us consider the following example. In wave soldering of circuit boards, an entire circuit board is run through the wave-soldering machine, and all solder joints are made. Suppose 5 major variables involved in the machine setup are measured for each run. A total of 25 separate runs of 5 boards each are made. (Each board contains 460 solder joints.) The soldered boards are subjected to visual and electrical inspection, and the number of defective solder joints per 100 joints inspected is recorded, with the following results:

Run No.	*Conveyor angle* x_1	*Solder temperature* x_2	*Flux concentration* x_3	*Conveyor speed* x_4	*Preheat temperature* x_5	*Faults per 100 solder joints* y
1	6.2	241	0.872	0.74	245	0.201
2	5.6	250	0.860	0.77	229	0.053
3	6.5	258	0.853	0.64	266	0.239
4	6.4	239	0.891	0.68	251	0.242
5	5.7	260	0.888	0.81	262	0.075
6	5.8	254	0.876	0.75	230	0.132
7	5.5	250	0.869	0.71	228	0.053
8	6.1	241	0.860	0.76	234	0.119
9	6.1	256	0.854	0.62	269	0.172
10	6.3	260	0.872	0.64	240	0.171
11	6.6	249	0.877	0.69	250	0.369
12	5.7	255	0.868	0.73	246	0.100
13	5.8	258	0.854	0.80	261	0.105
14	6.1	260	0.879	0.77	270	0.196
15	5.8	262	0.888	0.70	267	0.126
16	6.3	256	0.870	0.81	246	0.216
17	6.4	254	0.862	0.76	233	0.286
18	6.8	247	0.855	0.65	250	0.306
19	6.7	238	0.876	0.69	249	0.403
20	6.3	264	0.884	0.71	265	0.162
21	6.4	260	0.891	0.79	252	0.214
22	5.7	259	0.881	0.80	245	0.287
23	5.8	244	0.863	0.76	238	0.092
24	5.4	259	0.875	0.68	217	0.008
25	5.7	264	0.870	0.64	276	0.102

Using MINITAB software to perform a linear multiple regression analysis, we set the values of x_1 in column $C1, x_2$ in $C2, \ldots, x_5$ in C5, and y in C6, in the same order as the run numbers shown in the data table. Then, the "regress" command produces the results shown in Figure 7.

It is tempting to conclude that the coefficients in this, or any other, multiple-regression analysis represent the "effects" of the corresponding predictor variables

```
THE REGRESSION EQUATION IS
C6 = -1.79 + 0.214 C1 - 0.00096 C2 + 0.90 C3 + 0.122 C4 + 0.000169 C5

                              ST. DEV.     TRATIO =
COLUMN    COEFFICIENT        OF COEF.     COEF/S.D.
             -1.7885          0.9655        -1.85
C1            0.21357         0.03630        5.88
C2           -0.000959        0.001873      -0.51
C3            0.898           1.047          0.86
C4            0.1216          0.2167         0.56
C5            0.0001695       0.0009457      0.18

S = 0.05806
R-SQUARED = 73.6 PERCENT

              Y      PRED.Y   ST. DEV.
ROW   C1      C6     VALUE    PRED.Y     RESIDUAL     ST.RES.
22    6.70  0.2870  0.1104    0.0220     0.1766       3.29R

R DENOTES AN OBS. WITH A LARGE ST. RES.
```

Figure 7. Computer printout for the example above.

on the dependent variable. For example, it appears that the coefficient of x_1, having the value 0.214, is the estimated effect Y of increasing x_1 by 1 unit. But it probably is not true that Y, the number of faults per 100 solder joints, will increase by 0.214 when x_1, the conveyor angle, is increased by 1 unit. There are several reasons for making this statement.

Any estimate of a coefficient in a regression analysis is subject to random error. Using Theorem 8, a confidence interval can be found for such a coefficient when it can be assumed that the residuals are approximately normally distributed. Thus, the random error is relatively easily quantified, but it often plays only a small role relative to other sources of error.

A much more serious source of error in interpreting the coefficients of a multiple regression equation arises from multicollinearity among the independent variables in the multiple regression equation. When at least some of the independent variables are highly correlated with each other, it is not possible to separate their effects on the dependent variable. In such cases we say that the effects of the independent variables are **confounded** with each other. To investigate the degree of correlation among the independent variables, the following correlation matrix of pairwise correlation coefficients has been computed for the wave-solder data by giving the MINITAB command `CORRELATE C1-C5`:

	C1	C2	C3	C4
C2	-.328			
C3	-.039	.174		
C4	-.281	.030	.215	
C5	.251	.402	.117	-.207

(Only a portion of the full matrix is shown here, since the matrix is symmetrical; for example, the correlation of C1 with C2 equals the correlation of C2 with C1, and the correlation of any column with itself equals 1.) It can be seen that several of the data columns involving independent variables show evidence of multicollinearity.

The effect of multicollinearity in this example can be observed directly by performing a multiple linear-regression analysis of y on x_2, x_3, x_4, and x_5 only, that is, by omitting x_1 from the regression equation. The resulting multiple regression equation is

$$\hat{y} = 0.23 - 0.00617x_2 + 1.18x_3 - 0.150x_4 + 0.00238x_5$$

By comparison, the multiple regression equation previously obtained when all five independent variables were used was

$$\hat{y} = -1.79 + 0.214x_1 - 0.0096x_2 + 0.90x_3 + 0.122x_4 + 0.000169x_5$$

It is readily seen that the coefficients of x_2, x_3, x_4, and x_5 have changed by more than trivial amounts when the independent variable x_1 has been omitted from the analysis. For example, the coefficient of x_2, which was -0.0096 when x_1 was included in the regression equation, becomes-0.00617, an increase of 36%, when x_1 is not included, and the coefficient of x_4 actually changes sign.

Often in practice, nonlinear terms, such as x^2, x^3, x_1x_2, and so forth, are introduced into a multiple regression equation to fit curved surfaces to data. When nonlinear terms are added, however, there is a risk of introducing further multicollinearity, such as between x and x^2, for example. This difficulty may be avoided, or at least minimized, by standardizing the variables used in the regression analysis. (Standardization, in this case, consists of subtracting the mean of each variable from each value of that variable, and dividing the result by its standard deviation.)

The use of large multiple regression equations, containing many variables in both linear and nonlinear forms, can produce an equation with better predictive power than one containing only a few linear terms. However, this method often creates highly correlated independent variables, even when standardization is employed, thereby making the problems of multicollinearity even worse.

When normal multiple regression analysis is to be used, the **residuals** should be examined carefully. The quantity $e_i = y_i - \hat{y}_i$ is called the ***i*th residual** in the multiple regression. An analysis of the residuals is useful in checking if the data are adequately described by the form of the fitted equation, or by the variables included in the equation.

A normal-scores plot is used to check the assumption that the residuals are approximately normally distributed. While the *t*-tests associated with regression analysis are not highly sensitive to departures from normality, gross departures will invalidate the significance tests associated with the regression. (However, the regression equation remains useful for estimating values of the coefficients and for obtaining $\hat{y}$, a predicted value of y.)

In addition, a plot of the residuals against the predicted values of y can reveal errors in the assumptions leading to the form of the fitted equation. If the chosen equation adequately describes the data, such a plot will show a "random" pattern without trend or obvious curvilinearity. On the other hand, if a linear equation is fitted to data that are highly nonlinear, the residuals will show a curvilinear trend. When the data depart seriously enough from the assumed relationship, excessive errors in prediction will result, and estimates of the coefficients of the independent variables will be relatively meaningless.

Finally, a plot of the residuals against integers reflecting the order of taking the observations (or "run number") or the time each observation was taken also should show a random pattern, without trends. A trend in such a plot can be caused by the presence of one or more variables, not included in the regression analysis, whose values have a measurable influence on the value of Y over the time period of the experiment. (Ambient variables, such as temperature and humidity, are examples of such effects.) A time trend in the residuals may suggest that these (and possibly other) variables should be controlled or their values measured and included in the regression equation when performing further research.

To illustrate these methods for checking residuals, the residuals were computed for the wave-solder regression analysis. Standardized residuals can be found directly with MINITAB software by giving the command `REGRESS C6 ON 5 PREDICTORS C1-C5, PUT RESIDUALS IN C7`.

The normal-scores plot of the raw residuals is shown in Figure 8. The graph shows reasonable agreement with the assumption that the residuals are normally distributed. (There appears to be one "outlying" observation, run 22. If significance

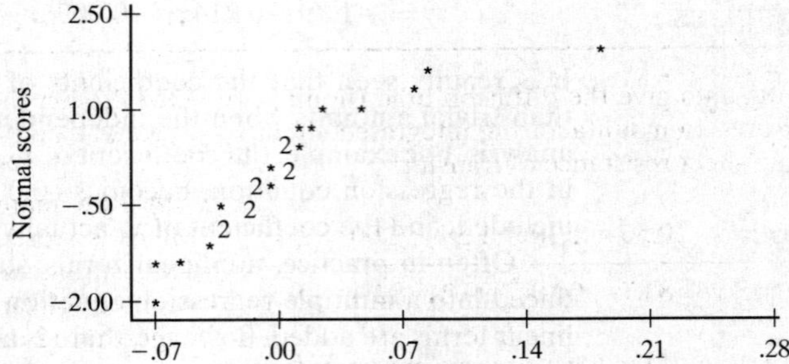

Figure 8. Normal-scores plot of wave-solder regression residuals.

tests are to be performed for the coefficients of the regression, it is recommended that the outlying observation be discarded and that the regression be rerun.)

A plot of the residuals against $\hat{y}$ is shown in Figure 9. Ignoring the outlier, this graph shows a random pattern with no obvious trends or curvilinearity. Thus, it appears that the linear multiple regression equation was adequate to describe the relationship between the dependent variable and the five independent variables over the range of observations.

These residuals are plotted against the run numbers in Figure 10. This graph, likewise, shows a random pattern, with no linear or curvilinear trends. It appears that no time-dependent extraneous variable has materially affected the value of y during the experiment.

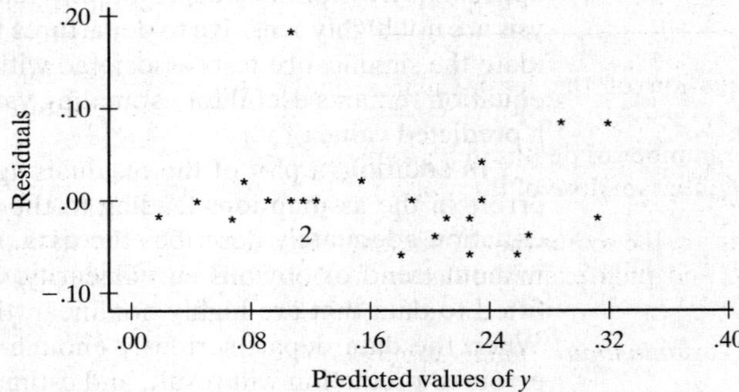

Figure 9. Plot of residuals against $\hat{y}$.

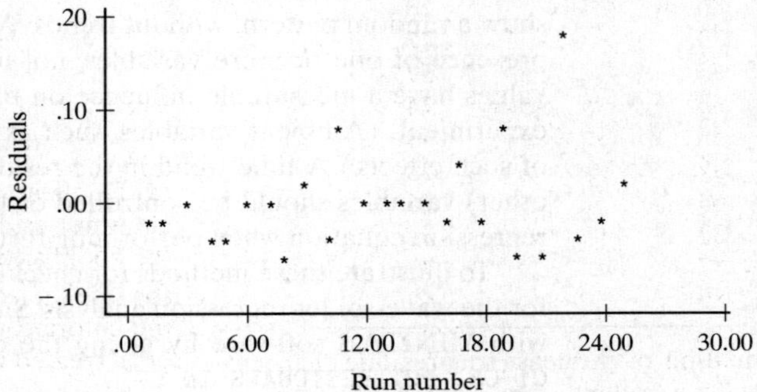

Figure 10. Plot of residuals against run numbers.

Applied Exercises

SECS. 1–3

41. The following data give the diffusion time (hours) of a silicon wafer used in manufacturing integrated circuits and the resulting sheet resistance of transfer:

Diffusion time, x	0.56	1.10	1.58	2.00	2.45
Sheet resistance, y	83.7	90.0	90.2	92.4	91.6

(a) Find the equation of the least squares line fit to these data.

(b) Predict the sheet resistance when the diffusion time is 1.3 hours.

42. Various doses of a poisonous substance were given to groups of 25 mice and the following results were observed:

Dose (mg) x	*Number of deaths* y
4	1
6	3
8	6
10	8
12	14
14	16
16	20

(a) Find the equation of the least squares line fit to these data.

(b) Estimate the number of deaths in a group of 25 mice that receive a 7-milligram dose of this poison.

43. The following are the scores that 12 students obtained on the midterm and final examinations in a course in statistics:

Midterm examination x	*Final examination* y
71	83
49	62
80	76
73	77
93	89
85	74
58	48
82	78
64	76
32	51
87	73
80	89

(a) Find the equation of the least squares line that will enable us to predict a student's final examination score in this course on the basis of his or her score on the midterm examination.

(b) Predict the final examination score of a student who received an 84 on the midterm examination.

44. Raw material used in the production of a synthetic fiber is stored in a place that has no humidity control. Measurements of the relative humidity and the moisture content of samples of the raw material (both in percentages) on 12 days yielded the following results:

Humidity	*Moisture content*
46	12
53	14
37	11
42	13
34	10
29	8
60	17
44	12
41	10
48	15
33	9
40	13

(a) Fit a least squares line that will enable us to predict the moisture content in terms of the relative humidity.

(b) Use the result of part (a) to estimate (predict) the moisture content when the relative humidity is 38 percent.

45. The following data pertain to the chlorine residual in a swimming pool at various times after it has been treated with chemicals:

Number of hours	*Chlorine residual (parts per million)*
2	1.8
4	1.5
6	1.4
8	1.1
10	1.1
12	0.9

(a) Fit a least squares line from which we can predict the chlorine residual in terms of the number of hours since the pool has been treated with chemicals.

(b) Use the equation of the least squares line to estimate the chlorine residual in the pool five hours after it has been treated with chemicals.

46. Use the coding of Exercise 15 to rework both parts of Exercise 42.

47. Use the coding of Exercise 15 to rework both parts of Exercise 45.

48. During its first five years of operation, a company's gross income from sales was 1.4, 2.1, 2.6, 3.5, and 3.7 million dollars. Use the coding of Exercise 15 to fit a least squares line and, assuming that the same linear trend continues, predict the company's gross income from sales during its sixth year of operation.

49. If a set of paired data gives the indication that the regression equation is of the form $\mu_{Y|x} = \alpha \cdot \beta^x$, it is customary to estimate α and β by fitting the line

$$\log \hat{y} = \log \hat{\alpha} + x \cdot \log \hat{\beta}$$

to the points $\{(x_i, \log y_i); i = 1, 2, \ldots, n\}$ by the method of least squares. Use this technique to fit an exponential curve of the form $\hat{y} = \hat{\alpha} \cdot \hat{\beta}^x$ to the following data on the growth of cactus grafts under controlled environmental conditions:

Weeks after grafting x	*Height (inches)* y
1	2.0
2	2.4
4	5.1
5	7.3
6	9.4
8	18.3

50. If a set of paired data gives the indication that the regression equation is of the form $\mu_{Y|x} = \alpha \cdot x^\beta$, it is customary to estimate α and β by fitting the line

$$\log \hat{y} = \log \hat{\alpha} + \hat{\beta} \cdot \log x$$

to the points $\{(\log x_i, \log y_i); i = 1, 2, \ldots, n\}$ by the method of least squares.

(a) Use this technique to fit a power function of the form $\hat{y} = \hat{\alpha} \cdot x^{\hat{\beta}}$ to the following data on the unit cost of producing certain electronic components and the number of units produced:

Lot size x	*Unit cost* y
50	\$108
100	\$53
250	\$24
500	\$9
1,000	\$5

(b) Use the result of part (a) to estimate the unit cost for a lot of 300 components.

SEC. 4

51. With reference to Exercise 42, test the null hypothesis $\beta = 1.25$ against the alternative hypothesis $\beta > 1.25$ at the 0.01 level of significance.

52. With reference to Exercise 44, test the null hypothesis $\beta = 0.350$ against the alternative hypothesis $\beta < 0.350$ at the 0.05 level of significance.

53. The following table shows the assessed values and the selling prices of eight houses, constituting a random sample of all the houses sold recently in a metropolitan area:

Assessed value (thousands) (of dollars)	*Selling price (thousands) (of dollars)*
170.3	214.4
202.0	269.3
162.5	206.2
174.8	225.0
157.9	199.8
181.6	232.1
210.4	274.2
188.0	243.5

(a) Fit a least squares line that will enable us to predict the selling price of a house in that metropolitan area in terms of its assessed value.

(b) Test the null hypothesis $\beta = 1.30$ against the alternative hypothesis $\beta > 1.30$ at the 0.05 level of significance.

54. With reference to Exercise 43, construct a 99% confidence interval for the regression coefficient β.

55. With reference to Exercise 45, construct a 98% confidence interval for the regression coefficient β.

56. With reference to Example 4, use the theory of Exercise 22 to test the null hypothesis $\alpha = 21.50$ against the alternative hypothesis $\alpha \neq 21.50$ at the 0.01 level of significance.

57. The following data show the advertising expenses (expressed as a percentage of total expenses) and the net operating profits (expressed as a percentage of total sales) in a random sample of six drugstores:

Advertising expenses	*Net operating profits*
1.5	3.6
1.0	2.8
2.8	5.4
0.4	1.9
1.3	2.9
2.0	4.3

(a) Fit a least squares line that will enable us to predict net operating profits in terms of advertising expenses.

(b) Test the null hypothesis $\alpha = 0.8$ against the alternative hypothesis $\alpha > 0.8$ at the 0.01 level of significance.

58. With reference to Exercise 42, use the theory of Exercise 22 to construct a 95% confidence interval for α.

59. With reference to Exercise 43, use the theory of Exercise 22 to construct a 99% confidence interval for α.

60. Use the theory of Exercises 24 and 26, as well as the quantities already calculated in Examples 4 and 5, to construct

(a) a 95% confidence interval for the mean test score of persons who have studied 14 hours for the test;

(b) 95% limits of prediction for the test score of a person who has studied 14 hours for the test.

61. Use the theory of Exercises 24 and 26, as well as the quantities already calculated in Exercise 51 for the data of Exercise 42, to find

(a) a 99% confidence interval for the expected number of deaths in a group of 25 mice when the dosage is 9 milligrams;

(b) 99% limits of prediction of the number of deaths in a group of 25 mice when the dosage is 9 milligrams.

62. Redo Exercise 61 when the dosage is 20 milligrams. Note the greatly increased width of the confidence limits for the expected number of deaths and of the limits of prediction. This example illustrates that **extrapolation**, estimating a value of Y for observations outside the range of the data, usually results in a highly inaccurate estimate.

63. The following table shows the elongation (in thousandths of an inch) of steel rods of nominally the same composition and diameter when subjected to various tensile forces (in thousands of pounds).

Force x	*Elongation* y
1.2	15.6
5.3	80.3
3.1	39.0
2.2	34.3
4.1	58.2
2.6	36.7
6.5	88.9
8.3	111.5
7.6	99.8
4.9	65.7

(a) Use appropriate computer software to fit a straight line to these data.

(b) Construct 99% confidence limits for the slope of the fitted line.

64. The following are loads (grams) put on the centers of like plastic rods with the resulting deflections (cm).

Load x	*Deflection* y
25	1.58
30	1.39
35	1.41
40	1.60
55	1.81
45	1.78
50	1.65
60	1.94

(a) Use an appropriate computer program to fit a straight line to these data.

(b) Using the 0.95 level of significance, test the null hypothesis that $\beta = 0.01$ against the alternative that $\beta > 0.01$.

SEC. 5

65. An achievement test is said to be reliable if a student who takes the test several times will consistently get high (or low) scores. One way of checking the reliability of a test is to divide it into two parts, usually the even-numbered problems and the odd-numbered problems, and observe the correlation between the scores that students get in both halves of the test. Thus, the following data represent the grades, x and y, that 20 students obtained for the even-numbered problems and the odd-numbered problems of a new objective test designed to test eighth grade achievement in general science:

x	y	x	y
27	29	33	42
36	44	39	31
44	49	38	38
32	27	24	22
27	35	33	34
41	33	32	37
38	29	37	38
44	40	33	35
30	27	34	32
27	38	39	43

Calculate r for these data and test its significance, that is, the null hypothesis $\rho = 0$ against the alternative hypothesis $\rho \neq 0$ at the 0.05 level of significance.

66. With reference to Exercise 65, use the formula obtained in Exercise 31 to construct a 95% confidence interval for ρ.

67. The following data pertain to x, the amount of fertilizer (in pounds) that a farmer applies to his soil, and y, his yield of wheat (in bushels per acre):

x	y	x	y	x	y
112	33	88	24	37	27
92	28	44	17	23	9
72	38	132	36	77	32
66	17	23	14	142	38
112	35	57	25	37	13
88	31	111	40	127	23
42	8	69	29	88	31
126	37	19	12	48	37
72	32	103	27	61	25
52	20	141	40	71	14
28	17	77	26	113	26

Assuming that the data can be looked upon as a random sample from a bivariate normal population, calculate r and test its significance at the 0.01 level of significance. Also, draw a scattergram of these paired data and judge whether the assumption seems reasonable.

68. With reference to Exercise 67, use the formula obtained in Exercise 31 to construct a 99% confidence interval for ρ.

69. Use the formula of Exercise 29 to calculate a 95% confidence interval for β for the numbers of hours studied and the test scores in the table in Section 3 and compare this interval with the one obtained in Example 6.

70. The calculation of r can often be simplified by adding the same constant to each x, adding the same constant to each y, or multiplying each x and/or y by the same positive constants. Recalculate r for the data of Example 7 by first multiplying each x and each y by 10 and then subtracting 70 from each x and 60 from each y.

71. The table at the bottom of the page shows how the history and economics scores of 25 students are distributed. Use the method of Exercise 32 to determine the value of r, replacing the row headings by the corresponding **class marks** (midpoints) 23, 28, 33, 38, 43, and 48 and the column headings by the corresponding class marks 23, 28, 33, 38, and 43. Use this value of r to test at the 0.05 level of significance whether there is a relationship between scores in the two subjects.

72. Rework Exercise 71, coding the class marks of the history scores −2, −1, 0, 1, and 2 and the class marks of the economics scores −2, −1, 0, 1, 2, and 3. (It follows from Exercise 70 that this kind of coding will not affect the value of r.)

Economics scores	*History scores* 21–25	26–30	31–35	36–40	41–45
21–25	1				
26–30		3	1		
31–35		2	5	2	
36–40			1	4	1
41–45			1	3	
46–50					1

73. This question has been intentionally omitted for this edition.

74. This question has been intentionally omitted for this edition.

75. (a) Use an appropriate computer program to obtain the sample correlation coefficient for the data of Exercise 63.

(b) Test whether r is significantly different from 0 using the 0.05 level.

76. (a) Use an appropriate computer program to obtain the sample correlation coefficient for the data of Exercise 64.

(b) Test whether this coefficient is significant using the 0.10 level.

SECS. 6–7

77. The following are sample data provided by a moving company on the weights of six shipments, the distances they were moved, and the damage that was incurred:

Weight (1,000 *pounds*) x_1	*Distance* (1,000 *miles*) x_2	*Damage* (*dollars*) y
4.0	1.5	160
3.0	2.2	112
1.6	1.0	69
1.2	2.0	90
3.4	0.8	123
4.8	1.6	186

(a) Assuming that the regression is linear, estimate β_0, β_1, and β_2.

(b) Use the results of part (a) to estimate the damage when a shipment weighing 2,400 pounds is moved 1,200 miles.

78. The following are data on the average weekly profits (in $1,000) of five restaurants, their seating capacities, and the average daily traffic (in thousands of cars) that passes their locations:

Seating capacity x_1	*Traffic count* x_2	*Weekly net profit* y
120	19	23.8
200	8	24.2
150	12	22.0
180	15	26.2
240	16	33.5

(a) Assuming that the regression is linear, estimate β_0, β_1, and β_2.

(b) Use the results of part (a) to predict the average weekly net profit of a restaurant with a seating capacity of 210 at a location where the daily traffic count averages 14,000 cars.

79. The following data consist of the scores that 10 students obtained in an examination, their I.Q.'s, and the numbers of hours they spent studying for the examination:

I.Q. x_1	*Number of hours studied* x_2	*Score* y
112	5	79
126	13	97
100	3	51
114	7	65
112	11	82
121	9	93
110	8	81
103	4	38
111	6	60
124	2	86

(a) Assuming that the regression is linear, estimate β_0, β_1, and β_2.

(b) Predict the score of a student with an I.Q. of 108 who studied 6 hours for the examination.

80. The following data were collected to determine the relationship between two processing variables and the hardness of a certain kind of steel:

Hardness (Rockwell 30-T) y	*Copper content (percent)* x_1	*Annealing temperature (degrees F)* x_2
78.9	0.02	1,000
55.2	0.02	1,200
80.9	0.10	1,000
57.4	0.10	1,200
85.3	0.18	1,000
60.7	0.18	1,200

Fit a plane by the method of least squares, and use it to estimate the average hardness of this kind of steel when the copper content is 0.14 percent and the annealing temperature is 1,100°F.

81. When the x_1's, x_2's, ..., and/or the x_k's are equally spaced, the calculation of the $\hat{\beta}$'s can be simplified by using the coding suggested in Exercise 15. Rework Exercise 80 coding the x_1-values -1, 0, and 1 and the x_2-values -1 and 1. (Note that for the coded x_1's and x_2's, call them z_1's and z_2's, we have not only $\Sigma z_1 = 0$ and $\Sigma z_2 = 0$, but also $\Sigma z_1 z_2 = 0$.)

82. The following are data on the percent effectiveness of a pain reliever and the amounts of three different medications (in milligrams) present in each capsule:

Medication A x_1	*Medication B* x_2	*Medication C* x_3	*Percent effective* y
15	20	10	47
15	20	20	54
15	30	10	58
15	30	20	66
30	20	10	59
30	20	20	67
30	30	10	71
30	30	20	83
45	20	10	72
45	20	20	82
45	30	10	85
45	30	20	94

Assuming that the regression is linear, estimate the regression coefficients after suitably coding each of the x's, and express the estimated regression equation in terms of the original variables.

83. The regression models that we introduced in Sections 2 and 6 are linear in the x's, but, more important, they are also linear in the β's. Indeed, they can be used in some problems where the relationship between the x's and y is not linear. For instance, when the regression is parabolic and of the form

$$\mu_{Y|x} = \beta_0 + \beta_1 x + \beta_2 x^2$$

we simply use the regression equation $\mu_{Y|x} = \beta_0 + \beta_1 x_1 + \beta_2 x_2$ with $x_1 = x$ and $x_2 = x^2$. Use this method to fit a parabola to the following data on the drying time of a varnish and the amount of a certain chemical that has been added:

Amount of additive (grams) x	*Drying time (hours)* y
1	8.5
2	8.0
3	6.0
4	5.0
5	6.0
6	5.5
7	6.5
8	7.0

Also, predict the drying time when 6.5 grams of the chemical is added.

84. The following data pertain to the demand for a product (in thousands of units) and its price (in cents) charged in five different market areas:

Price x	*Demand* y
20	22
16	41
10	120
11	89
14	56

Fit a parabola to these data by the method suggested in Exercise 83.

85. To judge whether it was worthwhile to fit a parabola in Exercise 84 and not just a straight line, test the null hypothesis $\beta_2 = 0$ against the alternative hypothesis $\beta_2 \neq 0$ at the 0.05 level of significance.

86. Use the results obtained for the data of Example 9 to construct a 90% confidence interval for the regression coefficient β_2 (see Exercise 38).

87. With reference to Exercise 77, test the null hypothesis $\beta_2 = 10.0$ against the alternative hypothesis $\beta_2 \neq 10.0$ at the 0.05 level of significance.

88. With reference to Exercise 77, construct a 95% confidence interval for the regression coefficient β_1.

89. With reference to Exercise 78, test the null hypothesis $\beta_1 = 0.12$ against the alternative hypothesis $\beta_1 < 0.12$ at the 0.05 level of significance.

90. With reference to Exercise 78, construct a 98% confidence interval for the regression coefficient β_2.

91. Use the results obtained for the data of Example 9 and the result of part (b) of Exercise 39 to construct a 95% confidence interval for the mean sales price of a three-bedroom house with two baths in the given housing development.

92. Use the results obtained for the data of Example 9 and the result of part (b) of Exercise 40 to construct 99% limits of prediction for the sales price of a three-bedroom house with two baths in the given housing development.

93. With reference to Exercise 77, use the result of part (b) of Exercise 39 to construct a 98% confidence interval for the mean damage of 2,400-pound shipments that are moved 1,200 miles.

94. With reference to Exercise 77, use the result of part (b) of Exercise 40 to construct 95% limits of prediction for the damage that will be incurred by a 2,400-pound shipment that is moved 1,200 miles.

95. With reference to Exercise 78, use the result of part (b) of Exercise 39 to construct a 99% confidence interval for the mean weekly net profit of restaurants with a seating capacity of 210 at a location where the daily traffic count averages 14,000 cars.

96. With reference to Exercise 78, use the result of part (b) of Exercise 40 to construct 98% limits of prediction for the average weekly net profit of a restaurant with a seating capacity of 210 at a location where the daily traffic count averages 14,000 cars.

97. Use an appropriate computer program to redo Exercise 82 without coding the x-values.

98. (a) Use an appropriate computer program to fit a plane to the following data relating the monthly water usage of a production plant (gallons) to its monthly production (tons), mean monthly ambient temperature (°F), and the monthly number of days of plant operation over a period of 12 months.

Water usage y	*Production* x_1	*Mean temperature* x_2	*Days of operation* x_3
2,228	98.5	67.4	19
2,609	108.2	70.3	20
3,088	109.6	82.1	21
2,378	101.0	69.2	21
1,980	83.3	64.5	19
1,717	70.0	63.7	21
2,723	144.7	58.0	19
2,031	84.4	58.1	20
1,902	97.4	36.6	17
1,721	131.8	49.6	23
2,254	82.1	44.3	18
2,522	64.5	44.1	19

(b) Estimate the water usage of the plant during a month when its production is 90.0 tons, the mean ambient temperature is 65°F, and it operates for 20 days.

SEC. 8

99. (a) Fit a linear surface to the following data:

y	x_1	x_2
118	41	−6
38	76	3
156	19	6
45	67	−3
31	62	−1
17	99	−3
109	27	−5
349	43	12
195	25	−8
72	24	2
94	48	5
118	3	4

(b) How good a fit is obtained?

(c) Plot the residuals against $\hat{y}$ and determine whether the pattern is "random."

(d) Check for multicollinearity among the independent variables.

100. The following data represent more extended measurements of monthly water usage at the plant referred to in Exercise 98 over a period of 20 months:

Water usage y	*Production* x_1	*Mean temperature* x_2	*Days of operation* x_3
2,609	108	70	20
2,228	97	68	19
2,559	113	66	19
2,723	144	58	19
3,088	109	82	21
2,522	64	44	19
2,012	91	61	20
2,254	82	44	18
2,436	126	59	21
2,460	111	62	21
2,147	85	54	18
2,378	101	69	21
2,031	84	58	20
1,717	70	64	21
2,117	107	51	22
1,902	97	36	17
2,251	98	56	22
2,357	96	85	19
1,721	132	49	23
1,980	84	64	19

(a) Use an appropriate computer program to fit a linear surface to these data.

(b) Use a computer program to make a normal-scores plot of the residuals. Does the assumption of normality appear to be satisfied at least approximately?

(c) Plot the residuals against $\hat{y}$ and determine whether the pattern is random.

(d) Check for excessive multicollinearity among the independent variables.

101. Using the data of Exercise 99,

(a) Create a new variable, $x_2{}^2$.

(b) Fit a surface of the form

$$y = b_0 + b_1x_1 + b_2x_2 + b_3x_2{}^2$$

(c) Find the correlation matrix of the three independent variables. Is there evidence of multicollinearity?

(d) Standardize each of the independent variables, x_1 and x_2, and create a new variable that is the square of the *standardized* value of x_2.

(e) Fit a surface of the same form as in part (b) to the standardized variables. Compare the goodness of fit of this surface to that of the linear surface fitted in Exercise 99.

(f) Plot the residuals of this regression analysis against the values of $\hat{y}$ and compare this plot to the one obtained in Exercise 99.

102. Using the data of Exercise 100,

(a) Create a new variable, x_1x_2.

(b) Fit a surface of the form

$$y = b_0 + b_1x_1 + b_2x_2 + b_3x_3 + b_4x_1x_2$$

(c) Find the correlation matrix of the four independent variables. Is there evidence of multicollinearity?

(d) Standardize each of the three independent variables x_1, x_2, and x_3, and create a new variable that is the product of the *standardized* values of x_1 and x_2.

(e) Fit a curved surface of the same form to the standardized variables. Compare the goodness of fit of this surface to that of the linear surface fitted in Exercise 100.

(f) Find the correlation matrix of the four standardized independent variables and compare with the results of part (c).

References

A proof of Theorem 3 and other mathematical details left out in the chapter may be found in Wilks, S. S., *Mathematical Statistics*. New York: John Wiley & Sons, Inc., 1962, and information about the distribution of $\frac{1}{2} \cdot \ln \frac{1+R}{1-R}$ may be found in Kendall, M. G., and Stuart, A., *The Advanced Theory of Statistics*, Vol. 1, 4th ed. New York: Macmillan Publishing Co., Inc., 1977. A derivation of the maximum likelihood estimates of σ_1, σ_2, and ρ is given in the third edition (but not in the fourth edition) of

HOEL, P., *Introduction to Mathematical Statistics*, 3rd ed. New York: John Wiley & Sons, Inc., 1962.

More detailed treatments of multiple regression may be found in numerous more advanced books, for instance, in

MORRISON, D. F., *Applied Linear Statistical Methods*. Upper Saddle River, N.J.: Prentice Hall, 1983,

WEISBERG, S., *Applied Linear Regression*, 2nd ed. New York: John Wiley & Sons, Inc., 1985,

WONNACOTT, T. H., and WONNACOTT, R. J., *Regression: A Second Course in Statistics*. New York: John Wiley & Sons, Inc., 1981.

Answers to Odd-Numbered Exercises

3 $\mu_{Y|x} = \frac{1+x}{2}$ and $\mu_{X|y} = \frac{2y}{3}$.

5 $\mu_{X|1} = \frac{4}{7}$ and $\mu_{Y|0} = \frac{9}{8}$.

13 $\hat{\beta} = \frac{\sum_{i=1}^{n} x_i y_i}{\sum_{i=1}^{n} x_i^2}$.

19 (a) $t = \frac{\hat{\beta} - \beta}{s_e/\sqrt{S_{xx}}}$; **(b)** $\hat{\beta} - t_{\alpha/2,n-2} \cdot \frac{S_e}{\sqrt{S_{xx}}} < \beta < \hat{\beta} + t_{\alpha/2,n-2} \cdot \frac{S_e}{\sqrt{S_{xx}}}$.

31 $\frac{1+r-(1-r)e^{-2z_{\alpha/2}/\sqrt{n-3}}}{1+r+(1-r)e^{-2z_{\alpha/2}/\sqrt{n-3}}} < \rho < \frac{1+r-(1-r)e^{2z_{\alpha/2}/\sqrt{n-3}}}{1+r+(1-r)e^{2z_{\alpha/2}/\sqrt{n-3}}}$.

39 (b) $B'X_0 \pm t_{\alpha/2,n-k}\hat{\sigma} \cdot \sqrt{\frac{n[X_0'(X'X)^{-1}X_0]}{n-k-1}}$.

41 (a) $\hat{y} = 83.46 + 3.98x$; **(b)** 88.63.

43 (a) $\hat{y} = 31.609 + 0.5816x$; **(b)** $\hat{y} = 88.63$.

45 (a) $\hat{y} = 1.8999 + 0.0857x$; **(b)** $\hat{y} = 1.4714$.

47 (a) $\hat{y} = 1.3 - 0.0857x$ (coded); **(b)** $\hat{y} = 1.4714$.

49 $\hat{y} = 1.371(1.383)^x$.

51 $t = 3.72$; the null hypothesis must be rejected.

53 (a) $\hat{y} = -37.02 + 1.4927x$; **(b)** $t = 3.413$; the null hypothesis must be rejected.

55 $-0.1217 < \beta < -0.0497$.

57 (a) $\hat{y} = 1.2594 + 1.4826x$; **(b)** $t = 3.10$; the null hypothesis cannot be rejected.

59 $-2.2846 < \alpha < 65.5026$.

61 (a) $6.452 < \mu_{Y|9} < 9.7634$; **(b)** 3.4777 and 12.7009.

63 (a) $\hat{y} = 2.20 + 13.3x$; **(b)** $11.5 < \beta < 15.1$.

65 $r = 0.55$; $z = 2.565$ and the value of r is significant.

67 $r = 0.727$; $z = 5.05$ and the value of r is significant.

69 $2.84 < \beta < 4.10$.

71 $r = 0.772$; $z = 4.81$ and the value of r is significant.

73 $r = 0.285$; $z = 5.55$ and the value of r is significant.

75 (a) 0.994; **(b)** $z = 7.68$; it is significantly different from 0 at the 0.05 level of significance.

77 (a) $\hat{\beta}_0 = 14.56$; $\hat{\beta}_1 = 30.109$ and $\hat{\beta}_2 = 12.16$; **(b)** $\hat{y} = \$101.41$.

79 (a) $\hat{\beta}_0 = -124.57$, $\hat{\beta}_1 = 1.659$ and $\hat{\beta}_2 = 1.439$; **(b)** $\hat{y} = 63.24$.

81 $\hat{y} = 69.73 + 2.975z_1 - 11.97z_2$ (coded); $\hat{y} = 71.2$.

83 $\hat{y} = 10.5 - 2.0x + 0.2x^2$; $\hat{y} = 5.95$.

85 $t = 2.94$; the null hypothesis cannot be rejected and there is no real evidence that it is worthwhile to fit a parabola rather than a straight line.

87 $t = 0.16$; the null hypothesis cannot be rejected.

89 $t = -4.18$; the null hypothesis must be rejected.

91 $\$288,650 < \mu_{Y|3,2} < \$296,920$.

93 $\$74.5 < \mu_{Y|2,4,1,2} < \$128.3$.

97 $\hat{y} = -2.33 + 0.900x_1 + 1.27x_2 + 0.900x_3$.

99 (a) $\hat{y} = 170 - 1.39x_1 + 6.07x_2$.

101 (b) $\hat{y} = 86.9 - 0.904x_1 + 0.508x_2 + 2.06x_2^2$; **(c)** $r_{x_1x_2} = -0.142$, $r_{x_1x_2^2} = -0.218$, $r_{x_2x_2^2} = 0.421$; **(e)** $\hat{y} = 47.5 - 24.8x_1' + 15.0x_2' + 70.2(x_2')^2$.

Appendix

SUMS AND PRODUCTS

1 Rules for Sums and Products

To simplify expressions involving sums and products, the $\sum$ and $\prod$ notations are widely used in statistics. In the usual notation we write

$$\sum_{i=a}^{b} x_i = x_a + x_{a+1} + x_{a+2} + \cdots + x_b$$

and

$$\prod_{i=a}^{b} x_i = x_a \cdot x_{a+1} \cdot x_{a+2} \cdot \ldots \cdot x_b$$

for any nonnegative integers a and b with $a \leqq b$.

When working with sums or products, it is often helpful to apply the following rules, which can all be verified by writing the respective expressions in full, that is, without the $\sum$ or $\prod$ notation:

THEOREM 1.

1. $\sum_{i=1}^{n} kx_i = k \cdot \sum_{i=1}^{n} x_i$

2. $\sum_{i=1}^{n} k = nk$

3. $\sum_{i=1}^{n} (x_i + y_i) = \sum_{i=1}^{n} x_i + \sum_{i=1}^{n} y_i$

4. $\prod_{i=1}^{n} kx_i = k^n \cdot \prod_{i=1}^{n} x_i$

5. $\prod_{i=1}^{n} k = k^n$

6. $\prod_{i=1}^{n} x_i y_i = \left(\prod_{i=1}^{n} x_i\right)\left(\prod_{i=1}^{n} y_i\right)$

7. $\ln \prod_{i=1}^{n} x_i = \sum_{i=1}^{n} \ln x_i$

Double sums, triple sums, ... are also widely used in statistics, and if we repeatedly apply the definition of $\sum$ given above, we have, for example,

$$\begin{aligned}\sum_{i=1}^{m}\sum_{j=1}^{n} x_{ij} &= \sum_{i=1}^{m}(x_{i1}+x_{i2}+\cdots+x_{in}) \\ &= (x_{11}+x_{12}+\cdots+x_{1n}) \\ &\quad + (x_{21}+x_{22}+\cdots+x_{2n}) \\ &\quad \cdots\cdots\cdots\cdots\cdots\cdots \\ &\quad + (x_{m1}+x_{m2}+\cdots+x_{mn})\end{aligned}$$

Note that when the x_{ij} are thus arranged in a rectangular array, the first subscript denotes the row to which a particular element belongs, and the second subscript denotes the column.

When we work with double sums, the following theorem is of special interest; it is an immediate consequence of the multinomial expansion of

$$(x_1+x_2+\cdots+x_n)^2$$

THEOREM 2.

$$\sum_{i<j}\sum x_i x_j = \frac{1}{2}\left[\left(\sum_{i=1}^{n} x_i\right)^2 - \sum_{i=1}^{n} x_i^2\right]$$

where

$$\sum_{i<j}\sum x_i x_j = \sum_{i=1}^{n-1}\sum_{j=i+1}^{n} x_i x_j$$

2 Special Sums

In the theory of nonparametric statistics, particularly when we deal with rank sums, we often need expressions for the sums of powers of the first n positive integers, that is, expressions for

$$S(n,r) = 1^r + 2^r + 3^r + \cdots + n^r$$

for $r = 0, 1, 2, 3, \ldots$. The following theorem, which the reader will be asked to prove in Exercise 1, provides a convenient way of obtaining these sums.

THEOREM 3.

$$\sum_{r=0}^{k-1}\binom{k}{r} S(n,r) = (n+1)^k - 1$$

for any positive integers n and k.

A disadvantage of this theorem is that we have to find the sums $S(n,r)$ one at a time, first for $r = 0$, then for $r = 1$, then for $r = 2$, and so forth. For instance, for $k = 1$ we get

$$\binom{1}{0} S(n,0) = (n+1) - 1 = n$$

and hence $S(n,0) = 1^0 + 2^0 + \cdots + n^0 = n$. Similarly, for $k = 2$ we get

$$\binom{2}{0} S(n,0) + \binom{2}{1} S(n,1) = (n+1)^2 - 1$$

$$n + 2S(n,1) = n^2 + 2n$$

and hence $S(n,1) = 1^1 + 2^1 + \cdots + n^1 = \frac{1}{2}n(n+1)$. Using the same technique, the reader will be asked to show in Exercise 2 that

$$S(n,2) = 1^2 + 2^2 + \cdots + n^2 = \frac{1}{6}n(n+1)(2n+1)$$

and

$$S(n,3) = 1^3 + 2^3 + \cdots + n^3 = \frac{1}{4}n^2(n+1)^2$$

Exercises

1. Prove Theorem 3 by making use of the fact that

$$(m+1)^k - m^k = \sum_{r=0}^{k-1} \binom{k}{r} m^r$$

which follows from the binomial expansion of $(m+1)^k$.

2. Verify the formulas for $S(n,2)$ and $S(n,3)$ given previously, and find an expression for $S(n,4)$.

3. Given $x_1 = 1, x_2 = 3, x_3 = -2, x_4 = 4, x_5 = -1, x_6 = 2, x_7 = 1$, and $x_8 = 2$, find

(a) $\sum_{i=1}^{8} x_i$; **(b)** $\sum_{i=1}^{8} x_i^2$.

4. Given $x_1 = 3, x_2 = 4, x_3 = 5, x_4 = 6, x_5 = 7, f_1 = 3, f_2 = 7, f_3 = 10, f_4 = 5$, and $f_5 = 2$, find

(a) $\sum_{i=1}^{5} x_i$; **(b)** $\sum_{i=1}^{5} f_i$;

(c) $\sum_{i=1}^{5} x_i f_i$; **(d)** $\sum_{i=1}^{5} x_i^2 f_i$.

5. Given $x_1 = 2, x_2 = -3, x_3 = 4, x_4 = -2, y_1 = 5, y_2 = -3, y_3 = 2$, and $y_4 = -1$, find

(a) $\sum_{i=1}^{4} x_i$; **(b)** $\sum_{i=1}^{4} y_i$;

(c) $\sum_{i=1}^{4} x_i^2$; **(d)** $\sum_{i=1}^{4} y_i^2$; **(e)** $\sum_{i=1}^{4} x_i y_i$.

6. Given $x_{11} = 3, x_{12} = 1, x_{13} = -2, x_{14} = 2, x_{21} = 1, x_{22} = 4, x_{23} = -2, x_{24} = 5, x_{31} = 3, x_{32} = -1, x_{33} = 2$, and $x_{34} = 3$, find

(a) $\sum_{i=1}^{3} x_{ij}$ separately for $j = 1, 2, 3$, and 4;

(b) $\sum_{j=1}^{4} x_{ij}$ separately for $i = 1, 2$, and 3.

7. With reference to Exercise 6, evaluate the double summation $\sum_{i=1}^{3}\sum_{j=1}^{4} x_{ij}$ using

(a) the results of part (a) of that exercise;

(b) the results of part (b) of that exercise.

Answers to Odd-Numbered Exercises

3 (a) 10; **(b)** 40.

5 (a) 1; **(b)** 3; **(c)** 33; **(d)** 39; **(e)** 29.

7 (a) 19; **(b)** 19.

Appendix

SPECIAL PROBABILITY DISTRIBUTIONS

1 Bernoulli Distribution

$$f(x;\theta) = \theta^x(1-\theta)^{1-x} \qquad \text{for } x = 0, 1$$

Parameter: $0 < \theta < 1$

Mean and variance: $\mu = \theta$ and $\sigma^2 = \theta(1-\theta)$

2 Binomial Distribution

$$b(x;n,\theta) = \binom{n}{x}\theta^x(1-\theta)^{n-x} \qquad \text{for } x = 0, 1, 2, \ldots, n$$

Parameters: n is a positive integer and $0 < \theta < 1$

Mean and variance: $\mu = n\theta$ and $\sigma^2 = n\theta(1-\theta)$

3 Discrete Uniform Distribution (Special Case)

$$f(x;k) = \frac{1}{k} \qquad \text{for } x = 1, 2, \ldots, k$$

Parameter: k is a positive integer

Mean and variance: $\mu = \frac{k+1}{2}$ and $\sigma^2 = \frac{k^2-1}{12}$

4 Geometric Distribution

$$g(x;\theta) = \theta(1-\theta)^{x-1} \qquad \text{for } x = 1, 2, 3, \ldots$$

Parameter: $0 < \theta < 1$

Mean and variance: $\mu = \frac{1}{\theta}$ and $\sigma^2 = \frac{1-\theta}{\theta^2}$

5 Hypergeometric Distribution

$$h(x; n, N, M) = \frac{\binom{M}{x}\binom{N-M}{n-x}}{\binom{N}{n}} \quad \text{for } x = 0, 1, 2, \ldots, n, \; x \leqq M, \quad \text{and} \quad n - x \leqq N - M$$

Parameters: n and N are positive integers with $n \leqq N$, and M is a nonnegative integer with $M \leqq N$

Mean and variance: $\mu = \frac{nM}{N}$ and $\sigma^2 = \frac{nM(N-M)(N-n)}{N^2(N-1)}$

6 Negative Binomial Distribution

$$b^*(x; k, \theta) = \binom{x-1}{k-1}\theta^k(1-\theta)^{x-k} \qquad \text{for } x = k, k+1, k+2, \ldots$$

Parameters: k is a positive integer and $0 < \theta < 1$

Mean and variance: $\mu = \frac{k}{\theta}$ and $\sigma^2 = \frac{k(1-\theta)}{\theta^2}$

7 Poisson Distribution

$$p(x; \lambda) = \frac{\lambda^x e^{-\lambda}}{x!} \qquad \text{for } x = 0, 1, 2, \ldots$$

Parameter: $\lambda > 0$

Mean and variance: $\mu = \lambda$ and $\sigma^2 = \lambda$

Appendix

SPECIAL PROBABILITY DENSITIES

1 Beta Distribution

$$f(x;\alpha,\beta)=\begin{cases}\dfrac{\Gamma(\alpha+\beta)}{\Gamma(\alpha)\cdot\Gamma(\beta)}x^{\alpha-1}(1-x)^{\beta-1} & \text{for } 0<x<1\\ 0 & \text{elsewhere}\end{cases}$$

Parameters: $\alpha > 0$ and $\beta > 0$

Mean and variance: $\mu = \dfrac{\alpha}{\alpha+\beta}$ and $\sigma^2 = \dfrac{\alpha\beta}{(\alpha+\beta)^2(\alpha+\beta+1)}$

2 Cauchy Distribution

$$p(x;\alpha,\beta)=\frac{\frac{\beta}{\pi}}{(x-\alpha)^2+\beta^2}$$

Parameters: $-\infty<\alpha<\infty$ and $\beta>0$

Mean and variance: Do not exist

3 Chi-Square Distribution

$$f(x;\nu)=\begin{cases}\dfrac{1}{2^{\nu/2}\Gamma(\nu/2)}x^{\frac{\nu-2}{2}}e^{-\frac{x}{2}} & \text{for } x>0\\ 0 & \text{elsewhere}\end{cases}$$

Parameter: ν is a positive integer

Mean and variance: $\mu=\nu$ and $\sigma^2=2\nu$

4 Exponential Distribution

$$g(x;\theta) = \begin{cases} \dfrac{1}{\theta}e^{-x/\theta} & \text{for } x > 0 \\ 0 & \text{elsewhere} \end{cases}$$

Parameter: $\theta > 0$

Mean and variance: $\mu = \theta$ and $\sigma^2 = \theta^2$

5 *F* Distribution

$$g(f) = \begin{cases} \dfrac{\Gamma\left(\dfrac{\nu_1+\nu_2}{2}\right)}{\Gamma\left(\dfrac{\nu_1}{2}\right)\Gamma\left(\dfrac{\nu_2}{2}\right)}\left(\dfrac{\nu_1}{\nu_2}\right)^{\frac{\nu_1}{2}} \cdot f^{\frac{\nu_1}{2}-1}\left(1+\dfrac{\nu_1}{\nu_2}f\right)^{-\frac{1}{2}(\nu_1+\nu_2)} & \text{for } f > 0 \\ 0 & \text{elsewhere} \end{cases}$$

Parameters: $\nu_1 > 0$ and $\nu_2 > 0$

Mean: $\mu = \dfrac{\nu_2}{\nu_2 - 2}$

6 Gamma Distribution

$$f(x) = \begin{cases} \dfrac{1}{\beta^\alpha \Gamma(\alpha)} x^{\alpha-1}e^{-x/\beta} & \text{for } x > 0 \\ 0 & \text{elsewhere} \end{cases}$$

Parameters: $\alpha > 0$ and $\beta > 0$

Mean and variance: $\mu = \alpha\beta$ and $\sigma^2 = \alpha\beta^2$

7 Normal Distribution

$$n(x;\mu,\sigma) = \frac{1}{\sigma\sqrt{2\pi}}e^{-\frac{1}{2}\left(\frac{x-\mu}{\sigma}\right)^2} \quad \text{for } -\infty < x < \infty$$

Parameters: μ and $\sigma > 0$

Mean and variance: $\mu = \mu$ and $\sigma^2 = \sigma^2$

8 *t* Distribution (Student's *t* Distribution)

$$f(t;\nu) = \frac{\Gamma\left(\dfrac{\nu+1}{2}\right)}{\sqrt{\pi\nu}\,\Gamma\left(\dfrac{\nu}{2}\right)} \cdot \left(1+\frac{t^2}{\nu}\right)^{-\frac{\nu+1}{2}} \quad \text{for } -\infty < t < \infty$$

Parameter: ν is a positive integer

Mean and variance: $\mu = 0$ and $\sigma^2 = \dfrac{\nu}{\nu - 2}$ for $\nu > 2$

9 Uniform Distribution (Rectangular Distribution)

$$u(x; \alpha, \beta) = \begin{cases} \dfrac{1}{\beta - \alpha} & \text{for } \alpha < x < \beta \\ 0 & \text{elsewhere} \end{cases}$$

Parameters: $-\infty < \alpha < \beta < \infty$

Mean and variance: $\mu = \dfrac{\alpha + \beta}{2}$ and $\sigma^2 = \dfrac{1}{12}(\beta - \alpha)^2$

Statistical Tables

Table I: Binomial Probabilities†

		θ									
n	x	.05	.10	.15	.20	.25	.30	.35	.40	.45	.50
1	0	.9500	.9000	.8500	.8000	.7500	.7000	.6500	.6000	.5500	.5000
	1	.0500	.1000	.1500	.2000	.2500	.3000	.3500	.4000	.4500	.5000
2	0	.9025	.8100	.7225	.6400	.5625	.4900	.4225	.3600	.3025	.2500
	1	.0950	.1800	.2550	.3200	.3750	.4200	.4550	.4800	.4950	.5000
	2	.0025	.0100	.0225	.0400	.0625	.0900	.1225	.1600	.2025	.2500
3	0	.8574	.7290	.6141	.5120	.4219	.3430	.2746	.2160	.1664	.1250
	1	.1354	.2430	.3251	.3840	.4219	.4410	.4436	.4320	.4084	.3750
	2	.0071	.0270	.0574	.0960	.1406	.1890	.2389	.2880	.3341	.3750
	3	.0001	.0010	.0034	.0080	.0156	.0270	.0429	.0640	.0911	.1250
4	0	.8145	.6561	.5220	.4096	.3164	.2401	.1785	.1296	.0915	.0625
	1	.1715	.2916	.3685	.4096	.4219	.4116	.3845	.3456	.2995	.2500
	2	.0135	.0486	.0975	.1536	.2109	.2646	.3105	.3456	.3675	.3750
	3	.0005	.0036	.0115	.0256	.0469	.0756	.1115	.1536	.2005	.2500
	4	.0000	.0001	.0005	.0016	.0039	.0081	.0150	.0256	.0410	.0625
5	0	.7738	.5905	.4437	.3277	.2373	.1681	.1160	.0778	.0503	.0312
	1	.2036	.3280	.3915	.4096	.3955	.3602	.3124	.2592	.2059	.1562
	2	.0214	.0729	.1382	.2048	.2637	.3087	.3364	.3456	.3369	.3125
	3	.0011	.0081	.0244	.0512	.0879	.1323	.1811	.2304	.2757	.3125
	4	.0000	.0004	.0022	.0064	.0146	.0284	.0488	.0768	.1128	.1562
	5	.0000	.0000	.0001	.0003	.0010	.0024	.0053	.0102	.0185	.0312
6	0	.7351	.5314	.3771	.2621	.1780	.1176	.0754	.0467	.0277	.0156
	1	.2321	.3543	.3993	.3932	.3560	.3025	.2437	.1866	.1359	.0938
	2	.0305	.0984	.1762	.2458	.2966	.3241	.3280	.3110	.2780	.2344
	3	.0021	.0146	.0415	.0819	.1318	.1852	.2355	.2765	.3032	.3125
	4	.0001	.0012	.0055	.0154	.0330	.0595	.0951	.1382	.1861	.2344
	5	.0000	.0001	.0004	.0015	.0044	.0102	.0205	.0369	.0609	.0938
	6	.0000	.0000	.0000	.0001	.0002	.0007	.0018	.0041	.0083	.0156
7	0	.6983	.4783	.3206	.2097	.1335	.0824	.0490	.0280	.0152	.0078
	1	.2573	.3720	.3960	.3670	.3115	.2471	.1848	.1306	.0872	.0547
	2	.0406	.1240	.2097	.2753	.3115	.3177	.2985	.2613	.2140	.1641
	3	.0036	.0230	.0617	.1147	.1730	.2269	.2679	.2903	.2918	.2734
	4	.0002	.0026	.0109	.0287	.0577	.0972	.1442	.1935	.2388	.2734
	5	.0000	.0002	.0012	.0043	.0115	.0250	.0466	.0774	.1172	.1641
	6	.0000	.0000	.0001	.0004	.0013	.0036	.0084	.0172	.0320	.0547
	7	.0000	.0000	.0000	.0000	.0001	.0002	.0006	.0016	.0037	.0078
8	0	.6634	.4305	.2725	.1678	.1001	.0576	.0319	.0168	.0084	.0039
	1	.2793	.3826	.3847	.3355	.2670	.1977	.1373	.0896	.0548	.0312
	2	.0515	.1488	.2376	.2936	.3115	.2965	.2587	.2090	.1569	.1094
	3	.0054	.0331	.0839	.1468	.2076	.2541	.2786	.2787	.2568	.2188
	4	.0004	.0046	.0185	.0459	.0865	.1361	.1875	.2322	.2627	.2734
	5	.0000	.0004	.0026	.0092	.0231	.0467	.0808	.1239	.1719	.2188
	6	.0000	.0000	.0002	.0011	.0038	.0100	.0217	.0413	.0703	.1094
	7	.0000	.0000	.0000	.0001	.0004	.0012	.0033	.0079	.0164	.0312
	8	.0000	.0000	.0000	.0000	.0000	.0001	.0002	.0007	.0017	.0039

†Based on *Tables of the Binomial Probability Distribution*, National Bureau of Standards Applied Mathematics Series No. 6. Washington, D.C.: U.S. Government Printing Office, 1950.

Table I: *(continued)*

n	x	.05	.10	.15	.20	.25	.30	.35	.40	.45	.50
		θ									
9	0	.6302	.3874	.2316	.1342	.0751	.0404	.0207	.0101	.0046	.0020
	1	.2985	.3874	.3679	.3020	.2253	.1556	.1004	.0605	.0339	.0176
	2	.0629	.1722	.2597	.3020	.3003	.2668	.2162	.1612	.1110	.0703
	3	.0077	.0446	.1069	.1762	.2336	.2668	.2716	.2508	.2119	.1641
	4	.0006	.0074	.0283	.0061	.1168	.1715	.2194	.2508	.2600	.2461
	5	.0000	.0008	.0050	.0165	.0389	.0735	.1181	.1672	.2128	.2461
	6	.0000	.0001	.0006	.0028	.0087	.0210	.0424	.0743	.1160	.1641
	7	.0000	.0000	.0000	.0003	.0012	.0039	.0098	.0212	.0407	.0703
	8	.0000	.0000	.0000	.0000	.0001	.0004	.0013	.0035	.0083	.0176
	9	.0000	.0000	.0000	.0000	.0000	.0000	.0001	.0003	.0008	.0020
10	0	.5987	.3487	.1969	.1074	.0563	.0282	.0135	.0060	.0025	.0010
	1	.3151	.3874	.3474	.2684	.1877	.1211	.0725	.0403	.0207	.0098
	2	.0746	.1937	.2759	.3020	.2816	.2335	.1757	.1209	.0763	.0439
	3	.0105	.0574	.1298	.2013	.2503	.2668	.2522	.2150	.1665	.1172
	4	.0010	.0112	.0401	.0881	.1460	.2001	.2377	.2508	.2384	.2051
	5	.0001	.0015	.0085	.0264	.0584	.1029	.1536	.2007	.2340	.2461
	6	.0000	.0001	.0012	.0055	.0162	.0368	.0689	.1115	.1596	.2051
	7	.0000	.0000	.0001	.0008	.0031	.0090	.0212	.0425	.0746	.1172
	8	.0000	.0000	.0000	.0001	.0004	.0014	.0043	.0106	.0229	.0439
	9	.0000	.0000	.0000	.0000	.0000	.0001	.0005	.0016	.0042	.0098
	10	.0000	.0000	.0000	.0000	.0000	.0000	.0000	.0001	.0003	.0016
11	0	.5688	.3138	.1673	.0859	.0422	.0198	.0088	.0036	.0014	.0005
	1	.3293	.3835	.3248	.2362	.1549	.0932	.0518	.0266	.0125	.0054
	2	.0867	.2131	.2866	.2953	.2581	.1998	.1395	.0887	.0513	.0269
	3	.0137	.0710	.1517	.2215	.2581	.2568	.2254	.1774	.1259	.0806
	4	.0014	.0158	.0536	.1107	.1721	.2201	.2428	.2365	.2060	.1611
	5	.0001	.0025	.0132	.0388	.0803	.1321	.1830	.2207	.2360	.2256
	6	.0000	.0003	.0023	.0097	.0268	.0566	.0985	.1471	.1931	.2256
	7	.0000	.0000	.0003	.0017	.0064	.0173	.0379	.0701	.1128	.1611
	8	.0000	.0000	.0000	.0002	.0011	.0037	.0102	.0234	.0462	.0806
	9	.0000	.0000	.0000	.0000	.0001	.0005	.0018	.0052	.0126	.0269
	10	.0000	.0000	.0000	.0000	.0000	.0000	.0002	.0007	.0021	.0054
	11	.0000	.0000	.0000	.0000	.0000	.0000	.0000	.0000	.0002	.0005
12	0	.5404	.2824	.1422	.0687	.0317	.0138	.0057	.0022	.0008	.0002
	1	.3413	.3766	.3012	.2062	.1267	.0712	.0368	.0174	.0075	.0029
	2	.0988	.2301	.2924	.2835	.2323	.1678	.1088	.0639	.0339	.0161
	3	.0173	.0852	.1720	.2362	.2581	.2397	.1954	.1419	.0923	.0537
	4	.0021	.0213	.0683	.1329	.1936	.2311	.2367	.2128	.1700	.1208
	5	.0002	.0038	.0193	.0532	.1032	.1585	.2039	.2270	.2225	.1934
	6	.0000	.0005	.0040	.0155	.0401	.0792	.1281	.1766	.2124	.2256
	7	.0000	.0000	.0006	.0033	.0115	.0291	.0591	.1009	.1489	.1934
	8	.0000	.0000	.0001	.0005	.0024	.0078	.0199	.0420	.0762	.1208
	9	.0000	.0000	.0000	.0001	.0004	.0015	.0048	.0125	.0277	.0537
	10	.0000	.0000	.0000	.0000	.0000	.0002	.0008	.0025	.0068	.0161
	11	.0000	.0000	.0000	.0000	.0000	.0000	.0001	.0003	.0010	.0029
	12	.0000	.0000	.0000	.0000	.0000	.0000	.0000	.0000	.0001	.0002
13	0	.5133	.2542	.1209	.0550	.0238	.0097	.0037	.0013	.0004	.0001
	1	.3512	.3672	.2774	.1787	.1029	.0540	.0259	.0113	.0045	.0016
	2	.1109	.2448	.2937	.2680	.2059	.1388	.0836	.0453	.0220	.0095

Table I: *(continued)*

n	x	θ .05	.10	.15	.20	.25	.30	.35	.40	.45	.50
13	3	.0214	.0997	.1900	.2457	.2517	.2181	.1651	.1107	.0660	.0349
	4	.0028	.0277	.0838	.1535	.2097	.2337	.2222	.1845	.1350	.0873
	5	.0003	.0055	.0266	.0691	.1258	.1803	.2154	.2214	.1989	.1571
	6	.0000	.0008	.0063	.0230	.0559	.1030	.1546	.1968	.2169	.2095
	7	.0000	.0001	.0011	.0058	.0186	.0442	.0833	.1312	.1775	.2095
	8	.0000	.0000	.0001	.0011	.0047	.0142	.0336	.0656	.1089	.1571
	9	.0000	.0000	.0000	.0001	.0009	.0034	.0101	.0243	.0495	.0873
	10	.0000	.0000	.0000	.0000	.0001	.0006	.0022	.0065	.0162	.0349
	11	.0000	.0000	.0000	.0000	.0000	.0001	.0003	.0012	.0036	.0095
	12	.0000	.0000	.0000	.0000	.0000	.0000	.0000	.0001	.0005	.0016
	13	.0000	.0000	.0000	.0000	.0000	.0000	.0000	.0000	.0000	.0001
14	0	.4877	.2288	.1028	.0440	.0178	.0068	.0024	.0008	.0002	.0001
	1	.3593	.3559	.2539	.1539	.0832	.0407	.0181	.0073	.0027	.0009
	2	.1229	.2570	.2912	.2501	.1802	.1134	.0634	.0317	.0141	.0056
	3	.0259	.1142	.2056	.2501	.2402	.1943	.1366	.0845	.0462	.0222
	4	.0037	.0349	.0998	.1720	.2202	.2290	.2022	.1549	.1040	.0611
	5	.0004	.0078	.0352	.0860	.1468	.1963	.2178	.2066	.1701	.1222
	6	.0000	.0013	.0093	.0322	.0734	.1262	.1759	.2066	.2088	.1833
	7	.0000	.0002	.0019	.0092	.0280	.0618	.1082	.1574	.1952	.2095
	8	.0000	.0000	.0003	.0020	.0082	.0232	.0510	.0918	.1398	.1833
	9	.0000	.0000	.0000	.0003	.0018	.0066	.0183	.0408	.0762	.1222
	10	.0000	.0000	.0000	.0000	.0003	.0014	.0049	.0136	.0312	.0611
	11	.0000	.0000	.0000	.0000	.0000	.0002	.0010	.0033	.0093	.0222
	12	.0000	.0000	.0000	.0000	.0000	.0000	.0001	.0005	.0019	.0056
	13	.0000	.0000	.0000	.0000	.0000	.0000	.0000	.0001	.0002	.0009
	14	.0000	.0000	.0000	.0000	.0000	.0000	.0000	.0000	.0000	.0001
15	0	.4633	.2059	.0874	.0352	.0134	.0047	.0016	.0005	.0001	.0000
	1	.3658	.3432	.2312	.1319	.0668	.0305	.0126	.0047	.0016	.0005
	2	.1348	.2669	.2856	.2309	.1559	.0916	.0476	.0219	.0090	.0032
	3	.0307	.1285	.2184	.2501	.2252	.1700	.1110	.0634	.0318	.0139
	4	.0049	.0428	.1156	.1876	.2252	.2186	.1792	.1268	.0780	.0417
	5	.0006	.0105	.0449	.1032	.1651	.2061	.2123	.1859	.1404	.0916
	6	.0000	.0019	.0132	.0430	.0917	.1472	.1906	.2066	.1914	.1527
	7	.0000	.0003	.0030	.0138	.0393	.0811	.1319	.1771	.2013	.1964
	8	.0000	.0000	.0005	.0035	.0131	.0348	.0710	.1181	.1647	.1964
	9	.0000	.0000	.0001	.0007	.0034	.0116	.0298	.0612	.1048	.1527
	10	.0000	.0000	.0000	.0001	.0007	.0030	.0096	.0245	.0515	.0916
	11	.0000	.0000	.0000	.0000	.0001	.0006	.0024	.0074	.0191	.0417
	12	.0000	.0000	.0000	.0000	.0000	.0001	.0004	.0016	.0052	.0139
	13	.0000	.0000	.0000	.0000	.0000	.0000	.0001	.0003	.0010	.0032
	14	.0000	.0000	.0000	.0000	.0000	.0000	.0000	.0000	.0001	.0005
	15	.0000	.0000	.0000	.0000	.0000	.0000	.0000	.0000	.0000	.0000
16	0	.4401	.1853	.0743	.0281	.0100	.0033	.0010	.0003	.0001	.0000
	1	.3706	.3294	.2097	.1126	.0535	.0228	.0087	.0030	.0009	.0002
	2	.1463	.2745	.2775	.2111	.1336	.0732	.0353	.0150	.0056	.0018
	3	.0359	.1423	.2285	.2463	.2079	.1465	.0888	.0468	.0215	.0085
	4	.0061	.0514	.1311	.2001	.2252	.2040	.1553	.1014	.0572	.0278
	5	.0008	.0137	.0555	.1201	.1802	.2099	.2008	.1623	.1123	.0667
	6	.0001	.0028	.0180	.0550	.1101	.1649	.1982	.1983	.1684	.1222

Table I: *(continued)*

n	x	.05	.10	.15	.20	.25	.30	.35	.40	.45	.50
						θ					
16	7	.0000	.0004	.0045	.0197	.0524	.1010	.1524	.1889	.1969	.1746
	8	.0000	.0001	.0009	.0055	.0197	.0487	.0923	.1417	.1812	.1964
	9	.0000	.0000	.0001	.0012	.0058	.0185	.0442	.0840	.1318	.1746
	10	.0000	.0000	.0000	.0002	.0014	.0056	.0167	.0392	.0755	.1222
	11	.0000	.0000	.0000	.0000	.0002	.0013	.0049	.0142	.0337	.0667
	12	.0000	.0000	.0000	.0000	.0000	.0002	.0011	.0040	.0115	.0278
	13	.0000	.0000	.0000	.0000	.0000	.0000	.0002	.0008	.0029	.0085
	14	.0000	.0000	.0000	.0000	.0000	.0000	.0000	.0001	.0005	.0018
	15	.0000	.0000	.0000	.0000	.0000	.0000	.0000	.0000	.0001	.0002
	16	.0000	.0000	.0000	.0000	.0000	.0000	.0000	.0000	.0000	.0000
17	0	.4181	.1668	.0631	.0225	.0075	.0023	.0007	.0002	.0000	.0000
	1	.3741	.3150	.1893	.0957	.0426	.0169	.0060	.0019	.0005	.0001
	2	.1575	.2800	.2673	.1914	.1136	.0581	.0260	.0102	.0035	.0010
	3	.0415	.1556	.2359	.2393	.1893	.1245	.0701	.0341	.0144	.0052
	4	.0076	.0605	.1457	.2093	.2209	.1868	.1320	.0796	.0411	.0182
	5	.0010	.0175	.0668	.1361	.1914	.2081	.1849	.1379	.0875	.0472
	6	.0001	.0039	.0236	.0680	.1276	.1784	.1991	.1839	.1432	.0944
	7	.0000	.0007	.0065	.0267	.0668	.1201	.1685	.1927	.1841	.1484
	8	.0000	.0001	.0014	.0084	.0279	.0644	.1134	.1606	.1883	.1855
	9	.0000	.0000	.0003	.0021	.0093	.0276	.0611	.1070	.1540	.1855
	10	.0000	.0000	.0000	.0004	.0025	.0095	.0263	.0571	.1008	.1484
	11	.0000	.0000	.0000	.0001	.0005	.0026	.0090	.0242	.0525	.0944
	12	.0000	.0000	.0000	.0000	.0001	.0006	.0024	.0081	.0215	.0472
	13	.0000	.0000	.0000	.0000	.0000	.0001	.0005	.0021	.0068	.0182
	14	.0000	.0000	.0000	.0000	.0000	.0000	.0001	.0004	.0016	.0052
	15	.0000	.0000	.0000	.0000	.0000	.0000	.0000	.0001	.0003	.0010
	16	.0000	.0000	.0000	.0000	.0000	.0000	.0000	.0000	.0000	.0001
	17	.0000	.0000	.0000	.0000	.0000	.0000	.0000	.0000	.0000	.0000
18	0	.3972	.1501	.0536	.0180	.0056	.0016	.0004	.0001	.0000	.0000
	1	.3763	.3002	.1704	.0811	.0338	.0126	.0042	.0012	.0003	.0001
	2	.1683	.2835	.2556	.1723	.0958	.0458	.0190	.0069	.0022	.0006
	3	.0473	.1680	.2406	.2297	.1704	.1046	.0547	.0246	.0095	.0031
	4	.0093	.9700	.1592	.2153	.2130	.1681	.1104	.0614	.0291	.0117
	5	.0014	.0218	.0787	.1507	.1988	.2017	.1664	.1146	.0666	.0327
	6	.0002	.0052	.0301	.0816	.1436	.1873	.1941	.1655	.1181	.0708
	7	.0000	.0010	.0091	.0350	.0820	.1376	.1792	.1892	.1657	.1214
	8	.0000	.0002	.0022	.0120	.0376	.0811	.1327	.1734	.1864	.1669
	9	.0000	.0000	.0004	.0033	.0139	.0386	.0794	.1284	.1694	.1855
	10	.0000	.0000	.0001	.0008	.0042	.0149	.0385	.0771	.1248	.1669
	11	.0000	.0000	.0000	.0001	.0010	.0046	.0151	.0374	.0742	.1214
	12	.0000	.0000	.0000	.0000	.0002	.0012	.0047	.0145	.0354	.0708
	13	.0000	.0000	.0000	.0000	.0000	.0002	.0012	.0045	.0134	.0327
	14	.0000	.0000	.0000	.0000	.0000	.0000	.0002	.0011	.0039	.0117
	15	.0000	.0000	.0000	.0000	.0000	.0000	.0000	.0002	.0009	.0031
	16	.0000	.0000	.0000	.0000	.0000	.0000	.0000	.0000	.0001	.0006
	17	.0000	.0000	.0000	.0000	.0000	.0000	.0000	.0000	.0000	.0001
	18	.0000	.0000	.0000	.0000	.0000	.0000	.0000	.0000	.0000	.0000
19	0	.3774	.1351	.0456	.0144	.0042	.0011	.0003	.0001	.0000	.0000
	1	.3774	.2852	.1529	.0685	.0268	.0093	.0029	.0008	.0002	.0000

Table I: *(continued)*

n	x	.05	.10	.15	.20	.25	.30	.35	.40	.45	.50
		θ									
19	2	.1787	.2852	.2428	.1540	.0803	.0358	.0138	.0046	.0013	.0003
	3	.0533	.1796	.2428	.2182	.1517	.0869	.0422	.0175	.0062	.0018
	4	.0112	.0798	.1714	.2182	.2023	.1491	.0909	.0467	.0203	.0074
	5	.0018	.0266	.0907	.1636	.2023	.1916	.1468	.0933	.0497	.0222
	6	.0002	.0069	.0374	.0955	.1574	.1916	.1844	.1451	.0949	.0518
	7	.0000	.0014	.0122	.0443	.0974	.1525	.1844	.1797	.1443	.0961
	8	.0000	.0002	.0032	.0166	.0487	.0981	.1489	.1797	.1771	.1442
	9	.0000	.0000	.0007	.0051	.0198	.0514	.0980	.1464	.1771	.1762
	10	.0000	.0000	.0001	.0013	.0066	.0220	.0528	.0976	.1449	.1762
	11	.0000	.0000	.0000	.0003	.0018	.0077	.0233	.0532	.0970	.1442
	12	.0000	.0000	.0000	.0000	.0004	.0022	.0083	.0237	.0529	.0961
	13	.0000	.0000	.0000	.0000	.0001	.0005	.0024	.0085	.0233	.0518
	14	.0000	.0000	.0000	.0000	.0000	.0001	.0006	.0024	.0082	.0222
	15	.0000	.0000	.0000	.0000	.0000	.0000	.0001	.0005	.0022	.0074
	16	.0000	.0000	.0000	.0000	.0000	.0000	.0000	.0001	.0005	.0018
	17	.0000	.0000	.0000	.0000	.0000	.0000	.0000	.0000	.0001	.0003
	18	.0000	.0000	.0000	.0000	.0000	.0000	.0000	.0000	.0000	.0000
	19	.0000	.0000	.0000	.0000	.0000	.0000	.0000	.0000	.0000	.0000
20	0	.3585	.1216	.0388	.0115	.0032	.0008	.0002	.0000	.0000	.0000
	1	.3774	.2702	.1368	.0576	.0211	.0068	.0020	.0005	.0001	.0000
	2	.1887	.2852	.2293	.1369	.0669	.0278	.0100	.0031	.0008	.0002
	3	.0596	.1901	.2428	.2054	.1339	.0716	.0323	.0123	.0040	.0011
	4	.0133	.0898	.1821	.2182	.1897	.1304	.0738	.0350	.0139	.0046
	5	.0022	.0319	.1028	.1746	.2023	.1789	.1272	.0746	.0365	.0148
	6	.0003	.0089	.0454	.1091	.1686	.1916	.1712	.1244	.0746	.0370
	7	.0000	.0020	.0160	.0545	.1124	.1643	.1844	.1659	.1221	.0739
	8	.0000	.0004	.0046	.0222	.0609	.1144	.1614	.1797	.1623	.1201
	9	.0000	.0001	.0011	.0074	.0271	.0654	.1158	.1597	.1771	.1602
	10	.0000	.0000	.0002	.0020	.0099	.0308	.0686	.1171	.1593	.1762
	11	.0000	.0000	.0000	.0005	.0030	.0120	.0336	.0710	.1185	.1602
	12	.0000	.0000	.0000	.0001	.0008	.0039	.0136	.0355	.0727	.1201
	13	.0000	.0000	.0000	.0000	.0002	.0010	.0045	.0146	.0366	.0739
	14	.0000	.0000	.0000	.0000	.0000	.0002	.0012	.0049	.0150	.0370
	15	.0000	.0000	.0000	.0000	.0000	.0000	.0003	.0013	.0049	.0148
	16	.0000	.0000	.0000	.0000	.0000	.0000	.0000	.0003	.0013	.0046
	17	.0000	.0000	.0000	.0000	.0000	.0000	.0000	.0000	.0002	.0011
	18	.0000	.0000	.0000	.0000	.0000	.0000	.0000	.0000	.0000	.0002
	19	.0000	.0000	.0000	.0000	.0000	.0000	.0000	.0000	.0000	.0000
	20	.0000	.0000	.0000	.0000	.0000	.0000	.0000	.0000	.0000	.0000

Table II: Poisson Probabilities†

x	λ									
	0.1	0.2	0.3	0.4	0.5	0.6	0.7	0.8	0.9	1.0
0	.9048	.8187	.7408	.6703	.6065	.5488	.4966	.4493	.4066	.3679
1	.0905	.1637	.2222	.2681	.3033	.3293	.3476	.3595	.3659	.3679
2	.0045	.0164	.0333	.0536	.0758	.0988	.1217	.1438	.1647	.1839
3	.0002	.0011	.0033	.0072	.0126	.0198	.0284	.0383	.0494	.0613
4	.0000	.0001	.0002	.0007	.0016	.0030	.0050	.0077	.0111	.0153
5	.0000	.0000	.0000	.0001	.0002	.0004	.0007	.0012	.0020	.0031
6	.0000	.0000	.0000	.0000	.0000	.0000	.0001	.0002	.0003	.0005
7	.0000	.0000	.0000	.0000	.0000	.0000	.0000	.0000	.0000	.0001

x	λ									
	1.1	1.2	1.3	1.4	1.5	1.6	1.7	1.8	1.9	2.0
0	.3329	.3012	.2725	.2466	.2231	.2019	.1827	.1653	.1496	.1353
1	.3662	.3614	.3543	.3452	.3347	.3230	.3106	.2975	.2842	.2707
2	.2014	.2169	.2303	.2417	.2510	.2584	.2640	.2678	.2700	.2707
3	.0738	.0867	.0998	.1128	.1255	.1378	.1496	.1607	.1710	.1804
4	.0203	.0260	.0324	.0395	.0471	.0551	.0636	.0723	.0812	.0902
5	.0045	.0062	.0084	.0111	.0141	.0176	.0216	.0260	.0309	.0361
6	.0008	.0012	.0018	.0026	.0035	.0047	.0061	.0078	.0098	.0120
7	.0001	.0002	.0003	.0005	.0008	.0011	.0015	.0020	.0027	.0034
8	.0000	.0000	.0001	.0001	.0001	.0002	.0003	.0005	.0006	.0009
9	.0000	.0000	.0000	.0000	.0000	.0000	.0001	.0001	.0001	.0002

x	λ									
	2.1	2.2	2.3	2.4	2.5	2.6	2.7	2.8	2.9	3.0
0	.1225	.1108	.1003	.0907	.0821	.0743	.0672	.0608	.0550	.0498
1	.2572	.2438	.2306	.2177	.2052	.1931	.1815	.1703	.1596	.1494
2	.2700	.2681	.2652	.2613	.2565	.2510	.2450	.2384	.2314	.2240
3	.1890	.1966	.2033	.2090	.2138	.2176	.2205	.2225	.2237	.2240
4	.0992	.1082	.1169	.1254	.1336	.1414	.1488	.1557	.1622	.1680
5	.0417	.0476	.0538	.0602	.0668	.0735	.0804	.0872	.0940	.1008
6	.0146	.0174	.0206	.0241	.0278	.0319	.0362	.0407	.0455	.0504
7	.0044	.0055	.0068	.0083	.0099	.0118	.0139	.0163	.0188	.0216
8	.0011	.0015	.0019	.0025	.0031	.0038	.0047	.0057	.0068	.0081
9	.0003	.0004	.0005	.0007	.0009	.0011	.0014	.0018	.0022	.0027
10	.0001	.0001	.0001	.0002	.0002	.0003	.0004	.0005	.0006	.0008
11	.0000	.0000	.0000	.0000	.0000	.0001	.0001	.0001	.0002	.0002
12	.0000	.0000	.0000	.0000	.0000	.0000	.0000	.0000	.0000	.0001

x	λ									
	3.1	3.2	3.3	3.4	3.5	3.6	3.7	3.8	3.9	4.0
0	.0450	.0408	.0369	.0334	.0302	.0273	.0247	.0224	.0202	.0183
1	.1397	.1304	.1217	.1135	.1057	.0984	.0915	.0850	.0789	.0733
2	.2165	.2087	.2008	.1929	.1850	.1771	.1692	.1615	.1539	.1465
3	.2237	.2226	.2209	.2186	.2158	.2125	.2087	.2046	.2001	.1954
4	.1734	.1781	.1823	.1858	.1888	.1912	.1931	.1944	.1951	.1954
5	.1075	.1140	.1203	.1264	.1322	.1377	.1429	.1477	.1522	.1563
6	.0555	.0608	.0662	.0716	.0771	.0826	.0881	.0936	.0989	.1042
7	.0246	.0278	.0312	.0348	.0385	.0425	.0466	.0508	.0551	.0595
8	.0095	.0111	.0129	.0148	.0169	.0191	.0215	.0241	.0269	.0298
9	.0033	.0040	.0047	.0056	.0066	.0076	.0089	.0102	.0116	.0132

†Based on E. C. Molina, *Poisson's Exponential Binomial Limit*, 1973 Reprint, Robert E. Krieger Publishing Company, Melbourne, Fla., by permission of the publisher.

Table II: *(continued)*

x	λ = 3.1	3.2	3.3	3.4	3.5	3.6	3.7	3.8	3.9	4.0
10	.0010	.0013	.0016	.0019	.0023	.0028	.0033	.0039	.0045	.0053
11	.0003	.0004	.0005	.0006	.0007	.0009	.0011	.0013	.0016	.0019
12	.0001	.0001	.0001	.0002	.0002	.0003	.0003	.0004	.0005	.0006
13	.0000	.0000	.0000	.0000	.0001	.0001	.0001	.0001	.0002	.0002
14	.0000	.0000	.0000	.0000	.0000	.0000	.0000	.0000	.0000	.0001

x	λ = 4.1	4.2	4.3	4.4	4.5	4.6	4.7	4.8	4.9	5.0
0	.0166	.0150	.0136	.0123	.0111	.0101	.0091	.0082	.0074	.0067
1	.0679	.0630	.0583	.0540	.0500	.0462	.0427	.0395	.0365	.0337
2	.1393	.1323	.1254	.1188	.1125	.1063	.1005	.0948	.0894	.0842
3	.1904	.1852	.1798	.1743	.1687	.1631	.1574	.1517	.1460	.1404
4	.1951	.1944	.1933	.1917	.1898	.1875	.1849	.1820	.1789	.1755
5	.1600	.1633	.1662	.1687	.1708	.1725	.1738	.1747	.1753	.1755
6	.1093	.1143	.1191	.1237	.1281	.1323	.1362	.1398	.1432	.1462
7	.0640	.0686	.0732	.0778	.0824	.0869	.0914	.0959	.1002	.1044
8	.0328	.0360	.0393	.0428	.0463	.0500	.0537	.0575	.0614	.0653
9	.0150	.0168	.0188	.0209	.0232	.0255	.0280	.0307	.0334	.0363
10	.0061	.0071	.0081	.0092	.0104	.0118	.0132	.0147	.0164	.0181
11	.0023	.0027	.0032	.0037	.0043	.0049	.0056	.0064	.0073	.0082
12	.0008	.0009	.0011	.0014	.0016	.0019	.0022	.0026	.0030	.0034
13	.0002	.0003	.0004	.0005	.0006	.0007	.0008	.0009	.0011	.0013
14	.0001	.0001	.0001	.0001	.0002	.0002	.0003	.0003	.0004	.0005
15	.0000	.0000	.0000	.0000	.0001	.0001	.0001	.0001	.0001	.0002

x	λ = 5.1	5.2	5.3	5.4	5.5	5.6	5.7	5.8	5.9	6.0
0	.0061	.0055	.0050	.0045	.0041	.0037	.0033	.0030	.0027	.0025
1	.0311	.0287	.0265	.0244	.0225	.0207	.0191	.0176	.0162	.0149
2	.0793	.0746	.0701	.0659	.0618	.0580	.0544	.0509	.0477	.0446
3	.1348	.1293	.1239	.1185	.1133	.1082	.1033	.0985	.0938	.0892
4	.1719	.1681	.1641	.1600	.1558	.1515	.1472	.1428	.1383	.1339
5	.1753	.1748	.1740	.1728	.1714	.1697	.1678	.1656	.1632	.1606
6	.1490	.1515	.1537	.1555	.1571	.1584	.1594	.1601	.1505	.1606
7	.1086	.1125	.1163	.1200	.1234	.1267	.1298	.1326	.1353	.1377
8	.0692	.0731	.0771	.0810	.0849	.0887	.0925	.0962	.0998	.1033
9	.0392	.0423	.0454	.0486	.0519	.0552	.0586	.0620	.0654	.0688
10	.0200	.0220	.0241	.0262	.0285	.0309	.0334	.0359	.0386	.0413
11	.0093	.0104	.0116	.0129	.0143	.0157	.0173	.0190	.0207	.0225
12	.0039	.0045	.0051	.0058	.0065	.0073	.0082	.0092	.0102	.0113
13	.0015	.0018	.0021	.0024	.0028	.0032	.0036	.0041	.0046	.0052
14	.0006	.0007	.0008	.0009	.0011	.0013	.0015	.0017	.0019	.0022
15	.0002	.0002	.0003	.0003	.0004	.0005	.0006	.0007	.0008	.0009
16	.0001	.0001	.0001	.0001	.0001	.0002	.0002	.0002	.0003	.0003
17	.0000	.0000	.0000	.0000	.0000	.0001	.0001	.0001	.0001	.0001

x	λ = 6.1	6.2	6.3	6.4	6.5	6.6	6.7	6.8	6.9	7.0
0	.0022	.0020	.0018	.0017	.0015	.0014	.0012	.0011	.0010	.0009
1	.0137	.0126	.0116	.0106	.0098	.0090	.0082	.0076	.0070	.0064
2	.0417	.0390	.0364	.0340	.0318	.0296	.0276	.0258	.0240	.0223

Table II: *(continued)*

x	λ 6.1	6.2	6.3	6.4	6.5	6.6	6.7	6.8	6.9	7.0
3	.0848	.0806	.0765	.0726	.0688	.0652	.0617	.0584	.0552	.0521
4	.1294	.1249	.1205	.1162	.1118	.1076	.1034	.0992	.0952	.0912
5	.1579	.1549	.1519	.1487	.1454	.1420	.1385	.1349	.1314	.1277
6	.1605	.1601	.1595	.1586	.1575	.1562	.1546	.1529	.1511	.1490
7	.1399	.1418	.1435	.1450	.1462	.1472	.1480	.1486	.1489	.1490
8	.1066	.1099	.1130	.1160	.1188	.1215	.1240	.1263	.1284	.1304
9	.0723	.0757	.0791	.0825	.0858	.0891	.0923	.0954	.0985	.1014
10	.0441	.0469	.0498	.0528	.0558	.0588	.0618	.0649	.0679	.0710
11	.0245	.0265	.0285	.0307	.0330	.0353	.0377	.0401	.0426	.0452
12	.0124	.0137	.0150	.0164	.0179	.0194	.0210	.0227	.0245	.0264
13	.0058	.0065	.0073	.0081	.0089	.0098	.0108	.0119	.0130	.0142
14	.0025	.0029	.0033	.0037	.0041	.0046	.0052	.0058	.0064	.0071
15	.0010	.0012	.0014	.0016	.0018	.0020	.0023	.0026	.0029	.0033
16	.0004	.0005	.0005	.0006	.0007	.0008	.0010	.0011	.0013	.0014
17	.0001	.0002	.0002	.0002	.0003	.0003	.0004	.0004	.0005	.0006
18	.0000	.0001	.0001	.0001	.0001	.0001	.0001	.0002	.0002	.0002
19	.0000	.0000	.0000	.0000	.0000	.0000	.0000	.0001	.0001	.0001

x	λ 7.1	7.2	7.3	7.4	7.5	7.6	7.7	7.8	7.9	8.0
0	.0008	.0007	.0007	.0006	.0006	.0005	.0005	.0004	.0004	.0003
1	.0059	.0054	.0049	.0045	.0041	.0038	.0035	.0032	.0029	.0027
2	.0208	.0194	.0180	.0167	.0156	.0145	.0134	.0125	.0116	.0107
3	.0492	.0464	.0438	.0413	.0389	.0366	.0345	.0324	.0305	.0286
4	.0874	.0836	.0799	.0764	.0729	.0696	.0663	.0632	.0602	.0573
5	.1241	.1204	.1167	.1130	.1094	.1057	.1021	.0986	.0951	.0916
6	.1468	.1445	.1420	.1394	.1367	.1339	.1311	.1282	.1252	.1221
7	.1489	.1486	.1481	.1474	.1465	.1454	.1442	.1428	.1413	.1396
8	.1321	.1337	.1351	.1363	.1373	.1382	.1388	.1392	.1395	.1396
9	.1042	.1070	.1096	.1121	.1144	.1167	.1187	.1207	.1224	.1241
10	.0740	.0770	.0800	.0829	.0858	.0887	.0914	.0941	.0967	.0993
11	.0478	.0504	.0531	.0558	.0585	.0613	.0640	.0667	.0695	.0722
12	.0283	.0303	.0323	.0344	.0366	.0388	.0411	.0434	.0457	.0481
13	.0154	.0168	.0181	.0196	.0211	.0227	.0243	.0260	.0278	.0296
14	.0078	.0086	.0095	.0104	.0113	.0123	.0134	.0145	.0157	.0169
15	.0037	.0041	.0046	.0051	.0057	.0062	.0069	.0075	.0083	.0090
16	.0016	.0019	.0021	.0024	.0026	.0030	.0033	.0037	.0041	.0045
17	.0007	.0008	.0009	.0010	.0012	.0013	.0015	.0017	.0019	.0021
18	.0003	.0003	.0004	.0004	.0005	.0006	.0006	.0007	.0008	.0009
19	.0001	.0001	.0001	.0002	.0002	.0002	.0003	.0003	.0003	.0004
20	.0000	.0000	.0001	.0001	.0001	.0001	.0001	.0001	.0001	.0002
21	.0000	.0000	.0000	.0000	.0000	.0000	.0000	.0000	.0001	.0001

x	λ 8.1	8.2	8.3	8.4	8.5	8.6	8.7	8.8	8.9	9.0
0	.0003	.0003	.0002	.0002	.0002	.0002	.0002	.0002	.0001	.0001
1	.0025	.0023	.0021	.0019	.0017	.0016	.0014	.0013	.0012	.0011
2	.0100	.0092	.0086	.0079	.0074	.0068	.0063	.0058	.0054	.0050
3	.0269	.0252	.0237	.0222	.0208	.0195	.0183	.0171	.0160	.0150
4	.0544	.0517	.0491	.0466	.0443	.0420	.0398	.0377	.0357	.0337

Table II: *(continued)*

x	λ 8.1	8.2	8.3	8.4	8.5	8.6	8.7	8.8	8.9	9.0
5	.0882	.0849	.0816	.0784	.0752	.0722	.0692	.0663	.0635	.0607
6	.1191	.1160	.1128	.1097	.1066	.1034	.1003	.0972	.0941	.0911
7	.1378	.1358	.1338	.1317	.1294	.1271	.1247	.1222	.1197	.1171
8	.1395	.1392	.1388	.1382	.1375	.1366	.1356	.1344	.1332	.1318
9	.1256	.1269	.1280	.1290	.1299	.1306	.1311	.1315	.1317	.1318
10	.1017	.1040	.1063	.1084	.1104	.1123	.1140	.1157	.1172	.1186
11	.0749	.0776	.0802	.0828	.0853	.0878	.0902	.0925	.0948	.0970
12	.0505	.0530	.0555	.0579	.0604	.0629	.0654	.0679	.0703	.0728
13	.0315	.0334	.0354	.0374	.0395	.0416	.0438	.0459	.0481	.0504
14	.0182	.0196	.0210	.0225	.0240	.0256	.0272	.0289	.0306	.0324
15	.0098	.0107	.0116	.0126	.0136	.0147	.0158	.0169	.0182	.0194
16	.0050	.0055	.0060	.0066	.0072	.0079	.0086	.0093	.0101	.0109
17	.0024	.0026	.0029	.0033	.0036	.0040	.0044	.0048	.0053	.0058
18	.0011	.0012	.0014	.0015	.0017	.0019	.0021	.0024	.0026	.0029
19	.0005	.0005	.0006	.0007	.0008	.0009	.0010	.0011	.0012	.0014
20	.0002	.0002	.0002	.0003	.0003	.0004	.0004	.0005	.0005	.0006
21	.0001	.0001	.0001	.0001	.0001	.0002	.0002	.0002	.0002	.0003
22	.0000	.0000	.0000	.0000	.0001	.0001	.0001	.0001	.0001	.0001

x	λ 9.1	9.2	9.3	9.4	9.5	9.6	9.7	9.8	9.9	10
0	.0001	.0001	.0001	.0001	.0001	.0001	.0001	.0001	.0001	.0000
1	.0010	.0009	.0009	.0008	.0007	.0007	.0006	.0005	.0005	.0005
2	.0046	.0043	.0040	.0037	.0034	.0031	.0029	.0027	.0025	.0023
3	.0140	.0131	.0123	.0115	.0107	.0100	.0093	.0087	.0081	.0076
4	.0319	.0302	.0285	.0269	.0254	.0240	.0226	.0213	.0201	.0189
5	.0581	.0555	.0530	.0506	.0483	.0460	.0439	.0418	.0398	.0378
6	.0881	.0851	.0822	.0793	.0764	.0736	.0709	.0682	.0656	.0631
7	.1145	.1118	.1091	.1064	.1037	.1010	.0982	.0955	.0928	.0901
8	.1302	.1286	.1269	.1251	.1232	.1212	.1191	.1170	.1148	.1126
9	.1317	.1315	.1311	.1306	.1300	.1293	.1284	.1274	.1263	.1251
10	.1198	.1210	.1219	.1228	.1235	.1241	.1245	.1249	.1250	.1251
11	.0991	.1012	.1031	.1049	.1067	.1083	.1098	.1112	.1125	.1137
12	.0752	.0776	.0799	.0822	.0844	.0866	.0888	.0908	.0928	.0948
13	.0526	.0549	.0572	.0594	.0617	.0640	.0662	.0685	.0707	.0729
14	.0342	.0361	.0380	.0399	.0419	.0439	.0459	.0479	.0500	.0521
15	.0208	.0221	.0235	.0250	.0265	.0281	.0297	.0313	.0330	.0347
16	.0118	.0127	.0137	.0147	.0157	.0168	.0180	.0192	.0204	.0217
17	.0063	.0069	.0075	.0081	.0088	.0095	.0103	.0111	.0119	.0128
18	.0032	.0035	.0039	.0042	.0046	.0051	.0055	.0060	.0065	.0071
19	.0015	.0017	.0019	.0021	.0023	.0026	.0028	.0031	.0034	.0037
20	.0007	.0008	.0009	.0010	.0011	.0012	.0014	.0015	.0017	.0019
21	.0003	.0003	.0004	.0004	.0005	.0006	.0006	.0007	.0008	.0009
22	.0001	.0001	.0002	.0002	.0002	.0002	.0003	.0003	.0004	.0004
23	.0000	.0001	.0001	.0001	.0001	.0001	.0001	.0001	.0002	.0002
24	.0000	.0000	.0000	.0000	.0000	.0000	.0000	.0001	.0001	.0001

x	λ 11	12	13	14	15	16	17	18	19	20
0	.0000	.0000	.0000	.0000	.0000	.0000	.0000	.0000	.0000	.0000
1	.0002	.0001	.0000	.0000	.0000	.0000	.0000	.0000	.0000	.0000

Table II: *(continued)*

x	λ = 11	12	13	14	15	16	17	18	19	20
2	.0010	.0004	.0002	.0001	.0000	.0000	.0000	.0000	.0000	.0000
3	.0037	.0018	.0008	.0004	.0002	.0001	.0000	.0000	.0000	.0000
4	.0102	.0053	.0027	.0013	.0006	.0003	.0001	.0001	.0000	.0000
5	.0224	.0127	.0070	.0037	.0019	.0010	.0005	.0002	.0001	.0001
6	.0411	.0255	.0152	.0087	.0048	.0026	.0014	.0007	.0004	.0002
7	.0646	.0437	.0281	.0174	.0104	.0060	.0034	.0018	.0010	.0005
8	.0888	.0655	.0457	.0304	.0194	.0120	.0072	.0042	.0024	.0013
9	.1085	.0874	.0661	.0473	.0324	.0213	.0135	.0083	.0050	.0029
10	.1194	.1048	.0859	.0663	.0486	.0341	.0230	.0150	.0095	.0058
11	.1194	.1144	.1015	.0844	.0663	.0496	.0355	.0245	.0164	.0106
12	.1094	.1144	.1099	.0984	.0829	.0661	.0504	.0368	.0259	.0176
13	.0926	.1056	.1099	.1060	.0956	.0814	.0658	.0509	.0378	.0271
14	.0728	.0905	.1021	.1060	.1024	.0930	.0800	.0655	.0514	.0387
15	.0534	.0724	.0885	.0989	.1024	.0992	.0906	.0786	.0650	.0516
16	.0367	.0543	.0719	.0866	.0960	.0992	.0963	.0884	.0772	.0646
17	.0237	.0383	.0550	.0713	.0847	.0934	.0963	.0936	.0863	.0760
18	.0145	.0256	.0397	.0554	.0706	.0830	.0909	.0936	.0911	.0844
19	.0084	.0161	.0272	.0409	.0557	.0699	.0814	.0887	.0911	.0888
20	.0046	.0097	.0177	.0286	.0418	.0559	.0692	.0798	.0866	.0888
21	.0024	.0055	.0109	.0191	.0299	.0426	.0560	.0684	.0783	.0846
22	.0012	.0030	.0065	.0121	.0204	.0310	.0433	.0560	.0676	.0769
23	.0006	.0016	.0037	.0074	.0133	.0216	.0320	.0438	.0559	.0669
24	.0003	.0008	.0020	.0043	.0083	.0144	.0226	.0328	.0442	.0557
25	.0001	.0004	.0010	.0024	.0050	.0092	.0154	.0237	.0336	.0446
26	.0000	.0002	.0005	.0013	.0029	.0057	.0101	.0164	.0246	.0343
27	.0000	.0001	.0002	.0007	.0016	.0034	.0063	.0109	.0173	.0254
28	.0000	.0000	.0001	.0003	.0009	.0019	.0038	.0070	.0117	.0181
29	.0000	.0000	.0001	.0002	.0004	.0011	.0023	.0044	.0077	.0125
30	.0000	.0000	.0000	.0001	.0002	.0006	.0013	.0026	.0049	.0083
31	.0000	.0000	.0000	.0000	.0001	.0003	.0007	.0015	.0030	.0054
32	.0000	.0000	.0000	.0000	.0001	.0001	.0004	.0009	.0018	.0034
33	.0000	.0000	.0000	.0000	.0000	.0001	.0002	.0005	.0010	.0020
34	.0000	.0000	.0000	.0000	.0000	.0000	.0001	.0002	.0006	.0012
35	.0000	.0000	.0000	.0000	.0000	.0000	.0000	.0001	.0003	.0007
36	.0000	.0000	.0000	.0000	.0000	.0000	.0000	.0001	.0002	.0004
37	.0000	.0000	.0000	.0000	.0000	.0000	.0000	.0000	.0001	.0002
38	.0000	.0000	.0000	.0000	.0000	.0000	.0000	.0000	.0000	.0001
39	.0000	.0000	.0000	.0000	.0000	.0000	.0000	.0000	.0000	.0001

Table III: Standard Normal Distribution

z	.00	.01	.02	.03	.04	.05	.06	.07	.08	.09
0.0	.0000	.0040	.0080	.0120	.0160	.0199	.0239	.0279	.0319	.0359
0.1	.0398	.0438	.0478	.0517	.0557	.0596	.0636	.0675	.0714	.0753
0.2	.0793	.0832	.0871	.0910	.0948	.0987	.1026	.1064	.1103	.1141
0.3	.1179	.1217	.1255	.1293	.1331	.1368	.1406	.1443	.1480	.1517
0.4	.1554	.1591	.1628	.1664	.1700	.1736	.1772	.1808	.1844	.1879
0.5	.1915	.1950	.1985	.2019	.2054	.2088	.2123	.2157	.2190	.2224
0.6	.2257	.2291	.2324	.2357	.2389	.2422	.2454	.2486	.2517	.2549
0.7	.2580	.2611	.2642	.2673	.2704	.2734	.2764	.2794	.2823	.2852
0.8	.2881	.2910	.2939	.2967	.2995	.3023	.3051	.3078	.3106	.3133
0.9	.3159	.3186	.3212	.3238	.3264	.3289	.3315	.3340	.3365	.3389
1.0	.3413	.3438	.3461	.3485	.3508	.3531	.3554	.3577	.3599	.3621
1.1	.3643	.3665	.3686	.3708	.3729	.3749	.3770	.3790	.3810	.3830
1.2	.3849	.3869	.3888	.3907	.3925	.3944	.3962	.3980	.3997	.4015
1.3	.4032	.4049	.4066	.4082	.4099	.4115	.4131	.4147	.4162	.4177
1.4	.4192	.4207	.4222	.4236	.4251	.4265	.4279	.4292	.4306	.4319
1.5	.4332	.4345	.4357	.4370	.4382	.4394	.4406	.4418	.4429	.4441
1.6	.4452	.4463	.4474	.4484	.4495	.4505	.4515	.4525	.4535	.4545
1.7	.4554	.4564	.4573	.4582	.4591	.4599	.4608	.4616	.4625	.4633
1.8	.4641	.4649	.4656	.4664	.4671	.4678	.4686	.4693	.4699	.4706
1.9	.4713	.4719	.4726	.4732	.4738	.4744	.4750	.4756	.4761	.4767
2.0	.4772	.4778	.4783	.4788	.4793	.4798	.4803	.4808	.4812	.4817
2.1	.4821	.4826	.4830	.4834	.4838	.4842	.4846	.4850	.4854	.4857
2.2	.4861	.4864	.4868	.4871	.4875	.4878	.4881	.4884	.4887	.4890
2.3	.4893	.4896	.4898	.4901	.4904	.4906	.4909	.4911	.4913	.4916
2.4	.4918	.4920	.4922	.4925	.4927	.4929	.4931	.4932	.4934	.4936
2.5	.4938	.4940	.4941	.4943	.4945	.4946	.4948	.4949	.4951	.4952
2.6	.4953	.4955	.4956	.4957	.4959	.4960	.4961	.4962	.4963	.4964
2.7	.4965	.4966	.4967	.4968	.4969	.4970	.4971	.4972	.4973	.4974
2.8	.4974	.4975	.4976	.4977	.4977	.4978	.4979	.4979	.4980	.4981
2.9	.4981	.4982	.4982	.4983	.4984	.4984	.4985	.4985	.4986	.4988
3.0	.4987	.4987	.4987	.4988	.4988	.4989	.4989	.4989	.4990	.4990

Also, for $z = 4.0$, 5.0, and 6.0, the probabilities are 0.49997, 0.4999997, and 0.499999999.

Table IV: Values of $t_{\alpha,\nu}$†

ν	$\alpha = .10$	$\alpha = .05$	$\alpha = .025$	$\alpha = .01$	$\alpha = .005$	ν
1	3.078	6.314	12.706	31.821	63.657	1
2	1.886	2.920	4.303	6.965	9.925	2
3	1.638	2.353	3.182	4.541	5.841	3
4	1.533	2.132	2.776	3.747	4.604	4
5	1.476	2.015	2.571	3.365	4.032	5
6	1.440	1.943	2.447	3.143	3.707	6
7	1.415	1.895	2.365	2.998	3.499	7
8	1.397	1.860	2.306	2.896	3.355	8
9	1.383	1.833	2.262	2.821	3.250	9
10	1.372	1.812	2.228	2.764	3.169	10
11	1.363	1.796	2.201	2.718	3.106	11
12	1.356	1.782	2.179	2.681	3.055	12
13	1.350	1.771	2.160	2.650	3.012	13
14	1.345	1.761	2.145	2.624	2.977	14
15	1.341	1.753	2.131	2.602	2.947	15
16	1.337	1.746	2.120	2.583	2.921	16
17	1.333	1.740	2.110	2.567	2.898	17
18	1.330	1.734	2.101	2.552	2.878	18
19	1.328	1.729	2.093	2.539	2.861	19
20	1.325	1.725	2.086	2.528	2.845	20
21	1.323	1.721	2.080	2.518	2.831	21
22	1.321	1.717	2.074	2.508	2.819	22
23	1.319	1.714	2.069	2.500	2.807	23
24	1.318	1.711	2.064	2.492	2.797	24
25	1.316	1.708	2.060	2.485	2.787	25
26	1.315	1.706	2.056	2.479	2.779	26
27	1.314	1.703	2.052	2.473	2.771	27
28	1.313	1.701	2.048	2.467	2.763	28
29	1.311	1.699	2.045	2.462	2.756	29
inf.	1.282	1.645	1.960	2.326	2.576	inf.

†Based on Richard A. Johnson and Dean W. Wichern, *Applied Multivariate Statistical Analysis*, 2nd ed., © 1988, Table 2, p. 592. By permission of Prentice Hall, Upper Saddle River, N.J.

Table V: Values of $\chi^2_{\alpha,\nu}$ †

ν	$\alpha = .995$	$\alpha = .99$	$\alpha = .975$	$\alpha = .95$	$\alpha = .05$	$\alpha = .025$	$\alpha = .01$	$\alpha = .005$	ν
1	.0000393	.000157	.000982	.00393	3.841	5.024	6.635	7.879	1
2	.0100	.0201	.0506	.103	5.991	7.378	9.210	10.597	2
3	.0717	.115	.216	.352	7.815	9.348	11.345	12.838	3
4	.207	.297	.484	.711	9.488	11.143	13.277	14.860	4
5	.412	.554	.831	1.145	11.070	12.832	15.086	16.750	5
6	.676	.872	1.237	1.635	12.592	14.449	16.812	18.548	6
7	.989	1.239	1.690	2.167	14.067	16.013	18.475	20.278	7
8	1.344	1.646	2.180	2.733	15.507	17.535	20.090	21.955	8
9	1.735	2.088	2.700	3.325	16.919	19.023	21.666	23.589	9
10	2.156	2.558	3.247	3.940	18.307	20.483	23.209	25.188	10
11	2.603	3.053	3.816	4.575	19.675	21.920	24.725	26.757	11
12	3.074	3.571	4.404	5.226	21.026	23.337	26.217	28.300	12
13	3.565	4.107	5.009	5.892	22.362	24.736	27.688	29.819	13
14	4.075	4.660	5.629	6.571	23.685	26.119	29.141	31.319	14
15	4.601	5.229	6.262	7.261	24.996	27.488	30.578	32.801	15
16	5.142	5.812	6.908	7.962	26.296	28.845	32.000	34.267	16
17	5.697	6.408	7.564	8.672	27.587	30.191	33.409	35.718	17
18	6.265	7.015	8.231	9.390	28.869	31.526	34.805	37.156	18
19	6.844	7.633	8.907	10.117	30.144	32.852	36.191	38.582	19
20	7.434	8.260	9.591	10.851	31.410	34.170	37.566	39.997	20
21	8.034	8.897	10.283	11.591	32.671	35.479	38.932	41.401	21
22	8.643	9.542	10.982	12.338	33.924	36.781	40.289	42.796	22
23	9.260	10.196	11.689	13.091	35.172	38.076	41.638	44.181	23
24	9.886	10.856	12.401	13.848	36.415	39.364	42.980	45.558	24
25	10.520	11.524	13.120	14.611	37.652	40.646	44.314	46.928	25
26	11.160	12.198	13.844	15.379	38.885	41.923	45.642	48.290	26
27	11.808	12.879	14.573	16.151	40.113	43.194	46.963	49.645	27
28	12.461	13.565	15.308	16.928	41.337	44.461	48.278	50.993	28
29	13.121	14.256	16.047	17.708	42.557	45.722	49.588	52.336	29
30	13.787	14.953	16.791	18.493	43.773	46.979	50.892	53.672	30

† Based on Table 8 of *Biometrika Tables for Statisticians*, Vol. 1, Cambridge University Press, 1954, by permission of the *Biometrika* trustees.

Table VI: Values of $f_{0.05,\nu_1,\nu_2}$ †

ν_2 = Degrees of freedom for denominator	ν_1 = Degrees of freedom for numerator																		
	1	2	3	4	5	6	7	8	9	10	12	15	20	24	30	40	60	120	∞
1	161	200	216	225	230	234	237	239	241	242	244	246	248	249	250	251	252	253	254
2	18.5	19.0	19.2	19.2	19.3	19.3	19.4	19.4	19.4	19.4	19.4	19.4	19.4	19.5	19.5	19.5	19.5	19.5	19.5
3	10.1	9.55	9.28	9.12	9.01	8.94	8.89	8.85	8.81	8.79	8.74	8.70	8.66	8.64	8.62	8.59	8.57	8.55	8.53
4	7.71	6.94	6.59	6.39	6.26	6.16	6.09	6.04	6.00	5.96	5.91	5.86	5.80	5.77	5.75	5.72	5.69	5.66	5.63
5	6.61	5.79	5.41	5.19	5.05	4.95	4.88	4.82	4.77	4.74	4.68	4.62	4.56	4.53	4.50	4.46	4.43	4.40	4.37
6	5.99	5.14	4.76	4.53	4.39	4.28	4.21	4.15	4.10	4.06	4.00	3.94	3.87	3.84	3.81	3.77	3.74	3.70	3.67
7	5.59	4.74	4.35	4.12	3.97	3.87	3.79	3.73	3.68	3.64	3.57	3.51	3.44	3.41	3.38	3.34	3.30	3.27	3.23
8	5.32	4.46	4.07	3.84	3.69	3.58	3.50	3.44	3.39	3.35	3.28	3.22	3.15	3.12	3.08	3.04	3.01	2.97	2.93
9	5.12	4.26	3.86	3.63	3.48	3.37	3.29	3.23	3.18	3.14	3.07	3.01	2.94	2.90	2.86	2.83	2.79	2.75	2.71
10	4.96	4.10	3.71	3.48	3.33	3.22	3.14	3.07	3.02	2.98	2.91	2.85	2.77	2.74	2.70	2.66	2.62	2.58	2.54
11	4.84	3.98	3.59	3.36	3.20	3.09	3.01	2.95	2.90	2.85	2.79	2.72	2.65	2.61	2.57	2.53	2.49	2.45	2.40
12	4.75	3.89	3.49	3.26	3.11	3.00	2.91	2.85	2.80	2.75	2.69	2.62	2.54	2.51	2.47	2.43	2.38	2.34	2.30
13	4.67	3.81	3.41	3.18	3.03	2.92	2.83	2.77	2.71	2.67	2.60	2.53	2.46	2.42	2.38	2.34	2.30	2.25	2.21
14	4.60	3.74	3.34	3.11	2.96	2.85	2.76	2.70	2.65	2.60	2.53	2.46	2.39	2.35	2.31	2.27	2.22	2.18	2.13
15	4.54	3.68	3.29	3.06	2.90	2.79	2.71	2.64	2.59	2.54	2.48	2.40	2.33	2.29	2.25	2.20	2.16	2.11	2.07

†Reproduced from M. Merrington and C. M. Thompson, "Tables of percentage points of the inverted beta (F) distribution," *Biometrika*, Vol. 33 (1943), by permission of the *Biometrika* trustees.

Table VI: *(continued)* Values of $f_{0.05,\nu_1,\nu_2}$

ν_2 = Degrees of freedom for denominator	ν_1 = Degrees of freedom for numerator																		
	1	2	3	4	5	6	7	8	9	10	12	15	20	24	30	40	60	120	∞
16	4.49	3.63	3.24	3.01	2.85	2.74	2.66	2.59	2.54	2.49	2.42	2.35	2.28	2.24	2.19	2.15	2.11	2.06	2.01
17	4.45	3.59	3.20	2.96	2.81	2.70	2.61	2.55	2.49	2.45	2.38	2.31	2.23	2.19	2.15	2.10	2.06	2.01	1.96
18	4.41	3.55	3.16	2.93	2.77	2.66	2.58	2.51	2.46	2.41	2.34	2.27	2.19	2.15	2.11	2.06	2.02	1.97	1.92
19	4.38	3.52	3.13	2.90	2.74	2.63	2.54	2.48	2.42	2.38	2.31	2.23	2.16	2.11	2.07	2.03	1.98	1.93	1.88
20	4.35	3.49	3.10	2.87	2.71	2.60	2.51	2.45	2.39	2.35	2.28	2.20	2.12	2.08	2.04	1.99	1.95	1.90	1.84
21	4.32	3.47	3.07	2.84	2.68	2.57	2.49	2.42	2.37	2.32	2.25	2.18	2.10	2.05	2.01	1.96	1.92	1.87	1.81
22	4.30	3.44	3.05	2.82	2.66	2.55	2.46	2.40	2.34	2.30	2.23	2.15	2.07	2.03	1.98	1.94	1.89	1.84	1.78
23	4.28	3.42	3.03	2.80	2.64	2.53	2.44	2.37	2.32	2.27	2.20	2.13	2.05	2.01	1.96	1.91	1.86	1.81	1.76
24	4.26	3.40	3.01	2.78	2.62	2.51	2.42	2.36	2.30	2.25	2.18	2.11	2.03	1.98	1.94	1.89	1.84	1.79	1.73
25	4.24	3.39	2.99	2.76	2.60	2.49	2.40	2.34	2.28	2.24	2.16	2.09	2.01	1.96	1.92	1.87	1.82	1.77	1.71
30	4.17	3.32	2.92	2.69	2.53	2.42	2.33	2.27	2.21	2.16	2.09	2.01	1.93	1.89	1.84	1.79	1.74	1.68	1.62
40	4.08	3.23	2.84	2.61	2.45	2.34	2.25	2.18	2.12	2.08	2.00	1.92	1.84	1.79	1.74	1.69	1.64	1.58	1.51
60	4.00	3.15	2.76	2.53	2.37	2.25	2.17	2.10	2.04	1.99	1.92	1.84	1.75	1.70	1.65	1.59	1.53	1.47	1.39
120	3.92	3.07	2.68	2.45	2.29	2.18	2.09	2.02	1.96	1.91	1.83	1.75	1.66	1.61	1.55	1.50	1.43	1.35	1.25
∞	3.84	3.00	2.60	2.37	2.21	2.10	2.01	1.94	1.88	1.83	1.75	1.67	1.57	1.52	1.46	1.39	1.32	1.22	1.00

Table VI: (*continued*) Values of $f_{0.01,\nu_1,\nu_2}$

ν_2 = Degrees of freedom for denominator	ν_1 = Degrees of freedom for numerator																		
	1	2	3	4	5	6	7	8	9	10	12	15	20	24	30	40	60	120	∞
1	4,052	5,000	5,403	5,625	5,764	5,859	5,928	5,982	6,023	6,056	6,106	6,157	6,209	6,235	6,261	6,287	6,313	6,339	6,366
2	98.5	99.0	99.2	99.2	99.3	99.3	99.4	99.4	99.4	99.4	99.4	99.4	99.4	99.5	99.5	99.5	99.5	99.5	99.5
3	34.1	30.8	29.5	28.7	28.2	27.9	27.7	27.5	27.3	27.2	27.1	26.9	26.7	26.6	26.5	26.4	26.3	26.2	26.1
4	21.2	18.0	16.7	16.0	15.5	15.2	15.0	14.8	14.7	14.5	14.4	14.2	14.0	13.9	13.8	13.7	13.7	13.6	13.5
5	16.3	13.3	12.1	11.4	11.0	10.7	10.5	10.3	10.2	10.1	9.89	9.72	9.55	9.47	9.38	9.29	9.20	9.11	9.02
6	13.7	10.9	9.78	9.15	8.75	8.47	8.26	8.10	7.98	7.87	7.72	7.56	7.40	7.31	7.23	7.14	7.06	6.97	6.88
7	12.2	9.55	8.45	7.85	7.46	7.19	6.99	6.84	6.72	6.62	6.47	6.31	6.16	6.07	5.99	6.91	5.82	5.74	5.65
8	11.3	8.65	7.59	7.01	6.63	6.37	6.18	6.03	5.91	5.81	5.67	5.52	5.36	5.28	5.20	5.12	5.03	4.95	4.86
9	10.6	8.02	6.99	6.42	6.06	5.80	5.61	5.47	5.35	5.26	5.11	4.96	4.81	4.73	4.65	4.57	4.48	4.40	4.31
10	10.0	7.56	6.55	5.99	5.64	5.39	5.20	5.06	4.94	4.85	4.71	4.56	4.41	4.33	4.25	4.17	4.08	4.00	3.91
11	9.65	7.21	6.22	5.67	5.32	5.07	4.89	4.74	4.63	4.54	4.40	4.25	4.10	4.02	3.94	3.86	3.78	3.69	3.60
12	9.33	6.93	5.95	5.41	5.06	4.82	4.64	4.50	4.39	4.30	4.16	4.01	3.86	3.78	3.70	3.62	3.54	3.45	3.36
13	9.07	6.70	5.74	5.21	4.86	4.62	4.44	4.30	4.19	4.10	3.96	3.82	3.66	3.59	3.51	3.43	3.34	3.25	3.17
14	8.86	6.51	5.56	5.04	4.70	4.46	4.28	4.14	4.03	3.94	3.80	3.66	3.51	3.43	3.35	3.27	3.18	3.09	3.00
15	8.68	6.36	5.42	4.89	4.56	4.32	4.14	4.00	3.89	3.80	3.67	3.52	3.37	3.29	3.21	3.13	3.05	2.96	2.87

Table VI: (*continued*) Values of $f_{0.01,\nu_1,\nu_2}$

ν_2 = Degrees of freedom for denominator	ν_1 = Degrees of freedom for numerator																		
	1	2	3	4	5	6	7	8	9	10	12	15	20	24	30	40	60	120	∞
16	8.53	6.23	5.29	4.77	4.44	4.20	4.03	3.89	3.78	3.69	3.55	3.41	3.26	3.18	3.10	3.02	2.93	2.84	2.75
17	8.40	6.11	5.19	4.67	4.34	4.10	3.93	3.79	3.68	3.59	3.46	3.31	3.16	3.08	3.00	2.92	2.83	2.75	2.65
18	8.29	6.01	5.09	4.58	4.25	4.01	3.84	3.71	3.60	3.51	3.37	3.23	3.08	3.00	2.92	2.84	2.75	2.66	2.57
19	8.19	5.93	5.01	4.50	4.17	3.94	3.77	3.63	3.52	3.43	3.30	3.15	3.00	2.92	2.84	2.76	2.67	2.58	2.49
20	8.10	5.85	4.94	4.43	4.10	3.87	3.70	3.56	3.46	3.37	3.23	3.09	2.94	2.86	2.78	2.69	2.61	2.52	2.42
21	8.02	5.78	4.87	4.37	4.04	3.81	3.64	3.51	3.40	3.31	3.17	3.03	2.88	2.80	2.72	2.64	2.55	2.46	2.36
22	7.95	5.72	4.82	4.31	3.99	3.76	3.59	3.45	3.35	3.26	3.12	2.98	2.83	2.75	2.67	2.58	2.50	2.40	2.31
23	7.88	5.66	4.76	4.26	3.94	3.71	3.54	3.41	3.30	3.21	3.07	2.93	2.78	2.70	2.62	2.54	2.45	2.35	2.26
24	7.82	5.61	4.72	4.22	3.90	3.67	3.50	3.36	3.26	3.17	3.03	2.89	2.74	2.66	2.58	2.49	2.40	2.31	2.21
25	7.77	5.57	4.68	4.18	3.86	3.63	3.46	3.32	3.22	3.13	2.99	2.85	2.70	2.62	2.53	2.45	2.36	2.27	2.17
30	7.56	5.39	4.51	4.02	3.70	3.47	3.30	3.17	3.07	2.98	2.84	2.70	2.55	2.47	2.39	2.30	2.21	2.11	2.01
40	7.31	5.18	4.31	3.83	3.51	3.29	3.12	2.99	2.89	2.80	2.66	2.52	2.37	2.29	2.20	2.11	2.02	1.92	1.80
60	7.08	4.98	4.13	3.65	3.34	3.12	2.95	2.82	2.72	2.63	2.50	2.35	2.20	2.12	2.03	1.94	1.84	1.73	1.60
120	6.85	4.79	3.95	3.48	3.17	2.96	2.79	2.66	2.56	2.47	2.34	2.19	2.03	1.95	1.86	1.76	1.66	1.53	1.38
∞	6.63	4.61	3.78	3.32	3.02	2.80	2.64	2.51	2.41	2.32	2.18	2.04	1.88	1.79	1.70	1.59	1.47	1.32	1.00

Table VII: Factorials and Binomial Coefficients

Factorials

n	$n!$	$\log n!$
0	1	0.0000
1	1	0.0000
2	2	0.3010
3	6	0.7782
4	24	1.3802
5	120	2.0792
6	720	2.8573
7	5,040	3.7024
8	40,320	4.6055
9	362,880	5.5598
10	3,628,800	6.5598
11	39,916,800	7.6012
12	479,001,600	8.6803
13	6,227,020,800	9.7943
14	87,178,291,200	10.9404
15	1,307,674,368,000	12.1165

Binomial Coefficients

n	$\binom{n}{0}$	$\binom{n}{1}$	$\binom{n}{2}$	$\binom{n}{3}$	$\binom{n}{4}$	$\binom{n}{5}$	$\binom{n}{6}$	$\binom{n}{7}$	$\binom{n}{8}$	$\binom{n}{9}$	$\binom{n}{10}$
0	1										
1	1	1									
2	1	2	1								
3	1	3	3	1							
4	1	4	6	4	1						
5	1	5	10	10	5	1					
6	1	6	15	20	15	6	1				
7	1	7	21	35	35	21	7	1			
8	1	8	28	56	70	56	28	8	1		
9	1	9	36	84	126	126	84	36	9	1	
10	1	10	45	120	210	252	210	120	45	10	1
11	1	11	55	165	330	462	462	330	165	55	11
12	1	12	66	220	495	792	924	792	495	220	66
13	1	13	78	286	715	1287	1716	1716	1287	715	286
14	1	14	91	364	1001	2002	3003	3432	3003	2002	1001
15	1	15	105	455	1365	3003	5005	6435	6435	5005	3003
16	1	16	120	560	1820	4368	8008	11440	12870	11440	8008
17	1	17	136	680	2380	6188	12376	19448	24310	24310	19448
18	1	18	153	816	3060	8568	18564	31824	43758	48620	43758
19	1	19	171	969	3876	11628	27132	50388	75582	92378	92378
20	1	20	190	1140	4845	15504	38760	77520	125970	167960	184756

Table VIII: Values of e^x and e^{-x}

x	e^x	e^{-x}	x	e^x	e^{-x}
0.0	1.000	1.000	**5.0**	148.4	0.0067
0.1	1.105	0.905	**5.1**	164.0	0.0061
0.2	1.221	0.819	**5.2**	181.3	0.0055
0.3	1.350	0.741	**5.3**	200.3	0.0050
0.4	1.492	0.670	**5.4**	221.4	0.0045
0.5	1.649	0.607	**5.5**	244.7	0.0041
0.6	1.822	0.549	**5.6**	270.4	0.0037
0.7	2.014	0.497	**5.7**	298.9	0.0033
0.8	2.226	0.449	**5.8**	330.3	0.0030
0.9	2.460	0.407	**5.9**	365.0	0.0027
1.0	2.718	0.368	**6.0**	403.4	0.0025
1.1	3.004	0.333	**6.1**	445.9	0.0022
1.2	3.320	0.301	**6.2**	492.8	0.0020
1.3	3.669	0.273	**6.3**	544.6	0.0018
1.4	4.055	0.247	**6.4**	601.8	0.0017
1.5	4.482	0.223	**6.5**	665.1	0.0015
1.6	4.953	0.202	**6.6**	735.1	0.0014
1.7	5.474	0.183	**6.7**	812.4	0.0012
1.8	6.050	0.165	**6.8**	897.8	0.0011
1.9	6.686	0.150	**6.9**	992.3	0.0010
2.0	7.389	0.135	**7.0**	1,096.6	0.0009
2.1	8.166	0.122	**7.1**	1,212.0	0.0008
2.2	9.025	0.111	**7.2**	1,339.4	0.0007
2.3	9.974	0.100	**7.3**	1,480.3	0.0007
2.4	11.023	0.091	**7.4**	1,636.0	0.0006
2.5	12.18	0.082	**7.5**	1,808.0	0.00055
2.6	13.46	0.074	**7.6**	1,998.2	0.00050
2.7	14.88	0.067	**7.7**	2,208.3	0.00045
2.8	16.44	0.061	**7.8**	2,440.6	0.00041
2.9	18.17	0.055	**7.9**	2,697.3	0.00037
3.0	20.09	0.050	**8.0**	2,981.0	0.00034
3.1	22.20	0.045	**8.1**	3,294.5	0.00030
3.2	24.53	0.041	**8.2**	3,641.0	0.00027
3.3	27.11	0.037	**8.3**	4,023.9	0.00025
3.4	29.96	0.033	**8.4**	4,447.1	0.00022
3.5	33.12	0.030	**8.5**	4,914.8	0.00020
3.6	36.60	0.027	**8.6**	5,431.7	0.00018
3.7	40.45	0.025	**8.7**	6,002.9	0.00017
3.8	44.70	0.022	**8.8**	6,634.2	0.00015
3.9	49.40	0.020	**8.9**	7,332.0	0.00014
4.0	54.60	0.018	**9.0**	8,103.1	0.00012
4.1	60.34	0.017	**9.1**	8,955.3	0.00011
4.2	66.69	0.015	**9.2**	9,897.1	0.00010
4.3	73.70	0.014	**9.3**	10,938	0.00009
4.4	81.45	0.012	**9.4**	12,088	0.00008
4.5	90.02	0.011	**9.5**	13,360	0.00007
4.6	99.48	0.010	**9.6**	14,765	0.00007
4.7	109.95	0.009	**9.7**	16,318	0.00006
4.8	121.51	0.008	**9.8**	18,034	0.00006
4.9	134.29	0.007	**9.9**	19,930	0.00005

Table IX: Values of r_p for $\alpha = 0.01$†

d.f. \ p	2	3	4	5	6	7	8	9	10
1	90.02								
2	14.04	14.04							
3	8.26	8.32	8.32						
4	6.51	6.68	6.74	6.76					
5	5.70	5.90	5.99	6.04	6.07				
6	5.24	5.44	5.55	5.62	5.66	5.68			
7	4.95	5.15	5.26	5.33	5.38	5.42	5.44		
8	4.74	4.94	5.06	5.13	5.19	5.23	5.26	5.28	
9	4.60	4.79	4.91	4.99	5.04	5.09	5.12	5.14	5.16
10	4.48	4.67	4.79	4.88	4.93	4.98	5.01	5.04	5.06
11	4.39	4.58	4.70	4.78	4.84	4.89	4.92	4.95	4.97
12	4.32	4.50	4.62	4.71	4.77	4.81	4.85	4.88	4.91
13	4.26	4.44	4.56	4.64	4.71	4.75	4.79	4.82	4.85
14	4.21	4.39	4.51	4.59	4.66	4.70	4.74	4.77	4.80
15	4.17	4.34	4.46	4.55	4.61	4.66	4.70	4.73	4.76
16	4.13	4.31	4.43	4.51	4.57	4.62	4.66	4.70	4.72
17	4.10	4.27	4.39	4.47	4.54	4.59	4.63	4.66	4.69
18	4.07	4.25	4.36	4.45	4.51	4.56	4.60	4.64	4.66
19	4.05	4.22	4.33	4.42	4.48	4.53	4.57	4.61	4.64
20	4.02	4.20	4.31	4.40	4.46	4.51	4.55	4.59	4.62
24	3.96	4.13	4.24	4.32	4.39	4.44	4.48	4.52	4.55
30	3.89	4.06	4.17	4.25	4.31	4.36	4.41	4.45	4.48
40	3.82	3.99	4.10	4.18	4.24	4.29	4.33	4.38	4.41
60	3.76	3.92	4.03	4.11	4.18	4.23	4.37	4.31	4.34
120	3.70	3.86	3.97	4.04	4.11	4.16	4.20	4.24	4.27
∞	3.64	3.80	3.90	3.98	4.04	4.09	4.13	4.17	4.21

†This table is reproduced from H. L. Harter, "Critical Values for Duncan's New Multiple Range Test." It contains some corrected values to replace those given by D. B. Duncan in "Multiple Range and Multiple F Tests," *Biometrics*, Vol. 11 (1955). The above table is reproduced with the permission of the author and the Biometric Society.

Table IX: (*continued*) Values of r_p for $\alpha = 0.05$

d.f. \ p	2	3	4	5	6	7	8	9	10
1	17.97								
2	6.09	6.09							
3	4.50	4.52	4.52						
4	3.93	4.01	4.03	4.03					
5	3.64	3.75	3.80	3.81	3.81				
6	3.46	3.59	3.65	3.68	3.69	3.70			
7	3.34	3.48	3.55	3.59	3.61	3.62	3.63		
8	3.26	3.40	3.48	3.52	3.55	3.57	3.57	3.58	
9	3.20	3.34	3.42	3.47	3.50	3.52	3.54	3.54	3.55
10	3.15	3.29	3.38	3.43	3.47	3.49	3.51	3.52	3.52
11	3.11	3.26	3.34	3.40	3.44	3.46	3.48	3.49	3.50
12	3.08	3.23	3.31	3.37	3.41	3.44	3.46	3.47	3.48
13	3.06	3.20	3.29	3.35	3.39	3.42	3.46	3.46	3.47
14	3.03	3.18	3.27	3.33	3.37	3.40	3.43	3.44	3.46
15	3.01	3.16	3.25	3.31	3.36	3.39	3.41	3.43	3.45
16	3.00	3.14	3.23	3.30	3.34	3.38	3.40	3.42	3.44
17	2.98	3.13	3.22	3.28	3.33	3.37	3.39	3.41	3.43
18	2.97	3.12	3.21	3.27	3.32	3.36	3.38	3.40	3.42
19	2.96	3.11	3.20	3.26	3.31	3.35	3.38	3.40	3.41
20	2.95	3.10	3.19	3.25	3.30	3.34	3.37	3.39	3.41
24	2.92	3.07	3.16	3.23	3.28	3.31	3.35	3.37	3.39
30	2.89	3.03	3.13	3.20	3.25	3.29	3.32	3.35	3.37
40	2.86	3.01	3.10	3.17	3.22	3.27	3.30	3.33	3.35
60	2.83	2.98	3.07	3.14	3.20	3.24	3.28	3.31	3.33
120	2.80	2.95	3.04	3.12	3.17	3.22	3.25	3.29	3.31
∞	2.77	2.92	3.02	3.09	3.15	3.19	3.23	3.27	3.29

Table X: Critical Values for the Signed-Rank Test†

n	$T_{0.10}$	$T_{0.05}$	$T_{0.02}$	$T_{0.01}$
4				
5	1			
6	2	1		
7	4	2	0	
8	6	4	2	0
9	8	6	3	2
10	11	8	5	3
11	14	11	7	5
12	17	14	10	7
13	21	17	13	10
14	26	21	16	13
15	30	25	20	16
16	36	30	24	19
17	41	35	28	23
18	47	40	33	28
19	54	46	38	32
20	60	52	43	37
21	68	59	49	43
22	75	66	56	49
23	83	73	62	55
24	92	81	69	61
25	101	90	77	68

†From F. Wilcoxon and R. A. Wilcox, *Some Rapid Approximate Statistical Procedures*, American Cyanamid Company, Pearl River, N. Y., 1964. Reproduced with permission of American Cyanamid Company.

Table XI: Critical Values for the U Test†

Values of $U_{0.10}$

n_1 \ n_2	2	3	4	5	6	7	8	9	10	11	12	13	14	15
2				0	0	0	1	1	1	1	2	2	3	3
3		0	0	1	2	2	3	4	4	5	5	6	7	7
4		0	1	2	3	4	5	6	7	8	9	10	11	12
5	0	1	2	4	5	6	8	9	11	12	13	15	16	18
6	0	2	3	5	7	8	10	12	14	16	17	19	21	23
7	0	2	4	6	8	11	13	15	17	19	21	24	26	28
8	1	3	5	8	10	13	15	18	20	23	26	28	31	33
9	1	4	6	9	12	15	18	21	24	27	30	33	36	39
10	1	4	7	11	14	17	20	24	27	31	34	37	41	44
11	1	5	8	12	16	19	23	27	31	34	38	42	46	50
12	2	5	9	13	17	21	26	30	34	38	42	47	51	55
13	2	6	10	15	19	24	28	33	37	42	47	51	56	61
14	3	7	11	16	21	26	31	36	41	46	51	56	61	66
15	3	7	12	18	23	28	33	39	44	50	55	61	66	72

Values of $U_{0.05}$

n_1 \ n_2	2	3	4	5	6	7	8	9	10	11	12	13	14	15
2							0	0	0	0	1	1	1	1
3				0	1	1	2	2	3	3	4	4	5	5
4			0	1	2	3	4	4	5	6	7	8	9	10
5		0	1	2	3	5	6	7	8	9	11	12	13	14
6		1	2	3	5	6	8	10	11	13	14	16	17	19
7		1	3	5	6	8	10	12	14	16	18	20	22	24
8	0	2	4	6	8	10	13	15	17	19	22	24	26	29
9	0	2	4	7	10	12	15	17	20	23	26	28	31	34
10	0	3	5	8	11	14	17	20	23	26	29	30	36	39
11	0	3	6	9	13	16	19	23	26	30	33	37	40	44
12	1	4	7	11	14	18	22	26	29	33	37	41	45	49
13	1	4	8	12	16	20	24	28	30	37	41	45	50	54
14	1	5	9	13	17	22	26	31	36	40	45	50	55	59
15	1	5	10	14	19	24	29	34	39	44	49	54	59	64

Values of $U_{0.02}$

n_1 \ n_2	2	3	4	5	6	7	8	9	10	11	12	13	14	15
2												0	0	0
3						0	0	1	1	1	2	2	2	3
4				0	1	1	2	3	3	4	5	5	6	7
5			0	1	2	3	4	5	6	7	8	9	10	11
6			1	2	3	4	6	7	8	9	11	12	13	15
7		0	1	3	4	6	7	9	11	12	14	16	17	19
8		0	2	4	6	7	9	11	13	15	17	20	22	24
9		1	3	5	7	9	11	14	16	18	21	23	26	28
10		1	3	6	8	11	13	16	19	22	24	27	30	33
11		1	4	7	9	12	15	18	22	25	28	31	34	37
12		2	5	8	11	14	17	21	24	28	31	35	38	42
13	0	2	5	9	12	16	20	23	27	31	35	39	43	47
14	0	2	6	10	13	17	22	26	30	34	38	43	47	51
15	0	3	7	11	15	19	24	28	33	37	42	47	51	56

†This table is based on D. Auble, "Extended Tables for the Mann–Whitney Statistics," *Bulletin of the Institute of Educational Research at Indiana University*, Vol. 1, 1953. By permission of the author.

Table XI: *(continued)*

Values of $U_{0.01}$

n_1 \ n_2	3	4	5	6	7	8	9	10	11	12	13	14	15
3							0	0	0	1	1	1	2
4				0	0	1	1	2	2	3	3	4	5
5			0	1	1	2	3	4	5	6	7	7	8
6		0	1	2	3	4	5	6	7	9	10	11	12
7		0	1	3	4	6	7	9	10	12	13	15	16
8		1	2	4	6	7	9	11	13	15	17	18	20
9	0	1	3	5	7	9	11	13	16	18	20	22	24
10	0	2	4	6	9	11	13	16	18	21	24	26	29
11	0	2	5	7	10	13	16	18	21	24	27	30	33
12	1	3	6	9	12	15	18	21	24	27	31	34	37
13	1	3	7	10	13	17	20	24	27	31	34	38	42
14	1	4	7	11	15	18	22	26	30	34	38	42	46
15	2	5	8	12	16	20	24	29	33	37	42	46	51

Table XII: Critical Values for the Runs Test†

Values of $u'_{0.025}$

n_1 \ n_2	2	3	4	5	6	7	8	9	10	11	12	13	14	15
2											2	2	2	2
3					2	2	2	2	2	2	2	2	2	3
4				2	2	2	3	3	3	3	3	3	3	3
5			2	2	3	3	3	3	3	4	4	4	4	4
6		2	2	3	3	3	3	4	4	4	4	5	5	5
7		2	2	3	3	3	4	4	5	5	5	5	5	6
8		2	3	3	3	4	4	5	5	5	6	6	6	6
9		2	3	3	4	4	5	5	5	6	6	6	7	7
10		2	3	3	4	5	5	5	6	6	7	7	7	7
11		2	3	4	4	5	5	6	6	7	7	7	8	8
12	2	2	3	4	4	5	6	6	7	7	7	8	8	8
13	2	2	3	4	5	5	6	6	7	7	8	8	9	9
14	2	2	3	4	5	5	6	7	7	8	8	9	9	9
15	2	3	3	4	5	6	6	7	7	8	8	9	9	10

Values of $u_{0.025}$

n_1 \ n_2	4	5	6	7	8	9	10	11	12	13	14	15
4		9	9									
5	9	10	10	11	11							
6	9	10	11	12	12	13	13	13	13			
7		11	12	13	13	14	14	14	14	15	15	15
8		11	12	13	14	14	15	15	16	16	16	16
9			13	14	14	15	16	16	16	17	17	18
10			13	14	15	16	16	17	17	18	18	18
11			13	14	15	16	17	17	18	19	19	19
12			13	14	16	16	17	18	19	19	20	20
13				15	16	17	18	19	19	20	20	21
14				15	16	17	18	19	20	20	21	22
15				15	16	18	18	19	20	21	22	22

†This table is adapted, by permission, from F. S. Swed and C. Eisenhart, "Tables for testing randomness of grouping in a sequence of alternatives," *Annals of Mathematical Statistics*, Vol. 14.

Table XII: *(continued)*

Values of $u'_{0.005}$

n_1 \ n_2	3	4	5	6	7	8	9	10	11	12	13	14	15
3										2	2	2	2
4						2	2	2	2	2	2	2	3
5				2	2	2	2	3	3	3	3	3	3
6			2	2	2	3	3	3	3	3	3	4	4
7			2	2	3	3	3	3	4	4	4	4	4
8		2	2	3	3	3	3	4	4	4	5	5	5
9		2	2	3	3	3	4	4	5	5	5	5	6
10		2	3	3	3	4	4	5	5	5	5	6	6
11		2	3	3	4	4	5	5	5	6	6	6	7
12	2	2	3	3	4	4	5	5	6	6	6	7	7
13	2	2	3	3	4	5	5	5	6	6	7	7	7
14	2	2	3	4	4	5	5	6	6	7	7	7	8
15	2	3	3	4	4	5	6	6	7	7	7	8	8

Values of $u_{0.005}$

n_1 \ n_2	5	6	7	8	9	10	11	12	13	14	15
5		11									
6	11	12	13	13							
7		13	13	14	15	15	15				
8		13	14	15	15	16	16	17	17	17	
9			15	15	16	17	17	18	18	18	19
10			15	16	17	17	18	19	19	19	20
11			15	16	17	18	19	19	20	20	21
12				17	18	19	19	20	21	21	22
13				17	18	19	20	21	21	22	22
14				17	18	19	20	21	22	23	23
15					19	20	21	22	22	23	24

INDEX

Page references followed by "f" indicate illustrated figures or photographs; followed by "t" indicates a table.

E

F

G

H

I

J

L

M